## METRIC PREFIXES

| PREFIX | ABBREVIATION | VALUE |
|---|---|---|
| tera | T | $10^{12}$ |
| giga | G | $10^{9}$ |
| **mega** | **M** | $\mathbf{10^{6}}$ |
| **kilo** | **k** | $\mathbf{10^{3}}$ |
| hecto | h | $10^{2}$ |
| deka | da | $10^{1}$ |
| deci | d | $10^{-1}$ |
| **centi** | **c** | $\mathbf{10^{-2}}$ |
| **milli** | **m** | $\mathbf{10^{-3}}$ |
| **micro** | **μ** | $\mathbf{10^{-6}}$ |
| nano | n | $10^{-9}$ |
| pico | p | $10^{-12}$ |

## MATHEMATICAL SYMBOLS

| | |
|---|---|
| $\equiv$ | defined by |
| $=$ | equal to |
| $\neq$ | not equal to |
| $\approx$ | approximately equal to |
| $\propto$ | proportional to |
| $>$ | greater than |
| $\gg$ | much greater than |
| $<$ | less than |
| $\ll$ | much less than |
| $\geq$ | greater than or equal to |
| $\leq$ | less than or equal to |
| $\rightarrow$ | implying, yielding, approaching |
| $\Delta$ | change in |
| $\infty$ | infinity |

# MODERN PHYSICS
## for Science and Engineering

# MODERN PHYSICS for Science and Engineering

Marshall L. Burns

*Tuskegee University*

Harcourt Brace Jovanovich, Publishers

and its subsidary, Academic Press

*San Diego New York Chicago Austin Washington, D.C.*
*London Sydney Tokyo Toronto*

ISBN: 0-15-562351-6
Library of Congress Catalog Card Number: 87-83020
Printed in the United States of America

*To my dearest friend and closest companion—*
*my wife, June Anne*

# PREFACE

This book provides an introduction to modern physics for students who have completed an academic year of general physics. A continuation of introductory general physics, it includes the subject areas of classical relativity (Chapter 1), Einstein's *special theory of relativity* (Chapters 2–4), the *old* quantum theory (Chapters 5–7), an introduction to quantum mechanics (Chapters 8–10), and introductory classical and quantum statistical mechanics (Chapters 11–12). In a two-term course, Chapters 1–7 may be covered in the first term and Chapters 8–12 in the second. For schools offering a one-term course in modern physics, many of the topics in Chapters 1–7 may have previously been covered; consequently, the portions of this textbook to be covered might include parts of the *old* quantum theory, all of quantum mechanics, and possibly some of the topics in statistical mechanics.

It is important to recognize that mathematics is only a tool in the development of physical theories and that the mathematical skills of students at the sophomore level are often limited. Accordingly, algebra and basic trigonometry are primarily used in Chapters 1–7, with *elementary* calculus being introduced either as an alternative approach *or* when necessary to preserve the integrity and rigor of the subject. The math review provided in Appendix A is more than sufficient for a study of the *entire* book. On occasions when higher mathematics is required, as with the solution to a second-order partial differential equation in Chapter 8, the mathematics is sufficiently detailed to allow understanding with only a knowledge of elementary calculus. Even quantum theory and statistical mechanics are easily managed with this approach through the introduction of *operator algebra* and with the occasional use of one of the five definite integrals provided in Appendix A. This reduced mathematical emphasis allows students to concentrate on the more important underlying physical concepts and not be distracted or intimidated by unfamiliar mathematics.

A major objective of this book is to enhance student understanding and appreciation of the fundamentals of physics by illustrating the necessary physical and quantitative *reasoning with fundamentals* that is essential for theoretical modeling of phenomena in science and engineering. The majority of physics textbooks at both the introductory and the intermediate level concentrate on introducing the basic concepts, formulas, and associated terminology of a broad spectrum of physics topics, leaving little space for the development of mathematical logic and physical reasoning from first principles. Certainly, students must first learn the fundamentals of the subject before intricate, detailed logic and reasoning are possible. But most intermediate and advanced books follow the lead of introductory textbooks and seldom elaborate in sufficient detail the development of physical theories. Students are expected somehow to develop the necessary physical and quantitative reasoning either on their own or from classroom lectures. The result is that many students simply memorize physical formulas and stereotyped problems in their initial study of physics and continue the practice in intermediate and advanced courses. Students entering college are often accomplished at rote memorization but poorly prepared in reasoning skills. They must *learn* how to reason and how to employ logic with a set of fundamentals to obtain insights and results that are not obvious or commonly recognized. Developing understanding and reasoning is difficult in the qualitative nonscience courses and supremely challenging in such highly quantitative courses as physics and engineering. The objective is, however, most desirable in these areas, since memorized equations and problems are rapidly forgotten by even the best students.

In this textbook, a *deliberate* and *detailed* approach has been employed. All of the topics presented are developed from first principles. In fact, *all but three equations are rigorously derived via physical reasoning* before being applied to problems or used in the discussion of other topics. Thus, the order of topics throughout the text is dictated by the requirement that fundamentals and physical derivations be carefully and judiciously introduced. And there is a gradual increase in the complexity of topics being considered to allow students to mature steadily in *physical* and *quantitative reasoning* as they progress through the book. For example, relativity is discussed early, since it depends on only a small number of physical fundamentals from kinematics and dynamics of general classical mechanics.

Chapter 1 allows students to review pertinent fundamental equations of classical mechanics and to apply them to *classical relativity* before they are employed in the development of Einstein's *special theory of relativity* in Chapters 2–4. This allows students time to develop the necessary quantitative skills and gain an *overview* of relativity before considering the conceptually subtle points of Einsteinian relativity. This basic approach, of reviewing the classical point of view before developing that of modern physics, continues throughout the text, to allow students to build upon what they already know and to develop strong connections between classical and modern physics. With this approach—where later subject areas are dependent on the fundamentals and results of earlier sections—students are led to develop greater insights as they apply previously gained knowledge to new physical situations. They also see how concepts of classical and modern physics are tied together, rather than seeing them as confused, isolated areas of interest.

This development of reasoning skills and fundamental understanding better prepares students for all higher level courses. This book does not therefore pretend to be a survey of all modern physics topics. The pace of developing scientific understanding requires that some topics be omitted. For example, since a rigorous development of nuclear physics requires relativistic quantum mechanics, only a few basic topics (e.g., the size of the nucleus, nuclear binding energy, etc.) merit development within the pedagogic framework of the text. The goal of this book is to provide the background required for meaningful future studies and not to be a catalog of modern physics topics. Thus, the traditional coverage of nuclear physics has been displaced by the extremely useful subject of statistical mechanics. The fundamentals of statistical mechanics are carefully developed and applied to numerous topics in solid state physics and engineering, topics which themselves are so very important for many courses at the intermediate and advanced levels.

The following pedagogic features appear throughout this textbook:

1. Each chapter begins with an *introductory overview* of the direction and objectives of the chapter.
2. Boldface type is used to emphasize important concepts, principles, postulates, equation titles, and new terminology when they are first introduced; thereafter, they may be italicized to reemphasize their importance.
3. Verbal definitions are set off by the use of italics.
4. Reference titles (and comments) for important equations appear in the margin of the text.
5. Fundamental defining equations and important results from deri-

vations are highlighted in color. Furthermore, a defining symbol is used with fundamental defining equations in place of an equality sign.

6. A logical and comprehensive list of the fundamental and derived equations in each chapter appears in a *review* section. It will assist students in the assimilation of fundamental equations (and associated reference terminology) and test their quantitative reasoning ability.
7. Formal *solutions* for the odd-numbered problems are provided at the end of each chapter, and *answers* are given for the even-numbered problems. A student's efficiency in assimilating fundamentals and developing quantitative reasoning is greatly enhanced by making solutions an integral part of the text. The problems generally require students to be deliberate, reflective, and straightforward in their logic with physical fundamentals.
8. Examples and applications of physical theories are limited in order not to distract students from the primary aim of understanding the physical reasoning, fundamentals, and objectives of each section or chapter. Having solutions to problems at the end of a chapter reduces the number of examples required within the text, since many of the problems complement the chapter sections with subtle concepts being further investigated and discussed.
9. Endpapers provide a *quick reference* of frequently used quantities: the Greek alphabet, metric prefixes, mathematical symbols, calculus identities, and physical constants.

Marshall L. Burns

# CONTENTS

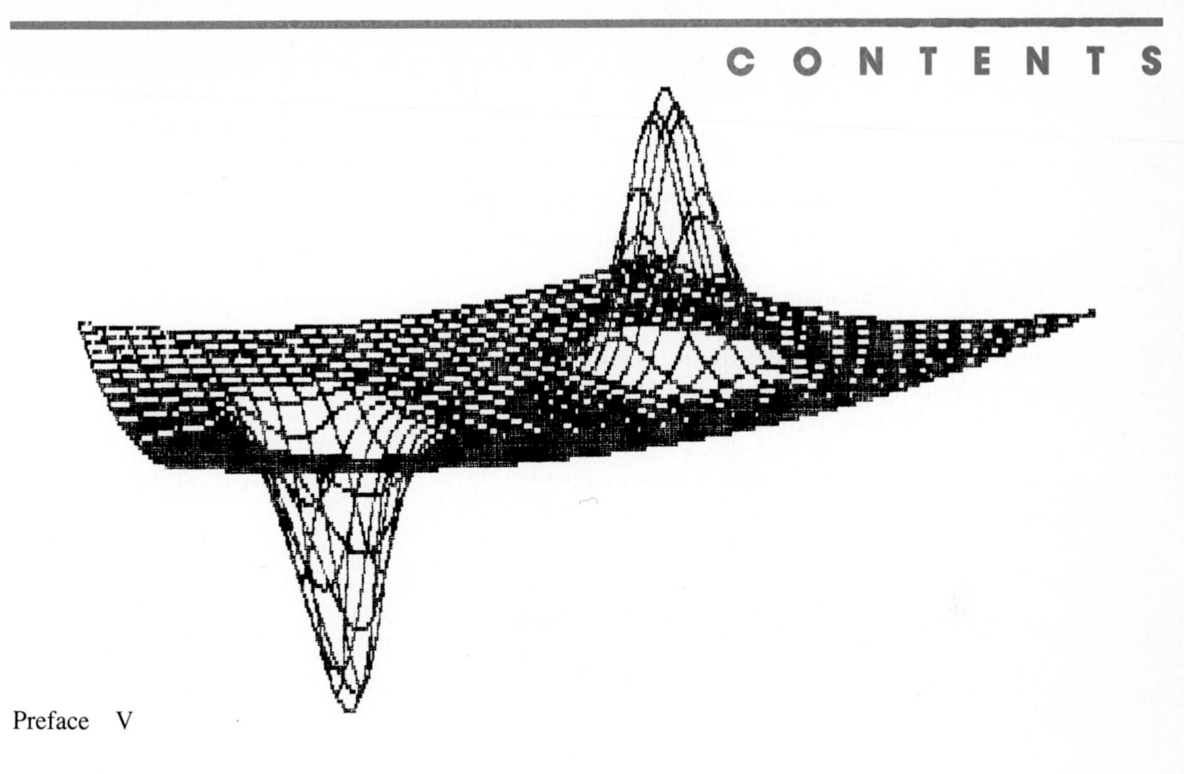

## Classical Statistical Mechanics 439

CHAPTER 11



## Quantum Statistical Mechanics 510

CHAPTER 12

# MODERN PHYSICS
## for Science and Engineering

CHAPTER 1

# Classical Transformations

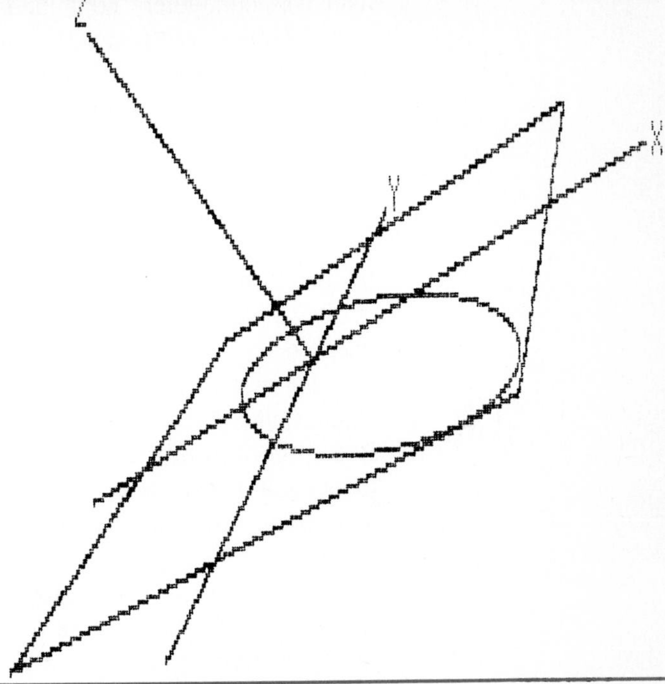


In experimental philosophy we are to look upon propositions obtained by general induction from phenomena as accurately or very nearly true . . . till such time as other phenomena occur, by which they may either be made accurate, or liable to exception.
SIR ISAAC NEWTON, *Principia* (1686)

## Introduction

Before the turn of the twentieth century, classical physics was fully developed within the three major disciplines—mechanics, thermodynamics, and electromagnetism. At that time the concepts, fundamental principles, and theories of classical physics were generally in accord with common sense and highly developed in precise, sophisticated mathematical formalisms. Alternative formulations to Newtonian mechanics were available through Lagrangian dynamics, Hamilton's formulation, and the Hamilton-Jacobi theory, which were equivalent physical descriptions of nature but differed mathematically and philosophically. By 1864 the theory of electromagnetism was completely contained in a set of four partial differential equations. Known as Maxwell's equations, they embodied all of the laws of electricity, magnetism, optics, and the propagation of electromagnetic radiation. The applicability and degree of sophistication of theoretical physics by the end of the nineteenth century was such that it was considered to be practically a closed subject. In fact, during the early 1890s some physicists purported that future accomplishments in physics would be limited to improving the accuracy of physical measurements. But, by the turn of the century, they realized classical physics was limited in its ability to accurately and completely describe many physical phenomena.

For nearly 200 years after Newton's contributions to classical mechanics, the disciplines of physics enjoyed an almost flawless existence. But at the turn of the twentieth century there was considerable turmoil in theoretical physics, instigated in 1900 by Max Planck's theory for the quantization of atoms regarded as electromagnetic oscillators and in 1905 by Albert Einstein's publication of the special theory of relativity. The latter work appeared in a paper entitled "On the Electrodynamics of Moving Bodies," in the German scholarly periodical, *Annalen der Physik*. This theory shattered the Newtonian view of nature and brought about an intellectual revelation concerning the concepts of space, time, matter, and energy.

The major objective of the following three chapters is to develop an understanding of Einsteinian relativity. It should be noted that the basic concept of relativity, namely that the *laws of physics assume the same form in many different reference frames,* is as old as the mechanics of Galileo Galilei (1564–1642) and Isaac Newton (1642–1727). The immediate task, however, is to review a few fundamental principles and defining equations of classical mechanics, which will be utilized in the development of relativistic transformation equations. In particular, the classical transformation equations for space, time, velocity, and acceleration are developed for two inertial reference frames, along with the appropriate frequency and wavelength equations for the classical Doppler effect. By this review and development of classical transformations, we will obtain an overview of the fundamental principles of classical relativity, which we are going to modify, in order that the relationship between the old theory and the new one can be fully understood and appreciated.

## 1.1 Fundamental Units

A philosophical approach to the study of natural phenomena might lead one to the acceptance of a few basic concepts in terms of which all physical quantities can be expressed. The concepts of space, time, and matter appear to be the most fundamental quantities in nature that allow for a description of physical reality. Certainly, reflection dictates space and time to be the more basic of the three, since they can exist independently of matter in what would constitute an empty universe. In this sense our philosophical and commonsense construction of the physical universe begins with space and time as given primitive, indefinable concepts and allows for the distribution of matter here and there in space and now and then in time.

A classical scientific description of the basic quantities of nature departs slightly from the philosophical view. Since space is regarded as three-dimensional, a spatial quantity like volume can be expressed by a length measurement cubed. Further, the existence of matter gives rise to gravitational, electric, and magnetic fields in nature. These fundamental fields in the universe are associated with the basic quantities of mass, electric charge, and state of motion of matter, respectively, with the latter being expressed in terms of length, time, and charge. Thus, the scientific view suggests four basic or fundamental quantities in nature: *length*, *mass*, *time*, and *electric charge*. It should be realized that an electrically charged body has an associated electric field according to an observer at rest with respect to the charged body. However, if relative motion exists between an observer and the charged body, the observer will detect not only an electric field, but also a magnetic field associated with the charged body. As the constituents of the universe are considered to be in a state of motion, the fourth fundamental quantity in nature is commonly taken to be *electric current* as opposed to *electric charge*.

The conventional scientific description of the physical universe, according to classical physics, is in terms of the four fundamental quantities: *length*, *mass*, *time*, and *electric current*. It should be noted that these four fundamental or primitive concepts have been somewhat arbitrarily chosen, as a matter of convenience. For example, all physical concepts of classical mechanics can be expressed in terms of the first three basic quantities, whereas electromagnetism requires the inclusion of the fourth. Certainly, these four fundamental quantities are convenient choices for the disciplines of mechanics and electromagnetism; however, in thermodynamics it proves convenient to define temperature as a fundamental or primitive concept. The point is that the number of *basic quantities* selected to describe physical reality is arbitrary, to a certain extent, and can be increased or decreased for convenience in the description of physical concepts in different areas.

Just as important as the number of basic quantities used in describing nature is the selection of a *system of units*. Previously, the systems most commonly utilized by scientists and engineers included the MKS *(meter-kilogram-second)*, Gaussian or CGS *(centimeter-gram-second)*, and British engineering or FPS *(foot-pound-second)* systems. Fortunately, an international system of units, called the Système internationale (SI), has been adopted as the preferred system by scientists in most countries. It is based upon the original MKS rationalized metric system and will probably become universally adopted by scientists and engineers in all countries, even those in the United States. For this reason it will be primarily utilized as the system of units in this textbook, although other special units (e.g., Angstrom (Å) for length and electron volt (eV) for energy) will be used in some instances for emphasis and convenience. In addition to the fundamental units of length, mass, time, and electric current, the SI system includes units for temperature, amount of substance, and luminous intensity. In the SI (MKS) system the *basic units* associated with these seven fundamental quantities are the *meter* (m), *kilogram* (kg), *second* (s), *ampere* (A), *kelvin* (K), *mole* (mol), and *candela* (cd), respectively. The units associated with every physical quantity in this textbook will be expressed as some combination of these seven basic units, with frequent reference to their equivalence in the CGS metric system. Since the CGS system is in reality a sub-system of the SI, knowledge of the metric prefixes, listed on the inside cover, allows for the easy conversion of physical units from one system to the other.

## 1.2 Review of Classical Mechanics

Before developing the transformation equations of classical relativity, it will prove prudent to review a few of the fundamental principles and defining equations of classical mechanics. In *kinematics* we are primarily concerned with the motion and path of a particle represented as a mathematical point. The motion of the particle is normally described by the position of its representative point in space as a function of time, relative to some chosen *reference frame* or *coordinate system*. Using the usual *Cartesian* coordinate system, the position of a particle at time $t$ in three dimensions is described by its **displacement vector r,**

$$\mathbf{r} = x\mathbf{i} + y\mathbf{j} + z\mathbf{k}, \tag{1.1}$$

relative to the *origin of coordinates*, as illustrated in Figure 1.1. Assuming we know the spatial coordinates as a function of time,

$$x = x(t) \qquad y = y(t) \qquad z = z(t), \tag{1.2}$$

then the **instantaneous translational velocity** of the particle is defined by

$$\mathbf{v} \equiv \frac{d\mathbf{r}}{dt}, \tag{1.3}$$

with fundamental units of m/s in the SI system of units. The three-dimensional velocity vector can be expressed in terms of its rectangular components as

$$\mathbf{v} = v_x\mathbf{i} + v_y\mathbf{j} + v_z\mathbf{k}, \tag{1.4}$$

where the components of velocity are defined by

$$v_x \equiv \frac{dx}{dt}, \tag{1.5a}$$

$$v_y \equiv \frac{dy}{dt}, \tag{1.5b}$$

$$v_z \equiv \frac{dz}{dt}. \tag{1.5c}$$

Although these equations for the instantaneous translational components of velocity will be utilized in Einsteinian relativity, the defining equations for **average translational velocity** and its components, given by

$$\bar{\mathbf{v}} \equiv \frac{\Delta\mathbf{r}}{\Delta t}, \tag{1.6a}$$

$$\bar{v}_x \equiv \frac{\Delta x}{\Delta t}, \tag{1.6b}$$

$$\bar{v}_y \equiv \frac{\Delta y}{\Delta t}, \tag{1.6c}$$

$$\bar{v}_z \equiv \frac{\Delta z}{\Delta t}, \tag{1.6d}$$

will be primarily used in the derivations of classical relativity. As is customary, the Greek letter **delta** ($\Delta$) in these equations is used to denote the *change* in a quantity. For example, $\Delta x = x_2 - x_1$ indicates the displacement of the particle along the $X$-axis from its *initial* position $x_1$ to its *final* position $x_2$.

To continue with our review of kinematics, recall that the definition of acceleration is the *time rate of change of velocity*. Thus, **instantaneous translational acceleration** can be defined mathematically by the equation

$$\begin{aligned}\mathbf{a} &\equiv \frac{d\mathbf{v}}{dt} = \frac{d^2r}{dt^2} \\ &= a_x\mathbf{i} + a_y\mathbf{j} + a_z\mathbf{k},\end{aligned} \tag{1.7}$$

having components given by

$$a_x \equiv \frac{dv_x}{dt} = \frac{d^2x}{dt^2}, \tag{1.8a}$$

$$a_y \equiv \frac{dv_y}{dt} = \frac{d^2y}{dt^2}, \tag{1.8b}$$

$$a_z \equiv \frac{dv_z}{dt} = \frac{d^2z}{dt^2}. \tag{1.8c}$$

Likewise, **average translational acceleration** is defined by

$$\bar{\mathbf{a}} \equiv \frac{\Delta\mathbf{v}}{\Delta \mathrm{t}} = \bar{a}_x\mathbf{i} + \bar{a}_y\mathbf{j} + \bar{a}_z\mathbf{k} \tag{1.9}$$

with Cartesian components

$$\bar{a}_x \equiv \frac{\Delta v_x}{\Delta t}, \tag{1.10a}$$

$$\bar{a}_y \equiv \frac{\Delta v_y}{\Delta t}, \tag{1.10b}$$

$$\bar{a}_z \equiv \frac{\Delta v_z}{\Delta t}. \tag{1.10c}$$

The basic units of acceleration in the SI system are $m/s^2$, which should be obvious from the second equality in Equation 1.7.

The kinematical representation of the motion and path of a *system of particles* is normally described by the position of the system's **center of mass point** as a function of time, as defined by

$$\mathbf{r}_c \equiv \frac{1}{M}\sum_i m_i\mathbf{r}_i. \tag{1.11}$$

In this equation the Greek letter **sigma** ($\Sigma$) denotes a sum over the $i$-particles, $m_i$ is the mass of the $i$th particle having the position vector $\mathbf{r}_1$, and $M = \Sigma m_i$ is the total mass of the system of *discrete* particles. For a *continuous* distribution of mass, the position vector for the center of mass is defined in terms of the *integral* expression

$$r_c \equiv \frac{1}{M} \int \mathbf{r}\, dm. \tag{1.12}$$

From these definitions, the velocity and acceleration of the center of mass of a system are obtained by taking the first and second order time derivatives, respectively. That is, for a *discrete* system of particles,

$$\mathbf{v}_c = \frac{1}{M} \sum_i m_i \mathbf{v}_i \tag{1.13}$$

for the velocity and

$$\mathbf{a}_c = \frac{1}{M} \sum_i m_i \mathbf{a}_i \tag{1.14}$$

for the acceleration of the center of mass point.

Whereas kinematics is concerned *only* with the motion and path of particles, classical *dynamics* is concerned with the effect that external forces have on the state of motion of a particle or system of particles. Newton's three laws of motion are by far the most important and complete formulation of dynamics and can be stated as follows:

1. A body in a state of rest or uniform motion will continue in that state unless acted upon by an external unbalanced force.
2. The net external force acting on a body is equal to the time rate of change of the body's momentum.
3. For every force acting on a body there exists a reaction force, equal in magnitude and oppositely directed, acting on another body.

With *liner momentum* defined by

$$\mathbf{p} \equiv m\mathbf{v}, \tag{1.15}$$

Newton's second law of motion can be represented by the mathematical equation

$$\mathbf{F} \equiv \frac{d\mathbf{p}}{dt} \tag{1.16}$$

for the net external *force* acting on a body. If the mass of a body is time independent, then substitution of Equation 1.15 into Equation 1.16 and using Equation 1.7 yields

$$\mathbf{F} = m\mathbf{a}. \tag{1.17}$$

From this equation it is obvious that the gravitational force acting on a body, or the **weight** of a body $F_g$, is given by

$$F_g = mg, \tag{1.18}$$

where $g$ is the acceleration due to gravity. In the SI system the defined unit of force (or weight) is the *Newton* (N), which has fundamental units given by

$$\text{N} = \frac{\text{kg} \cdot \text{m}}{\text{s}^2}. \tag{1.19}$$

In the Gaussian or CGS system of units, force has the defined unit *dyne* (dy) and fundamental units of $\text{g} \cdot \text{cm/s}^2$.

Another fundamental concept of classical dynamics that is of particular importance in Einsteinian relativity is that of **infinitesimal work** $dW$, which is defined as the *dot* or *scalar product* of a force $\mathbf{F}$ and an infinitesimal displacement vector $d\mathbf{r}$, as given by the equation

$$dW \equiv \mathbf{F} \cdot d\mathbf{r}. \tag{1.20}$$

Work has the defined unit of a *Joule* (J) in SI units (an *erg* in CGS units), with corresponding fundamental units of

$$\text{J} = \frac{\text{kg} \cdot \text{m}^2}{\text{s}^2}. \tag{1.21}$$

These are the same units that are associated with **kinetic energy,**

$$T \equiv \tfrac{1}{2}mv^2, \tag{1.22}$$

and **gravitational potential energy**

$$V_g = mgy, \tag{1.23}$$

since it can be shown that the work done on or by a body is equivalent to the change in mechanical energy of the body.

Although there are a number of other fundamental principles, concepts, and defining equations of classical mechanics that will be utilized in this textbook, those presented in this review will more than satisfy our needs for the next few chapters. A review of a general physics textbook of the defining equations, defined and derived units, basic SI units, and conventional symbols for fundamental quantities of classical physics might be prudent. Appendix A contains a review of the mathematics (symbols, algebra, trigonometry, and calculus) necessary for a successful study of intermediate level modern physics.

## 1.3 Classical Space-Time Transformations

The classical or *Galilean-Newtonian* transformation equations for space and time are easily obtained by considering *two* inertial frames of reference, similar to the coordinate system depicted in Figure 1.1. An **inertial**

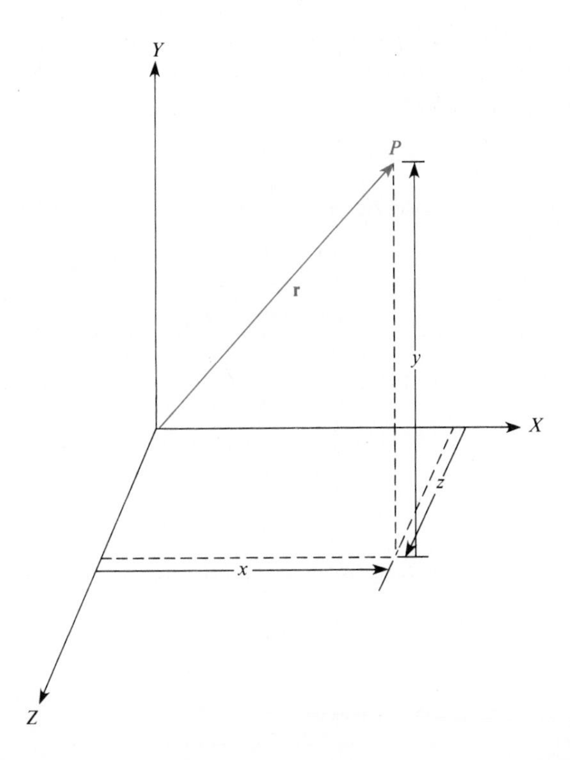


Figure 1.1
The position of a particle specified by a displacement vector in Cartesian coordinates.

**frame of reference** can be thought of as a *nonaccelerating coordinate system*, where Newton's laws of motion are valid. Further, all frames of reference moving at a constant velocity relative to an inertial one are themselves inertial and in principle *equivalent* for the formulation of physical laws.

Consider two inertial systems S and S′, as depicted in Figure 1.2, that are separating from one another at a constant speed $u$. We consider the *axis of relative motion* between S and S′ to coincide with their respective $X$, $X'$ axis and that their *origin of coordinates* coincided at time $t = t' \equiv 0$. Generality is not sacrificed by regarding system S as being at rest and system S′ to be moving in the positive $X$ direction with a uniform speed $u$ relative to S. Further, the uniform separation of two systems need not be along their common $X$, $X'$ axes. However, they can be so chosen without any loss in generality, since the selection of an origin of coordinates and the orientation of the coordinate axes in each system is entirely arbitrary. This requirement essentially simplifies the mathematical details, while maximizing the readability and understanding of classical and Einsteinian relativistic kinematics. Further, the requirement that S and S′ coincide at a time defined to be zero means that identical clocks in the two

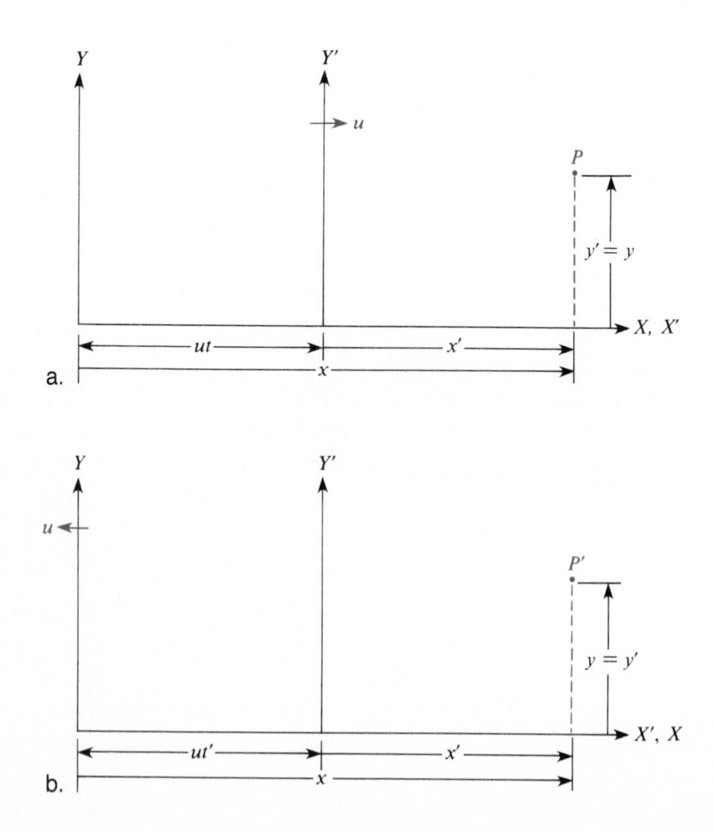


Figure 1.2
The classical coordinate transformations from *(a)* S to S′ and *(b)* S′ to S.

systems are started *simultaneously* at that *instant* in time. This requirement is essentially an **assumption of absolute time,** since classical common sense dictates that for all time thereafter $t = t'$.

Consider a particle $P$ ($P'$ in S$'$) moving about with a velocity at every instant in time and tracing out some kind of path. At an instant in time $t = t' > 0$, the position of the particle can be denoted by the coordinates $x$, $y$, $z$ in system S or, alternatively, by the coordinates $x'$, $y'$, $z'$ in system S$'$, as illustrated in Figure 1.2. The immediate problem is to deduce the relation between these two sets of coordinates, which should be clear from the figure. From the geometry below the $X$-$X'$ axis of Figure 1.2a and the assumption of absolute time, we have

$$x' = x - ut, \quad (1.24a)$$
$$y' = y, \quad (1.24b)$$
$$z' = z, \quad (1.24c)$$
$$t' = t, \quad (1.24d)$$

S $\rightarrow$ S$'$

for the classical transformation equations for **space-time coordinates,** according to an observer in system S. These equations indicate how an observer in the S system relates his coordinates of particle $P$ to the S$'$ coordinates of the particle, that *he* measures for *both* systems. From the point of view of an observer in the S$'$ system, the transformations are given by

$$x = x' + ut', \quad (1.25a)$$
$$y = y', \quad (1.25b)$$
$$z = z', \quad (1.25c)$$
$$t = t', \quad (1.25d)$$

S$'$ $\rightarrow$ S

where the relation between the $x$ and $x'$ coordinates is suggested by the geometry below the $X'$-axis in Figure 1.2b. These equations are just the *inverse* of Equations 1.24 and show how an observer in S$'$ relates the coordinates that he measures in both systems for the position of the particle at time $t'$. These sets of equations are known as *Galilean transformations*. The space-time coordinate relations for the case where the uniform relative motion between S and S$'$ is along the $Y$-$Y'$ axis or the $Z$-$Z'$ axis should be obvious by analogy.

The space-time transformation equations deduced above are for *coordinates* and are not appropriate for *length* and *time interval* calculations. For example, consider two particles $P_1$ ($P_1'$) and $P_2$ ($P_2'$) a fixed distance $y = y'$ above the $X$-$X'$ axis at an instant $t = t' > 0$ in time. The horizontal coordinates of these particles at time $t = t'$ are $x_1$ and $x_2$ in systems S and $x_1'$ and $x_2'$ in system S$'$. The relation between these four coordinates, according to Equation 1.24a, is

$$x_1' = x_1 - ut,$$
$$x_2' = x_2 - ut.$$

The *distance* between the two particles as measured with respect to the S′ system is $x_2' - x_1'$. Thus, from the above two equations we have

$$x_2' - x_1' = x_2 - x_1, \tag{1.26}$$

which shows that *length* measurements made at an *instant* in time are **invariant** (i.e., constant) for inertial frames of reference under a Galilean transformation.

Equations 1.24, 1.25, and 1.26 are called *transformation equations* because they transform physical measurements from one coordinate system to another. The basic problem in relativistic kinematics is to deduce the motion and path of a particle relative to the S′ system, when we know the kinematics of the particle relative to system S. More generally, the problem is that of relating any physical measurement in S with the corresponding measurement in S′. This central problem is of crucial importance, since an inability to solve it would mean that much of theoretical physics is a hopeless endeavor.

## 1.4 Classical Velocity and Acceleration Transformations

In the last section we considered the static effects of classical relativity by comparing a particle's position coordinates at an instant in time for two inertial frames of reference. *Dynamic* effects can be taken into account by considering how velocity and acceleration transform between inertial systems. To simplify our mathematical arguments, we assume all displacements, velocities, and accelerations to be *collinear*, in the same direction, and parallel to the $X$-$X'$ axis of relative motion. Further, systems S and S′ coincided at time $t = t' \equiv 0$ and S′ is considered to be receding from S at the constant speed $u$.

Our simplified view allows us to deduce the classical velocity transformation equation for *rectilinear* motion by commonsense arguments. For example, consider yourself to be standing at a train station, watching a jogger running due east at 5 m/s relative to and in front of you. Now, if you observe a train to be traveling due east at 15 m/s relative to and behind you, then you conclude that the relative speed between the jogger and the train is 10 m/s. Because all motion is assumed to be collinear and in the same direction, then the train must be approaching the jogger with a relative *velocity* of 10 m/s due east. A commonsense interpretation of these velocities (speeds and corresponding directions) can easily be associated with the symbolism adopted for our two inertial systems. From your point of view, you are a stationary observer in system S, the jogger represents an observer in system S′, and the train represents a particle in rectilinear motion. Consequently, a reasonable symbolic representation of the observed velocities would be $u = 5$ m/s, $v_x = 15$ m/s, and $v'_x = 10$ m/s, which would obey the mathematical relation

$$v'_x = v_x - u. \tag{1.27}$$

This equation represents the classical or Galilean transformation of velocities and is expressed as a *scalar* equation, because of our simplifying assumptions on rectilinear motion.

For those not appreciating the above commonsense arguments used for obtaining the velocity transformation equation, perhaps the following quantitative derivation will be more palatable. Consider the situation indicated in Figure 1.3, where a particle is moving in the $X$-$Y$ plane for some reasonable time interval $\Delta t = \Delta t'$. As the particle moves from position $P_1$ at time $t_1$ to position $P_2$ at time $t_2$, its rectilinear displacement is measured by an observer in $S$ to be $x_2 - x_1$. According to this observer, this *distance* is also given by his measurements of $x'_2 + u(t_2 - t_1) - x'_1$, as suggested in Figure 1.3. By comparing these two sets of measurements, the observer in system S concludes that

$$x'_2 - x'_1 = x_2 - x_1 - u(t_2 - t_1) \tag{1.28}$$

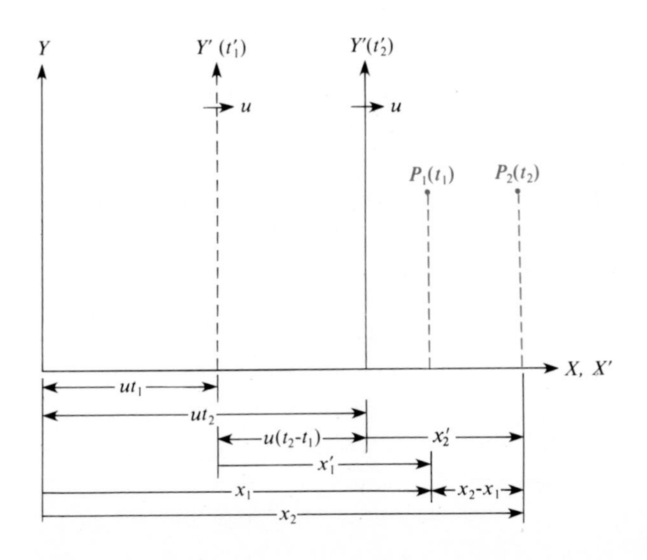


Figure 1.3
The displacement geometry of a particle at two different instants $t_1$ and $t_2$, as viewed by an observer in system S.

for *distance traveled* by the particle in the S′ system. It should be noted that for classical systems a displacement occurring over a *nonzero* time interval is *not invariant*, although previously we found that a length measurement made at an instant in time was invariant. Also, since the time interval for the particle's rectilinear displacement is

$$t_2' - t_1' = t_2 - t_1, \tag{1.29}$$

the division of Equation 1.28 by the time interval equation yields the expected velocity transformation given in Equation 1.27. This result is also easily produced by considering the coordinate transformations given by Equations 1.24a and 1.24d for the two positions of the particle in space and time. Further, the generalization to three-dimensional motion, where the particle has $x$, $y$, and $z$ components of velocity, should be obvious from the classical space-time transformation equations. The results obtained for the *Galilean velocity transformations* in three dimensions are

S′ → S

$$v_x' = v_x - u, \tag{1.30a}$$
$$v_y' = v_y, \tag{1.30b}$$
$$v_z' = v_z. \tag{1.30c}$$

Observe that the $y$- and $z$-components of the particle's velocity are invariant, while the $x$-components, measured by different inertial observers, are not invariant under a transformation between classical coordinate systems. We shall later realize that the $y$- and $z$-components of velocity are observed to be the same in both systems because of our commonsense assumption of *absolute time*. Further, note that the velocities expressed in Equations 1.30a to 1.30c should be denoted as *average* velocities (e.g., $\bar{v}_x'$, $\bar{v}_x$, etc.), because of the manner in which the derivations were performed. However, transformation equations for *instantaneous* velocities are directly obtained by taking the first order time derivative of the transformation equations for rectangular *coordinates* (Equations 1.24a to 1.24c). Clearly, the results obtained are identical to those given in Equations 1.30a to 1.30c, so we can consider all velocities in these equations as representing either *average* or *instantaneous* quantities. Further, a similar set of velocity transformation equations could have been obtained by taking the point of view of an observer in system S′. From Equations 1.25a through 1.25d we obtain

S → S′

$$v_x = v_x' + u, \tag{1.31a}$$
$$v_y = v_y', \tag{1.31b}$$
$$v_z = v_z', \tag{1.31c}$$

which are just the inverse of Equations 1.30a to 1.30c.

To finish our kinematical considerations, we consider taking a first order time derivative of Equations 1.30a through 1.30c. *or* Equations 1.31a through 1.31c. The same results

$$a'_x = a_x, \tag{1.32a}$$

$$a'_y = a_y, \tag{1.32b}$$

$$a'_z = a_z \tag{1.32c}$$

are obtained, irrespective of which set of velocity transformation equations we differentiate. These three equations for the components of acceleration are more compactly represented by

$$\mathbf{a}' = \mathbf{a}, \tag{1.33}$$

which indicates acceleration is invariant under a classical transformation. Whether **a** and **a**$'$ are regarded as *average* or *instantaneous* accelerations is immaterial, as Equation 1.33 is obtained by either operational derivation.

At the beginning of our discussion of classical transformations, we stated that an inertial frame of reference is one in which Newton's laws of motion are valid and that all inertial systems are equivalent for a description of physical reality. It is immediately apparent from Equation 1.33 that Newton's second law of motion is invariant with respect to a Galilean transformation. That is, since classical common sense dictates that mass is an invariant quantity, or

$$m' = m \tag{1.34}$$

for the mass of a particle as measured relative to system $S'$ or $S$, then from Equations 1.33 and 1.17 we have

$$\mathbf{F}' = \mathbf{F}. \tag{1.35}$$

Thus, the net external force acting on a body to cause its uniform acceleration will have the same magnitude and direction to all inertial observers. Since mass, time, acceleration, and Newton's second law of motion are invariant under a Galilean coordinate transformation, there is no *preferred* frame of reference for the measurement of these quantities.

We could continue our study of Galilean-Newtonian relativity by developing other transformation equations for classical dynamics (i.e., momentum, kinetic energy, etc.), but these would not contribute to our study of modern physics. There is, however, one other classical relation that deserves consideration, which is the transformation of *sound frequencies*. The classical **Doppler effect** for sound waves is developed in the next section from first principles of classical mechanics. An analogous pedagogic treatment for *electromagnetic waves* is presented in Chapter 3, with the inclusion of Einsteinian relativistic effects. As always, we consider only inertial systems that are moving relative to one another at a constant speed.

## 1.5 Classical Doppler Effect

It is of interest to know how the frequency of sound waves transforms between inertial reference frames. Sound waves are recognized as *longitudinal* waves and, unlike *transverse* light waves, they require a material medium for their propagation. In fact the speed of sound waves depends strongly on the physical properties (i.e., temperature, mass density, etc.) of the material medium through which they propagate. Assuming a uniform material medium, the speed of sound, or the speed at which the waves propagate through a *stationary* material medium, is constant. The basic relation

$$v_s = \lambda\nu, \tag{1.36}$$

requires that the product of the wavelength $\lambda$ and frequency $\nu$ of the waves be equal to their uniform speed $v_s$ of propagation. Classical physics requires that the relation expressed by Equation 1.36 is true for all observers who are at *rest* with respect to the transmitting material medium. That is, once sound waves have been produced by a vibrating source, which can either be at rest or moving with respect to the propagating medium, the speed of sound measured by *different spatial observers* will be identical, provided they are all stationary with respect to and in the same uniform material medium. Certainly, the measured values of frequency and wavelength in a system that is stationary with respect to the transmitting medium need not be the same as the measured values of frequency and wavelength in a *moving* system.

In this section the unprimed variables (e.g., $x$, $t$, $\lambda$, etc.) are associated with an observer in the *receiver* R system while the primed variables (e.g., $x'$, $t'$, etc.) are associated with the source of sound or *emitter* E′ system. In all cases the *transmitting material medium*, assumed to be air, is considered to be *stationary*, whereas the emitter E′ and receiver R may be either stationary or moving, relative to the transmitting medium. For the situation where the receiver R is stationary with respect to air, and the emitter E′ is receding or approaching the receiver, the speed of sound $v_s$ as perceived by R is given by Equation 1.36.

To deduce the classical frequency transformation, consider the emitter E′ of sound waves to be positioned at the origin of coordinates of the S′ reference frame. Let the sound waves be emitted in the direction of the receiver R, which is located at the origin of coordinates of the unprimed system and is *stationary with respect to air*. This situation, depicted in Figure 1.4, corresponds to the case where the emitter and detector *recede* from each other with a uniform speed $u$. In Figure 1.4 the *wave pulses* of the emitted sound are depicted by *arcs*. It should be noted that the *first wave pulse* received at R occurs at a time $\Delta t$ *after* the emitter E′ was activated (indicated by the dashed $Y'$-axis in the figure). The emitter E′ can be thought of as being activated by a pulse of light from R at time $t_1 = t_1'$. A continuous emission of sound waves traveling at approximately 330 m/s is assumed until the *first* sound wave is perceived by R at time $t_2 = t_2'$. As illustrated in Figure 1.4, E′ has moved through the distance $u\Delta t$ during the time $t_2 - t_1$ required for the first sound wave to travel the distance $v_s(t_2 - t_1)$ to R. When R detects the first sound wave, it transmits a light pulse traveling at the constant speed of essentially $3 \times 10^8$ m/s to E′, thereby stopping the emission of sound waves almost instantaneously. Consequently, the number of wave pulses $N'$ emitted by E′ in the time interval $\Delta t' = \Delta t$ is exactly the number of wave pulses $N$ that will be perceived *eventually* by R. With $x$ being defined as the distance between R and E′ at that instant in time when R detects the very *first* sound wave emitted by E′, we have

$$\lambda = \frac{x}{N}, \tag{1.37}$$

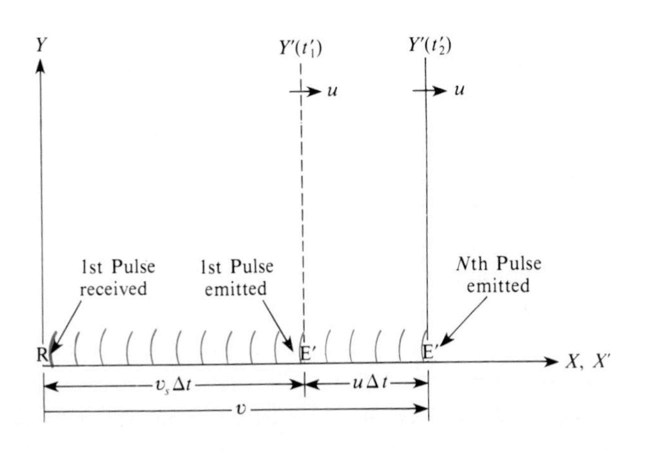


Figure 1.4
An emitter E′ of sound waves receding from a detector R, which is stationary with respect to air. E′ is activated at time $t_1'$ and deactivated at time $t_2'$, when R receives the first wave pulse.

where $\lambda$ is the wavelength of the sound waves according to an observer in the receiving system. Solving Equation 1.36 for $\nu$ and substituting from Equation 1.37 gives

$$\nu = \frac{v_s N}{x} \tag{1.38}$$

for the frequency of sound waves as observed in system R.

From Figure 1.4

$$x = (v_s + u)\Delta t, \tag{1.39}$$

thus Equation 1.38 can be rewritten as

$$\nu = \frac{v_s N}{(v_s + u)\Delta t}. \tag{1.40}$$

Substituting

$$N = N' = \nu' \Delta t' \tag{1.41}$$

into Equation 1.40 and using the Greek letter **kappa** ($\kappa$) to represent the ratio $u/v_s$,

$$\kappa \equiv \frac{u}{v_s}, \tag{1.42}$$

we obtain the relation

E′ receding from R

$$\nu = \frac{\nu'}{1 + \kappa}, \tag{1.43}$$

where the identity $\Delta t = \Delta t'$ has been utilized. Since the denominator of Equation 1.43 is always greater than one (i.e., $1 + \kappa > 1$), the detected frequency $\nu$ is always lower than the emitted or *proper frequency* $\nu'$ (i.e., $\nu < \nu'$). With musical pitch being related to frequency, in a subjective sense, then this phenomenon could be referred to as a *down-shift*. To appreciate the rationale of this reference terminology, realize that as a train recedes from you the pitch of its emitted sound is noticeably lower than when it was approaching. The appropriate wavelength transformation is obtained by using Equation 1.36 with Equation 1.43 and is of the form

$$\lambda = \frac{v_s}{\nu'}(1 + \kappa) \equiv \lambda'(1 + \kappa). \quad (1.44)$$

E′ receding from R

Since $1 + \kappa > 1$, $\lambda > \lambda'$ and there is a shift to larger wavelengths when an emitter E′ of sound waves recedes from an observer R who is stationary with respect to air.

What about the case where the emitter is *approaching* a receiver that is stationary with respect to air? We should expect the sound waves to be *bunched* together, thus resulting in an *up-shift* phenomenon. To quantitatively develop the appropriate transformation equations for the frequency and wavelength, consider the situation as depicted in Figure 1.5. Again, let the emitter E′ be at the origin of coordinates of the primed reference system and the receiver R at the origin of coordinates of the unprimed system. As viewed by observers in the receiving system R, a time interval $\Delta t = t_2 - t_1 = t_2' - t_1'$ is required for the very *first* wave pulse emitted by E′ to reach the receiver R, at which time the emission by E′ is terminated. During this time interval the emitter E′ has moved a distance $u\Delta t$ *closer* to the receiver R. Hence, the total number of wave pulses $N'$, emitted by E′ in the elapsed time $\Delta t'$, will be *bunched together* in the distance $x$, as illustrated in Figure 1.5. By comparing this situation with the previous one, we find that Equations 1.37 and 1.38 are still valid. But now,

$$x = (v_s - u)\Delta t \quad (1.45)$$

and substitution into Equation 1.38 yields

$$\nu = \frac{v_s N}{(v_s - u)\Delta t}. \quad (1.46)$$

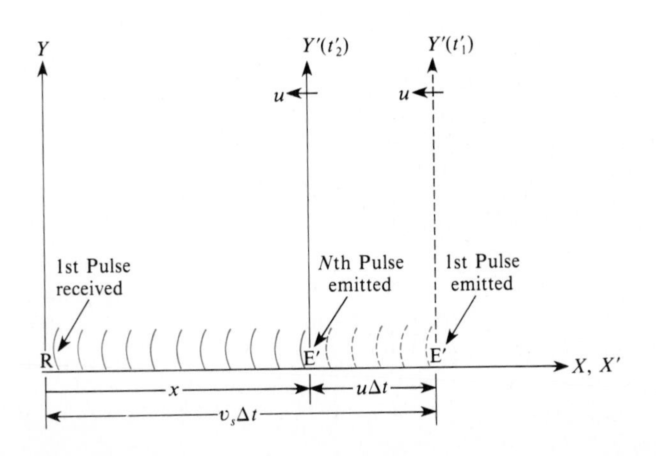


Figure 1.5
An emitter E′ of sound waves approaching a detector R, which is stationary with respect to air. E′ is activated at time $t_1'$ and deactivated at time $t_2'$, at the instant when R receives the first wave pulse.

Using Equations 1.41 and 1.42 with Equation 1.46 results in

E′ approaching R

$$\nu = \frac{\nu'}{1 - \kappa}. \tag{1.47}$$

Since $1 - \kappa < 1$, $\nu > \nu'$ and we have an *up-shift* phenomenon. Utilization of Equation 1.36 will transform Equation 1.47 from the domain of frequencies to that of wavelengths. The result obtained is

E′ approaching R

$$\lambda = \lambda'(1 - \kappa), \tag{1.48}$$

where, obviously, $\lambda < \lambda'$, since $1 - \kappa < 1$.

In the above cases the *receiver* R was considered to be *stationary* with respect to the *transmitting* material medium. If, instead, the *source* of the sound waves is *stationary* with respect to the material medium, then the transformation equations for frequency and wavelength take on a slightly different form. To obtain the correct set of equations, we need only perform the following inverse operations:

$$\nu \rightarrow \nu' \qquad \nu' \rightarrow \nu \qquad u \rightarrow -u. \tag{1.49}$$

Using these operations on Equations 1.43 and 1.47 gives

R receding from E′

$$\nu = \nu'(1 - \kappa) \tag{1.50}$$

R approaching E′

and

$$\nu = \nu'(1 + \kappa), \tag{1.51}$$

respectively. In the last two equations the receiver R is considered to be moving with respect to the transmitting medium of sound waves, while the emitter E′ is considered to be stationary with respect to the transmitting material medium. In all cases discussed above, $\nu'$ always represents the *natural* or *proper frequency* of the sound waves emitted by E′ in one system, while $\nu$ represents an *apparent frequency* detected by the receiver R in another inertial system. Clearly, the apparent frquency can be any one of *four* values for known values of $\nu'$, $v_s$, and $u$, as given by Equations 1.43, 1.44, 1.50, and 1.51.

For those wanting to derive Equation 1.50, you need only consider the situation as depicted in Figure 1.6. In this case the first wave pulse is perceived by R at time $t_1$, at which time the emission from E′ is terminated. R recedes from E′ at the constant speed $u$ while counting the $N'$ wave pulses. At time $t_2$ the last wave pulse emitted by E′ is detected by R and of course $N = N'$. Since the material medium is at rest with respect to E′

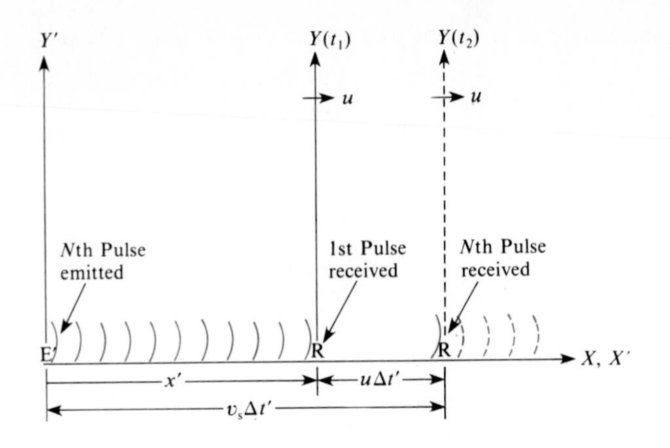


Figure 1.6
A detector R of sound waves receding from an emitter E′, which is stationary with respect to air. E′ is deactivated at time $t_1$, when R perceives the first wave pulse. R receives the last wave pulse at a later time $t_2$.

$$v_s = \lambda' \nu'. \tag{1.52}$$

Solving this equation for $\nu'$ and using

$$\lambda' = \frac{x'}{N'} \tag{1.53}$$

and

$$x' = (v_s - u)\,\Delta t', \tag{1.54}$$

we obtain

$$\nu' = \frac{v_s N'}{(v_s - u)\,\Delta t'}.$$

Realizing that

$$N' = N = \nu\,\Delta t \tag{1.55}$$

and, of course,

$$\Delta t = \Delta t',$$

we have the sought after result

$$\nu = \nu'(1 - \kappa), \tag{1.50}$$

R receding from E′

where Equation 1.42 has been used. A similar derivation can be employed to obtain the frequency transformation represented by Equation 1.51.

The wavelength $\lambda$ detected by R when R is receding from E′ is directly obtained by using Equation 1.50 and the fact that the speed of sound waves, as measured by R, is given by

R receding from E′

$$v = v_s - u = \lambda\nu. \tag{1.56}$$

Solving Equation 1.56 for the wavelength and substituting from Equation 1.50 for the frequency yields

$$\lambda = \frac{v_s - u}{\nu'(1 - \kappa)}, \tag{1.57}$$

which, in view of Equation 1.42, immediately reduces to

$$\lambda = \frac{v_s}{\nu'}. \tag{1.58}$$

Clearly, from Equations 1.52 and 1.58 we have

R receding from E′

$$\lambda = \lambda'. \tag{1.59}$$

This same result is obtained for the case where R is *approaching* the *stationary* emitter E′. By using Equations 1.51 and 1.52 and realizing that the speed of the sound waves as measured by R is given by

R approaching E′

$$v = v_s + u = \lambda\nu, \tag{1.60}$$

then we directly obtain the result

R approaching E′

$$\lambda = \lambda' \tag{1.61}$$

In each of the four cases presented either the receiver R or the emitter E′ was considered to be *stationary* with respect to *air*, the assumed transmitting material medium for sound waves. Certainly, the more general Doppler effect problem involves an emitter E′ and a receiver R both of which are *moving* with respect to air. Such a problem is handled by considering two of our cases *separately* for its complete solution. For example, consider a train traveling at 30 m/s due east relative to air, and approaching an eastbound car traveling at 15 m/s relative to air. If the train emits sound of 600 Hz, find the frequency and wavelength of the sound to

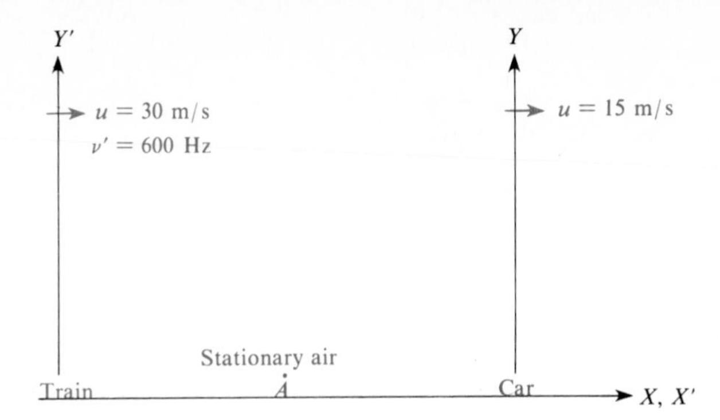


Figure 1.7
An emitter (train) and a receiver (car) of sound waves, both moving with respect to stationary air.

observers in the car for a speed of sound of $v_s = 330$ m/s. This situation is illustrated in Figure 1.7, where the reference frame of the train is denoted as the primed system and that of the car as the unprimed system. To employ the equations for one of our four cases, we must have a situation where either E′ or R is *stationary* with respect to air. In this example, we simply consider a point, such as *A* in Figure 1.7, between the emitter (the train) and the receiver (the car), that is stationary with respect to air. This point becomes the *receiver* of sound waves from the train and the *emitter* of sound waves to the observers in the car. In the first consideration, the emitter E′ (train) is *approaching* the receiver R (point *A*) and the frequency is determined by

$$\nu = \frac{\nu'}{1 - \kappa} = \frac{600\text{Hz}}{1 - \dfrac{30}{330}} = 660 \text{ Hz}.$$

Since the receiver R (point *A*) is stationary with respect to air, then the wavelength is easily calculated by

$$\lambda = \frac{v_s}{\nu} = \frac{330 \text{ m/s}}{660 \text{ Hz}} = 0.5 \text{ m}.$$

Indeed, the train's sound waves at any point between the train and the car have a 660 Hz frequency and a 0.5 m wavelength. Now, we can consider point *A* as the emitter E′ of 660 Hz sound waves to observers in the receding car. In this instance R is receding from E′, thus

$$\nu = \nu'(1 - \kappa) = (660 \text{ Hz})\left(1 - \frac{15}{330}\right) = 630 \text{ Hz}.$$

The wavelength is easily obtained, since for this case (R receding from E′)

$$\lambda = \lambda' = 0.5 \text{ m}.$$

Alternatively, the wavelength could be determined by

$$\lambda = \frac{v_s - u}{\nu} = \frac{330 \text{ m/s} - 15 \text{ m/s}}{630 \text{ Hz}} = 0.5 \text{ m},$$

since observers in the car are receding from a stationary emitter (point $A$) of sound waves. The passengers in the car will measure the frequency and wavelength of the train's sound waves to be 630 Hz and 0.5 m, respectively.

It should be understood that the velocity, acceleration, and frequency transformations are a direct and logical consequence of the space and time transformations. Therefore, any subsequent criticism of Equations 1.24a through 1.24d will necessarily affect all the aforementioned results. In fact there is an a priori criticism available! Is one entitled to assume that what is apparently true of one's own experience, is also absolutely, universally true? Certainly, when the speeds involved are within our domain of ordinary experience, the validity of the classical transformations is easily verified experimentally. But will the transformations be valid at speeds approaching the speed of light? Since even our fastest satellite travels approximately at a mere 1/13,000 the speed of light, we have no business assuming that $v_x' = v_x - u$ for all possible values of $u$. Our common sense (which a philosopher once defined as the total of all prejudices acquired by age seven) must be regarded as a handicap, and thus subdued, if we are to be successful in uncovering and understanding the fundamental laws of nature. As a last consideration before studying Einstein's theory on relative motion, we will review in the next section some historical events and conceptual crises of classical physics that made for the timely introduction of a consistent theory of special relativity.

## 1.6 Historical and Conceptual Perspective

The **classical principle of relativity** (CPR) has always been part of physics (once called natural philosophy) and its validity seems fundamental, unquestionable. Because it will be referred to many times in this section, and because it is one of the two basic postulates of Einstein's development of relativity, we will define it now by several equivalent statements:

1. The laws of physics are preserved in all inertial frames of reference.
2. There exists no preferred reference frame as physical reality contradicts the notion of absolute space.
3. An unaccelerated person is incapable of experimentally determining whether he is in a state of rest or uniform motion—he can only perceive relative motion existing between himself and other objects.

The last statement is perhaps the most informative. Imagine two astronauts in different spaceships traveling through space at constant but different velocities. Each can determine the velocity of the other relative to his system. But, neither astronaut can determine, *by any experimental measurement*, whether he is in a state of absolute rest or uniform motion. In fact, each astronaut will consider himself at rest and the other as moving. When you think about it, the *classical principle of relativity* is surprisingly subtle, yet it is completely in accord with common sense and classical physics.

The role played by the classical principle of relativity in the crises of theoretical physics that occurred in the years from 1900 to 1905 is schematically presented in Figure 1.8. Here, the relativistic space-time transformations (RT) were developed in 1904 by H. A. Lorentz, and for now

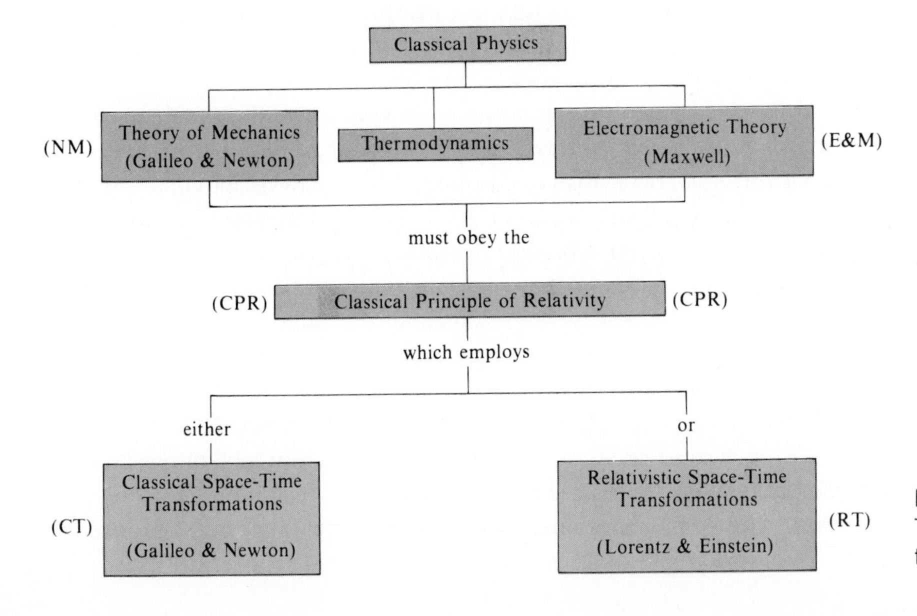


Figure 1.8
Theoretical physics at the turn of the 20th century.

we will simply accept it without elaborating. There are four points that should be emphasized about the consistency of the mathematical formalism suggested in Figure 1.8. Using the abbreviations indicated in Figure 1.8, we now assert the following:

1. NM obeys the CPR under the CT.
2. E&M does not obey the CPR under the CT.
3. E&M obeys the CPR under the RT.
4. NM does not obey the CPR under the RT.

The first statement asserts that NM, the CPR, and the CT are all compatible and in agreement with *common sense*. But the second statement indicates that Maxwell's equations are *not* **covariant** (invariant in form) when subjected to the CPR and the CT. New terms appeared in the mathematical expressions of Maxwell's equations when they were subjected to the classical transformations (CT). These *new terms* involved the relative speed of the two reference frames and predicted the existence of new electromagnetic phenomena. Unfortunately, such phenomena were never experimentally confirmed. This might suggest that the laws of electromagnetism should be revised to be covariant with the CPR and the CT. When this was attempted, not even the simplest electromagnetic phenomena could be described by the resulting laws.

Around 1903 Lorentz, understanding the difficulties in resolving the problem of the first and second statements, decided to retain E&M and the CPR and to replace the CT. He sought to mathematically develop a set of space-time transformation equations that would leave Maxwell's laws of electromagnetism invariant under the CPR. Lorentz succeeded in 1904, but saw merely the formal validity for the new RT equations and as applicable to only the theory of electromagnetism.

During this same time Einstein was working independently on this problem and succeeded in developing the RT equations, but his reasoning was quite different from that of Lorentz. Einstein was convinced that the propagation of light was invariant—a direct consequence of Maxwell's equations of E&M. The Michelson-Morley experiment, which was conducted prior to this time, also supported this supposition that electromagnetic waves (e.g., light waves) propagate at the same speed $c = 3 \times 10^8$ m/s relative to *any inertial reference frame*. One way of maintaining the invariance of $c$ was to require Maxwell's equations of electromagnetism (E&M) to be covariant under a transformation from S to S′. He also reasoned that such a set of space and time transformation equations should be the correct ones for NM as well as E&M. But, according to the fourth statement, Newtonian mechanics (NM) is incompatible with the principle of relativity (CPR), if the Lorentz (Einstein) transformation (RT) is used. Realizing this, Einstein considered that if the RT is universally applicable, and if the CPR is universally true, then the laws of NM cannot be completely valid at all allowable speeds of uniform separation between two inertial reference frames. He was then led to modify the laws of NM in order to make them compatible with the CPR under the RT. However, he was always guided by the requirement that these *new laws of mechanics* must reduce exactly to the classical laws of Galilean-Newtonian mechanics, when the uniform relative speed between two inertial reference frames is much less than the speed of light (i.e., $u \ll c$). This requirement will be referred to as the **correspondence principle,** which was formally proposed by Niels Bohr in 1924. Bohr's principle simply states that *any new theory must yield the same result as the corresponding classical theory, when the domains of the two theories converge or overlap*. Thus, when $u \ll c$, Einsteinian relativity must reduce to the well-established laws of classical physics. It is in this sense, and this sense only, that Newton's celebrated laws of motion are *incorrect*. Obviously, Newton's laws of motion are easily validated for our fastest rocket; however, we must always be on guard against unwarranted extrapolation, lest we predict incorrectly nature's phenomena.

## Review of Fundamental and Derived Equations

A listing of the fundamental and derived equations for sections concerned with classical relativity and the Doppler effect is presented below. Also indicated are the fundamental postulates defined in this chapter.

### GALILEAN TRANSFORMATION (S → S′)

$$\left.\begin{aligned} x' &= x - ut \\ y' &= y \\ z' &= z \\ t' &= t \end{aligned}\right\} \quad \text{Space-Time Transformations}$$

$$\left.\begin{aligned} v'_x &= v_x - u \\ v'_y &= v_y \\ v'_z &= v_z \end{aligned}\right\} \quad \text{Velocity Transformations}$$

$$m' = m \qquad \text{Mass Transformation}$$

$$\mathbf{a}' = \mathbf{a} \qquad \text{Acceleration Transformation}$$

$$\mathbf{F}' = \mathbf{F} \qquad \text{Force Transformation}$$

## CLASSICAL DOPPLER EFFECT

$v_s = \lambda\nu$ — $R$ stationary

$$\left.\begin{aligned} \nu &= \frac{\nu'}{1+\kappa} \\ \lambda &= \lambda'(1+\kappa) \end{aligned}\right\} \quad \text{E}' \text{ receding from R}$$

$$\left.\begin{aligned} \nu &= \frac{\nu'}{1-\kappa} \\ \lambda &= \lambda'(1-\kappa) \\ v_s &= \lambda'\nu' \end{aligned}\right\} \quad \begin{aligned} &\text{E}' \text{ approaching R} \\ &\text{E}' \text{ stationary} \end{aligned}$$

$$\left.\begin{aligned} \nu &= \nu'(1-\kappa) \\ v &= v_s - \text{u} = \lambda\nu \\ \lambda &= \lambda' \end{aligned}\right\} \quad \text{R receding from E}'$$

$$\left.\begin{aligned} \nu &= \nu'(1+\kappa) \\ v &= v_s + u = \lambda\nu \\ \lambda &= \lambda' \end{aligned}\right\} \quad \text{R approaching E}'$$

## FUNDAMENTAL POSTULATES

1. Classical Principle of Relativity
2. Bohr's Correspondence Principle

# Problems

**1.1** Starting with the defining equation for average velocity and assuming uniform translation acceleration, derive the equation $\Delta x = v_1 \Delta \text{t} + \frac{1}{2}a(\Delta t)^2$.

*Solution:*
For one-dimensional motion with constant acceleration, average velocity can be expressed as the **arithmetic mean** of the final velocity $v_2$ and initial velocity $v_1$. Assuming motion along the $X$-axis, we have

$$\bar{v} \equiv \frac{\Delta x}{\Delta t} = \frac{v_2 + v_1}{2}$$

and from the defining equation for average acceleration (Equation 1.9) we obtain

$$v_2 = v_1 + a\Delta t,$$

where the average sign has been dropped. Substitution of the second equation into the first equation gives

$$\frac{\Delta x}{\Delta t} = \frac{v_1 + a\Delta t + v_1}{2},$$

which is easily solved for $\Delta x$,

$$\Delta x = v_1 \Delta t + \tfrac{1}{2}a (\Delta t)^2.$$

**1.2** Starting with the defining equation for average velocity and assuming uniform translational acceleration, derive an equation for the final velocity $v_2$ in terms of the initial velocity $v_1$, the constant acceleration $a$, and the displacement $\Delta x$.

*Answer:* $v_2^2 = v_1^2 + 2a\Delta x$

**1.3** Do Problem 1.1 starting with Equation 1.5a and using calculus.

*Solution:*
Dropping the subscript notation in Equation 1.5a and solving it for $dx$ gives

$$dx = v dt.$$

By integrating both sides of this equation and interpreting $v$ as the final velocity $v_2$ we have

$$\int_{x_1}^{x_2} dx = \int_{t_1}^{t_2} v_2 \, dt.$$

Since $v_2 = v_1 + at$, substitution into and integration of the last equation yields

$$\Delta x = \int_{t_1}^{t_2} (v_1 + at)\, dt = v_1 \Delta t + \tfrac{1}{2}a(\Delta t)^2.$$

**1.4** Do Problem 1.2 starting with Equation 1.7 and using calculus.

*Answer:* $v_2^2 = v_1^2 + 2a\Delta x$

**1.5** Starting with $W = \mathbf{F} \cdot \mathbf{\Delta x}$ and assuming translational motion, show that $W = \Delta T$ by using the defining equations for *average* velocity and acceleration.

*Solution:*

$$
\begin{aligned}
W &= \mathbf{F} \cdot \mathbf{\Delta x} \\
&= F \cos \theta \, \Delta x && \text{definition of a dot product} \\
&= F \, \Delta x && \text{assuming } \theta = 0° \\
&= ma \, \Delta x && \text{assuming m} \neq \text{m(t)} \\
&= m \frac{\Delta v}{\Delta t} \Delta x && \text{from Eq. 1.9} \\
&= m\Delta v \frac{\Delta x}{\Delta t} \\
&= m\Delta v \left( \frac{v_2 + v_1}{2} \right) && \text{average velocity definition} \\
&= \tfrac{1}{2}\text{m}(v_2 - v_1)(v_2 + v_1) \\
&= \tfrac{1}{2}mv_2^2 - \tfrac{1}{2}mv_1^2 \\
&= \Delta T && \text{from Equation 1.22}
\end{aligned}
$$

**1.6** Starting with the defining equation for work (Equation 1.20) and using calculus, derive the work-energy theorem.

*Answer:* $W = \Delta T$

**1.7** Consider two cars traveling due east and separating from one another. Let the first car be moving at 20 m/s and the second car at 30 m/s relative to the highway. If a passenger in the second car measures the speed of an eastbound bus to be 15 m/s, find the speed of the bus relative to observers in the first car.

*Solution:*
Thinking of the first car as system S and the second as system S′, then

$$u = (30 - 20) \text{ m/s} = 10 \text{ m/s}.$$

With the speed of the bus denoted accordingly as $v_x' = 15$ m/s, $v_x$ is given by Equation 1.30a or Equation 1.31a as

$$v_x = v_x' + u = 15 \text{ m/s} + 10 \text{ m/s} = 25 \text{ m/s}.$$

**1.8** Consider a system S′ to be moving at a uniform rate of 30 m/s relative to system S, and a system S″ to be receding at a constant speed of 20 m/s relative to system S′. If observers in S″ measure the translational speed of a particle to be 50 m/s, what will observers in S′ and S measure for the speed of the particle? Assume all motion to be in the positive $x$-direction along the *common* axis of relative motion.

*Answer:* 70 m/s, 100 m/s

**1.9** A passenger on a train traveling at 20 m/s passes a train station attendant. Ten seconds after the train passes, the attendant observes a plane

500 m away horizontally and 300 m high moving in the same direction as the train. Five seconds after the first observation, the attendant notes the plane to be 700 m away and 450 m high. What are the space-time coordinates of the plane to the passenger on the train?

*Solution:*
For the train station attendant

$$x_1 = 500 \text{ m} \qquad y_1 = 300 \text{ m} \qquad t_1 = 10 \text{ s}$$
$$x_2 = 700 \text{ m} \qquad y_2 = 450 \text{ m} \qquad t_2 = 15 \text{ s}.$$

For the passenger on the train

$$x_1' = x_1 - ut_1 = 500 \text{ m} - \left(20 \frac{\text{m}}{\text{s}}\right)(10 \text{ s}) = 300 \text{ m},$$
$$y_1' = y_1 = 300 \text{ m},$$
$$t_1' = t_1 = 10 \text{ s},$$
$$x_2' = x_2 - ut_2 = 700 \text{ m} - \left(20 \frac{\text{m}}{\text{s}}\right)(15 \text{ s}) = 400 \text{ m},$$
$$y_2' = y_2 = 450 \text{ m},$$
$$t_2' = t_2 = 15 \text{ s}.$$

**1.10** From the results of Problem 1.9, find the velocity of the plane as measured by both the attendant and the passenger on the train.

*Answer:* 50 m/s at 36.9°, 36.1 m/s at 56.3°

**1.11** A tuning fork of 660 Hz frequency is *receding* at 30 m/s from a stationary (with respect to air) observer. Find the apparent frequency and wavelength of the sound waves as measured by the observer for $v_s = 330$ m/s.

*Solution:*
With $\nu' = 660$ Hz and $u = 30$ m/s for the case where E′ is *receding* from R,

$$\nu = \frac{\nu'}{1 + \kappa} = \frac{660 \text{ Hz}}{\frac{12}{11}} = 605 \text{ Hz}$$

and

$$\lambda = \frac{v_s}{\nu} = \frac{330 \text{ m/s}}{605 \text{ Hz}} = 0.545 \text{ m}.$$

**1.12** Consider Problem 1.11 for the case where the tuning fork is *approaching* the stationary observer.

*Answer:* 726 Hz, 0.455 m

**1.13** Consider Problem 1.11 for the case where the observer is *approaching* the stationary tuning fork.

*Solution:*
With $\nu' = 660$ Hz and $u = 30$ m/s for the case where R is *approaching* E′, we have

$$\nu = \nu' (1 + \kappa) = 660 \text{ Hz} \left(\frac{12}{11}\right) = 720 \text{ Hz},$$

$$\lambda = \frac{v_s + u}{\nu} = \frac{330 \text{ m/s} + 30 \text{ m/s}}{720 \text{ Hz}} = 0.5 \text{ m}$$

or

$$\lambda = \lambda' = 0.5 \text{ m}.$$

**1.14** Draw the appropriate schematic and derive the frequency transformation equation for the case where the emitter E′ is stationary with respect to air and the receiver R is approaching the emitter.

*Answer:* $\nu = \nu' (1 + \kappa)$

**1.15** Consider Problem 1.11 for the case where the observer is *receding* from the stationary tuning fork.

*Solution:*
Given that $\nu' = 660$ Hz, $u = 30$ m/s, $v_s = 330$ m/s, and R is receding from E′, then

$$\nu = \nu' (1 - \kappa) = 660 \text{ Hz} \left(\frac{10}{11}\right) = 600 \text{ Hz},$$

$$\lambda = \frac{v_s - u}{\nu} = \frac{330 \text{ m/s} - 30 \text{ m/s}}{600 \text{ Hz}} = 0.5 \text{ m}$$

or

$$\lambda = \lambda' = 0.5 \text{ m}.$$

**1.16** Consider a train to be traveling at a uniform rate of 25 m/s relative to stationary air and a plane to be in front of the train traveling at 40 m/s relative to and in the same direction as the train. If the engines of the plane produce sound waves of 800 Hz frequency, what is the frequency and wavelength of the sound wave to a ground observer located behind the plane for $v_s = 335$ m/s.

*Answer:* 670 Hz, 0.5 m

**1.17** What is the apparent frequency and wavelength of the plane's engines of Problem 1.16 to passengers on the train?

*Solution:*
Any stationary point in air between the plane and the train serves as a receiver of sound waves from the plane and an emitter of sound waves to the

passengers on the train. Thus, from the previous problem we have $\nu' =$ 670 Hz, $\lambda' = 0.5$ m, $u = 25$ m/s, and $v_s = 335$ m/s, where the receiver (train) is *approaching* the emitter (stationary point). For this case the frequency becomes

$$\nu = \nu' (1 + \kappa) = 670 \text{ Hz} \left(1 + \frac{5}{67}\right) = 720 \text{ Hz}$$

and the wavelength is given by

$$\lambda = \frac{v_s + u}{\nu} = \frac{360}{720} = 0.5 \text{ m} \qquad \text{or} \qquad \lambda = \lambda' = 0.5 \text{ m}.$$

**1.18** A train traveling at 30 m/s due east, relative to stationary air, is approaching an east bound car traveling at 15 m/s, relative to air. If the train emits sound of 600 Hz, find the frequency and wavelength of the sound to a passenger in the car for $v_s = 330$ m/s.

*Answer:* 630 Hz, 0.5 m

**1.19** A train traveling due west at 30 m/s emits 500 Hz sound waves while approaching a train station attendant. A driver of an automobile traveling due east at 15 m/s and emitting sound waves of 460 Hz is directly approaching the attendant, who is at rest with respect to air. For $v_s = 330$ m/s, find the frequency and wavelength of the train's sound waves to the driver of the automobile.

*Solution:*
From the train to the attendant we have $\nu' = 500$ Hz, $u = 30$ m/s, $v_s =$ 330 m/s, and E′ *approaching* R:

$$\nu = \frac{\nu'}{1 - \kappa} = \frac{500 \text{ Hz}}{1 - 1/11} = 550 \text{ Hz},$$

$$\lambda = \frac{v_s}{\nu} = \frac{330 \text{ m/s}}{550 \text{ Hz}} = \frac{3}{5} \text{ m}.$$

From the attendant to the automobile we have $\nu' = 550$ Hz, $u = 15$ m/s, $v_s = 330$ m/s, and R *approaching* E′:

$$\nu = \nu' (1 + \kappa) = 550 \text{ Hz} \left(1 + \frac{1}{22}\right) = 575 \text{ Hz},$$

$$\lambda = \lambda' = \frac{3}{5} \text{ m} \qquad \text{or} \qquad \lambda = \frac{v_s + u}{\nu} = \frac{345}{575} = \frac{3}{5} \text{ m}.$$

**1.20** After the automobile and train of Problem 1.19 pass the train station attendant, what is the frequency of the automobile's sound waves to passengers on the train?

*Answer:* 400 Hz

CHAPTER 2

# Basic Concepts of Einsteinian Relativity

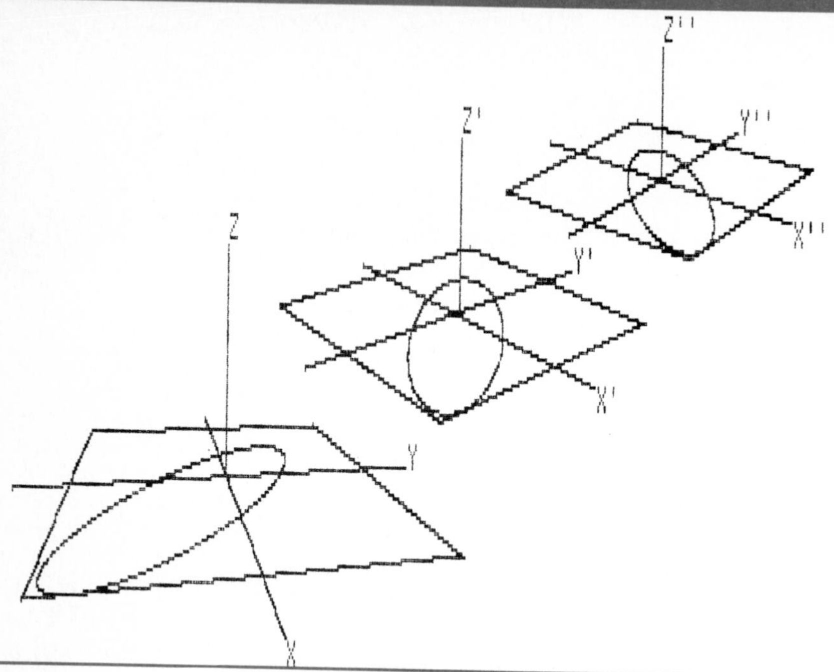


The relativity theory arose from necessity, from serious and deep contradictions in the old theory from which there seemed no escape. The strength of the new theory lies in the consistency and simplicity with which it solves all these difficulties, using only a few very convincing assumptions. . . . The old mechanics is valid for small velocities and forms the limiting case of the new one.

A. EINSTEIN AND L. INFELD, *The Evolution of Physics* (1938)

## Introduction

The discussion of Galilean relativity in the last chapter was mostly in accord with common sense. The results obtained were intuitive and in agreement with everyday experience for inertial systems separating from one another at relatively low speeds. The problems associated with comparing physical measurements made by different inertial observers were easily handled, once the Galilean space-time coordinate transformations were obtained. Central to Galilean relativity was the assumption of *absolute time*, which suggests that two clocks initially synchronized at $t = t' \equiv 0$ will remain *synchronized* when they are moving relative to one another at a constant speed. A direct consequence of this assumption is that *time interval measurements are invariant*, $\Delta t = \Delta t'$, for observers in different inertial systems, since temporal coordinates would be identical, $t = t'$, for all of time. Thus, *simultaneous* measurements of two spatial positions at an instant in time results in an *invariance of length*, $\Delta x = \Delta x'$, in Galilean relativity.

This chapter is primarily concerned with *time interval* and *length* measurements, along with the concepts of *synchronization* and *simultaneity*, within the framework of Einstein's **special theory of relativity.** Einsteinian relativity is an elegant theory that arises logically and naturally from two fundamental postulates: (1) the classical principle of relativity and (2) the invariance of the speed of light. Like Galilean relativity, Einstein's theory is concerned with problems involving the comparison of physical measurements made by observers in different inertial frames of reference. It differs significantly and fundamentally from Galilean relativity in that the two postulates of relativity are devoid of any *temporal* assumption. It will be shown that the *invariance of time*, as well as the *invariance of length*, must in general be abandoned, when the speed of light is assumed invariant. The results of Galilean relativity for time interval and length measurements are, however, directly obtained from Einsteinian relativity under the *correspondence principle*, for the situation where the uniform speed of separation between two inertial systems is *small* compared to the speed of light.

Einsteinian relativity has the completely undeserved reputation of being mathematically intimidating and conceptually mystifying to all but a few students. This misconception will be laid to rest in this and the next two chapters, as Einstein's special theory of relativity is fully developed utilizing only elementary mathematics of algebra, trigonometry, and occasionally introductory calculus. In developing Einsteinian relativity from the two fundamental postulates, *gedanken* (German for *thought*) experiments will be utilized in illustrating particular concepts that are not intuitive from everyday experience. Although some concepts will be introduced that are not in accord with common sense and may defy visualization, they will be intellectually stimulating and exciting to the imagination. These new concepts and results present a far reaching and nonclassical view of the intimate relationships between space, time, matter, and energy that is essential for the understanding of *microscopic* phenomena. Consequently, the study of special relativity is very important for students of contemporary (atomic, nuclear, solid state, etc.) physics and electrical engineering.

## 2.1 Einstein's Postulates of Special Relativity

The incompatibility of the laws of electromagnetism, the classical principle of relativity, and the Galilean space-time transformation led Einstein to a critical reevaluation of the concepts of space, time, and simultaneity. He decided to abandon the Galilean transformation of relativity and adopt a more fundamental principle of relativity that would be applicable to all physical laws—electromagnetism and mechanics. Einstein developed the *special theory of relativity* from the following two fundamental postulates:

1. the classical principle of relativity,
2. the invariance of the speed of light.

Einstein was the first to recognize the profound nature and universal applicability of the classical principle of relativity and to raise it to the status of a postulate. This postulate suggests that *all* physical laws, including those of electromagnetism and mechanics, are covariant in *all* inertial frames of reference. Not only are the mathematical interrelations of physical laws preserved, but also the values associated with all physical constants are identical in all inertial reference frames. Thus, the notion of an *absolute frame of reference* is forever discarded, and the concept of invariance is assumed for all of physics. The second postulate is a direct consequence of Maxwell's equations and a fact resulting from the Michelson-Morley experiment. It is *incompatible* with the first postulate under the Galilean transformation but toally *compatible* with the classical principle of relativity under the Lorentz transformation, as will be seen in the next chapter. For now, we will attempt to develop an understanding of the physical implications of Einstein's postulates, by considering some simple and intuitive examples involving two identical luxury vans traveling on a straight smooth road.

To elaborate on Einstein's first postulate, consider that you are riding in one of the vans at a constant velocity of 25 m/s due west. You lie back in the captain's chair against the headrest and feel very comfortable, not experiencing any bumps or accelerations. In fact, if you slump down in the chair and look far out at the horizon, you will not have any physical sensation of motion. Now, let a second identical van, being driven by your brother, approach you from behind at a uniform speed of 30 m/s relative to the ground. When your brother's van comes into view, you make some measurements and conclude that he is traveling relative to you at 5 m/s due west. On the other hand, if your brother measured your speed, he would conclude that you are *backing up toward him* with a relative speed of 5 m/s. You and your brother each determine that the other is moving with a relative speed of 5 m/s, but you can not make physical measurements that would determine whether you are moving, he is moving, or both of you are moving. This result is consistent with the classical principle of relativity and carries over to all possible physical measurements in each van. Further, the results of an experiment performed on any physical system in your van must yield identical numerical results to those obtained by your brother, when he conducts the same experiment on an identical system in his van. For example, if you measure the frequency of sound waves produced by your van's horn to be 550 Hz, then your brother would measure the fundamental frequency of sound from his van's horn to be 550 Hz. Also, you would measure the speed of light from your van's dome light to be $3 \times 10^8$ m/s (approximately) and your brother would obtain the same result for the speed of light from his van's dome light.

What about the situation where you measure the frequency of sound produced by your brother's horn and he measures the frequency of sound from your horn? From the results of the classical Doppler effect in the last chapter, you know that the two frequency measurements will be dissimilar, since your brother is approaching and you are receding from sound waves in *stationary air*. The relative speed between the sound waves in air and your brother would be $v_s + 30$ m/s, while it would be $v_s - 25$ m/s between the sound in air and you. In this situation the transporting material medium (air) for sound waves is another frame of reference in addition to the reference frames of the two vans.

What about the situation where you measure the speed of light from your brother's headlights and he measures the speed of light from your taillights? Unlike sound waves, light does not require any material medium for its transmission, and you need only consider your frame of reference and that of your brother. In this case the two vans are viewed as approaching each other with a constant speed of $u = 5$ m/s. As such, common sense dictates that the relative speed between you and the light from your brother's headlights would be $c + u$ and, likewise, between your brother and the light from your taillights. Although this result is intuitive and in agreement with Galilean relativity, it is *not consistent* with Einstein's second postulate. According to Einstein, you and your brother would measure the relative speed to be $c = 3 \times 10^8$ m/s, since the speed of light is invariant and independent of the relative motion between the *source* and *observer*. To emphasize this point, suppose your brother's van is overtaking you at half the speed of light and after he passes, you flash your headlights at him (possibly to indicate to him that in your considered opinion he is violating a traffic law). Einstein's second postulate predicts that he would measure the speed of light from your headlights to pass him at the normal speed of $c = 3 \times 10^8$ m/s! In fact, he would observe the same speed of light for that emitted from your headlights as you would observe, *regardless of his speed relative to you*. Surely, you must find this hard to believe and most remarkable. After all, does it not violate our notions of *common sense*? Yet, every experimental examination of this phenomenon verifies the truth of Einstein's second postulate. Serious experimental verification of Einstein's second postulate was not possible until the technological advances of the twentieth century. Perhaps because of this, the invariance of light was not realized nor even seriously contemplated, until Einstein published his work on relativity. It was his remarkable insight and understanding that brought about an intellectual and philosophical revolution and made man recognize the limitations of his dimensional experience. If natural phenomena exist that are conceptually beyond the grasp of man's reasoning—beyond deductive realization or physical verification—then we must unshackle our imaginations and arrest our beliefs, if we are to visualize the subtle laws of nature.

We will now deduce the basic results of Einstein's special theory of relativity from some *gedanken* experiments. As always, we consider two frames of reference, S and S′, to be separating from each other along their common $X$-$X'$ axis at the uniform speed $u$, after having coincided at time $t = t' \equiv 0$. Further, it is conceptually convenient to allow observers to exist at many different spatial positions at the same instant in time in both the S and S′ frames of reference. Thus, observers spatially separated and at rest in a particular frame of reference can synchronize their clocks and simplify their measurements of spatial and temporal intervals for a particle (or system) traveling at a relativistic speed.

## 2.2 Lengths Perpendicular to the Axis of Relative Motion

As a beginning to the development of Einsteinian relativity from the two fundamental postulates, consider whether coordinate axes that are perpendicular to an axis of relative motion in one frame of reference will be viewed by other inertial observers as being perpendicular. The gedanken experiment for this query is depicted in Figure 2.1, where a meter stick is aligned with the $Y$-$Z$ plane in system S and another with the $Y'$-$Z'$ plane in system S′. At the top of each meter stick is located a small plane mirror, labeled M in system S and M′ in the S′ system, adjusted so the mirror surfaces face each other. At some instant in time, a beam of light is emitted from M parallel to the $X$-axis at a distance $y$ above it and in the direction of the M′ mirror.

How does the beam of light appear to an observer in the S′ frame of reference? For consistency with Einstein's second postulate, he will view the beam of light as traveling at the uniform speed $c$ and incident on his

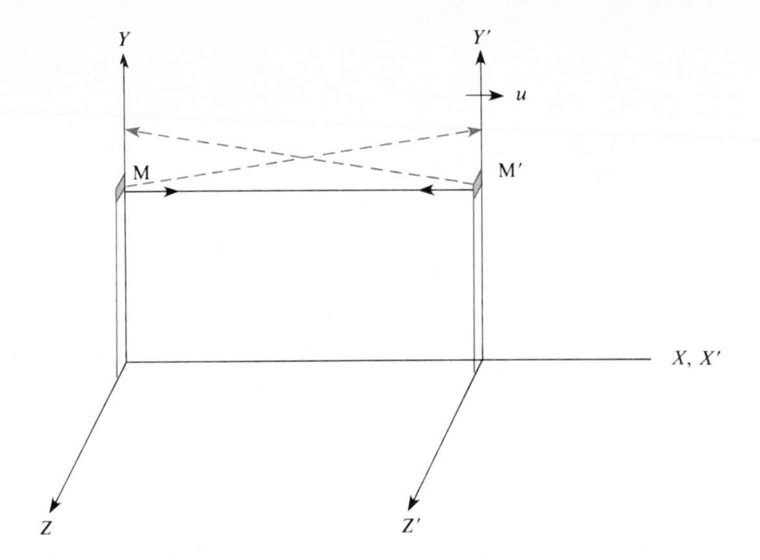


Figure 2.1
The reflection of a light beam between two plane mirrors set up parallel to one another in systems S and S′.

M′ mirror at some distance $y'$ above his $X'$-axis. When the beam of light strikes the M′ mirror, the observer in S′ notices the beam of light is reflected upon its incident path concluding that the beam of light is parallel to his $X'$-axis. This result is consistent with the initial requirements that the $X'$-axis coincides with the $X$-axis and that the beam of light was adjusted to be parallel to the $X$-axis. An observer in system S will also notice the beam of light to be reflected from M′ upon its incident path, in accordance with Einstein's first postulate. He concludes that the M′ mirror must have been perpendicular to the beam of light and, thus, M′ must be parallel to his mirror M.

The conclusion that the two mirrors are parallel would also be obtained by the observer in S′, if he were to initiate a beam of light from M′ parallel to his $X'$-axis and in the direction of M. As yet we do not know if $y = y'$, but the observers in both reference frames conclude that the two mirrors are parallel to one another and perpendicular to their common $X$-$X'$ axis. Since the mirrors are parallel, then so are the $Y$-$Z$ and $Y'$-$Z'$ planes. Thus, in general, observers in different inertial frames of reference see coordinate axes $Y$, $Y'$, $Z$, $Z'$ as being perpendicular to the $X$-$X'$ axis. Since our systems S and S′ coincided at time $t = t' \equiv 0$, in addition to being perpendicular to the axis of relative motion $Y$ is parallel to $Y'$ and, likewise, $Z$ and $Z'$ are parallel. In general, this need not be true for inertial systems that are not moving along a common axis; however, all inertial observers having a common axis of relative motion will view any length measurement made perpendicular to their common axis as being *normal* to that axis.

From the above discussion it is easy to argue that the distance of the light beam above the $X$-$X'$ axis is the same for observers in both systems (i.e., $y = y'$). Obviously, since the light beam strikes M′ and is reflected upon its incident path a distance $y'$ above the $X'$-axis, the value of $y'$ measured in S′ must be the same as the value $y$ measured in S. Suppose, however, that the light reflected from M′ strikes above the mirror M in system S. An observer in the S frame of reference would surmise that his meter stick was smaller than the meter stick in the S′ frame of reference. But, according to Einstein's first postulate, if an observer in S′ were to send out a beam of light from M′ parallel to the $X$-$X'$ axis so as to be reflected from M, he would conclude that *his* meter stick was smaller than the meter stick in system S. As illustrated in Figure 2.1, the two reflected rays for this hypothetical case would necessarily cross. However, since light, propagating at a constant velocity, travels in a straight line, parallel incident rays (in our case they are coincident) are reflected from parallel plane mirrors such that the reflected rays are parallel! Thus, there is a contradiction between Einstein's first and second postulates, which means that our initial supposition is in violation of nature's laws and, therefore, incorrect.

According to Einstein's first postulate, any supposition made by an observer in system S would necessarily be the same supposition made by an observer in the S′ frame of reference. If the supposition violates any known laws, like the first gedanken experiment or the fact that parallel light rays never cross, then the supposition is wrong. An analogous argument shows the impossibility of a beam of light being emitted from M, parallel to the $X$-$X'$ axis, and reflected from M′ such as to return *below* M. Consequently, the only possible conclusion is the one stated initially: *any length or coordinate measurement made perpendicular to the axis of relative motion has the same value for all inertial observers*. This generalization can be expressed as

$$y' = y \qquad \Delta y' = \Delta y \tag{2.1}$$

and

$$z' = z \qquad \Delta z' = \Delta z, \tag{2.2}$$

which are two of the three **Lorentz space-coordinate transformations.** Clearly, these are identical to the *Galilean space transformations* for the $y$, $y'$ and $z$, $z'$ coordinates. As yet, we do not know the relativistic spatial transformation for the $x$, $x'$ coordinates and we should not, at this time, make any assumptions regarding its form.

## 2.3 Time Interval Comparisons

In this gedanken experiment, illustrated in Figure 2.2, we let the plane of a mirror M′ be perpendicular to the $Y'$-axis at some distance $\Delta y'$ above the origin of coordinates. An observer in S′ sends a light pulse up the $Y'$-axis, where it is reflected upon its incident path by the mirror M′ and eventually absorbed at its point of origin O′. In accordance with Figure 2.2, any S′ observer will measure the distance from O′ to M′ to be given by

$$\Delta y' = c\left(\frac{\Delta t'}{2}\right) = \tfrac{1}{2}c\Delta t', \tag{2.3}$$

where $\Delta t'$ is defined as the time it takes the light pulse to travel from O′ to M′ and back to O′.

Observers in the S frame of reference do not see the motion and path of the light pulse in S′ as being vertical, since M′ (at rest relative to S′) is moving at a uniform speed $u$ relative to their reference frame. Instead, they observe the motion and path of the light pulse in S′ to be something like the isosceles triangle depicted in Figure 2.2, where $\Delta t \equiv t_B - t_A$ is the time interval, according to S observers, for the light pulse to go from $A$ to M′ to $B$. It should be obvious that, whereas S′ observers need only one clock to measure the time interval $\Delta t'$, S observers need *two* clocks for their measurement of the corresponding time interval $\Delta t$. In system S,

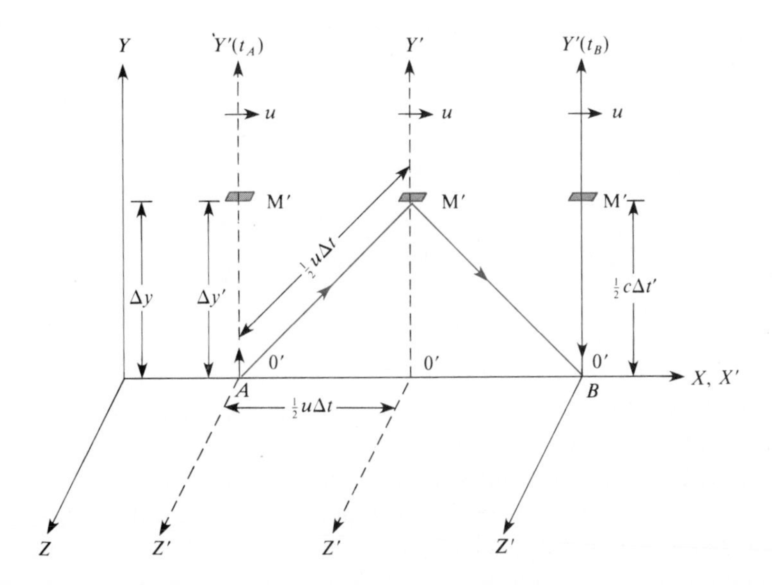


Figure 2.2
The path of a vertical pulse of light being reflected from a mirror in S′, as viewed by observers in S′ and in S.

we need one clock at $A$ and another at $B$, and since the two clocks are *spatially separated*, it is essential that they be *synchronized*. Later, we will consider a method by which clocks can be synchronized, but for now, we simply assume the synchronization of clocks at $A$ and $B$ has been effected. Clearly, $\Delta t$ corresponds to the difference between the reading of the two clocks at $A$ and $B$ in system S.

Viewing the left triangle of Figure 2.2 and invoking the Pythagorean theorem, we have

$$(\Delta y)^2 + (\tfrac{1}{2}u\,\Delta t)^2 = (\tfrac{1}{2}c\,\Delta t)^2,$$

which is easily solved for $\Delta y$ in the form

$$\Delta y = \tfrac{1}{2}c\,\Delta t\sqrt{1-\beta^2}, \tag{2.4}$$

where we have set

$$\beta \equiv \frac{u}{c}. \tag{2.5}$$

Here, $\Delta y$ is the vertical displacement of the mirror M$'$, as measured by observers in system S. Now, using Equation 2.1 from the previous gedanken experiment ($\Delta y = \Delta y'$) along with Equations 2.3 and 2.4, we have

$$\tfrac{1}{2}c\,\Delta t\sqrt{1-\beta^2} = \tfrac{1}{2}c\,\Delta t'. \tag{2.6}$$

By defining

$$\gamma \equiv \frac{1}{\sqrt{1-\beta^2}} \tag{2.7}$$

and solving Equation 2.6 for $\Delta t$, we obtain

Time Dilation

$$\Delta t = \gamma\,\Delta t'. \tag{2.8}$$

The result expressed by Equation 2.8 gives a comparison of time intervals measured in two different inertial reference frames. The meaning and implications of this result may need some elaboration. First, consider that in the limit as $u$ approaches $c$ in Equation 2.7 (remember $\beta = u/c$), $\gamma$ approaches $\infty$. However, for $u \ll c$, $\gamma$ approaches 1, and Equation 2.8 reduces to the Galilean transformation (Equation 1.29) in accordance with the *correspondence principle*. Thus, the range of values for $\gamma$ can be expressed as

$$1 \leq \gamma \leq \infty, \tag{2.9}$$

which implies that $\Delta t > \Delta t'$ in Equation 2.8 for $u$ *not small*. That is, the time interval $\Delta t$ measured between two events (emission and absorption of the light pulse) occurring at spatially *different* positions in system S is greater than the **proper time** $\Delta t'$, which is *the time interval measured in system S′ between two events occurring at the same position*. Because of the factor $\gamma$ in Equation 2.8, $\Delta t$ is *greater* than the *proper time* and is referred to as a *dilated time*. Hence, the name **time dilation** that is normally associated with Equation 2.8.

It is important to keep in mind Einstein's first postulate, because if the experiment were performed in system S, observers in S′ would conclude that $\Delta t' = \gamma \Delta t$. The significance of the time dilation result is that it emphasizes how time interval measurements differ between inertial observers in relative motion, because of differing physical measurements. *A time interval between two events is always shortest in a system where the events occur at the same position and dilated by the factor* $\gamma$ *in all other inertial systems*. In solving problems associated with the time dilation equation, it is most convenient and less confusing to identify the *proper time* as occurring in the *primed* system, with the *dilated time* then given by Equation 2.8 for the *unprimed* inertial system. As the concept of time dilation is most incredible, a few examples will be presented in an atttempt to clarify the subtleties of the phenomenon.

If you believe the secret of eternal youth is in keeping on the move, you are not far from wrong. In fact, time dilation predicts that if you have a twin, your *biological clocks* will be different, if one twin is traveling uniformly at a relativistic speed (e.g., $u > 0.4c$) relative to the other. For example, consider that at age twenty you take off in a rocket ship traveling at 0.866 the speed of light. You leave behind a younger brother of age ten and travel through space for twenty years (according to your clock). When you return home, at age forty, you will find your *kid brother* to be fifty years old! He will have aged by $\Delta t = 40$ years, while you have aged by only $\Delta t' = 20$ yrs. That is, time dilation gives

$$\Delta t = \gamma \Delta t' = (20 \text{ yrs})/\sqrt{1 - (.866c/c)^2} = 40 \text{ yrs}.$$

As compared to your brother, your biological clock and aging process was slowed by the time dilation effect. It should be emphasized that time dilation is a *real effect* that applies not just to clocks but to time itself—time *flows* at different rates to different inertial observers.

Other time dilation effects have been observed in laboratories for the lifetimes of radioactive particles. As a particular example, consider the decay of unstable elementary particles called **muons** (or **mu-mesons**). A muon is observed in a laboratory to decay into an electron in an average time of $2.20 \times 10^{-6}$ sec, after it comes into being. Normally, muons are created in the upper atmosphere by cosmic ray particles and travel with a uniform mean speed of $2.994 \times 10^8$ m/s $= 0.9987c$ toward the earth's surface. In their lifetime, the muons should be able to travel a distance of

$$s = u\,\Delta t = \left(2.994 \times 10^8 \frac{\text{m}}{\text{s}}\right)(2.20 \times 10^{-6}\ \text{s}) = 659\ \text{m},$$

according to the laws of *classical kinematics*. Since they are created at altitudes exceeding 6000 m, they should rarely reach the earth's surface. But they do reach the earth's surface and in profusion—approximately 207 muons per square meter per second are detected at sea level.

The *muon paradox* is immediately resolved, if we take *time dilation* into account. According to observers in the earth's reference frame, the muon's *proper mean lifetime* is $\Delta t' = 2.20 \times 10^{-6}$ s. The *mean* time that *we* observe the muons in motion is a time *dilated* to the value

$$\Delta t = \gamma\,\Delta t' = \frac{2.20 \times 10^{-6}\ \text{s}}{\sqrt{1 - \left(\dfrac{0.9987c}{c}\right)^2}}$$

$$= 43.1 \times 10^{-6}\ \text{s}.$$

Thus, according to Einstein's postulates, the muons should (and do) travel a mean distance of

$$\Delta x = u\,\Delta t = \left(2.994 \times 10^8 \frac{\text{m}}{\text{s}}\right)(43.1 \times 10^{-6}\ \text{s}) = 1.29 \times 10^4\ \text{m}$$

in *our reference frame*.

## 2.4 Lengths Parallel to the Axis of Relative Motion

Consider the situation schematically illustrated in Figure 2.3, where observers in S′ have placed a mirror M′ perpendicular to their $X'$-axis at

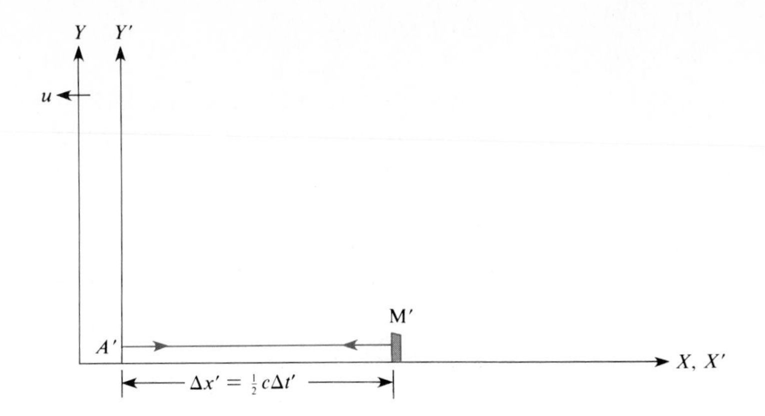


Figure 2.3
A horizontal pulse of light being reflected from a vertical mirror in S′, as viewed by observers in S′.

some distance $\Delta x'$ from the origin of their reference frame. In order to measure the distance $\Delta x'$, they send out a light pulse from $A'$ parallel to the $X'$-axis such that it will be reflected by M′ back to $A'$. For generality we will denote this distance by $\Delta x'$ and, thus, observers in S′ reason that

$$c = \frac{2\Delta x'}{\Delta t'},$$

where $\Delta t'$ is the time interval required for the light pulse to travel from $A'$ to M′ and back to $A'$. Thus, the distance they measure between $A'$ and M′ is given by

$$\Delta x' = \tfrac{1}{2}c\,\Delta t', \tag{2.10}$$

where $\Delta x'$ should be interpreted as a *length measurement*.

What do observers in system S measure, and how do they operationally perform the necessary measurements for the *length* in question? First consider the motion of the light pulse from $A'$ to M′ as viewed by observers in the S-frame and depicted in Figure 2.4. Observers in system S note the origination of the light pulse at point $A$ at time $t_A$. It propagates to the right at the uniform speed $c$, while the mirror M′ moves to the right uniformly at a speed $u$. At some later time $t_B$, the observers in S will note the light pulse striking the mirror M′. The two clocks in S record a time difference of $\Delta t_{A-B} \equiv t_B - t_A$ and, thus, the light pulse must have traveled the distance $c\Delta t_{A-B}$. With the distance from the origin of the $Y'$-axis to the mirror M′ (as measured by S-observers) denoted by $\Delta x$, then Figure

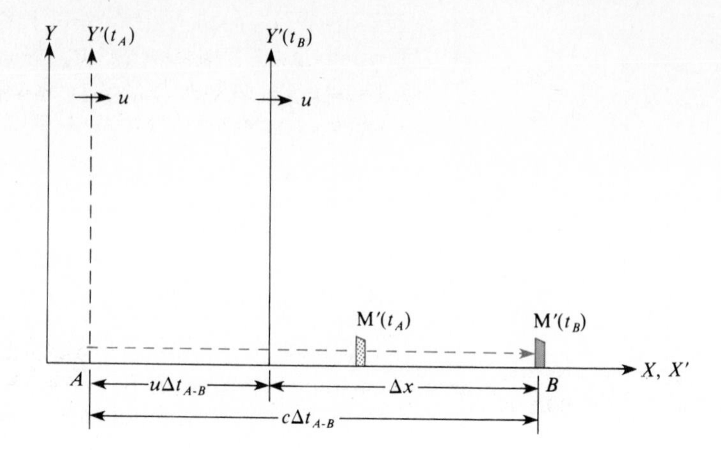


Figure 2.4
A horizontal pulse of light propagating towards a vertical mirror in S′, as viewed by observers in S.

2.4 suggests that

$$\Delta x = (c - u)\Delta t_{A-B}.$$

Thus, the time interval $\Delta t_{A-B}$ is given by

$$\Delta t_{A-B} = \frac{\Delta x}{c - u}. \tag{2.11}$$

Now, consider the motion of the light pulse from M′ back to the $Y'$-axis, as viewed by observers in system S. The situation, as depicted in Figure 2.5, suggests that the light pulse is reflected from M′ at point $B$ at the time $t_B$. It then propagates to the *left* at the speed of light $c$, arriving at the $Y'$-axis at point $C$ and time $t_C$. Of course, during the time interval $\Delta t_{B-C} = t_C - t_B$, the S′-frame and mirror M′ were moving to the *right* with constant speed $u$. In this case, the distance $\Delta x$ in question is immediately obtained from Figure 2.5 as

$$\Delta x = (c + u)\Delta t_{B-C}.$$

Solving the above equation for the time interval $\Delta t_{B-C}$ gives

$$\Delta t_{B-C} = \frac{\Delta x}{c + u}, \tag{2.12}$$

which is *not* the same as $\Delta t_{A-B}$, as far as observers in system S are concerned. These results also agree with classical common sense. That is, if

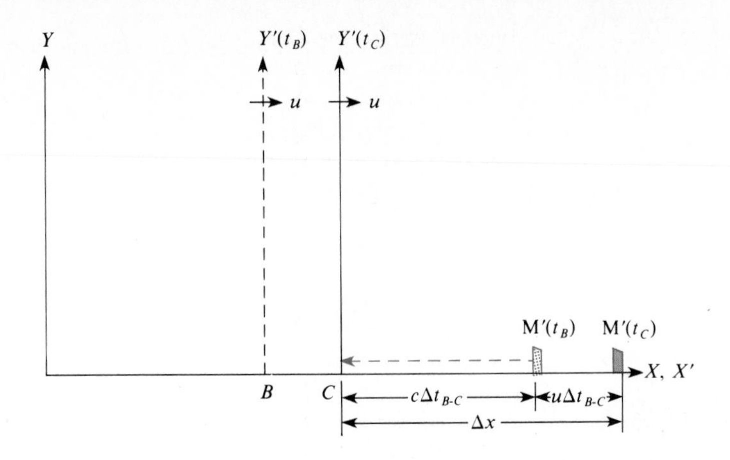


Figure 2.5
A horizontal pulse of light being reflected from a vertical mirror in S′, as viewed by observers in S.

the mirror and light pulse are traveling to the right, then the relative speed between the light pulse and the mirror is $c - u$ and Equation 2.11 is immediately obtained. On the other hand, if the mirror is moving to the right while the light pulse is traveling to the left, then the relative speed of separation between the mirror and the light pulse is $c + u$, which leads to Equation 2.12.

Let observers in S define the time interval $\Delta t$ as that time required for the pulse of light in S′ to travel from $A'$ to M′ and back to $A'$, as viewed in their reference frame.

Clearly, then

$$\Delta t = \Delta t_{A-B} + \Delta t_{B-C} = \frac{\Delta x}{c - u} + \frac{\Delta x}{c + u}$$

which, when solved for $\Delta x$, yields

$$\Delta x = \tfrac{1}{2}c\,\Delta t\left(1 - \frac{u^2}{c^2}\right). \tag{2.13}$$

Substitution of Equation 2.8 into Equation 2.13 gives

$$\Delta x = \tfrac{1}{2}c\,\Delta t'\gamma(1 - \beta^2), \tag{2.14}$$

where Equation 2.5 has been utilized for $\beta$. Now, from Equations 2.10

and 2.14 we have the famous Lorentz-Fitzgerald **length contraction** equation in the form

$$\Delta x = \Delta x' \sqrt{1 - \beta^2},$$

which simplifies to

Length Contraction

$$\Delta x = \frac{\Delta x'}{\gamma}, \tag{2.15}$$

by using the definition of $\gamma$ given by Equation 2.7. This equation describes how length measurements made parallel to the axis of relative motion compare between the two inertial frames of reference. In view of the result in Equation 2.9, the length measurement obtained by S-observers on a *moving* object is *always less* than the corresponding *proper length* measured by S′-observers on the object at rest (i.e., $\Delta x < \Delta x'$), when $u$ is *not small* and the length of the object is *parallel* to the axis of relative motion. Since $\Delta x < \Delta x'$, the terminology *contraction* is associated with Equation 2.15 and the name **proper length** *is always associated with a length measured in the rest frame of an object*. For $u \ll c$ we have $\gamma \approx 1$, and Equation 2.15 reduces to the Galilean transformation (Equation 1.26) in accordance with the correspondence principle.

The length contraction phenomenon has also been verified in actual laboratory experiments. For example, we can consider the muon paradox by taking into account *contracted displacements*. From the point of view of the muon, which is at rest relative to itself, it views the earth as traveling toward it with a speed of $u = 0.9987c$ for an average time of $\Delta t = 2.20 \times 10^{-6}$ s. Thus, the resulting displacement of the earth relative to the muon is

$$\Delta x = u\,\Delta t = 659 \text{ m},$$

which is a contracted length as measured by the muon. The proper length ($\Delta x'$ in the earth's reference frame) would correspond to

$$\Delta x' = \frac{\Delta x}{\sqrt{1 - \beta^2}} = 1.29 \times 10^4 \text{ m}.$$

This result is in perfect agreement with that obtained by the time dilation arguments, except for the interpretive meanings of $\Delta x$ and $\Delta x'$ being reversed. Note that the primed variables still refer to proper time intervals $\Delta t'$ and length measurements $\Delta x'$, and the problem can be solved by Equations 2.8 or 2.15.

## 2.5 Simultaneity and Clock Synchronization

Consider the situation as depicted in the schematic of Figure 2.6, where observers in S′ set up two identical clocks on the $X'$-axis that are spatially separated by the distance $\Delta x'$. The origin of the S′ reference frame is located at the midpoint between the clocks that are positioned at points $A'$ and $B'$. A flash bulb at the origin of S′ is used to send out simultaneous light pulses in the direction of $A'$ and $B'$ respectively, so that the clocks can be started at the instant the light pulses strike. In this manner, observers in system S′ can be assured of their clocks being started at the same instant in time (simultaneously started) and thus **synchronized** for all time.

The inquiry now is as to whether two events that appear to occur simultaneously to observers in S′ will be viewed as occurring simultaneously to observers in S. Clearly, when the flash bulb goes off, light pulses are simultaneously propagating at the speed $c$ in the direction of $A'$ and $B'$. But according to observers in S, $A'$ is approaching a light pulse at the uniform speed $u$, while $B'$ is receding from a light pulse at the speed $u$. As depicted in Figure 2.7a, the observers in S notice the clock at $A'$ being started at a time $t_A$, as recorded on *their* clock at position $A$. But at that particular instant in time, the observers in S do *not* notice the light

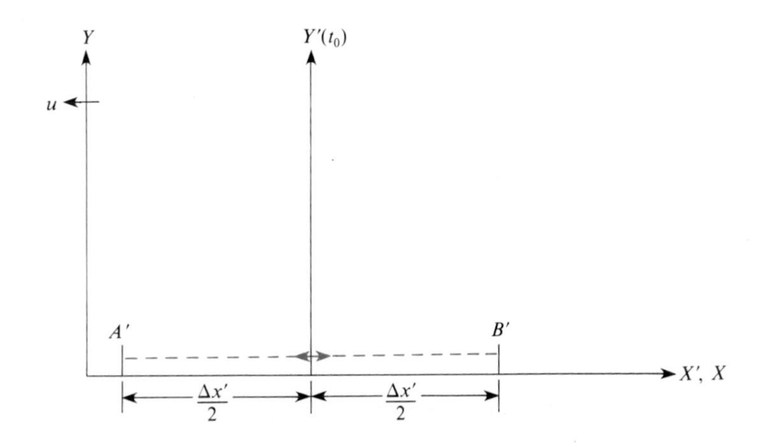


Figure 2.6
The synchronization of two clocks at rest in the S′ frame of reference, according to observers in system S′.

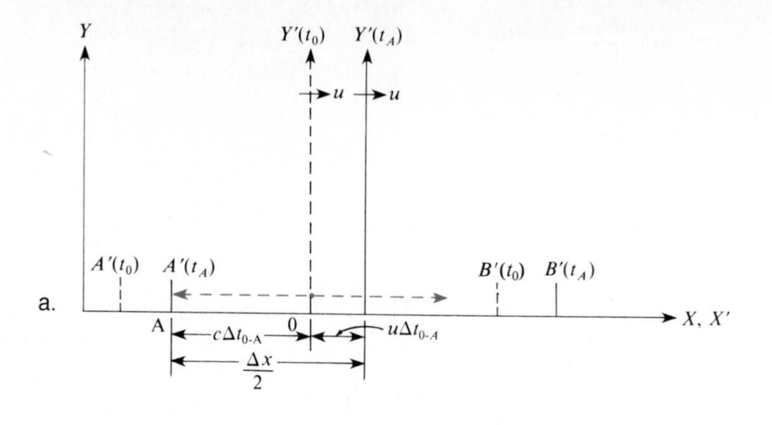


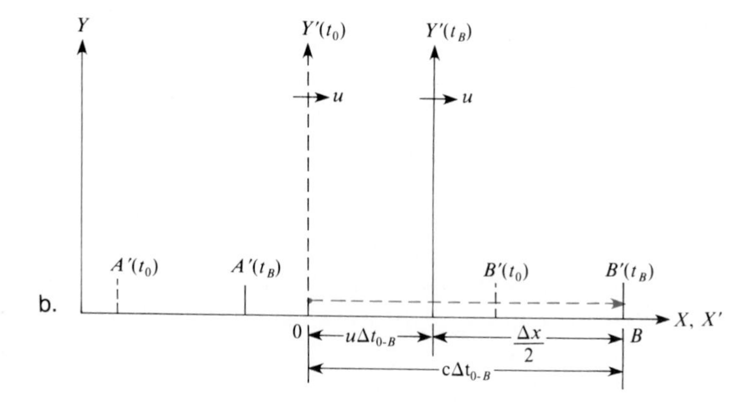


Figure 2.7
The synchronization of two clocks at rest in the S′ frame of reference, according to observers in system S.

pulse striking the clock at $B'$, since the clock at $B'$ has been moving to the right at a constant speed $u$. Instead, as suggested by Figure 2.7b, the clock at $B'$ will be started at a time later than $t_A \equiv t_A - t_0$, say $t_B \equiv t_B - t_0$, according to observers in S. From Figure 2.7,

$$\tfrac{1}{2}\Delta x = (c + u)t_A,$$

and observers in S conclude that the clock $A'$ in S′ starts at a time

$$t_A = \frac{\frac{1}{2}\Delta x}{c + u};$$

whereas, the clock at $B'$ in S′ starts at a later time

$$t_B = \frac{\frac{1}{2}\Delta x}{c - u}.$$

We can call this discrepancy in time, as viewed only by observers in S on events (the starting of the two clocks) occurring simultaneously in S′, a **synchronization correction** term, $\tau_s$, and define it by

$$\tau_s \equiv t_B - t_A = \tfrac{1}{2}\Delta x \left(\frac{1}{c - u} - \frac{1}{c + u}\right). \tag{2.16}$$

The algebraic simplification of Equation 2.16 gives the *synchronization correction* as

$$\tau_s = \frac{\Delta x\, u/c^2}{1 - \beta^2}. \tag{2.17}$$

Taking into account length contraction, as given by Equation 2.15, $\tau_s$ can be expressed in the more convenient form

Synchronization Correction

$$\tau_s = \frac{\Delta x' u/c^2}{\sqrt{1 - \beta^2}} = \gamma \frac{\Delta x' u}{c^2}, \tag{2.18}$$

where the definition given in Equation 2.7 has been used in the last representation. In Equation 2.18, $\Delta x'$ should be recognized as the *proper distance* between the two clocks in S′. It should also be noted that the *synchronization correction* of Equation 2.18 is entirely different from the previously discussed *time dilation* effect. Whereas synchronization is related to the nonuniqueness of simultaneity to different inertial observers, time dilation has to do with differing time interval measurements between two events by different inertial observers. For the meaning of the former, consider clocks in both reference frames to be identical and initially started at the same *instant* in time (say at the instant when both reference frames were coincident), such that all clocks are *initially* synchronized. Later on in time, we can say that all clocks in S are still synchronized and likewise for all clocks in S′, but the clocks in S are no longer synchronized with the clocks in S′. There is a *synchronization correction* resulting from the motion of separation between the two sets of clocks.

The form of Equation 2.18 can be varied to a still more convenient one by taking out the *time dilation* effect. That is, since

$$\Delta t = \gamma \Delta t',$$

we can write

$$\tau_s = \gamma \tau_s', \tag{2.19}$$

where, in accordance with Equation 2.18,

$$\tau_s' \equiv \frac{\Delta x' u}{c^2}. \tag{2.20}$$

$\tau_s'$ must be interpreted as the time discrepancy *expected* by observers in S to *exist* between the two S′ clocks according to observers in S′! It should be emphasized that from the point of view of the S frame of reference, observers in S′ should see the clock at $A'$ leading the clock at $B'$ by the time $\tau_s'$; but, according to observers in S′, there is no time discrepancy between the two clocks in S′. Think about the meaning of this in relation to Equation 2.19, when you consider the problems at the end of the chapter.

## 2.6 Time Dilation Paradox

According to Einstein's first postulate, all inertial systems are equivalent for the formulation of physical laws. This principle, combined with time dilation considerations, means that observers in any one inertial system will consider their own clocks to *run faster* than clocks in any other inertial system. This presents an apparently paradoxical situation, since observers in two different inertial systems view the other's clocks as *running slower*. As a particular case in point, consider Homer and Triper to be identical twins. Homer remains on the earth (home), while Triper travels to a distant planet and back at a relativistic speed. Because of time dilation, Homer considers his biological clock as being faster than Triper's and concludes that he will be older than Triper when Triper returns from his voyage. But, while Triper was traveling at a constant speed away from or toward the earth, he regarded himself as stationary and the earth as moving. As such, he would consider Homer's clock to be slower than his and would expect Homer to be younger than he is when he returns home. It is paradoxical to Triper to find that Homer is older than he when he returns to earth. The paradox arises from the seemingly symmetric roles played by the twins as contrasted with their asymmetric aging.

To particularize this example, let planet P be 20 light years (abbreviated as $c$-yrs) from, and stationary with respect to, the earth. Further, let the acceleration and deceleration times for Triper be negligible in comparison to his coasting times, where he has the uniform speed of $0.8c$. Actually, a detailed treatment of this problem would require the inclusion of accelerating reference frames, a topic requiring the *general theory of relativity*. However, we will attempt to gain some insight into this problem by using the synchronization disparity of moving clocks.

Consider the situation as depicted in Figure 2.8, where Triper is illustrated in system S′ as either receding or approaching the earth at the uniform speed of $0.8c$. Homer considers himself at rest and calculates the time required for Triper to go from the earth (E) to the planet (P) as

$$\Delta t_{E-P} = \frac{\Delta x}{u} = \frac{20\ c\text{-yrs}}{0.8c} = 25\ \text{yrs}.$$

Since the return trip would take the same time, $\Delta t_{P-E} = 25$ yrs, then Homer would age by

$$\Delta t_H = \Delta t_{E-P} + \Delta t_{P-E} = 50\ \text{yrs}$$

during Triper's voyage. But, according to Homer, Triper's clock would register a time change of only

$$\Delta t'_T = \Delta t_H\sqrt{1 - \beta^2} = (50\ \text{yrs})(0.6) = 30\ \text{yrs}.$$

Thus, Homer concludes that he will be 20 yrs older than his twin when Triper returns from the voyage.

From Triper's point of view, the distance from the earth to the planet is *contracted* to the value

$$\Delta x = \frac{\Delta x'}{\gamma} = \frac{20\ c\text{-yrs}}{5/3} = 12\ c\text{-yrs},$$

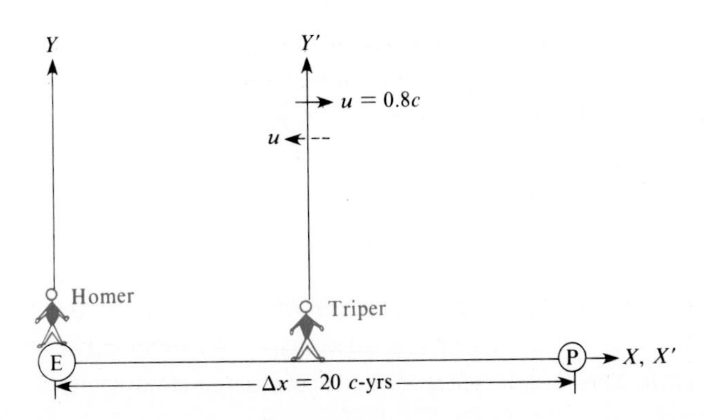


Figure 2.8
The aging of Triper as viewed by Homer.

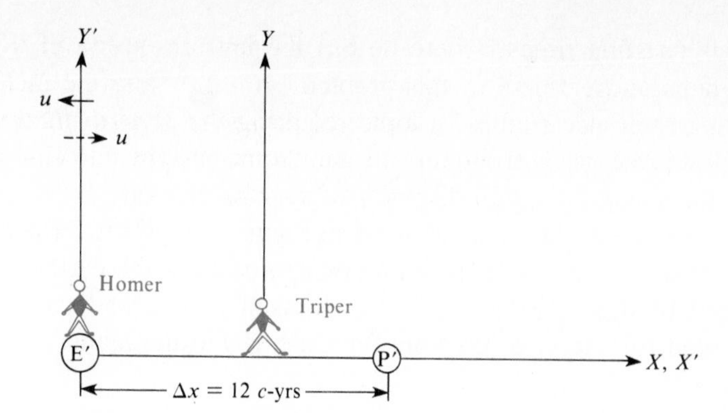


Figure 2.9
The aging of Homer as viewed by Triper.

as suggested by Figure 2.9. Here, of course, Triper considers himself in the stationary system S and Homer in the moving system S′. Accordingly, for $u = 0.8c$ we have

$$\Delta t_{E-P} = \frac{\Delta x}{u} = \frac{12\ c\text{-yrs}}{0.8c} = 15\ \text{yrs},$$

and the total time required for his trip would be

$$\Delta t_T = \Delta t_{E-P} + \Delta t_{P-E} = 30\ \text{yrs},$$

which is exactly what Homer predicted for Triper's time. But, the real problem is that Triper does not understand why Homer would age by 50 yrs. After all, Triper considers himself at rest in his system and Homer to be moving, which means that Homer's clock measures proper time. Thus, for 30 yrs to pass on Triper's clock, Homer's clock should record only

$$\Delta t'_H = \Delta t_T \sqrt{1 - \beta^2} = (30\ \text{yrs})(0.6) = 18\ \text{yrs},$$

which is of course the real paradox.

To understand this apparent discrepancy, let there be a clock on the planet P that is synchronized with Homer's clock on the earth E. To Triper these clocks are *unsynchronized* by the amount.

$$\tau'_s = \frac{\Delta x' u}{c^2} = \frac{(20\ c\text{-yrs})(0.8c)}{c^2} = 16\ \text{yrs}.$$

This means that when Triper is *approaching the planet*, the planet-clock *leads* the earth-clock by 16 yrs. When Triper momentarily stops at P, he is in the same reference system as the earth-planet system and observes the E-clock and P-clock to be synchronized. This means that in the decelerating time required for Triper to slow down to a stop (assumed negligible for Triper), the E-clock must have gained a time of 16 yrs. Of course, as Triper is returning home, he is *approaching* the earth and the E-clock *leads* the P-clock by 16 yrs, suggesting to Triper that the E-clock gained another 16 yrs while he was accelerating up to the $0.8c$ speed. When he lands on the earth, the P-clock is synchronized with the E-clock, indicating that the P-clock gained 16 yrs while Triper was decelerating to a stop. When Triper takes the total *synchronization correction time* into account, he realizes that Homer must have aged by

$$\Delta t'_{\mathrm{H}} = 18 \text{ yrs} + 16 \text{ yrs} + 16 \text{ yrs} = 50 \text{ yrs},$$

while he has aged by only 30 years. This result predicted by Einsteinian relativity becomes nonparadoxical only when the asymmetric roles of the twins is properly taken into account. Homer and Triper realize that the one (Homer) who remains in an inertial frame will age more than the one (Triper) who accelerates.

## Review of Derived Equations

A listing of the derived equations in this chapter is presented below, along with new defined units and symbols. Not included are the well-known definitions of kinematics.

### EINSTEIN'S POSTULATES

1. Classical Principle of Relativity
2. Invariance of the Speed of Light

### DEFINED UNITS

| | |
|---|---|
| $c$-yrs | Distance |

### SPECIAL SYMBOLS

$$\beta \equiv \frac{u}{c}$$

$$\gamma \equiv \frac{1}{\sqrt{1 - \beta^2}}$$

## DERIVED EQUATIONS

$$\Delta t = \gamma \Delta t' \qquad \text{Time Dilation}$$

$$\Delta x = \frac{\Delta x'}{\gamma} \qquad \text{Length Contraction}$$

$$\tau_s = \gamma \frac{\Delta x' u}{c^2} \qquad \text{Synchronization Correction}$$

## Problems

**2.1** Find the value of $\gamma$ for $u = (0.84)^{1/2}c$, $u = 0.6c$, $u = 0.8c$, and $u = 0.866c$.

*Solution:*
From Equations 2.5 and 2.7 we have

$$\gamma = \frac{1}{\sqrt{1 - u^2/c^2}},$$

so direct substitution for $u$ gives

$$\gamma = \frac{1}{\sqrt{1 - 0.84}} = \frac{1}{\sqrt{0.16}} = \frac{1}{0.4} = \frac{5}{2},$$

$$\gamma = \frac{1}{\sqrt{1 - 0.36}} = \frac{1}{\sqrt{0.64}} = \frac{1}{0.8} = \frac{5}{4},$$

$$\gamma = \frac{1}{\sqrt{1 - 0.64}} = \frac{1}{\sqrt{0.36}} = \frac{1}{0.6} = \frac{5}{3},$$

$$\gamma = \frac{1}{\sqrt{1 - 0.75}} = \frac{1}{\sqrt{0.25}} = \frac{1}{0.5} = 2$$

**2.2** Express $\gamma$ as a series, using the binomial expansion.

*Answer:* $\gamma = 1 + \frac{1}{2}\beta^2 + \frac{3}{8}\beta^4 + \cdots$

**2.3** Two inertial systems are receding from one another at a uniform speed of $0.6c$. In one system a sprinter runs 200 m in 20 s, according to his stopwatch. If the path of the sprinter is *perpendicular* to the axis of relative motion between the two systems, how far did the sprinter run and how long did it take him, according to observers in the other inertial system?

*Solution:*
Given $u = 0.6c$, $\Delta z' = 200$ m, $\Delta t' = 20$ s, and $\Delta z = \Delta z' = 200$ m, according to Equation 2.2. With $\beta \equiv u/c = 0.6$, $\gamma$ can be computed with Equation 2.7,

$$\gamma \equiv \frac{1}{\sqrt{1 - \beta^2}} = \frac{5}{4},$$

and Equation 2.8 immediately yields

$$\Delta t = \gamma \, \Delta t' = \left(\frac{5}{4}\right)(20 \text{ s}) = 25 \text{ s}.$$

**2.4** Consider Problem 2.3 for the situation where the path of the sprinter is *parallel* to the axis of relative motion between the two systems.

*Answer:* 160 m, 25 s

**2.5** An observer moves at $0.8c$ parallel to the edge of a cube having a proper volume of $15^3$ cm$^3$. What does the observer measure for the volume of the cube?

*Solution:*
With $u = 0.8c \rightarrow \gamma = 5/3$, $\Delta y = \Delta y' = 15$ cm, $\Delta z = \Delta z' = 15$ cm, and $\Delta x' = 15$ cm,

$$\Delta x = \frac{\Delta x'}{\gamma} = \frac{15 \text{ cm}}{5/3} = 9 \text{ cm}.$$

Accordingly, the volume measured by the observer is

$$V = \Delta x \, \Delta y \, \Delta z = (9 \text{ cm})(15 \text{ cm})(15 \text{ cm}) = 2025 \text{ cm}^3.$$

**2.6** Two inertial systems are uniformly separating at a speed of exactly $\sqrt{0.84}c$. In one system a jogger runs a mile (1609m) in 6 min along the axis of relative motion. How far in meters does he run and how long does it take, to observers in the other system?

*Answer:* 643.6 m, 15 min

**2.7** Consider two inertial systems separating at the uniform speed of $3c/5$. If a rod is parallel to the axis of relative motion and measures 1.5 m in its system, what is its length to observers in the other system?

*Solution:*
Using $u = 3c/5 \rightarrow \gamma = 5/4$ and the proper length as $\Delta x' = 1.5$ m,

$$\Delta x = \frac{\Delta x'}{\gamma} = \frac{1.5 \text{ m}}{5/4} = 1.2 \text{ m}.$$

**2.8** The proper mean lifetime of $\pi$-mesons with a speed of $0.90c$ is $2.6 \times 10^{-8}$ s. Compute their average lifetime as measured in a laboratory and the average distance they would travel before decaying.

*Answer:* $6.0 \times 10^{-8}$ s, 16 m

**2.9** A meterstick moves parallel to its length with a uniform speed of $0.6c$, relative to an observer. Compute the length of the meter stick as measured by the observer and the time it takes for the meter stick to pass him.

*Solution:*
With $u = 0.6c \rightarrow \gamma = 5/4$ and $\Delta x' = 1$ m,

$$\Delta x = \frac{\Delta x'}{\gamma} = \frac{1 \text{ m}}{5/4} = 0.8 \text{ m},$$

$$\Delta t = \frac{\Delta x}{u} = \frac{0.8 \text{ m}}{(0.6)(3 \times 10^8 \text{ m/s})} \approx 4.4 \times 10^{-9} \text{ s}.$$

**2.10** Two inertal systems are separating at the uniform rate of $0.6c$. If in one system a particle is observed to move parallel to the axis of relative motion between the two systems at a speed of $0.1c$ for $2 \times 10^{-5}$ s, how far does the particle move, according to observers in the other system?

*Answer:* 480 m

**2.11** How long must a satellite orbit the earth at the uniform rate of 6,711 mi/hr before its clock loses one second by comparison with an earth clock?

*Solution:*
With knowledge of $u = 6{,}711$ mi/hr $= 3 \times 10^3$ m/s and $\Delta t - \Delta t' = 1$ s, we need to find $\Delta t'$. Since

$$\Delta t - \Delta t' = \gamma \Delta t' - \Delta t' = \Delta t'(\gamma - 1),$$

the *proper time* can be expressed as

$$\Delta t' = \frac{\Delta t - \Delta t'}{\gamma - 1} = \frac{1 \text{ s}}{\gamma - 1}.$$

Unfortunately, since $u$ is very small compared to the speed of light, $\gamma$ in the last expression is very nearly one. To avoid this difficulty, we use the result obtained in Problem 2.2,

$$\gamma \approx 1 + \frac{1}{2}\beta^2,$$

and substitute into the previous expression to obtain

$$\Delta t' = \frac{1 \text{ s}}{\frac{1}{2}\beta^2} = \frac{2c^2}{u^2}(1 \text{ s}).$$

Using $c = 3 \times 10^8$ m/s and the value for $u$ gives

$$\Delta t' = 2 \times 10^{10} \text{ s} \approx 634 \text{ yrs}.$$

**2.12** What must be the relative speed of separation between two inertial observers, if their time interval measurements are to differ by ten percent? That is, for $(\Delta t - \Delta t')/\Delta t' = 0.10$, find $u$.

*Answer:* $0.417c$

**2.13** What must be the relative speed of separation between two inertial systems, for a length measurement to be contracted to 0.90 of its proper length?

*Solution:*
For this situation

$$\Delta x = 0.90\,\Delta x' \rightarrow \frac{\Delta x'}{\gamma} = 0.90\,\Delta x'.$$

Therefore, from the definition of $\gamma$ we have

$$\sqrt{1 - \beta^2} = 0.90 \rightarrow 1 - \beta^2 = 0.81 \rightarrow \beta^2 = 0.19.$$

Since $\beta = u/c$, $u = 0.436c$.

**2.14** A flying saucer passes a rocket ship traveling at $0.8c$ and the alien adjusts his clock to coincide with the rocket pilot's watch. Twenty minutes later, according to the alien, the flying saucer passes a space station that is stationary with respect to the rocket ship. What is the distance in meters between the rocket ship and the space station, according to (a) the alien and (b) the pilot of the rocket ship?

*Answer:* $2.88 \times 10^{11}$ m, $4.80 \times 10^{11}$ m

**2.15** Consider the situation described in Section 2.5, with the distance between $A'$ and $B'$ in S′ as 100 $c$-min. With $u = 0.6c$, compute the distance between the clocks at $A'$ and $B'$, as measured by observers in system S. Further, if stopwatches in S are started at the *instant* the flash-bulb in S′ goes off, show that a stopwatch in S reads 25 min when the light flash reaches $A'$ and another reads 100 min when the light flash reaches $B'$.

*Solution:*
With $u = 0.6c \rightarrow \gamma = 5/4$ and the proper length given as $\Delta x' = 100$ $c$-min, then

$$\Delta x = \frac{\Delta x'}{\gamma} = \frac{100\ c\text{-min}}{5/4} = 80\ c\text{-min},$$

$$t_A = \frac{\frac{1}{2}\Delta x}{c + u} = \frac{40\ c\text{-min}}{1.6c} = 25\ \text{min},$$

$$t_B = \frac{\frac{1}{2}\Delta x}{c - u} = \frac{40\ c\text{-min}}{0.4c} = 100\ \text{min}.$$

**2.16** According to observers in system S of Problem 2.15, how much time elapses between the activation of the two clocks in S′? How much time do they *expect* to have elapsed on the clock at $A'$ in S′, when the $B'$ clock is activated?

*Answer:* 75 min, 60 min

**2.17** Two explosions, separated by a distance of 200 $c$-min in space, occur simultaneously to an earth observer. How much time elapses between the two explosions, according to aliens traveling at $0.8c$ parallel to a line connecting the two events?

*Solution:*
With proper distance between the explosions being measured by the earth observer, we have $\Delta x' = 200$ $c$-min and $u = 0.8c \rightarrow \gamma = 5/3$. Thus, the aliens see the explosions occurring a distance of

$$\Delta x = \frac{\Delta x'}{\gamma} = \frac{200\ c\text{-min}}{5/3} = 120\ c\text{-min}$$

apart, during the time interval

$$\tau_s = \gamma\frac{\Delta x' u}{c^2} = \left(\frac{5}{3}\right)\frac{(200\ c\text{-min})(0.8c)}{c^2} = \frac{800}{3}\ \text{min}.$$

**2.18** How much time will the aliens of Problem 2.17 *expect* to have elapsed, on an earth clock, between the occurrence of the two explosions? Is this time interval equal to that measured by the earth observer?

*Answer:* 160 min, No

**2.19** Two inertial systems are uniformly separating at a speed of $0.8c$. A gun fired in one system is equidistant from two observers in that system. Both observers hear the shot 6 s after it was fired and each raises a flag. If the speed of sound in that system is 300 m/s, how much time elapses between the occurrence of the two events (raising the flags), according to observers in the *other* system?

*Solution:*
In this problem we know $u = 0.8c \rightarrow \gamma = 5/3$, $\Delta t' = 6$ s, and $v_s' = 300$ m/s. The question is answered by finding $\tau_s$, which requires that we first compute $\Delta x'$. Accordingly,

$$\tfrac{1}{2}\Delta x' = v_s' \, \Delta t' = \left(300 \, \frac{\text{m}}{\text{s}}\right)(6 \text{ s}) = 1800 \text{ m},$$

which results in

$$\Delta x' = 3600 \text{ m},$$
$$\tau_s = \gamma \frac{\Delta x' u}{c^2} = \left(\frac{5}{3}\right) \frac{(3600 \text{ m})(0.8c)}{(3 \times 10^8 \text{ m/s})(c)} = 16 \times 10^{-6} \text{ s}.$$

**2.20** Consider the situation described for Homer and Triper in Section 2.6, with $u = 0.6c$ and the distance between the earth and planet 9 $c$-yrs. How many years will Homer and Triper age, during Triper's voyage?

*Answer:* 30 yrs, 24 yrs

CHAPTER 3

# Transformations of Relativistic Kinematics

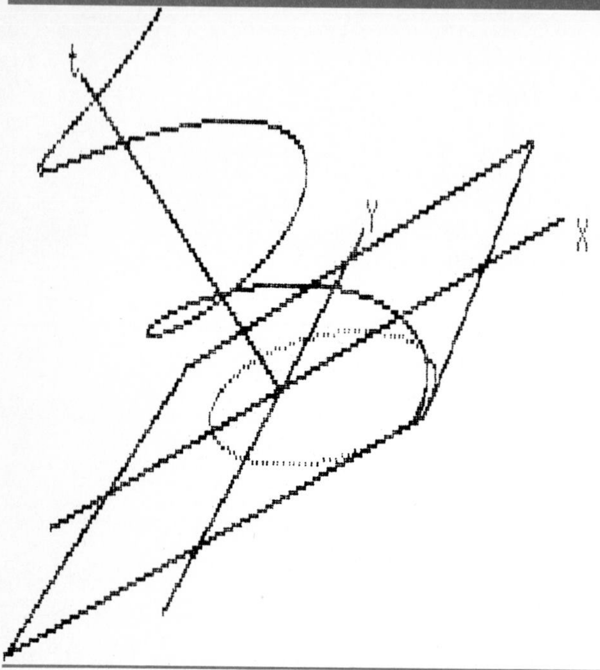


The velocity of light forms the upper limit of velocities for all material bodies. . . . The simple mechanical law of adding and subtracting velocities is no longer valid or, more precisely, is only approximately valid for small velocities, but not for those near the velocity of light.
A. EINSTEIN AND L. INFIELD, *The Evolution of Physics* (1938)

## Introduction

The initial consideration of Einsteinian relativity in the preceding chapter was based totally on two fundamental postulates and basic physical reasoning (logic) applied to several gedanken experiments. The results obtained for time dilation and length contraction were in stark contrast to the time interval ($\Delta t = \Delta t'$) and length measurement ($\Delta x = \Delta x'$) transformation equations predicted by Galilean relativity in Chapter 1. It was demonstrated, however, that these two relativistic effects reduced exactly to their classical counterparts under the *correspondence principle*, and we can expect any extensions of Einsteinian relativity in kinematics and dynamics to reduce to their corresponding classical transformations, when the relative speed between two inertial systems is *small* compared to the speed of light.

Since the classical time interval and length measurement transformation equations were a direct consequence of the Galilean space-time coordinate transformations, Equations 1.25a and 1.25d, then the results of the last chapter clearly illustrate the inconsistency of the Galilean transformations with the basic postulates of Einstein's special theory of relativity. As such, we need a new set of space-time coordinate transformation equations capable of relating the position and time variables $x$, $y$, $z$, and $t$ of an event measured in one coordinate system with the coordinates $x'$, $y'$, $z'$, and $t'$ of the same event as measured in another system, when there is uniform relative motion between the two systems. The correct spatial transformations are obtained in the next section by incorporating the length contraction effect with physical arguments similar to those presented in Section 1.4. The relativistic temporal transformation equation is then directly obtained from either the spatial transformations or by qualitative arguments combining time dilation and synchronization phenomena. After the relativistic coordinate transformations are fully developed and compared with the corresponding classical transformations, the relativistic kinematic transformation equations for velocity and acceleration are derived from first principles in a manner similar to that presented in Chapter 1. Finally, the relativistic Doppler effect for transverse electromagnetic waves is considered using arguments analogous to those presented in the development of the classical Doppler effect for longitudinal sound waves.

## 3.1 Relativistic Spatial Transformations

To develop the relativistic *spatial* transformation equations, we consider two inertial systems S and S$'$ to be separating from one another at a constant speed $u$ along their common $X$-$X'$ axis. As always, we allow that the origins of S and S$'$ coincided at time $t = t' \equiv 0$ and that identical clocks in both systems were started *simultaneously* at that instant. To avoid any conflict with the concept of simultaneity, the spatial coordinates of the clocks in both systems should be identical ($x = x'$, $y = y'$, and $z = z'$) at the instant $t = t' \equiv 0$, so that we are comparing two clocks at the same point in space. Now, consider a particle moving about in space being viewed by observers in both systems S and S$'$. The immediate problem is to deduce the relation between the S-coordinates ($x$, $y$, and $z$) and S$'$-coordinates ($x'$, $y'$, and $z'$) of the particle's position at an instant $t > 0$ in time. From the results given by Equations 2.1 and 2.2, we know the relations $y = y'$ and $z = z'$ are valid for all of time. This suggests that the physical considerations for the particle being viewed can be simplified by allowing $y = y' = 0$ and $z = z' = 0$ at some instant $t > 0$ in time. Accordingly, Figure 3.1 depicts the position of the particle on the $X$-$X'$ axis at point $P$ in system S and at point $P'$ in system S′. The relation between the coordinates $x$ and $x'$ will obviously depend on the frame of reference assumed, so each point of view will be considered separately.

The point of view of observers in system S is illustrated in Figure 3.1a by length measurements given below the $X$-$X'$ axis. The particle's nonzero space-time coordinates are denoted as $x$ and $t$, where the coordinate $t$ represents the time that has elapsed on a clock in system S after the two systems coincided. Consequently, observers in S would measure the distance of separation between their origin O and the origin O′ as $ut$, while the distance between O′ and $P'$ would be viewed as being contracted to the value $x'/\gamma$. Thus, from Figure 3.1a observers in S conclude that

S′ → S

$$x' = \gamma\,(x - ut), \tag{3.1a}$$
$$y' = y, \tag{3.1b}$$
$$z' = z, \tag{3.1c}$$

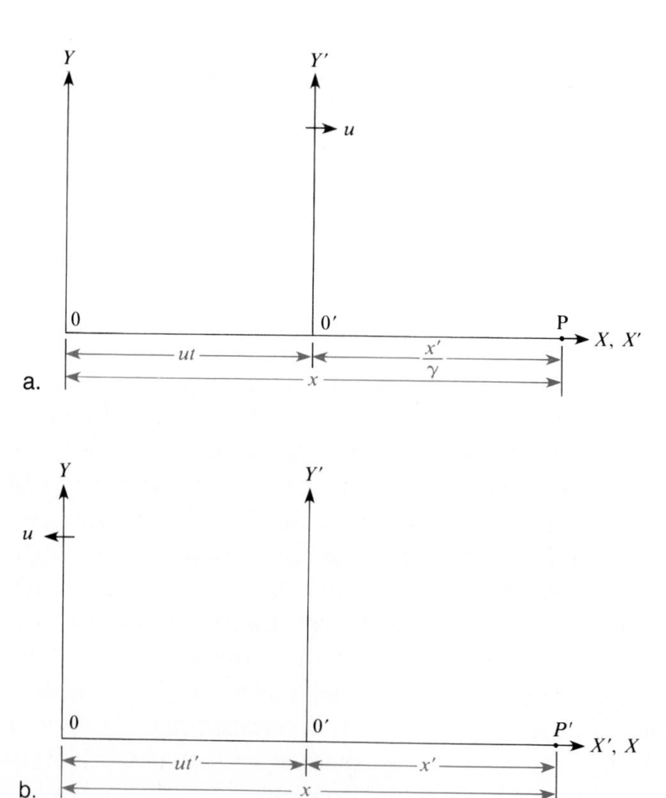


Figure 3.1
The spatial $x$-coordinate transformations from *(a)* S to S′ and *(b)* S′ to S.

are the correct set of *relativistic* transformation equations for *spatial coordinates*, where the latter two were previously obtained and noted as Equations 2.1 and 2.2, respectively. These equations constitute part of what is commonly called the **Lorentz transformation,** with the other part being the relativistic relation between $t'$ and $t$. They were first derived by H. A. Lorentz; however, it was not until a number of years later that their real significance was fully understood and explained by Einstein. Since $\gamma \approx 1$ in the limit of small $u$, these relativistic transformations reduce exactly to the Galilean transformation given by Equations 1.24a through 1.24c.

Observers in system S′ can also deduce a set of spatial transformation equations, as suggested by the geometry given in Figure 3.1b.. At that *instant* in time $t' > 0$ when the particle is on the $X'$ axis, observers in S′ view the distance from O′ to $P'$ as $x'$, the distance from O′ to O as $ut'$, and the distance from O to $P$ as *contracted* to the value $x/\gamma$. Accordingly, their spatial transformation equations are of the form

$$x = \gamma (x' + ut'), \tag{3.2a}$$

S → S′

$$y = y', \tag{3.2b}$$

$$z = z'. \tag{3.2c}$$

It should be noted that these equations are just the inverse of Equations 3.1a to 3.1c, which can be obtained from the first set by replacing primed with unprimed variables and vice versa, and by replacing $u$ with $-u$. Again, these equations reduce to the ordinary classical transformations, Equations 1.25a to 1.25c, when the relative speed of separation $u$ between S and S′ is *small* compared to the speed of light $c$. To complete our discussion of the space-time *Lorentz transformation*, the next section considers the manner in which time coordinates in S and S′ transform.

## 3.2 Relativistic Temporal Transformations

The correct relativistic relation between the time coordinates $t$ and $t'$ of systems S and S′ is easily obtained by direct *qualitative* arguments combining the effects of *time dilation* and *synchronization*. Since all clocks in both systems were *simultaneously* set to the value of zero, at that *instant* when the two systems coincided, then all clocks in both systems are *initially synchronized*. However, after that instant the clocks in system S are *no longer synchronized* with the clocks in system S′ (recall Section 2.5), because of the relativistic motion between the two systems. Now consider a stationary clock in system S′ to be located at point $P'$ in Figure 3.1, and another clock to be at point $P$ in system S, at that instant in time when the particle is on the $X$-$X'$ axis. The clock in S′ at $P'$ is *unsynchronized* with the clock in S at $P$ by the amount $\tau_s$. Further, any time measurement made in S′ will correspond to a dilated time in system S. Combining these arguments with the fact that $\tau_s = (\tau_s')_{\text{dilated}}$ suggests that a time measurement of $t$ in system S corresponds to a time measurement of $t'$ in S′ being increased by the amount $\tau_s'$ and the resulting total time being *dilated*. That is,

$$t = (t' + \tau_s')_{\text{dilated}}$$

or

$$t = \gamma\,(t' + \tau_s'). \tag{3.3}$$

Substitution of Equation 2.20 into Equation 3.3, with $\Delta x'$ being replaced appropriately with $x'$, gives the more fundamental result

S′ → S

$$t = \gamma\left(t' + \frac{x'u}{c^2}\right). \tag{3.4}$$

This is the correct relativistic *time coordinate* transformation equation that constitutes the other part of the *inverse Lorentz transformation* referred to in the previous section. It gives the relation between the measurement of the *time coordinate* of an event occurring in S′ to the corresponding measurement made on the *same event* occurring in S by an observer in system S′. From the point of view of an observer in system S, the time coordinate transformation is of the form

S → S′

$$t' = \gamma\left(t - \frac{xu}{c^2}\right) \tag{3.5}$$

and is just the inverse of the relation expressed by Equation 3.4. These results are clearly different from the classical results given by Equations 1.25d and 1.24d, respectively; however, they obviously reduce to their classical counterparts in the limit that $u$ is *small* compared to the speed of light $c$.

The relativistic time transformation equations, obtained above by *qualitative arguments*, can be easily derived by combining Equations 3.1a and 3.2a. From the point of view of an observer in system S, we desire a relativistic equation for $t'$ in terms of *unprimed* coordinates $x$ and $t$. This is directly accomplished by solving Equations 3.1a and 3.2a simultaneously to eliminate the variable $x'$. That is, substitution of Equation 3.1a into Equation 3.2a gives

$$\begin{aligned} x &= \gamma\,[\gamma\,(x - ut) + ut'] \\ &= \gamma^2\,(x - ut) + \gamma ut', \end{aligned}$$

which can be solved for $t'$ in terms of $x$ and $t$ as

$$
\begin{aligned}
t' &= \frac{x}{\gamma u} - \frac{\gamma}{u}(x - ut) \\
&= \gamma t + \frac{x}{\gamma u} - \frac{\gamma x}{u} \\
&= \gamma\left(t + \frac{x}{u\gamma^2} - \frac{x}{u}\right) \\
&= \gamma\left[t + \frac{x}{u}\left(\frac{1}{\gamma^2} - 1\right)\right].
\end{aligned} \tag{3.6}
$$

From Equations 2.5 and 2.7

$$\gamma = \frac{1}{\sqrt{1 - u^2/c^2}}, \tag{3.7}$$

thus $1/\gamma^2$ is simply given by

$$\frac{1}{\gamma^2} = 1 - \frac{u^2}{c^2}. \tag{3.8}$$

With Equation 3.8 substituted into Equation 3.6 we have

$$t' = \gamma\left[t + \frac{x}{u}\left(-\frac{u^2}{c^2}\right)\right],$$

which immediately reduces to the result given by Equation 3.5,

$$t' = \gamma\left(t - \frac{xu}{c^2}\right). \tag{3.5}$$

The result expressed by Equation 3.4 is obtained by a similar procedure, that is, Equation 3.2a is solved for $x$ and substituted into Equation 3.1a, and the resulting equation solved for $t$ in terms of $t'$ and $x'$.

## 3.3 Comparison of Classical and Relativistic Transformations

The relativistic space-time *coordinate* transformations have been developed in the previous two sections from fundamental considerations. These relations are known as the *Lorentz* or *Einstein transformations* and will be

TABLE 3.1
A comparison of the relativistic and classical space-time coordinate equations for the transformation of measurements from S′ to S.

| $S' \rightarrow S$ Coordinate Transformations | |
|---|---|
| Lorentz-Einstein | Galileo-Newton |
| $x = \gamma\,(x' + ut')$ | $x = x' + ut'$ |
| $y = y'$ | $y = y'$ |
| $z = z'$ | $z = z'$ |
| $t = \gamma\left(t' + \frac{x'u}{c^2}\right)$ | $t = t'$ |

tabulated below in a concise and informative manner and compared with the analogous *Galilean transformations* of classical mechanics.

Consider the occurrence of a single event in space and allow coordinate measurements to be made by an observer in system S and, likewise, by an observer in S′. The *Lorentz transformation* representing how an observer in S′ relates his coordinates ($x'$, $y'$, $z'$, and $t'$) of the event to the S coordinates ($x$, $y$, $z$, and $t$) of the same event, is compared with the *Galilean transformation* in Table 3.1. Although it should be obvious from our discussion of Einsteinian relativity, it merits emphasizing that it is the *relativistic* Lorentz transformations that are uniquely compatible with Einstein's postulates and, therefore, supersede the classical Galilean transformations. Another important observation is that for $u \ll c$, $\gamma \approx 1$, and the relativistic (Lorentzian-Einsteinian) equations reduce *exactly* to the classical (Galilean-Newtonian) transformations.

As indicated in the derivational sections, the inverse transformation equations to those presented in Table 3.1 are easily obtained by replacing unprimed coordinates with primed coordinates and vice versa, and by substituting $-u$ for $u$. That is,

$$\begin{array}{llllll} x \rightarrow x' & \text{while} & x' \rightarrow x, & y \rightarrow y' & \text{while} & y' \rightarrow y \\ z \rightarrow z' & \text{while} & z' \rightarrow z, & t \rightarrow t' & \text{while} & t' \rightarrow t \\ u \rightarrow -u. & & & & & \end{array}$$

Table 3.2 illustrates the results of applying these operations, thus giving the coordinate equations for the transformation of measurement on an event from S to S′. As before, the inverse Lorentz transformation equations reduce exactly to the inverse Galilean transformations for $u \ll c$.

The equations in Tables 3.1 and 3.2 represent the most *fundamental transformations* allowable by nature for problems involving the relative motion of inertial systems. It should be noted that these transformations are *universally* applicable in inertial systems; whereas the Galilean-Newtonian transformations are only good approximations to physical reality

| S → S′ Coordinate Transformations | |
|---|---|
| *Lorentz-Einstein* | *Galileo-Newton* |
| $x' = \gamma (x - ut)$ | $x' = x - ut$ |
| $y' = y$ | $y' = y$ |
| $z' = z$ | $z' = z$ |
| $t' = \gamma \left(t - \frac{xu}{c^2}\right)$ | $t' = t$ |

TABLE 3.2
A comparison of the relativistic and classical space-time coordinate equations for the transformation of measurements from S to S′.

when $u \ll c$. Unlike the Galilean-Newtonian transformations and classical mechanics, the Lorentzian-Einsteinian transformations predict an upper limit for the speed of a particle or system. This results from the mathematical form of $\gamma = \sqrt{1 - u^2/c^2}$ being *real* for $u \ll c$ and *imaginary* for $u > c$. Consequently, for $\gamma$ imaginary the transformation equations have no *physical* interpretation, and thus $c$ must be viewed as the upper limit for the speed of any physical entity. Furthermore, it should be emphasized that this *speed limitation* in nature is entirely consistent with Einstein's first postulate that required all inertial systems to be equivalent with respect to physical measurements. Otherwise, for $u > c$ an event, and the coordinate measurements associated with it, would be *real* and observable in one inertial system but *imaginary* or unobservable in another.

An appreciation and understanding of the Lorentz transformation equations are best attained by an application of these relations to basic problems. Having available two sets of tranformation equations tends to be *confusing* to beginning students, until it is realized that the two sets are in *essence* the *same* equations. Since all space-time coordinate variables for both systems S and S′ are contained in each set of transformations, knowledge of *either* set should be sufficient for solving any physical problem. To be specific, suppose you know the values of $x'$, $t'$, and $u$ and wish to find the values of $x$ and $t$. Certainly, the easiest approach would be to employ the Lorentz equations of Table 3.1; however, those in Table 3.2 will suffice. You need only replace $x'$, $t'$, $u$, and $\gamma$ in the first and last equation with their known values, and then solve the resulting equations simultaneously for $x$ and $t$. The results for $x$ and $t$ obtained by this procedure will be *identical* to those predicted by the equations of Table 3.1. To prove part of this last statement, we need only solve the equation

$$x' = \gamma (x - ut)$$

for

$$t = \frac{x}{u} - \frac{x'}{\gamma u},$$

substitute into the equation

$$t' = \gamma\left(t - \frac{xu}{c^2}\right),$$

and solve the resulting equation,

$$\begin{aligned} t' &= \gamma\left(\frac{x}{u} - \frac{x'}{\gamma u}\right) - \frac{\gamma x u}{c^2} \\ &= \frac{\gamma x}{u} - \frac{x'}{u} - \frac{\gamma x u}{c^2} \\ &= \frac{\gamma x}{u}\left(1 - \frac{u^2}{c^2}\right) - \frac{x'}{u} \\ &= \frac{x}{\gamma u} - \frac{x'}{u}, \end{aligned}$$

for $x$ to obtain

$$x = \gamma\,(x' + ut').$$

In a similar manner the equation

$$t = \gamma\left(t' + \frac{x'u}{c^2}\right)$$

is directly derived from the Lorentz relations for $x'$ and $t'$ given in Table 3.2. The important point to remember in using the Lorentz transformation is that *proper time* and *proper length* measurements were originally associated with the S′ system, where the *experiment of interest* was always considered to be *stationary*. Although it may be unnecessary, it tends to be more convenient and less confusing to identify *proper time* and *proper length* measurements with the S′ system. The following examples should help clarify this point and further decrease any confusion surrounding the two sets of transformations.

Imagine two events as occurring at the *same position* in a frame of reference at two different instances in time. Using the Lorentz *time coordinate* transformation equation, show that in any other frame of reference, the time interval between the events is greater than the proper time by the factor $\gamma$. In this problem, we choose S′ as the system where the two events occur at the *same position* at instants $t_1'$ and $t_2'$. Thus, with

$$x_1' = x_2' \equiv x' \rightarrow \Delta x' = 0$$

and the S′ → S (Table 3.1) Lorentz time transformation

$$t = \gamma\left(t' + \frac{x'u}{c^2}\right),$$

we immediately obtain

$$\Delta t = \gamma\left(\Delta t' + \frac{\Delta x' u}{c^2}\right) = \gamma\,\Delta t',$$

which is the *time dilation* result given by Equation 2.8. This same result can be obtained using the S → S′ space-time transformations (Table 3.2) in the form

$$x' = \gamma\,(x - ut) \;\rightarrow \Delta x' = \gamma\,(\Delta x - u\,\Delta t), \tag{3.9a}$$

$$t' = \gamma\left(t - \frac{xu}{c^2}\right) \rightarrow \Delta t' = \gamma\left(\Delta t - \frac{\Delta x\,u}{c^2}\right). \tag{3.9b}$$

Again, with *proper time* being defined in system S′, $\Delta x' = 0$ and Equation 3.9a gives

$$\Delta x = u\,\Delta t. \tag{3.10}$$

Now, substitution of Equation 3.10 into Equation 3.9b gives the expected *time dilation* equation, that is

$$\begin{aligned} \Delta t' &= \gamma\left(\Delta t - \frac{\Delta t\,u^2}{c^2}\right) \\ &= \gamma\,\Delta t\left(1 - \frac{u^2}{c^2}\right) \\ &= \frac{\gamma\,\Delta t}{\gamma^2} = \frac{\Delta t}{\gamma}. \end{aligned}$$

It should now be apparent that either set of Lorentz transformations can be utilized with equivalent and relative ease in solving physical problems. The results obtained by either set will be correct and entirely consistent with the point of view assumed. In the above example, we knew $\Delta t' = t_2' - t_1'$ would be the *proper time interval* and that $\Delta t = t_2 - t_1$ would be a *dilated time interval*, since we allowed S′ to be the system where the two events occurred at the same position. If we had assumed S to be the system where $x_1 = x_2$ for two events, then $t_2 - t_1$ would be the *proper time* and $t'_2 - t'_1$ would be *dilated* by the factor $\gamma$, as the equation

$$\Delta t' = \gamma \Delta t$$

would be directly obtained by *either* set of Lorentz transformations. Length *contraction* is also directly obtained from the Lorentz transformations, and is considered, along with other examples, in the problems at the end of the chapter.

## 3.4 Relativistic Velocity Transformations

In the last section we observed that the Lorentz transformation equations predict an upper limit for the speed of a particle. This result is also predicted by the relativistic transformation equations for velocity components, which will be derived by the same mathematical procedure used in obtaining the Galilean velocity transformations in Chapter 1. We consider a particle moving in a rectilinear path with constant velocity and being viewed by observers in two inertial systems S and S′. In system S the particle is observed to be at position $x_1$, $y_1$, and $z_1$ at time $t_1$ and at $x_2$, $y_2$, and $z_2$ at time $t_2$, while in system S′ the initial coordinates of the particle are denoted as $x'_1$, $y'_1$, $z'_1$, and $t'_1$ and the final coordinates as $x'_2$, $y'_2$, $z'_2$, and $t'_2$. The relativistic velocity transformations can be derived from the Lorentz space-time transformation by using the definition of either *average velocity* or *instantaneous velocity,* and each derivation will be considered separately.

To observers at rest in system S, the $x$-component of average velocity is defined as the ratio of displacement $x_2 - x_1$ to the corresponding time interval $t_2 - t_1$. But the position $x_1$ of the particle at time $t_1$ in S corresponds to the position

$$x'_1 = \gamma(x_1 - ut_1) \tag{3.11}$$

at time

$$t'_1 = \gamma\left(t_1 - \frac{x_1 u}{c^2}\right) \tag{3.12}$$

in system S′, according to the Lorentz transformation of Table 3.2. Likewise, the particle's position $x_2$ at time $t_2$ in S corresponds to position

$$x'_2 = \gamma\,(x_2 - ut_2) \tag{3.13}$$

at the instant in time

$$t_2' = \gamma\left(t_2 - \frac{x_2 u}{c^2}\right) \tag{3.14}$$

in S′, according to measurements made by observers in system S. Subtracting Equation 3.11 from Equation 3.13 gives the horizontal displacement of the particle in S′ as

$$\Delta x' = \gamma(\Delta x - u\,\Delta t), \tag{3.15}$$

which occurs during the time interval

$$\Delta t' = \gamma\left(\Delta t - \frac{\Delta x\, u}{c^2}\right) \tag{3.16}$$

obtained by subtracting Equation 3.12 from Equation 3.14. The particle's $x$-component of average velocity in system S′ is measured by observers in S to be

$$\bar{v}_x' = \frac{\Delta x - u\,\Delta t}{\Delta t - \dfrac{\Delta x\, u}{c^2}}, \tag{3.17}$$

which is obtained by dividing Equation 3.15 by Equation 3.16 and using the definition

$$\bar{v}_x' \equiv \frac{\Delta x'}{\Delta t'}. \tag{3.18}$$

As the $x$-component of average velocity observed in S is simply $\bar{v}_x \equiv \Delta x/\Delta t$, division of the numerator and denominator of Equation 3.17 by $\Delta t$ gives the relation between $\bar{v}_x'$ and $\bar{v}_x$ as

$$\bar{v}_x' = \frac{\bar{v}_x - u}{1 - \dfrac{\bar{v}_x u}{c^2}}. \tag{3.19}$$

S → S′

This is the relativistic velocity transformation equation for the motion of a particle parallel to the common $X$-$X'$ axis, as measured by observers in system S. The relativistic transformations for the $y$ and $z$ components of velocity are easily obtained by this same procedure. It is obvious from the

Lorentz transformation of Table 3.2 that during the time interval $\Delta t$ the particle's displacements parallel to the $y$ and $z$ axes in S are related to its displacements in system S′ by the equations

$$\Delta y' = \Delta y \quad \text{and} \quad \Delta z' = \Delta z.$$

Division of these relations by Equation 3.16 immediately yields

S → S′

$$\bar{v}_y' = \frac{\bar{v}_y/\gamma}{1 - \frac{\bar{v}_x u}{c^2}} \tag{3.20a}$$

S → S′ and

$$\bar{v}_z' = \frac{\bar{v}_z/\gamma}{1 - \frac{\bar{v}_x u}{c^2}}, \tag{3.20b}$$

where the definition of *average velocity* for each of the spatial coordinates has been appropriately utilized.

The *velocity-component* transformations given by Equations 3.19, 3.20a, and 3.20b were derived by using the definition of *average* velocity. However, similar relations (with average velocities replaced by instantaneous velocities) are obtainable by employing the definition of *instantaneous* velocity, after taking the *differential* of each Lorentz transformation equation in Table 3.2. The results obtained are

S → S′

$$v_x' = \frac{v_x - u}{1 - \frac{v_x u}{c^2}}, \tag{3.21a}$$

S → S′

$$v_y' = \frac{v_y/\gamma}{1 - \frac{v_x u}{c^2}}, \tag{3.21b}$$

S → S′

$$v_z' = \frac{v_z/\gamma}{1 - \frac{v_x u}{c^2}}, \tag{3.21c}$$

which are identical to the previous transformations with average velocities being replaced by instantaneous velocities. To illustrate the procedure used in obtaining these results, we will derive the *inverse* velocity transformation equations by adopting the point of view of observers in system S′. The derivation is based on the idea that for any inertial system S or S′, *instantaneous* velocity is defined as the ratio of an *infinitesimal displacement* to the corresponding *infinitesimal time interval dt* or *dt′*, in the limit that the time interval goes to zero. Accordingly, we differentiate the coordinates of Table 3.1 to obtain

$$dx = \gamma(dx' + u\,dt'), \tag{3.22a}$$

$$dy = dy', \tag{3.22b}$$

$$dz = dz', \tag{3.22c}$$

$$dt = \gamma\left(dt' + \frac{dx'u}{c^2}\right). \tag{3.22d}$$

From the definition of *instantaneous velocity,* we have for the *infinitesimal* displacements in S′

$$v'_x \equiv \frac{dx'}{dt'} \rightarrow dx' = v'_x\,dt',$$

$$v'_y \equiv \frac{dy'}{dt'} \rightarrow dy' = v'_y\,dt',$$

$$v'_z \equiv \frac{dz'}{dt'} \rightarrow dz' = v'_z\,dt',$$

which upon substitution into Equations 3.22 a-d give

$$dx = \gamma(v'_x + u)\,dt', \tag{3.23a}$$

$$dy = v'_y\,dt', \tag{3.23b}$$

$$dz = v'_z\,dt', \tag{3.23c}$$

$$dt = \gamma\left(1 + \frac{v'_x u}{c^2}\right) dt'. \tag{3.23d}$$

Using the same definition for instantaneous velocity in system S ($v_x \equiv dx/dt$, $v_y \equiv dy/dt$, and $v_z \equiv dz/dt$) we divide the equations for $dx$, $dy$, and $dz$ by the equation for $dt$ and immediately obtain

$$v_x = \frac{v'_x + u}{1 + \frac{v'_x u}{c^2}}, \qquad \text{(3.24a)} \quad S' \rightarrow S$$

$$v_y = \frac{v'_y/\gamma}{1 + \frac{v'_x u}{c^2}}, \qquad \text{(3.24b)} \quad S' \rightarrow S$$

$$v_z = \frac{v'_z/\gamma}{1 + \frac{v'_x u}{c^2}}. \qquad \text{(3.24c)} \quad S' \rightarrow S$$

These velocity component transformations are the inverse of those given in Equations 3.21 a-c and contain quantities measured by observers in system S'.

The two sets of relativistic velocity transformation equations have a surprising and interesting result in common. With respect to the axis of relative motion, a *transverse* component of velocity in one system is dependent on *both* the corresponding transverse and longitudinal components of velocity in the other system. Further, both sets of transformations reduce exactly to their classical counterparts under the *correspondence principle*. For example, in Equations 3.21 and 3.24

$$1 - \frac{v_x u}{c^2} \approx 1$$

and

$$1 + \frac{v'_x u}{c^2} \approx 1$$

for $u \ll c$. Consequently, both Equations 3.21a and 3.24a reduce to

$$v'_x = v_x - u,$$

which is exactly the classical or Galilean transformation equation given by Equation 1.30a or Equation 1.31a. It is also interesting to note that unlike the Galilean transformations, *none* of the relativistic velocity transformations are *invariant*. This is a direct result of the *dilation* of time in Einsteinian relativity.

To illustrate the consistency between Einstein's second postulate and the above results, consider a situation where a particle moves along the $X'$-axis with a speed $c$ relative to S' (i.e., $v'_x = c$). The problem is to calculate the particle's speed relative to observers in system S. According to the classical view, the particle's speed relative to S is $v_x = c + u$—it has a speed exceeding the speed of light as far as observers in S are concerned. However, according to Einstein and Equation 3.24a

$$v_x = \frac{c + u}{1 + \dfrac{cu}{c^2}}$$

$$= \frac{c + u}{\dfrac{1}{c}(c + u)} = c.$$

That is, if observers in S' measure a particle's speed to be $c$, then observers in system S will also measure $c$ for the speed of the particle, irrespective of the speed of separation between the inertial reference frames S and S′. This result is totally compatible with Einstein's postulates and it tends to suggest that particle velocities can not exceed the speed of light.

As another example of the application of Einstein's *velocity addition formula*, Equation 3.24a, consider three inertial reference frames separating from one another along a common axis of relative motion. Let S′ be moving to the right of S with a speed relative to S of $u = 3c/4$. Allow the third system S″ to be moving to the right of S′ with a speed $u' = 3c/4$ relative to S′. Now, consider a particle to be moving parallel to the axis of relative motion with a speed $v_x'' = 3c/4$, as measured by observers in S″. The problem is to obtain the speed of the particle $v_x$, as measured by observers in system S. Clearly, observers in S′ obtain

$$v_x' = \frac{v_x'' + u'}{1 + \frac{v_x'' u'}{c^2}} = \frac{\frac{3c}{4} + \frac{3c}{4}}{1 + \frac{9}{16}} = \frac{\frac{6c}{4}}{\frac{25}{16}} = \frac{24}{25}c,$$

while to observers in system S the particle's speed is

$$v_x = \frac{v_x' + u}{1 + \frac{v_x' u}{c^2}} = \frac{\frac{24c}{25} + \frac{3c}{4}}{1 + \left(\frac{24}{25}\right)\left(\frac{3}{4}\right)} = \frac{\left(\frac{96}{100} + \frac{75}{100}\right)c}{1 + \frac{72}{100}} = \frac{171}{172}c.$$

The result of this example suggests that *a particle's speed can be viewed as approaching the speed of light, but it can never really attain the exact speed c*. By comparison with the previous example, if a particle's speed is exactly $c$, as measured by *any* inertial observer, then its speed is $c$ according to all inertial observers. The difference between these two conclusions is subtle, but an important one to keep in mind!

## 3.5 Relativistic Acceleration Transformations

To complete our kinematical considerations, we need to develop the appropriate transformation equations for acceleration components in Einsteinian relativity. Frequently, you may hear a statement that the special theory of relativity is incapable of considering the acceleration of particles. Clearly this is a misconception, since we can surely consider the Lorentzian spatial coordinate transformations and all of the their time derivatives. It is only a special and restrictive class of problems that deal with rectilinear motion and constant velocity. The more general problem is concerned with a particle moving about in space and exhibiting **curvilinear** motion, which necessarily requires the particle to have a *nonzero* acceleration.

The derivational procedure for obtaining the relativistic acceleration transformations is based on the definition of *instantaneous acceleration* as being the ratio of an *infinitesimal* change in velocity to the corresponding time interval. To simplify the mathematics of this section, we introduce

$$\alpha \equiv 1 - \frac{v_x u}{c^2} \tag{3.25a}$$

and note that

$$\frac{d\alpha^{-1}}{dt} = \frac{a_x u}{c^2}\alpha^{-2}. \tag{3.25b}$$

Using these relations, the differential of Equation 3.21a becomes

$$\begin{aligned} dv_x' &= \frac{d}{dt}[(v_x - u)\alpha^{-1}]\,dt \\ &= \left[a_x\alpha^{-1} + (v_x - u)\frac{d\alpha^{-1}}{dt}\right]dt \\ &= \left[a_x\alpha^{-1} + (v_x - u)\frac{a_x u}{c^2}\alpha^{-2}\right]dt \end{aligned}$$

$$= \left[ a_x \alpha + (v_x - u) \frac{a_x u}{c^2} \right] \alpha^{-2} dt$$

$$= \left[ a_x - \frac{v_x a_x u}{c^2} + (v_x - u) \frac{a_x u}{c^2} \right] \alpha^{-2} dt$$

$$= a_x \frac{1 - u^2}{c^2} \alpha^{-2} dt$$

$$= \frac{a_x}{\gamma^2 \alpha^2} dt. \qquad (3.26)$$

Since in system S′, $a'_x \equiv dv'_x/dt'$, then Equation 3.26 can be divided by the differential of Equation 3.5 in the form

$$dt' = \gamma \left( 1 - \frac{v_x u}{c^2} \right) dt = \gamma \alpha \, dt \qquad (3.27)$$

to easily obtain the transformation equation for the $x$-component of acceleration as

$$a'_x = \frac{a_x}{\gamma^3 \alpha^3}. \qquad (3.28a) \quad S \rightarrow S'$$

By using similar reasoning to that above, we can derive the $y$-component of acceleration by differentiating Equation 3.21b:

$$dv'_y = \frac{d}{dt} (v_y \gamma^{-1} \alpha^{-1}) \, dt$$

$$= \left[ a_y \gamma^{-1} \alpha^{-1} + v_y \gamma^{-1} \frac{d\alpha^{-1}}{dt} \right] dt$$

$$= \left[ a_y \gamma^{-1} \alpha^{-1} + \frac{v_y a_x u}{c^2} \gamma^{-1} \alpha^{-2} \right] dt$$

$$= \left[ a_y \alpha + \frac{v_y a_x u}{c^2} \right] \gamma^{-1} \alpha^{-2} dt$$

$$= \left[ a_y - \frac{v_x a_y u}{c^2} + \frac{v_y a_x u}{c^2} \right] \gamma^{-1} \alpha^{-2} dt$$

$$= \left[ a_y - (v_x a_y - v_y a_x) \frac{u}{c^2} \right] \gamma^{-1} \alpha^{-2} dt.$$

Again, dividing this relation for the differential velocity by the expression for differential time (Equation 3.27), we obtain

S - S′

$$a'_y = \frac{a_y - (v_x a_y - v_y a_x)\dfrac{u}{c^2}}{\gamma^2 \alpha^3}. \tag{3.28b}$$

Of course, the relativistic transformation equation for the $z$-component of acceleration is identical to Equation 3.28b, except for the $y$-subscripts being replaced by $z$-subscripts:

S → S′

$$a'_z = \frac{a_z - (v_x a_z - v_z a_x)\dfrac{u}{c^2}}{\gamma^2 \alpha^3}. \tag{3.28c}$$

For the inverse acceleration transformations we obtain

S′ → S

$$a_x = \frac{a'_x}{\gamma^2}\left(1 + \frac{v'_x u}{c^2}\right)^3 \equiv \frac{a'_x}{\gamma^3 \alpha'^3}, \tag{3.29a}$$

S′ → S

$$a_y = \frac{a'_y - (v'_x a'_y - v'_y a'_x)\dfrac{u}{c^2}}{\gamma^2 \alpha'^3}, \tag{3.29b}$$

S′ → S

$$a_z = \frac{a'_z - (v'_x a'_z - v'_z a'_x)\dfrac{u}{c^2}}{\gamma^2 \alpha'^3}. \tag{3.29c}$$

Whereas the Galilean-Newtonian acceleration transformations (Equations 1.32 a-c) were invariant, acceleration is certainly not invariant in Einsteinian relativity. Not only are the acceleration transformations mathematically intimidating, but the *transverse* components transform differently than the *longitudinal* component. Interestingly, however, they do reduce exactly to their classical counterparts under the *correspondence principle*.

## 3.6 Relativistic Frequency Transformations

To complete our discussion relativistic kinematics, we will develop the transformation equations for the frequency and wavelength of electromagnetic waves. Unlike sound waves considered in the classical Doppler effect, electromagnetic waves (x-rays, visible light, etc.) do *not require* a physical medium for their propagation. Further, all electromagnetic waves in free space travel at the speed of light and obey the basic relation

$$c = \lambda\nu = \lambda'\nu'. \tag{3.30}$$

Invariance of Light

This equation reflects the requirement of Einstein's second postulate that the product of wavelength and frequency of an electromagnetic wave be equal to the universal constant $c$ for *all inertial observers*. Clearly, the values of $\lambda$ and $\nu$ measured by observers in system S need not be identical to $\lambda'$ and $\nu'$ measured in S′; however, Equation 3.30 must always be valid.

The relativistic transformation equations for the frequency and wavelength of electromagnetic waves are easily derived by using arguments similar to those presented for the classical Doppler effect. By analogy with the first case considered there, our relativistic gedanken experiment considers an *emitter* E′ of *monochromatic* light waves to be positioned at the origin of coordinates in system S′, and a *receiver* or *detector* to be located at the origin of system S. The first situation to be considered is depicted in Figure 3.2, where the emitter and detector are *receding* from each other with a uniform speed $u$. The schematic represents the point of view for observers in inertial system S, who are at rest relative to themselves and view system S′ as receding. Accordingly, a time interval $\Delta t = t_2 - t_1$ is required for the first electromagnetic wave emitted by E′ to travel the distance $c\Delta t$ to reach R. As illustrated in Figure 3.2, system S′ has *receded* through the distance $u\Delta t$ during the time of *wave emission* $\Delta t$. With $x$ being the distance between R and E′ at the instant $t_2$ when R *first detects* a wave, then the wavelength measured by an observer in system S is

$$\lambda = \frac{x}{N}, \tag{3.31}$$

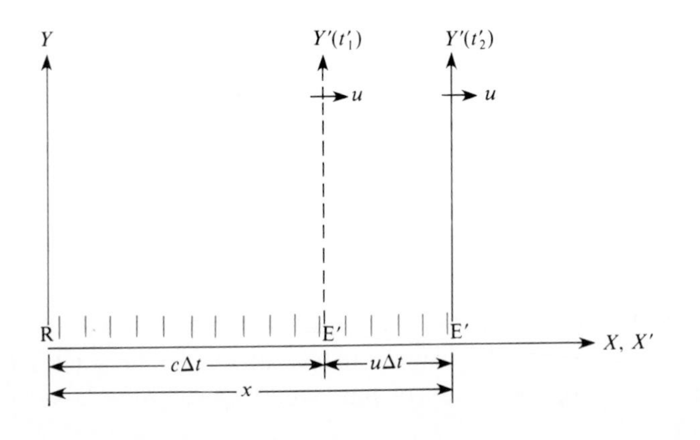


Figure 3.2
An emitter E′ of electromagnetic waves receding from a detector (receiver) R at a uniform speed $u$.

where $N$ is the number of waves perceived *eventually* by R. Solving Equation 3.30a for $\nu$ and using Equation 3.31 gives

$$\nu = \frac{cN}{x} \tag{3.32}$$

for the frequency as measured by an observer in system S. From the geometry of Fig. 3.2

$$x = (c + u)\,\Delta t, \tag{3.33}$$

which allows Equation 3.32 to take the form

$$\nu = \frac{N}{\Delta t\,(1 + u/c)}. \tag{3.34}$$

Since
$$N = N' \equiv \nu'\,\Delta t' \tag{3.35}$$

and $\beta \equiv u/c$, then Equation 3.34 becomes

$$\nu = \frac{\nu'\,\Delta t'}{\Delta t\,(1 + \beta)}. \tag{3.36}$$

Up to this point our derivation has been *exactly* like the first case considered for the classical Doppler effect leading to Equation 1.43; however, in Einsteinian relativity $\Delta t \neq \Delta t'$. Since the first and last wave emitted by E′ occurred at the *same position* in S′ at instants $t_1'$ and $t_2'$, respectively, then the time interval $\Delta t'$ must be recognized as the *proper time* in our equation. Thus, taking *time dilation* (Equation 2.8) into account allows our frequency transformation to be expressed as

Receding Case

$$\nu = \frac{\nu'}{\gamma\,(1 + \beta)}. \tag{3.37}$$

The wavelength transformation is directly obtained from this result by realizing that $\nu = c/\lambda$ and $\nu' = c/\lambda'$ from Equation 3.30:

Receding Case

$$\lambda = \lambda'\gamma(1 + \beta). \tag{3.38}$$

These two equations represent the **relativistic Doppler effect** of electromagnetic waves for *receding systems*. Except for the presence of $\gamma$ in the denominators, these results are of the same form as the classical Doppler effect given by Equations 1.43 and 1.44. This difference in form is solely precipitated by the absence of *absolute time* in Einsteinian relativity.

There is yet another significant difference between the classical and relativistic Doppler effects. In our classical considerations we had *two separate cases* for the situation where S and S′ were receding from each other. The second case (Equation 1.50) was the *inverse* of the first obtained by allowing

$$\nu \rightarrow \nu' \qquad \nu' \rightarrow \nu \qquad u \rightarrow -u. \tag{3.39}$$

Performing these operations on Equation 3.37 gives

$$\nu = \nu'\gamma(1 - \beta), \tag{3.40}$$

which corresponds to the situation where the emitter E′ is stationary in S′ and the receiver R in S is viewed as *receding* from observers in S′ with the uniform speed $u$. Unlike the classical effect, where the two equations (Equations 1.43 and 1.50) predicted different physical phenomena, Equation 3.40 can be shown to reduce exactly to Equation 3.37. That is,

$$\begin{aligned}
\nu &= \nu'\gamma(1 - \beta) \\
&= \frac{\nu'(1 - \beta)}{\sqrt{1 - \beta^2}} \\
&= \frac{\nu'\sqrt{1 - \beta}\,\sqrt{1 - \beta}}{\sqrt{1 - \beta}\,\sqrt{1 + \beta}} \\
&= \frac{\nu'\sqrt{1 - \beta}}{\sqrt{1 + \beta}} \\
&= \frac{\nu'\sqrt{1 - \beta}\,\sqrt{1 + \beta}}{\sqrt{1 + \beta}\,\sqrt{1 + \beta}} \\
&= \frac{\nu'\sqrt{1 - \beta^2}}{1 + \beta} \\
&= \frac{\nu'}{\gamma(1 + \beta)}.
\end{aligned}$$

Whereas in Galilean relativity there are two frequency transformation equations required for the complete description of sound waves perceived by observers in *receding* inertial systems, there is only one such equation predicted by Einsteinian relativity for electromagnetic waves. This particular difference is *not* due to the absence of *absolute time* in the latter theory, but, rather, by the absence of a *material medium* being required in nature for the propagation of electromagnetic waves.

Before considering the situation where the emitter and receiver *approach* one another, let us look at the phenomenological implications of Equations 3.37 and 3.38. For $u$ relativistic $\gamma > 1$ and $\beta < 1$, which means that

$$\gamma(1 + \beta) > 1. \tag{3.41}$$

Consequently, in Equations 3.37 and 3.38

$$\nu < \nu' \qquad \text{and} \qquad \lambda > \lambda', \tag{3.42}$$

respectively. Since blue light has a *shorter* wavelength ($\lambda_b \approx 4.7 \times 10^{-7}$ m) than red light ($\lambda_r \approx 6.7 \times 10^{-7}$ m), this phenomenon is referred to as a **red shift** because $\lambda > \lambda'$. When an emitter of electromagnetic waves is receding from a detector, the shift toward longer wavelengths is called a *red shift*. As an example of this phenomenon, consider a distant star to be receding from the earth at a speed of $3c/5$. If the star emits electromagnetic radiation of $3.3 \times 10^{-7}$ m, then observers on the earth will measure the wavelength of the incident waves to be

$$\lambda = \lambda'\gamma(1 + \beta) = (3.3 \times 10^{-7}\ \text{m})\left(\frac{5}{4}\right)\left(1 + \frac{3}{5}\right) = 6.6 \times 10^{-7}\ \text{m}.$$

In this example the waves have been shifted from the ultraviolet to the red wavelength region of the electromagnetic spectrum. Frequently, electromagnetic wavelengths are specified in **Angstrom** units, where an Angstrom unit is simply defined by

Angstrom Unit

$$\text{Å} \equiv 10^{-10}\ \text{m}. \tag{3.43}$$

What about the case where a distant star is approaching the earth with a uniform speed $u$ relative to the earth? We might expect the wave pulses to be *bunched* together thus giving rise to a *blue shift*. To quantitatively develop the appropriate relativistic frequency and wavelength transformations, consider the situation as depicted in Figure 3.3. As before, let the emitter E′ be at the origin of coordinates of the S′ system and the receiver R be at the coordinate origin of system S. As viewed by observers in S, a time interval $\Delta t$ is required for the first wave pulse emitted by E′ to reach the receiver R. During this time E′ has moved a distance $u\Delta t$ *closer* to R.

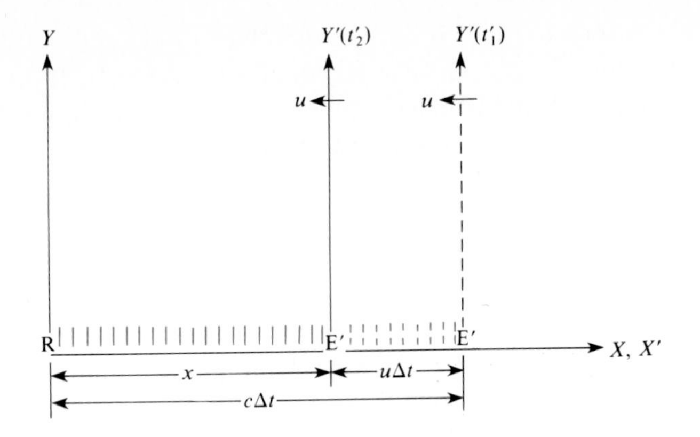


Figure 3.3
An emitter E′ of electromagnetic waves approaching a detector (receiver) R at a uniform speed $u$.

Hence, the total number of waves $N'$, emitted by E′ in the elapsed time $\Delta t$, will be *bunched together* in the distance $x$, as illustrated in Figure 3.3. By comparing this situation with the one previously discussed, we find that Equations 3.31 and 3.32 are still valid. But now

$$x = (c - u)\,\Delta t, \tag{3.44}$$

as seen from the geometry of Figure 3.3, and Equation 3.32 becomes

$$\nu = \frac{N}{\Delta t(1 - u/c)}. \tag{3.45}$$

Substitution of Equation 3.35 into Equation 3.45 results in

$$\nu = \frac{\nu'\,\Delta t'}{\Delta t\,(1 - \beta)}, \tag{3.46}$$

and after time dilation is properly accounted for, we have

$$\nu = \frac{\nu'}{\gamma\,(1 - \beta)}. \tag{3.47}$$

Approaching Case

Utilization of Equation 3.30 transforms Equation 3.47 from the domain of frequencies to that of wavelengths, resulting in

$$\lambda = \lambda'\gamma\,(1 - \beta). \tag{3.48}$$

Approaching Case

Since $\gamma\,(1 - \beta)$ is less than one for $u < c$, then $\lambda < \lambda'$ and we have what is called a **blue shift.** Also, it should be observed that Equations

3.47 and 3.48 are just the inverse of Equations 3.37 and 3.38, respectively, with $u$ being replaced by $-u$ ($\beta$ replaced by $-\beta$). If you were to consider this case from the point of view of observers in S′, who view themselves at rest and system S to be approaching, the frequency transformation obtained is of the form

$$\nu = \nu'\gamma\,(1 + \beta). \tag{3.49}$$

This result is easily deduced by performing the operations given in Equation 3.39 on Equation 3.47. However, it can readily be shown that this result reduces *exactly* to Equation 3.47, by using arguments similar to those presented following Equation 3.40. It is important to note that only *two* frequency transformation equations (Equations 3.37 and 3.47) are required for the *complete* description of the **relativistic Doppler effect** for electromagnetic waves; whereas, *four* such equations are required to describe the classical Doppler effect for sound waves.

## Review of Derived Equations

A listing of the derived Lorentz-Einstein transformation equations is presented below, along with the transformations for the frequency and wavelength of electromagnetic waves. Only the coordinate, velocity, and acceleration equations for the transformation of measurements on an event from S′ to S are listed, as the inverse transformations are easily obtained by replacing unprimed variables with primed variables and vice versa, and by substituting $-u$ for $u$. The velocity transformations can be derived by employing the definition of either average or instantaneous velocity with the space-time coordinate transformations.

### LORENTZ-EINSTEIN TRANSFORMATION (S′ → S)

#### *Space-Time Transformations*

$$x = \gamma(x' + ut')$$
$$y = y'$$
$$z = z'$$
$$t = \gamma\left(t' + \frac{x'u}{c^2}\right)$$

*Velocity Transformations*

$$v_x = \frac{v_x' + u}{1 + \frac{v_x' u}{c^2}}$$

$$v_y = \frac{v_y'/\gamma}{1 + \frac{v_x' u}{c^2}}$$

$$v_z = \frac{v_z'/\gamma}{1 + \frac{v_x' u}{c^2}}$$

*Acceleration Transformations*

$$\alpha' \equiv 1 + \frac{v_x' u}{c^2}$$

$$a_x = \frac{a_x'}{\gamma^3 \alpha'^3}$$

$$a_y = \frac{a_y' + (v_x' a_y' - v_y' a_x') \frac{u}{c^2}}{\gamma^2 \alpha'^3}$$

$$a_z = \frac{a_z' + (v_x' a_z' - v_z' a_x') \frac{u}{c^2}}{\gamma^2 \alpha'^3}$$

## RELATIVISTIC FREQUENCY TRANSFORMATIONS

$c = \lambda \nu = \lambda' \nu'$ Invariance of Light

Receding Cases:

$$\nu = \frac{\nu'}{\gamma(1 + \beta)}$$

$$\lambda = \lambda' \gamma (1 + \beta)$$

Approaching Cases:

$$\nu = \frac{\nu'}{\gamma(1 - \beta)}$$

$$\lambda = \lambda' \gamma (1 - \beta)$$

## Problems

**3.1** Consider inertial systems S and S′ to be separating along their common $X$-$X'$ axis at the uniform speed of $3c/5$. If an observer in S′ views an exploding flashbulb to occur 60 m from his origin of coordinates along the $X'$-axis at a time reading of $8 \times 10^{-8}$ s, what are the horizontal and time coordinates of the event according to an observer in system S?

*Solution:*
With $x' = 60$ m, $t' = 8 \times 10^{-8}$ s, and $u = 3c/5 \rightarrow \gamma = 5/4$, we obtain the $x$-coordinate from Equation 3.2a,

$$x = \gamma(x' + ut')$$
$$= \frac{5}{4}\left[60\text{ m} + \frac{3}{5}\left(3 \times 10^8 \frac{\text{m}}{\text{s}}\right)(8 \times 10^{-8}\text{ s})\right] = 93\text{ m},$$

and the time coordinate from Equation 3.4,

$$t = \gamma\left(t' + \frac{x'u}{c^2}\right)$$
$$= \frac{5}{4}\left[8 \times 10^{-8}\text{ s} + \frac{(60\text{ m})(3/5)}{3 \times 10^8\text{ m/s}}\right] = 2.5 \times 10^{-7}\text{ s}.$$

**3.2** Observers in system S measure the horizontal coordinate of an event to be 50 m at a time reading of $2 \times 10^{-7}$ s. What are the horizontal and time coordinates of the event to observers in S′, if the uniform speed of separation between S and S′ is $3c/5$?

*Answer:* $x' = 17.5$ m, $t' = 1.25 \times 10^{-7}$ s

**3.3** Consider measuring the length of an object moving relative to your reference frame by measuring the positions of each end $x_1$ and $x_2$ at the same instant in time. Using the Lorentz spatial coordinate transformation given by Equation 3.1a, show that the length you measure is smaller than the object's proper length.

*Solution:*
The *proper length* of an object is always measured in a frame of reference, say S′, in which the object is at rest. With $u$ being the speed of the object and system S′ relative to S, then from Equation 3.1a we have

$$x_1' = \gamma(x_1 - ut_1) \qquad \text{and} \qquad x_2' = \gamma(x_2 - ut_2).$$

Since $t_1 = t_2 \rightarrow \Delta t = 0$ in system S, $\Delta x' = x_2' - x_1'$ is

$$\Delta x' = \gamma(\Delta x - u\cancelto{0}{\Delta t}) = \gamma\,\Delta x.$$

Because $\gamma > 1$, the proper length $\Delta x' = x_2' - x_1'$ in S′ is greater than the length $\Delta x = x_2 - x_1$ you measure.

**3.4** Do Problem 3.3 using the Lorentz coordinate transformations given by Equations 3.2a and 3.4. You might wish to review Section 3.3 for a similar problem concerning time dilation.

*Answer:* $x_2 - x_1 = \dfrac{x_2' - x_1'}{\gamma}$

**3.5** Two clocks, positioned in system S′ at $x_A' = 25$ m and $x_B' = 75$ m, record the same time $t_0'$ for the occurrence of an event. What is the difference in time between the two clocks in S′, according to observers in system S, if $u = 3c/5$?

*Solution:*
With $\Delta x' \equiv x_B' - x_A' = 50$ m, $t_A' = t_B' \equiv t_0'$, and $u = 3c/5 \rightarrow \gamma = 5/4$, the Lorentz transformation gives

$$t_A = \gamma\left(t_0' + \frac{x_A' u}{c^2}\right) \quad \text{and} \quad t_B = \gamma\left(t_0' + \frac{x_B' u}{c^2}\right).$$

Subtracting $t_A$ from $t_B$ and substituting the data gives

$$\Delta t \equiv t_B - t_A = \gamma\left(0 + \frac{\Delta x' u}{c^2}\right)$$

$$= \left(\frac{5}{4}\right) \frac{(50 \text{ m})\left(\frac{3}{5}\right)}{3 \times 10^8 \frac{\text{m}}{\text{s}}} = 1.25 \times 10^{-7} \text{ s},$$

where $\Delta x' u/c^2$ is recognized as the quantity $\tau_s'$ of Chapter 2.

**3.6** Keeping in mind the physical coordinates associated with the event of Problem 3.2, if a second event occurs at 10 m, $3 \times 10^{-7}$ s as measured in system S, what is the time interval between the events measured by observers in system S′?

*Answer:* $\Delta t' = 2.25 \times 10^{-7}$ s

**3.7** Derive the time transformation equation $t = \gamma(t' + x'u/c^2)$ by using Equations 3.1a and 3.2a.

*Solution:*
Substitution of Equation 3.2a,

$$x = \gamma(x' + ut'),$$

into Equation 3.1a,

$$x' = \gamma(x - ut),$$

allows for the elimination of $x$:

$$\begin{aligned} x' &= \gamma[\gamma(x' + ut') - ut] \\ &= \gamma^2 x' + \gamma^2 ut' - \gamma ut. \end{aligned}$$

This result can be solved for $t$ in terms of $t'$ and $x'$ as

$$\begin{aligned} t &= \frac{1}{\gamma u}(\gamma^2 x' - x' + \gamma^2 ut') \\ &= \frac{\gamma x'}{u} - \frac{x'}{\gamma u} + \gamma t' \\ &= \gamma\left(t' + \frac{x'}{u} - \frac{x'}{\gamma^2 u}\right) \\ &= \gamma\left[t' + \frac{x'}{u}\left(1 - \frac{1}{\gamma^2}\right)\right] \\ &= \gamma\left[t' + \frac{x'}{u}\left(1 - \left(1 - \frac{u^2}{c^2}\right)\right)\right] \\ &= \gamma\left(t' + \frac{x'u}{c^2}\right). \end{aligned}$$

**3.8** Derive the $S' \rightarrow S$ time coordinate transformation equation by using Equations 3.1a and 3.5.

*Answer:* $t = \gamma\left(t' + \frac{x'u}{c^2}\right)$

**3.9** Observers in one inertial system measure the coordinates of two events and determine that they are separated in space and time by 1500 m and $7 \times 10^{-6}$ s, while observers in a second inertial system measure the two events to be separated by $5 \times 10^{-6}$ s. Find the relative speed of separation between the two systems.

*Solution:*

From the obvious dilation of time in the given data, we consider the second system to be moving relative to the first with an unknown speed $u$. Consequently, the data given can be identified as $\Delta x = 1500$ m, $\Delta t = 7 \times 10^{-6}$ s, and $\Delta t' = 5 \times 10^{-6}$ s. Using Equation 3.5 in the form (*Note:* $\beta \equiv u/c$)

$$\Delta t' = \gamma\left(\Delta t - \frac{\Delta x\, \beta}{c}\right),$$

direct substitution of the physical data gives

$$5 \times 10^{-6}\ \text{s} = \gamma\left[7 \times 10^{-6}\ \text{s} - \frac{(1500\ \text{m})\ \beta}{3 \times 10^{8}\ \text{m/s}}\right]$$

$$= \gamma(7 \times 10^{-6}\ \text{s} - \beta\ 5 \times 10^{-6}\ \text{s})$$

or, more simply

$$5 = \gamma(7 - 5\beta).$$

Squaring this equation and using the definition of $\gamma$ gives

$$50\beta^2 - 70\beta + 24 = 0,$$

which can be solved for $\beta$ using the general solution for a quadratic equation given in Appendix A, Section A.5. Accordingly

$$\beta = \frac{-(-70) \pm \sqrt{70^2 - (4)\ (50)\ (24)}}{(2)\ (50)}$$

$$= \frac{70 \pm \sqrt{4900 - 4800}}{100}$$

$$= \frac{70 \pm 10}{100} = 0.8 \quad \text{or} \quad 0.6,$$

and the two answers to this problem are $u = 0.8c$ and $u = 0.6c$.

**3.10** What must be the uniform speed of separation between two inertial systems, if in one system observers determine that two events are separated in space and time by 900 m and $1.8 \times 10^{-6}$ s, respectively, while in the other system the two events occur simultaneously?

*Answer:* $u = 0.6c$

**3.11** An alien in a flying saucer passes an astronaut in a space station at $0.6c$. Two-thirds second after the flying saucer passes, the astronaut observes a particle $3 \times 10^8$ m away moving in the same direction as the saucer. One second after the first observation, the astronaut notes the particle to be $5 \times 10^8$ m away. What are the particle's space-time coordinates for each position according to the alien?

*Solution:*
With the space station being identified as system S and the flying saucer as S′, then the physical data are denoted as $x_1 = 3 \times 10^8$ m, $t_1 = (2/3)$s, $x_2 = 5 \times 10^8$ m, $t_2 = (5/3)$s, and $u = 0.6c = 1.8 \times 10^8$ m/s $\rightarrow \gamma = 5/4$. Using the Lorentz transformations from S $\rightarrow$ S′ of Table 3.2, we have

$$x_1' = \gamma(x_1 - ut_1)$$

$$= \frac{5}{4}\left[3 \times 10^8\ \text{m} - \left(1.8 \times 10^8\ \frac{\text{m}}{\text{s}}\right)\left(\frac{2}{3}\ \text{s}\right)\right]$$

$$= \frac{5}{4}(3 - 1.2) \times 10^8 \text{ m} = 2.25 \times 10^8 \text{ m},$$

$$x_2' = \gamma(x_2 - ut_2)$$

$$= \frac{5}{4}\left[5 \times 10^8 \text{ m} - \left(1.8 \times 10^8 \frac{\text{m}}{\text{s}}\right)\left(\frac{5}{3}\text{ s}\right)\right]$$

$$= \frac{5}{4}(5 - 3) \times 10^8 \text{ m} = 2.5 \times 10^8 \text{ m},$$

$$t_1' = \gamma\left(t_1 - \frac{x_1 u}{c^2}\right)$$

$$= \frac{5}{4}\left[\frac{2}{3}\text{s} - \frac{(3 \times 10^8 \text{ m})(0.6)}{3 \times 10^8 \text{ m/s}}\right]$$

$$= \frac{5}{4}\left(\frac{2}{3} - 0.6\right)\text{ s} = \frac{5}{4}\left(\frac{1}{15}\right)\text{ s} = \frac{1}{12}\text{ s},$$

$$t_2' = \gamma\left(t_2 - \frac{x_2 u}{c^2}\right)$$

$$= \frac{5}{4}\left[\frac{5}{3}\text{ s} - \frac{(5 \times 10^8 \text{ m})(0.6)}{3 \times 10^8 \text{ m/s}}\right]$$

$$= \frac{5}{4}\left(\frac{5}{3} - 1\right)\text{ s} = \frac{5}{4}\left(\frac{2}{3}\right)\text{ s} = \frac{5}{6}\text{ s}.$$

**3.12** Referring to Problem 3.11, find the particle's $x$-component of velocity as measured by (a) the astronaut and (b) the alien.

*Answer:* $v_x = 2 \times 10^8 \frac{\text{m}}{\text{s}}, \; v_x' = \frac{1}{3} \times 10^8 \frac{\text{m}}{\text{s}}$

**3.13** Two spaceships are *receding* from each other at a uniform speed and in line with the earth. If the speed of each spaceship is $0.8c$ relative to the earth, find the speed of one relative to the other.

*Solution:*

The three systems in this problem can be identified in a simple manner. The earth is considered to be system S, and the spaceship that is *receding* from the earth is taken as system S′. Thus, its velocity relative to the earth is $u = 0.8c$, and the other spaceship must be *approaching* the earth with a velocity $v_x = -0.8c$. The problem now becomes one of finding $v_x'$ by using the appropriate velocity transformation equation. That is

$$v_x' = \frac{v_x - u}{1 - \dfrac{v_x u}{c^2}}$$

$$= \frac{-0.8c - 0.8c}{[1 - (-0.8)(0.8)]}$$

$$= \frac{-1.6c}{1 + 0.64} = -\frac{1.6c}{1.64}.$$

**3.14** Two spaceships are observed to have the same *speed* relative to the earth. If they are in line with the earth and *approaching* each other at a uniform speed of $1.2c/1.36$, what is the velocity of each relative to the earth?

*Answer:* $u = 0.6c,\ v_x = -0.6c$

**3.15** If a beam of light moves along the $Y'$-axis in system S', find (a) the components of velocity and (b) the magnitude of velocity for the light beam, as measured by observers in system S.

*Solution:*
With the data $v_x' = v_z' = 0$ and $v_y' = c$, then from the Lorentz velocity transformations we have

$$v_x = \frac{v_x' + u}{1 + \frac{v_x' u}{c^2}} = \frac{0 + u}{1 + 0} = u,$$

$$v_y = \frac{v_y'/\gamma}{1 + \frac{v_x' u}{c^2}} = \frac{c/\gamma}{1 + 0} = \frac{c}{\gamma},$$

$$v_z = \frac{v_z'/\gamma}{1 + \frac{v_x' u}{c^2}} = \frac{0/\gamma}{1 + 0} = 0$$

for the components of velocity observed in system S. The magnitude of the velocity is given by

$$v = \sqrt{v_x^2 + v_y^2 + v_z^2}$$

$$= \sqrt{u^2 + \frac{c^2}{\gamma^2} + 0}$$

$$= c$$

**3.16** Systems S and S' are separating uniformly at a speed of $0.8c$. Observers in S' view a particle moving in the positive $Y'$ direction, and at one instant the particle's instantaneous speed is measured to be $0.6c$. If the particle accelerates in the positive $Y'$ direction to a speed of $0.8c$ in six seconds, what is the particle's acceleration to observers in system S?

*Answer:* $a = a_y = \frac{9}{25} \times 10^7 \frac{\text{m}}{\text{s}^2}$

**3.17** Derive the wavelength transformation equation for the case of E′ receding from R, by expressing the distances in Figure 3.2 in terms of wavelength.

*Solution:*
From Figure 3.2 we have

$$x = c\,\Delta t + u\,\Delta t,$$

where each term can be related to wavelength as follows:

$$x = \lambda N = \lambda N',$$

$$c\,\Delta t = \lambda'\nu'\,\Delta t = \lambda'\nu'\,(\gamma\,\Delta t') = \gamma\lambda' N',$$

$$u\,\Delta t = \gamma u\,\Delta t' = \gamma u\,\frac{N'}{\nu'} = \gamma u\,\frac{N'}{c/\lambda'} = \gamma\lambda' N'\beta.$$

With these equalities substituted into the first equation above, we obtain

$$\lambda N' = \gamma\lambda' N' + \gamma\lambda' N'\beta,$$

which reduces to the familiar wavelength equation

$$\lambda = \lambda'\gamma(1 + \beta).$$

**3.18** Starting with Equation 3.49, show that Equation 3.47 is obtained.

*Answer:* $\nu = \dfrac{\nu'}{\gamma\,(1 - \beta)}$

**3.19** If a distant galaxy is approaching the earth at a uniform speed of $0.6c$, what is the ratio of $\nu'$ to $\nu$?

*Solution:*
Since $u = 0.6c$, $\beta = 3/5$ and $\gamma = 5/4$. From the *blue shift* frequency transformation equation

$$\nu = \frac{\nu'}{\gamma\,(1 - \beta)}$$

we immediately obtain

$$\frac{\nu'}{\nu} = \gamma(1 - \beta)$$

$$= \frac{5}{4}\left(1 - \frac{3}{5}\right) = \frac{5}{4}\left(\frac{2}{5}\right) = \frac{1}{2}.$$

**3.20** A distant galaxy is *receding* from the earth uniformly at a speed of $0.8c$. If the wavelength of electromagnetic radiation received by the earth measures 6600 Å, what is the wavelength of the emitted radiation?

*Answer:* $\lambda' = 2200$ Å

**3.21** How fast must you move toward light of *proper wavelength* 6400 Å for it to appear to have a wavelength of 3200 Å?

*Solution:*
This problem corresponds to a *blue shift* phenomenon with $\lambda = 3200$ Å and $\lambda' = 6400$ Å. From the wavelength transformation (Equation 3.48)

$$\lambda = \lambda'\gamma\,(1 - \beta)$$

and the fact that $\lambda/\lambda' = 1/2$, we obtain

$$\frac{1}{2} = \gamma\,(1 - \beta) = \frac{1 - \beta}{\sqrt{1 - \beta^2}}$$

$$= \frac{\sqrt{1 - \beta}}{\sqrt{1 + \beta}}.$$

Squaring this equation gives

$$\frac{1}{4} = \frac{1 - \beta}{1 + \beta},$$

which can be simplified to

$$\frac{1}{4} + \frac{\beta}{4} = 1 - \beta$$

and solved for $\beta$:

$$\beta = \frac{3/4}{5/4} = \frac{3}{5} \rightarrow u = \frac{3}{5}\,c.$$

**3.22** If a distant galaxy is receding from the earth such that the emitted radiation wavelength is shifted by a factor of two, what is the speed of the galaxy relative to the earth?

*Answer:* $u = \dfrac{3}{5}\,c$

CHAPTER 4

# Transformations of Relativistic Dynamics

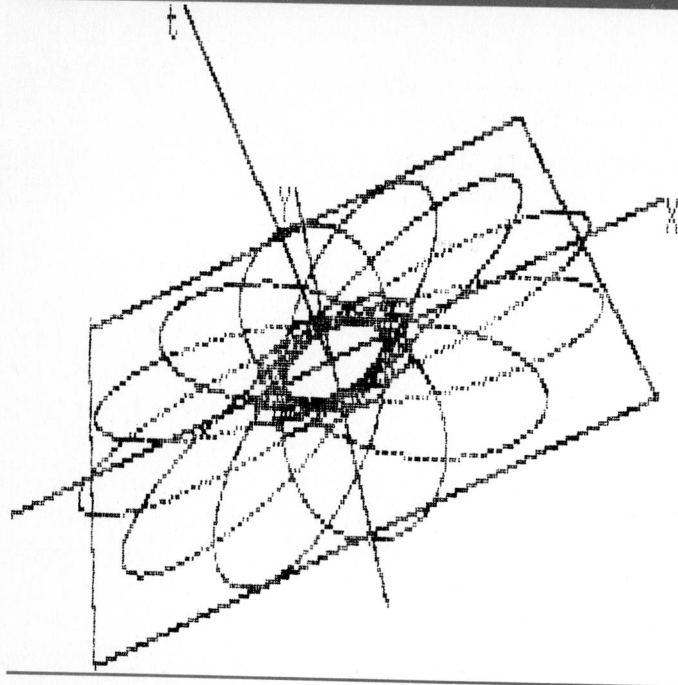


The most important result of a general character to which the special theory has led is concerned with the conception of mass. Before the advent of relativity, physics recognized two conservation laws of fundamental importance, namely, the law of the conservation of energy and the law of the conservation of mass; these two fundamental laws appeared to be quite independent of each other. By means of the theory of relativity they have been united into one law.

A. EINSTEIN, *Relativity* (1961)

## Introduction

The discussion of classical and Einsteinian relativity in the previous chapters illustrates how fundamental physical quantities, such as space-time coordinates, velocity, and acceleration, depend on the inertial system in which they are measured. In our investigation of relativistic dynamics, we will find how the mass, energy, and momentum of a body depend on the relativistic speed existing between the body and an inertial observer. We will also find it necessary to redefine basic quantities like total energy and kinetic energy, since their classical definitions are limited and become invalid for bodies traveling at relativistic speeds. However, these new relationships will reduce to their classical counterparts under the correspondence principle.

Up to this point, gedanken experiments have been viewed by observers in two different inertial systems, such that measurements of physical quantities in each system could be compared in obtaining transformation equations. The results obtained from this theoretical approach are most useful in that any physical quantity in *kinematics* can be transformed correctly from one inertial system to another by the appropriate use of the associated transformations. This same theoretical approach will be initially utilized herein to obtain a relativistic mass equation. For the most part, however, transformation equations per se are not obtained in our consideration of relativistic dynamics. Instead, equations for mass, force, energy, and momentum are developed that are appropriate for any particular inertial system. The derivations for relativistic force, energy, and momentum employ the fundamental defining equations of these quantities from classical mechanics, with a *relativistic mass relation* being appropriately incorporated. This allows for a logical development of these concepts in Einsteinian relativity, while capitalizing on our knowledge of the fundamentals of classical physics.

The classical conservation principles of energy and momentum are employed in obtaining new relativistic relationships, and the appropriateness of these conservation principles in special relativity is also investigated. As will be seen, conservation of momentum is assumed in both classical and relativistic mechanics; however, the classical conservation of energy principle becomes a *mass-energy conservation principle* in Einsteinian relativity. Our discussion of relativistic dynamics is concluded with a straight forward development of *momentum* and *energy* transformation equations for two inertial systems S and S′.

## 4.1 Relativistic Mass

It is of immediate interest to study the behavior of mass within the framework of Einstein's special theory of relativity. Unfortunately, we have a rather limited knowledge of mass and gravity, so a direct and logical development of the properties of a massive body moving at a relativistic speed is not available. Instead, we employ the fundamental conservation principle for *linear momentum* to deduce a relativistic mass relation for an isolated inertial system. As before, we will consider two inertial systems S and S′ to be separating from each other along their common $X$-$X'$ axis at a uniform relative speed $u$ and consider a gedanken experiment being viewed by observers in both systems.

Consider two *identical* massive bodies to be *approaching* each other on a *collision* course at constant speed parallel to the $X$-$X'$ axis, as illustrated in Figure 4.1a. Let observers in system S′ measure the speed of each body *before* the collision to be $u$, with $m_1'$ moving in the *positive* $x'$-direction and $m_2'$ traveling in the *negative* $x'$-direction. Further, allow the bodies to have a *perfectly inelastic collision*, such that observers in S′ will view the combined mass $m_1' + m_2' = 2m'$ to be at *rest* relative to S′ *after* the collision (see Figure 4.1b). We denote the velocities of $m_1'$ and $m_2'$ in S′ as

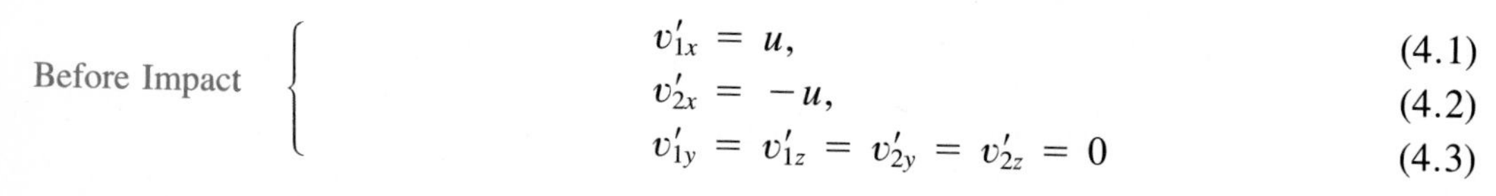

Before Impact

$$v_{1x}' = u, \tag{4.1}$$

$$v_{2x}' = -u, \tag{4.2}$$

$$v_{1y}' = v_{1z}' = v_{2y}' = v_{2z}' = 0 \tag{4.3}$$

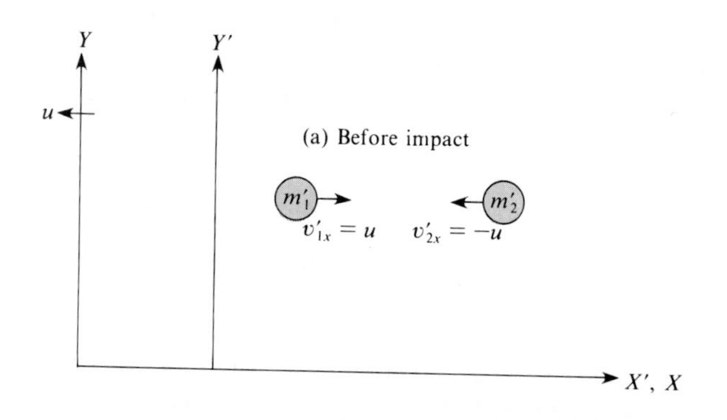


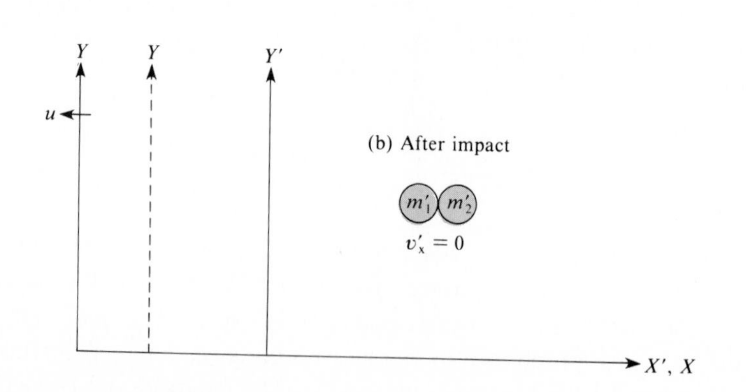


Figure 4.1
A perfectly inelastic collision of two identical bodies of mass $m_1' = m_2'$ traveling in opposite directions with identical speeds, according to observers in system S′.

*before* the collision and as

$$v'_x = v'_y = v'_z = 0 \tag{4.4}$$

After Impact

*after* the collision.

The physical data measured in S′ can now be transformed to system S, where observers consider themselves at rest and S′ to be moving in the positive $x$-direction with a speed $u$. To observers in system S, the situation *before* and *after* impact is similar to that depicted in Figures 4.2a and 4.2b. The velocity of each body can be obtained using the relativistic velocity component transformations given by Equations 3.24a, 3.24b, and 3.24c. Accordingly, we obtain

$$v_{1x} = \frac{v'_{1x} + u}{1 + \dfrac{v'_{1x}u}{c^2}} = \frac{u + u}{1 + \dfrac{uu}{c^2}}$$

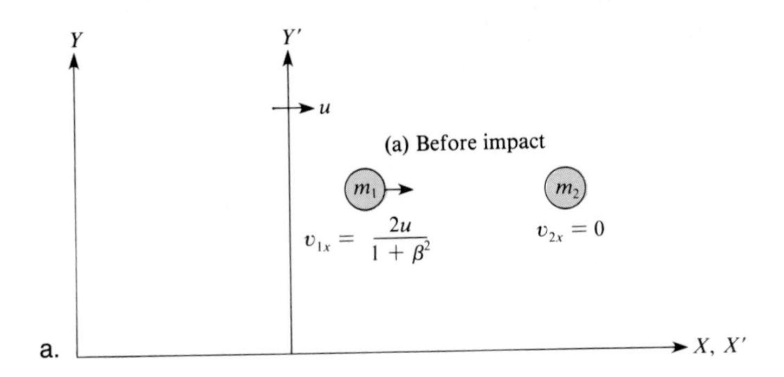


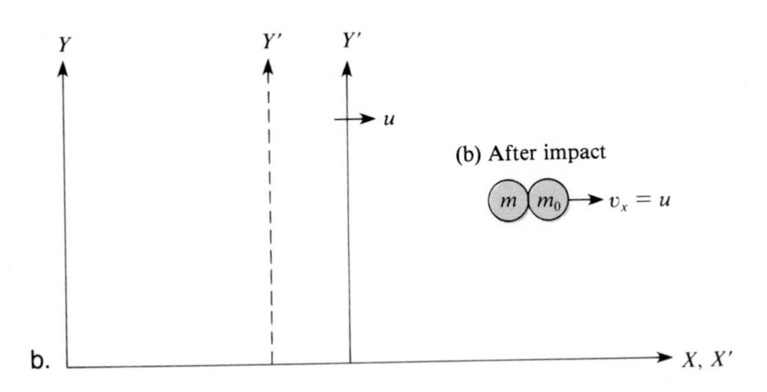


Figure 4.2
The perfectly inelastic collision of two massive bodies before *(a)* and after *(b)* impact, according to observers in system S.

Before Impact

$$= \frac{2u}{1 + \beta^2}, \tag{4.5}$$

$$v_{2x} = \frac{v'_{2x} + u}{1 + \dfrac{v'_{2x}u}{c^2}}$$

$$= \frac{-u + u}{1 + \dfrac{(-u)u}{c^2}}$$

Before Impact

$$= 0, \tag{4.6}$$

$$v_{1y} = v_{1z} = v_{2y} = v_{2z} = 0 \tag{4.7}$$

for the velocity components of $m_1$ and $m_2$ before the collision. *After* the collision, the two masses are viewed in S as stuck together with a total mass

$$M = m_1 + m_2 \tag{4.8}$$

and velocity components

$$v_x = \frac{v'_x + u}{1 + \dfrac{v'_x u}{c^2}}$$

$$= \frac{0 + u}{1 + \dfrac{(0)u}{c^2}}$$

After Impact

$$= u, \tag{4.9}$$

$$v_y = v_z = 0. \tag{4.10}$$

At this point we invoke the *conservation of linear momentum* principle, which says total linear momentum before impact must equal total linear momentum after impact. Since the $y$ and $z$ components of velocity are zero before (Equation 4.7) and after (Equation 4.10) impact, we are only concerned with the translational momentum in the $x$-direction. Thus, in system S the conservation of momentum can be expressed by

$$m_1 v_{1x} + m_2 v_{2x} = M v_x, \tag{4.11}$$

where $M$ is defined by Equation 4.8. Substitution from Equations 4.6, 4.8, and 4.9 allows the momentum equation to be expressed as

Momentum Conservation

$$m_1 v_{1x} = (m_1 + m_2)u. \tag{4.12}$$

It is now convenient to generalize the result expressed by Equation 4.12 and to simplify the symbolic notation. From Equation 4.6, it is apparent that mass $m_2$ is at *rest* in system S *before* the collision. We adopt the convention that the **rest** or **inertial mass** of a body be denoted as $m_0$ and, like proper length and proper time, allow it to be defined as that *mass measured in a frame of reference in which the body is at rest*. With this convention for $m_2$, there is no longer any need for the subscript on $m_1$ and, consequently, we have

$$m_1 \equiv m, \tag{4.13a}$$

$$m_2 \equiv m_0. \tag{4.13b}$$

Although the two masses are identical in S′, they are viewed differently in system S in that $m_0$ is at *rest* and $m$ is a *relativistic mass*, since it is in a state of *uniform relative motion* (Equation 4.5). To generalize Equation 4.12 we need only recognize that the *velocity* of $m$ ($m_1$ originally) *before* the collision corresponds to a *speed* in the positive $x$-direction of

$$v_1 = \sqrt{v_{1x}^2 + v_{1y}^2 + v_{1z}^2} = v_{1x}, \tag{4.14}$$

where Equation 4.7 has been used in obtaining the last equality. Again, there is no need for the subscript on $v_1$, so with

$$v_1 = v_{1x} \equiv v \tag{4.15}$$

and the identities of Equations 4.13a and 4.13b, Equation 4.12 becomes

$$mv = (m + m_0)u. \tag{4.16}$$

Momentum Conservation

Even though Equation 4.16 is expressed as a *scalar* equation in terms of speeds $v$ and $u$, it is equivalent to the *vector* equation expressed in terms of velocities **v** and **u**, since the directions of **v** and **u** are identical. For this reason we will refer to $v$ and $u$ as *velocities*, with the directions being understood to be in the positive $x$-direction. With this understanding, the *velocity* of $m$ *before* the collision, as given by Equation 4.5, can be rewritten, with $v_{1x}$ being replaced by $v$, as

$$v = \frac{2u}{1 + \beta^2}. \tag{4.17}$$

Referring to Equations 4.16 and 4.17, it is obvious that the relativistic mass $m$ can be expressed in terms of the rest mass $m_0$ and $v$, the velocity of $m$ relative to $m_0$, by solving the two equations simultaneously to eliminate $u$. To do so, we first solve Equation 4.16 for $m/m_0$ as

$$\frac{m}{m_0} = \frac{u}{v - u},$$

which can be rewritten in terms of $\beta$ ($\beta \equiv u/c$) as

Momentum Conservation

$$\frac{m}{m_0} = \frac{\beta}{\frac{v}{c} - \beta}. \tag{4.18}$$

Likewise, Equation 4.17 can be expressed in terms of $\beta$ as

$$\frac{v}{c} = \frac{2\beta}{1 + \beta^2} \tag{4.19}$$

and these last two equations can now be combined to eliminate $\beta$ and hence also $u$. The idea is to solve Equation 4.19 for $\beta$ and substitute into Equation 4.18. From Equation 4.19 we obtain a *quadratic* equation in terms of $\beta$,

$$\beta^2 - \frac{2c}{v}\beta + 1 = 0, \tag{4.20}$$

which has the solution

$$\beta = \frac{c}{v} \pm \sqrt{\frac{c^2}{v^2} - 1}. \tag{4.21}$$

Since $\beta \to 0$ for $u \ll c$, then the *negative* sign must clearly be chosen in Equation 4.21 with the result being

$$\beta = \frac{c}{v} - \sqrt{\frac{c^2}{v^2} - 1}$$

$$= \frac{c}{v} - \sqrt{\left(\frac{c^2}{v^2}\right)\left(1 - \frac{v^2}{c^2}\right)}$$

$$= \frac{c}{v}\left(1 - \sqrt{1 - \frac{v^2}{c^2}}\right). \tag{4.22}$$

By analogy with the definition of $\gamma$

$$\gamma \equiv \frac{1}{\sqrt{1 - \frac{u^2}{c^2}}}, \tag{2.7}$$

we define

$$\Gamma \equiv \frac{1}{\sqrt{1 - \frac{v^2}{c^2}}} \tag{4.23}$$

and immediately rewrite Equation 4.22 as

$$\beta = \frac{c}{v}(1 - \Gamma^{-1}). \tag{4.24}$$

Now, substitution of this expression for $\beta$ into the conservation of momentum equation (Equation 4.18) gives

$$\begin{aligned}
\frac{m}{m_0} &= \frac{\frac{c}{v}(1 - \Gamma^{-1})}{\frac{v}{c} - \frac{c}{v}(1 - \Gamma^{-1})} \\
&= \frac{1 - \Gamma^{-1}}{\frac{v^2}{c^2} - 1 + \Gamma^{-1}} \\
&= \frac{1 - \Gamma^{-1}}{-\Gamma^{-2} + \Gamma^{-1}} \\
&= \frac{1 - \Gamma^{-1}}{\Gamma^{-1}(1 - \Gamma^{-1})} \\
&= \Gamma.
\end{aligned}$$

This result can be rewritten in the more symmetrical form

$$m = \Gamma m_0, \tag{4.25}$$

Relativistic Mass

which is Einstein's *relativistic mass* equation. Since $\Gamma > 1$ for $v$ relativistic, then

$$m > m_0.$$

That is, the mass of a body is *not invariant* in Einsteinian relativity, but will be *observed* to increase when the body is in a *state of motion*. For $v \ll c$, this increase in mass is very small ($\Gamma \approx 1$) and $m \approx m_0$, which is exactly why we do not observe this phenomenon in everyday experiences. This result also explains why it is impossible to accelerate a body up to or beyond the speed of light $c$. As the body is accelerated toward the speed of light, its mass continuously increases and the externally applied accelerating force becomes less and less effective. At a speed very close to $c$, the mass of the body tends toward *infinity*, which would require an *infinite* external force for additional acceleration. Hence, acceleration of a body up to or beyond the speed of light is impossible by any *finite* force.

## 4.2 Relativistic Force

Frequently, in classical mechanics Newton's second law of motion is represented by

$$\mathbf{F} = m\mathbf{a}, \qquad m \neq m(t), \tag{1.17}$$

instead of the more *general* defining equation

Newton's Second Law

$$\mathbf{F} \equiv \frac{d\mathbf{p}}{dt}. \tag{1.16}$$

Often, the application of either equation to a problem will give the correct answer, since the mass of a body or a system of bodies is usually constant and independent of time in Newtonian dynamics. This is not the situation in relativistic dynamics, however, as the mass of an accelerating body experiences a *dilation* that is *time dependent*. This is immediately apparent from the *relativistic mass equation*, since the mass of a body is dependent on speed, which is ever changing for an accelerating body. From these arguments, it should be clear that $\mathbf{F} = m\mathbf{a} = \Gamma m_0\mathbf{a}$ is *not* valid in relativistic dynamics. The *defining equation* of Newton's second law is, however, applicable in relativistic dynamics and can be expressed as

$$\begin{aligned}\mathbf{F} &= \frac{d\mathbf{p}}{dt} = \frac{d(m\mathbf{v})}{dt}\\ &= m\frac{d\mathbf{v}}{dt} + \mathbf{v}\frac{dm}{dt}\\ &= m\mathbf{a} + \mathbf{v}m_0\frac{d\Gamma}{dt}\end{aligned}$$

or more fundamentally as

$$\mathbf{F} = \Gamma m_0 \mathbf{a} + m_0 \mathbf{v} \frac{d\Gamma}{dt}. \tag{4.26}$$

Relativistic Force

A relativistic relationship that is analogous to the classical $\mathbf{F} = m\mathbf{a}$ can be derived by using Newton's second law as expressed by Equation 4.26. The derivation can be greatly simplified by considering a body to be in a state of rectilinear motion and uniform acceleration, as viewed by observers in inertial system S. Under these conditions, Newton's second law can be expressed as

$$\begin{aligned} \mathbf{F} &= \frac{d\mathbf{p}}{dt} = \frac{dp}{dt}\,\mathbf{n} \\ &= \frac{dp}{dv}\frac{dv}{dt}\,\mathbf{n} \\ &= \frac{dp}{dv}\,a\mathbf{n} = \mathbf{a}\frac{dp}{dv}, \end{aligned} \tag{4.27}$$

where $\mathbf{n}$ is a *unit vector* in the direction of the momentum. It should be noted that we are considering a special case where the force, velocity, momentum, and acceleration vectors are all in the same direction. Substituting for linear momentum ($\mathbf{p} = m\mathbf{v}$), Equation 4.27 becomes

$$\begin{aligned} \mathbf{F} &= \mathbf{a}\frac{dmv}{dv} \\ &= m\mathbf{a} + v\mathbf{a}\frac{dm}{dv}, \end{aligned}$$

which from Equation 4.25 is rewritten as

$$\mathbf{F} = \Gamma m_0 \mathbf{a} + m_0 v \mathbf{a} \frac{d\Gamma}{dv}. \tag{4.28}$$

Relativistic Force

This result is also easily obtained from Equation 4.26, with $\mathbf{n}$ representing the assumed common direction for $\mathbf{v}$ and $\mathbf{a}$, that is

$$\begin{aligned} \mathbf{F} &= \left(\Gamma m_0 a + m_0 v \frac{d\Gamma}{dv}\frac{dv}{dt}\right)\mathbf{n} \\ &= \left(\Gamma m_0 a + m_0 v a \frac{d\Gamma}{dv}\right)\mathbf{n}. \end{aligned}$$

Further, Equation 4.28 for the relativistic force is easily reduced in form by obtaining the derivative of $\Gamma$ with respect to velocity. That is, differentiating Equation 4.23 yields

$$\frac{d\Gamma}{dv} = -\frac{1}{2}\left(1 - \frac{v^2}{c^2}\right)^{-3/2}\left(-\frac{2v}{c^2}\right)$$

$$= \Gamma^3 \frac{v}{c^2}, \tag{4.29}$$

which upon substitution into Equation 4.28 gives

$$\mathbf{F} = \Gamma m_0 \mathbf{a} + \Gamma^3 m_0 \frac{v^2}{c^2}\mathbf{a}$$

$$= \Gamma m_0 \mathbf{a}\left(1 + \Gamma^2 \frac{v^2}{c^2}\right)$$

$$= \Gamma^3 m_0 \mathbf{a}\left(\frac{1}{\Gamma^2} + \frac{v^2}{c^2}\right)$$

$$= \Gamma^3 m_0 \mathbf{a}\left(1 - \frac{v^2}{c^2} + \frac{v^2}{c^2}\right)$$

$$= \Gamma^3 m_0 \mathbf{a}.$$

Thus, for a body in rectilinear motion, the net accelerating force is given by

Relativistic Force

$$\mathbf{F} = \Gamma^3 m_0 \mathbf{a} = \Gamma^2 m \mathbf{a}. \tag{4.30}$$

This result is analogous to the classical equation $\mathbf{F} = m\mathbf{a}$ and in fact reduces to it as $\Gamma \rightarrow 1$. Although this result is limited to a *special class* of problems, it will prove useful for derivations in the next section.

## 4.3 Relativistic Kinetic and Total Energy

The *work-energy theorem* of classical mechanics states that the work done on a body is equivalent to its *change* in kinetic energy. Accordingly, if a body is at rest with inertial mass $m_0$, then the work done in accelerating it to a uniform velocity $\mathbf{v}$ is equivalent to its *final* kinetic energy $T$. For this situation the derivation of the body's relativistic kinetic energy can be simplified and previous results utilized, if we allow the net external force doing the work to be in the same direction as the body's displacement. This allows the *work* equation

$$dW \equiv \mathbf{F} \cdot d\mathbf{r} \tag{1.20}$$

Infinitesimal Work

to be written as

$$dT = F dr, \tag{4.31}$$

Since $\theta = 0°$ and $\cos\theta$ from the *dot product* becomes one. As the body is in rectilinear motion, the magnitude of the force expressed by Equation 4.30 can be substituted into Equation 4.31 and simplified as

$$\begin{aligned} dT &= \Gamma^2 ma\, dr \\ &= \Gamma^2 m \frac{dv}{dt}\, dr \\ &= \Gamma^2 m \frac{dr}{dt}\, dv \\ &= \Gamma^2 mv\, dv \\ &= m_0 \Gamma^3 v\, dv. \end{aligned} \tag{4.32}$$

This result is amenable to *integration by parts*; however, a further simplification is attained by solving Equation 4.29 for

$$c^2 d\Gamma = \Gamma^3 v\, dv$$

so that Equation 4.32 becomes

$$dT = m_0 c^2 d\Gamma. \tag{4.33}$$

Relativistic Kinetic Energy

This result represents the **relativistic kinetic energy** in differential form, which can be integrated once the *limits of integration* have been decided. In our present consideration, where work is done on a body initially at rest ($v_i = 0$, $v_f = v$), we need only recognize that for $v_i = 0$, $T_i = 0$ and $\Gamma_i = 1$. Thus, from Equation 4.33 we have

$$\int_0^T dT = m_0 c^2 \int_1^\Gamma d\Gamma \tag{4.34}$$

which immediately yields

Relativistic Kinetic Energy

$$T = m_0c^2(\Gamma - 1) \tag{4.35}$$

for the *relativistic kinetic energy*. This result has several interesting interpretations and is perfectly *general* and applicable in spite of our initial simplifying assumptions.

Under the *correspondence principle* the relativistic kinetic energy must reduce exactly to the classical kinetic energy. This is easily demonstrated by expanding $\Gamma$ in Equation 4.35 by use of the **binomial expansion**

$$(x + y)^n = x^n + nx^{n-1}y + \cdots \tag{4.36}$$

given in Appendix A, to obtain

$$\Gamma \approx 1 + \frac{1}{2}\frac{v^2}{c^2}, \qquad v << c. \tag{4.37}$$

With this relation for $\Gamma$ (remember $v << c$ and only the first couple of terms are significant) Equation 4.35 becomes

Classical Kinetic Energy

$$\begin{aligned} T &= m_0c^2\left(1 + \frac{1}{2}\frac{v^2}{c^2} - 1\right) \\ &= \tfrac{1}{2}m_0v^2 \end{aligned} \tag{1.22}$$

in agreement with classical mechanics. For a body moving at a relativistic speed, the error arising from using $\frac{1}{2}m_0v^2$ for $T$ is only 0.75 percent for $v = 0.1c$ but nearly 69 percent when $v = 0.9c$, for example

$$\begin{aligned} \frac{T - \frac{1}{2}m_0v^2}{T} &= 1 - \frac{\frac{1}{2}m_0v^2}{T} \\ &= 1 - \frac{\frac{1}{2}m_0v^2}{m_0c^2\,(\Gamma - 1)} \\ &= 1 - \frac{\frac{1}{2}(v/c)^2}{\Gamma - 1} \\ &= 1 - \frac{\frac{1}{2}(0.81)}{1.3} \approx 0.69. \end{aligned}$$

A more interesting consequence of Equation 4.35 becomes evident by rewriting it in the form

$$T = \Gamma m_0c^2 - m_0c^2 = mc^2 - m_0c^2. \quad (4.38)$$

Both terms on the right-hand side of this equation have, necessarily, dimensions of energy and they represent an *energy-mass equivalence* that are symbolically identified as

$$E_0 \equiv m_0c^2, \quad (4.39)$$

Rest Energy

$$E \equiv mc^2. \quad (4.40)$$

Relativistic Total Energy

The interpretation of these quantities is straightforward in that for a body at *rest* with *inertial mass* $m_0$, $E_0$ must correspond to its **rest energy,** while $E$ corresponds to its **total relativistic energy.** For a body at rest, *kinetic* energy, represented by

$$T = E - E_0, \quad (4.41a)$$

Relativistic Kinetic Energy

must be zero and $E$, given by

$$E = E_0 + T, \quad (4.41b)$$

Relativistic Total Energy

must equal $E_0$. Amazingly, a body at rest *possesses energy* $m_0c^2$ according to Einsteinian relativity. Equation 4.41a is practically always used as the fundamental equation for *relativistic kinetic energy* instead of Equation 4.35, since it is conceptually much simpler and logically much more direct. Further, the *relativistic energy* defined by Equation 4.40 is recognized as the *total energy* of a body from Equation 4.41b.

These results are surprising and have absolutely no counterpart in classical physics. The *energy mass equivalence* represented by Equation 4.40 is the single most important result of Einstein's *special theory* of relativity. It gives the energy equivalence of a 1 kg mass to be on the order of $c^2$ or $9 \times 10^{16}$ J. Consequently, even an extremely small mass has a relatively large energy equivalence. As an example, assuming the average caloric intake per person per day to be 3200 kcal, then the energy consumed per day by ten million people has a *mass equivalence* of approximately one and a half grams:

$$\begin{aligned} m_0 &= \frac{E_0}{c^2} \\ &= (3200 \text{ kcal})(10^7 \text{ people}) \frac{(4.186 \times 10^3 \text{ J/kcal})}{c^2} \end{aligned}$$

$$= \frac{1.340 \times 10^{14} \text{ J}}{9 \times 10^{16} \text{ m}^2/\text{s}^2}$$

$$= 1.489 \times 10^{-3} \text{ kg} \approx 1.5 \text{ g}.$$

Another interesting result is easily obtained from Equation 4.33. If a body has an initial velocity $v_i = v_1$ and final velocity $v_f = v_2$, then integration of Equation 4.33 gives

$$\begin{aligned} T_2 - T_1 &= m_0c^2(\Gamma_2 - \Gamma_1) \\ &= \Gamma_2 m_0 c^2 - \Gamma_1 m_0 c^2 \\ &= m_2c^2 - m_1c^2 \\ &= (m_2 - m_1)c^2. \end{aligned}$$

The result can be symbolically represented as

$$\Delta T = \Delta mc^2 = \Delta E, \tag{4.42}$$

where the last equality ($\Delta T = \Delta E$) is obvious from Equation 4.41. Consequently, any change in the *kinetic* or *total energy* of a body results in a corresponding change in its *mass*. Indeed, *hot* water has more *mass* than the same amount of *cold* water, and so forth. The reason we do not observe these changes in everyday experiences is because $\Delta m$ is *very* small as compared to the change in energy and, furthermore, commonly encountered values of $\Delta E$ are *relatively small*.

It needs to be emphasized that the concept of total energy in Einsteinian relativity differs from that of classical mechanics in that the former does not include *potential energy* $V$. The *conservation of energy* principle in a broader sense is, however, still valid in relativistic dynamics, provided the *rest energy* of a body or system of bodies is taken into account. The principle now becomes one of **mass-energy conservation,** which is represented as

Mass-Energy Conservation

$$E_0 + T + V = \text{CONSTANT} \tag{4.43}$$

for an isolated inertial system.

## 4.4 Relativistic Momentum

An expression for the relativistic momentum of a body is easily obtained from the classical definition of momentum and the equations representing relativistic mass and energy. That is,

$$\begin{aligned}\mathbf{p} &\equiv m\mathbf{v} \\ &= \Gamma m_0 \mathbf{v} = \frac{\Gamma m_0 c^2 \mathbf{v}}{c^2} \\ &= \frac{\Gamma E_0 \mathbf{v}}{c^2}\end{aligned} \tag{1.15}$$

Linear Momentum

or more simply

$$\mathbf{p} = \frac{E\mathbf{v}}{c^2} \tag{4.44}$$

Relativistic Momentum

One interesting interpretation of Equation 4.44 is for electromagnetic radiation (e.g., x-rays, $\gamma$-rays, visible light, etc.), since it propagates through free space at the speed $c$. In this case $v = c$ and the *total energy* of a *quantum* of radiation is given by

$$E = pc. \tag{4.45}$$

Energy of Photons

The result suggests a *particle-like* behavior for electromagnetic waves, which was originally proposed by Einstein in 1905. In explaining the *photoelectric effect* (see Chapter 6, Section 6.6), he postulated that electromagnetic radiation consisted of *quanta of light-energy* in the form of fundamental particles, later called **photons,** that propagate at the speed of light. This particle-like behavior of light will be the topic of considerable discussion in Chapter 6 as well as in the next section.

Another very useful relationship between momentum and energy can be obtained from either the *relativistic energy* (Equation 4.40) or *relativistic mass* (Equation 4.25) equations. From the former we have

$$E = mc^2 = \Gamma m_0 c^2 = \Gamma E_0,$$

which, when squared, gives

$$\begin{aligned}E^2 &= \Gamma^2 E_0^2 \\ &= \frac{E_0^2}{1 - \dfrac{v^2}{c^2}}\end{aligned}$$

$$= \frac{E_0^2}{1 - \dfrac{p^2}{m^2c^2}}$$

$$= \frac{E_0^2}{1 - \dfrac{p^2c^2}{E^2}},$$

where the definitions of momentum and relativistic energy have been utilized. The last equality, solved for $E^2$, gives

Energy-Momentum Invariant

$$E^2 = E_0^2 + p^2c^2, \tag{4.46}$$

which in Section 4.6 will be shown to be an *invariant* relationship to all inertial observers. At this time, however, additional insight into the properties of a particle traveling at the speed of light (i.e., a photon) can be obtained by combining Equations 4.45 and 4.46. Clearly, since for a photon $E = pc$ (Equation 4.45), then Equation 4.46 gives $E_0 = 0$, which means $m_0 = 0$ for a photon. That is, a body traveling at the speed of light *must* have a zero rest mass and, conversely, a particle of zero rest mass must be traveling at the speed of light. This result of Einsteinian relativity yields considerable insight into the behavior of *particles* and *waves* in nature, which was not fully appreciated nor understood for nearly two decades after Einstein's published work. His theory clearly predicts that particles having a nonzero rest mass can never be accelerated to the speed of light, while entities in nature traveling at such a speed must necessarily have a zero rest mass. This relativistic view is in sharp contrast to the predictions of classical physics, but, as will be seen, it is the correct one and accurately describes the properties of *photons*.

## 4.5 Energy and Inertial Mass Revisited

The results of the last section were totally surprising to the physics community of the early twentieth century; however, Equation 4.45 was known from classical electromagnetism for well overy thirty years (see Chapter 6, Section 6.4) before the publication of Einstein's theory. We will utilize that equation and the concept of a *photon* as an elementary particle or *quantum* of electromagnetic radiation to re-derive the energy-mass relationship, by considering a gedanken experiment originally developed by Einstein in 1906.

Consider two identical spheres, each of mass $\frac{1}{2}M$, separated a distance $L$ by a *rod of negligible mass*. This *dumbbell* system is assumed to be isolated from its surroundings and initially stationary with its *center of mass* (C.M.) located midway between the spheres on their common axis. At some instant in time a *burst of photons* is emitted from the right-hand sphere and propagates toward the left-hand sphere, as illustrated in Figure 4.3. If we think of these photons as possessing an *equivalent mass m*, then the radiant energy associated with the photons is

$$E = pc, \tag{4.45}$$

according to Equation 4.45. Assuming conservation of momentum to be valid, then the momentum of the photons to the left, $p$, is just equal to the momentum of the dumbbell system to the right, $(M - m)u$, where $m$ is the assumed *mass equivalent* of the emitted radiation. Thus,

$$p = (M - m)u \tag{4.47}$$

and Equation 4.45 becomes

$$E = (M - m)uc. \tag{4.48}$$

If $\Delta t$ is the time required for the photons to travel from the right-hand sphere to the left-hand sphere, then from the postulate of the constancy of the speed of light

$$\Delta t = \frac{L - \Delta x}{c}, \tag{4.49}$$

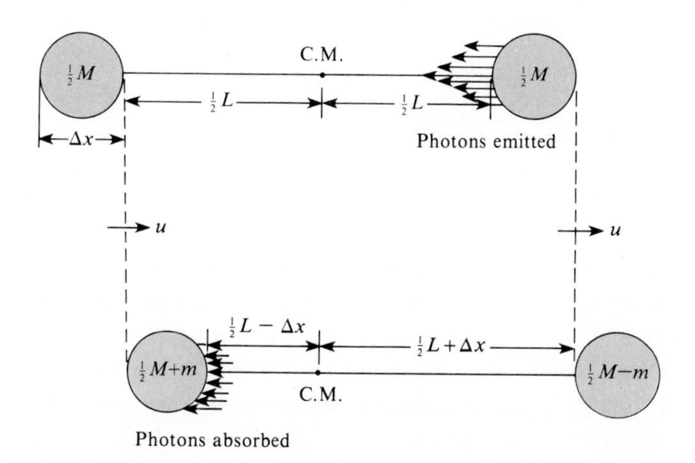


Figure 4.3
The emission of photons of equivalent mass $m$ from the right-hand sphere, the recoil of the dumb-bell with velocity $u$, and the final absorption of the photons by the left-hand sphere.

where the recoil distance $\Delta x$ of the dumbbell system has been taken into account. Accordingly, the average speed of the dumbbell is just

$$u \equiv \frac{\Delta x}{\Delta t} = \frac{c\,\Delta x}{L - \Delta x}, \tag{4.50}$$

which upon substitution into Equation 4.48 gives

$$\begin{aligned} E &= \frac{(M - m)c^2\,\Delta x}{L - \Delta x} \\ &= \frac{(M - m)c^2}{\dfrac{L}{\Delta x} - 1} \\ &= \frac{\left(\dfrac{M}{m} - 1\right)mc^2}{\dfrac{L}{\Delta x} - 1}. \end{aligned} \tag{4.51}$$

Since the sum of all *mass moments* on one side of the C.M. point must equal the sum of all mass moments on the other side (by definition of the center of mass), then

$$(\tfrac{1}{2}M + m)(\tfrac{1}{2}L - \Delta x) = (\tfrac{1}{2}M - m)(\tfrac{1}{2}L + \Delta x). \tag{4.52a}$$

Solving this equation for $\Delta x$,

$$\Delta x = \frac{mL}{M}, \tag{4.52b}$$

and substituting into Equation 4.51 immediately gives

Energy-Mass Equivalence

$$E = mc^2. \tag{4.40}$$

Although this equation has been obtained by a derivation that differs somewhat from Einstein's, the resulting implications are the same and are in agreement with the interpretations of Equation 4.40. The interpretation here, however, is that the *sphere emitting electromagnetic radiation* experiences an inertial mass *decrease* of $E/c^2$, while the other sphere's inertial mass *increases* by the amount $E/c^2$ upon absorption of the radiation.

Thus, *any* change in the energy $\Delta E$ of a body results in a corresponding change in its inertial mass $\Delta m$ in accordance with Equation 4.42.

## 4.6 Relativistic Momentum and Energy Transformations

The previous discussions of this chapter have been primarily concerned with the view of Einsteinian dynamics in one inertial system; however, it is desirable to transform momentum and energy measurements of a particle from one inertial system to another. Linear momentum has already been defined as the product of mass and velocity, so in system S the momentum of a particle is

$$\mathbf{p} \equiv m\mathbf{v} = \Gamma m_0 \mathbf{v}, \tag{4.53a}$$

while in system S′ it is denoted as

$$\mathbf{p}' \equiv m'\mathbf{v}' = \Gamma' m_0 \mathbf{v}'. \tag{4.53b}$$

Clearly, these are vector equations and can be expressed in terms of Cartesian components as

$$\mathbf{p} = p_x\mathbf{i} + p_y\mathbf{j} + p_z\mathbf{k} \tag{4.54a}$$

and

$$\mathbf{p}' = p'_x\mathbf{i}' + p'_y\mathbf{j}' + p'_z\mathbf{k}'. \tag{4.54b}$$

The immediate problem is to find out how these *components* of momentum transform between two inertial systems S and S′, when they are separating from each other at the constant speed $u$.

Consider a particle to be moving about in space and time with a velocity $\mathbf{v}$ measured in system S and $\mathbf{v}'$ measured in S′. In this context the velocity of the particle, as measured by observers in either system, need not be parallel to the common axis of relative motion between the two systems. According to observers in S, the particle has a longitudinal component of momentum given by

$$p_x = mv_x,$$

which, upon substitution of Equations 4.25 and 3.24a, can be expanded to the form

$$p_x = \Gamma m_0 \frac{(v'_x + u)}{1 + \dfrac{v'_x u}{c^2}}. \tag{4.55}$$

It will be shown that

$$\frac{\Gamma}{1 + \dfrac{v_x' u}{c^2}} = \gamma \Gamma', \tag{4.56}$$

where for observers in system S′

$$\Gamma' \equiv \frac{1}{\sqrt{1 - \dfrac{v'^2}{c^2}}} \tag{4.57}$$

by analogy with the definition given by Equation 4.23. To obtain this result, consider the expansion of $1 - v_x^2/c^2$,

$$\begin{aligned}
1 - \frac{v_x^2}{c^2} &= 1 - \frac{(v_x' + u)^2}{c^2\left(1 + \dfrac{v_x' u}{c^2}\right)^2} \\
&= \frac{\left(1 + \dfrac{v_x' u}{c^2}\right)^2 - \dfrac{(v_x' + u)^2}{c^2}}{\left(1 + \dfrac{v_x' u}{c^2}\right)^2} \\
&= \frac{1 - \dfrac{v_x'^2}{c^2} - \dfrac{u^2}{c^2} + \left(\dfrac{v_x' u}{c^2}\right)^2}{\left(1 + \dfrac{v_x'}{c^2}\right)^2} \\
&= \frac{\left(1 - \dfrac{v_x'^2}{c^2}\right)\left(1 - \dfrac{u^2}{c^2}\right)}{\left(1 + \dfrac{v_x' u}{c^2}\right)^2},
\end{aligned} \tag{4.58}$$

where Equation 3.24a has been used. From Equations 3.24b, 3.24c, and 2.7 we have

$$\frac{v_y^2}{c^2} = \frac{\left(\frac{v_y'}{c^2}\right)\left(1 - \frac{u^2}{c^2}\right)}{\left(1 + \frac{v_x'u}{c^2}\right)^2},$$

$$\frac{v_z^2}{c^2} = \frac{\left(\frac{v_z'}{c}\right)^2\left(1 - \frac{u^2}{c^2}\right)}{\left(1 + \frac{v_x'u}{c^2}\right)^2},$$

which upon subtraction from Equation 4.58 and rearrangement gives

$$\frac{1}{\left(1 - \frac{v^2}{c^2}\right)} = \frac{\left(1 + \frac{v_x'u}{c^2}\right)^2}{\left(1 - \frac{v'^2}{c^2}\right)\left(1 - \frac{u^2}{c^2}\right)}, \tag{4.59}$$

where the fundamental relations

$$v^2 = v_x^2 + v_y^2 + v_z^2, \tag{4.60a}$$

$$v'^2 = v_x'^2 + v_y'^2 + v_z'^2 \tag{4.60b}$$

have been used. Now, taking the square root of Equation 4.59 and using the defining equations for $\Gamma$, $\Gamma'$, and $\gamma$ gives

$$\Gamma = \gamma\Gamma'\left(1 + \frac{v_x'u}{c^2}\right),$$

which is equivalent to Equation 4.56. Using this result (actually Equation 4.56) allows the $x$-component of momentum, given by Equation 4.55, to be expressed in the more compact form

$$p_x = \gamma\Gamma' m_0(v_x' + u) = \gamma m'(v_x' + u)$$

or more simply

$$p_x = \gamma(p_x' + m'u). \tag{4.61}$$ $S' \rightarrow S$

In a similar manner the transformations for the transverse components of momentum are easily obtained. For example,

$$\begin{aligned}
p_y &\equiv mv_y \\
&= \Gamma m_0 \frac{(v_y'/\gamma)}{1 + \dfrac{v_x' u}{c^2}} \\
&= \frac{\Gamma}{1 + \dfrac{v_x' u}{c^2}} \frac{m_0 v_y'}{\gamma} \\
&= \gamma\Gamma' \frac{m_0 v_y'}{\gamma} \\
&= m' v_y',
\end{aligned}$$

which from Equation 4.53b yields

S′ → S

$$p_y = p_y' \tag{4.62}$$

and similarly

S′ → S

$$p_z = p_z'. \tag{4.63}$$

Equations 4.61, 4.62, and 4.63 represent the *relativistic momentum-component transformations* from system S′ to system S; whereas, the inverse transformations are given by

S → S′

$$\begin{cases} p_x' = \gamma(p_x - mu) & (4.64a) \\ p_y' = p_y & (4.64b) \\ p_z' = p_z. & (4.64c) \end{cases}$$

Frequently, the transformation for the longitudinal component of momentum is expressed in terms of energy. This is easily accomplished with Equations 4.61 and 4.64a by realizing that Einstein's energy-mass equivalence relationship is valid for *any* inertial system. That is, to observers in S

$$E = mc^2 = \Gamma m_0 c^2 = \Gamma E_0, \tag{4.65a}$$

while to observers in system S′

$$
\begin{aligned}
E' &= m'c^2 \\
&= \Gamma' m_0 c^2 \\
&= \Gamma' E_0 \qquad (4.65b)
\end{aligned}
$$

for a particle having an inertial mass $m_0$. Consequently, Equations 4.61 and 4.64a can be expressed as

$$p_x = \gamma\left(p_x' + \frac{E'u}{c^2}\right) \qquad (4.66) \quad S' \rightarrow S$$

and

$$p_x' = \gamma\left(p_x - \frac{Eu}{c^2}\right), \qquad (4.67) \quad S \rightarrow S'$$

which will prove useful in our next consideration.

To obtain the relativistic transformation for $E$ and $E'$, it is convenient to capitalize on the results expressed by Equations 4.66 and 4.67. If we desire an equation for $E$ in terms of primed quantities $p_x'$ and $E'$, then $p_x$ must be eliminated between the two equations. Thus, substitution of Equation 4.66 into Equation 4.67 gives

$$p_x' = \gamma\left[\gamma\left(p_x' + \frac{E'u}{c^2}\right) - \frac{Eu}{c^2}\right],$$

which can be solved for the term involving $E$ as

$$\frac{E\gamma u}{c^2} = \gamma^2 p_x' + \frac{\gamma^2 E' u}{c^2} - p_x'.$$

Multiplying both sides of this equation by $c^2/\gamma u$ gives

$$
\begin{aligned}
E &= \frac{\gamma p_x' c^2}{u} + \gamma E' - \frac{p_x' c^2}{\gamma u} \\
&= \gamma\left[E' + \frac{p_x' c^2}{u}(1 - \gamma^{-2})\right].
\end{aligned}
$$

Because $\gamma^{-2} = 1 - u^2/c^2$, the *energy transformation equation* from S′ to S becomes

$$E = \gamma(E' + p_x' u), \qquad (4.68) \quad S' \rightarrow S$$

while the inverse transformation is given by

$S \rightarrow S'$

$$E' = \gamma(E - p_x u). \tag{4.69}$$

This last expression is directly obtained in a similar manner by substituting $p'_x$ from Equation 4.67 directly into Equation 4.66 and solving for $E'$.

In this and previous sections it was clear that momentum and energy of a particle depend on the observer, and in general these quantities have different measured values for different inertial observers. Equation 4.46 solved in the form

$$E^2 - p^2c^2 = E_0^2$$

suggests that if $E_0$ for a particle is to have the same value to all inertial observers, then the particle's energy squared minus the square of the product of its momentum and the speed of light must be an *invariant* to all inertial observers. To verify this observation, we need only substitute Equations 4.69, 4.67, 4.66, and 4.64c into the expression $E'^2 - p'^2c^2$ to obtain

$$\begin{aligned}
E_0^2 &= E'^2 - p'^2c^2 \\
&= E'^2 - (p_x'^2 + p_y'^2 + p_z'^2)c^2 \\
&= E'^2 - \gamma^2\left(p_x - \frac{Eu}{c^2}\right)^2 c^2 - p_y^2c^2 - p_z^2c^2 \\
&= \gamma^2(E - p_x u)^2 - \gamma^2\left(p_x - \frac{Eu}{c^2}\right)^2 c^2 - p_y^2c^2 - p_z^2c^2 \\
&= \gamma^2\left(E^2 + p_x^2u^2 - p_x^2c^2 - \frac{E^2u^2}{c^2}\right) - p_y^2c^2 - p_z^2c^2 \\
&= \gamma^2\left[E^2\left(1 - \frac{u^2}{c^2}\right) - p_x^2c^2\left(1 - \frac{u^2}{c^2}\right)\right] - p_y^2c^2 - p_z^2c^2 \\
&= E^2 - p_x^2c^2 - p_y^2c^2 - p_z^2c^2 \\
&= E^2 - p^2c^2.
\end{aligned}$$

Thus, although our other relativistic transformations for momentum, energy, mass, and force, are *not invariant*, Equation 4.46

Energy-Momentum Invariant

$$E^2 = E_0^2 + p^2c^2, \tag{4.46}$$

represents a particular *invariant* combination of energy and momentum on which all inertial observers are in agreement.

Special relativity has profoundly altered our world view and raised a host of philosophical and scientific questions. Its exhaustively verified correctness suggests that we have been presumptuous in defining *nature's laws* to be consistent with our *common sense*. We should examine our fundamental view with the aim of removing inherent, prejudicious concepts, since we are three-dimensional creatures in, at least, a four-dimensional world. In particular, the concept of a *semi-infinite* time axis should be reexamined and our understanding of gravity (mass) could stand much improvement. At any rate, the developmental logic of special relativity illustrates the need to liberate our reasoning from physical prejudices and to rely only on pure logic within any carefully defined hypothetical framework. We may not be able to answer all the questions today, but we now know *how*, and *perhaps where*, to begin looking for at least some of the answers.

## Review of Derived Equations

A listing of the fundamental and derived equations of relativistic dynamics is presented below, along with the transformation equations for relativistic energy and momentum. Also included are the newly defined special symbols of this chapter.

### Special Symbols

$$\left.\begin{aligned} \gamma &\equiv \frac{1}{\sqrt{1-\frac{u^2}{c^2}}} \\ \Gamma &\equiv \frac{1}{\sqrt{1-\frac{v^2}{c^2}}} \\ \Gamma' &\equiv \frac{1}{\sqrt{1-\frac{v'^2}{c^2}}} \end{aligned}\right\} \quad \frac{\Gamma}{1+\frac{v_x' u}{c^2}} = \gamma\Gamma'$$

## DERIVED EQUATIONS

|  |  |
|---|---|
| $m = \Gamma m_0$ | Relativistic Mass |
| $\mathbf{F} \equiv \dfrac{d\mathbf{p}}{dt}$ | Newton's Second Law |
| $= \Gamma m_0\mathbf{a} + m_0\mathbf{v}\dfrac{d\Gamma}{dt}$ | |
| $= \Gamma m_0\mathbf{a} + m_0 v\mathbf{a}\dfrac{d\Gamma}{dv}$ | |
| $= \Gamma^3 m_0\mathbf{a}$ | Relativistic Force |
| $dT = F\,dr = m_0c^2 d\Gamma$; $T = m_0c^2(\Gamma - 1)$; $T = E - E_0$ | Relativistic Kinetic Energy |
| $E_0 \equiv m_0c^2$ | Rest Energy |
| $E \equiv mc^2$ | Relativistic Total Energy |
| $\mathbf{p} \equiv m\mathbf{v}$ | Classical Linear Momentum |
| $= \dfrac{E\mathbf{v}}{c^2}$ | Relativistic Momentum |
| $E = pc$ | Photon Energy |
| $E^2 = E_0^2 + p^2c^2$ | Energy-Momentum Invariant |

### *Relativistic Momentum Transformations*

| $\mathbf{S}' \to \mathbf{S}$ | $\mathbf{S} \to \mathbf{S}'$ |
|---|---|
| $p_x = \gamma(p_x' + m'u) = \gamma\left(p_x' + \dfrac{E'u}{c^2}\right)$ | $p_x' = \gamma(p_x - mu) = \gamma\left(p_x - \dfrac{Eu}{c^2}\right)$ |
| $p_y = p_y'$ | $p_y' = p_y$ |
| $p_z = p_z'$ | $p_z' = p_z$ |

### *Relativistic Energy Transformations*

$$E = \gamma(E' + p_x'u) \qquad E' = \gamma(E - p_xu)$$

## Problems

**4.1** Combining Equations 4.16 and 4.17, show that the ratio of $m$ to $m_0$ is given by $m/m_0 = (1 + \beta^2)/(1 - \beta^2)$.

*Solution:*
Solving Equation 4.16,

$$mv = (m + m_0)u,$$

for $m/m_0$ and substituting Equation 4.17,

$$v = \frac{2u}{1 + \beta^2},$$

immediately yields

$$\begin{aligned} \frac{m}{m_0} &= \frac{u}{v - u} \\ &= \frac{u}{\dfrac{2u}{1 + \beta^2} - u} \\ &= \frac{1}{\dfrac{2}{1 + \beta^2} - 1} \\ &= \frac{1 + \beta^2}{2 - (1 + \beta^2)} \\ &= \frac{1 + \beta^2}{1 - \beta^2}. \end{aligned}$$

**4.2** Starting with Equation 4.17 and using the result of Problem 4.1, show that $1 - v^2/c^2 = m_0^2/m^2$ and that Equation 4.25 for the relativistic mass is obtained.

*Answer:* $\Gamma = \dfrac{1 + \beta^2}{1 - \beta^2} = \dfrac{m}{m_0} \quad \rightarrow \quad m = \Gamma m_0$

**4.3** A particle of rest mass $1.60 \times 10^{-29}$ kg moves with a speed of $0.6c$ relative to some inertial system. Find its relativistic mass and momentum.

*Solution:*
With $m_0 = 1.60 \times 10^{-29}$ kg and $v = 3c/5 \rightarrow \Gamma = 5/4$ substituted directly into Equation 4.25,

$$m = \Gamma m_0,$$

the relativistic mass is

$$m = \frac{5}{4}(1.60 \times 10^{-29}\ \text{kg}) = 2.00 \times 10^{-29}\ \text{kg},$$

and the relativistic momentum is just

$$p = mv = (2.0 \times 10^{-29}\ \text{kg})\frac{3}{5}\left(3 \times 10^8\ \frac{\text{m}}{\text{s}}\right)$$

$$= 3.60 \times 10^{-21}\ \text{kg} \cdot \frac{\text{m}}{\text{s}}.$$

**4.4** A particle of relativistic mass $1.80 \times 10^{-29}$ kg is moving with a constant speed of $3c/5$. Find its relativistic momentum and rest mass.

*Answer:* $p = 3.24 \times 10^{-21}$ kg · m/s, $m = 1.44 \times 10^{-29}$ kg

**4.5** Find the rest energy, relativistic total energy, and relativistic kinetic energy for the particle of Problem 4.3.

*Solution:*

At this point we know $m_0 = 1.60 \times 10^{-29}$ kg, $v = 3c/5$, $\Gamma = 5/4$, $m = 2.00 \times 10^{-29}$ kg, $p = 3.60 \times 10^{-21}$ kg · m/s and we need to find $E_0$, $E$, and $T$. Direct substitution into Equations 4.39, 4.40, and 4.41a gives

$$E_0 = m_0c^2 = (1.60 \times 10^{-29}\ \text{kg})\left(9 \times 10^{16}\ \frac{\text{m}^2}{\text{s}^2}\right) = 1.44 \times 10^{-12}\ \text{J},$$

$$E = mc^2 = \Gamma E_0 = \frac{5}{4}(1.44 \times 10^{-12}\ \text{J}) = 1.80 \times 10^{-12}\ \text{J},$$

$$T = E - E_0 = (1.80 - 1.44) \times 10^{-12}\ \text{J} = 3.6 \times 10^{-13}\ \text{J}.$$

**4.6** Find the rest energy, total energy, and relativistic kinetic energy for the particle of Problem 4.4.

*Answer:*

$E_0 = 1.30 \times 10^{-12}$ J, $E = 1.62 \times 10^{-12}$ J, $T = 3.24 \times 10^{-13}$ J

**4.7** Express the answers to Problem 4.5 in units of MeV, where M $= 10^6$ and 1 eV $= 1.60 \times 10^{-19}$ J. Verify that the total energy in MeV is also given by Equation 4.46.

*Solution:*

For the conversion of units we have

$$E_0 = (1.44 \times 10^{-12}\ \text{J})\left(\frac{1\ \text{eV}}{1.60 \times 10^{-19}\ \text{J}}\right) = 9\ \text{MeV},$$

$$E = (1.80 \times 10^{-12}\ \text{J})\left(\frac{1\ \text{eV}}{1.60 \times 10^{-19}\ \text{J}}\right) \approx 11.25\ \text{MeV},$$

$$T = (3.6 \times 10^{-13}\ \text{J})\left(\frac{1\ \text{eV}}{1.60 \times 10^{-19}\ \text{J}}\right) = 2.25\ \text{MeV}.$$

Now, using Equation 4.46,

$$E^2 = E_0^2 + p^2c^2,$$

direct substitution yields

$$E^2 = (9 \text{ MeV})^2 + \left[(3.60 \times 10^{-21})\ (3 \times 10^8) \text{ J} \left(\frac{1 \text{ eV}}{1.6 \times 10^{-19} \text{ J}}\right)\right]^2$$
$$= (9 \text{ MeV})^2 + (6.75 \text{ MeV})^2.$$

Thus, the total energy is given by

$$E \approx 11.25 \text{ MeV}.$$

**4.8** Find the percentage of error arising from using the classical definition of kinetic energy ($T = \frac{1}{2}m_0v^2$) instead of the relativistic definition for a particle traveling at $3c/5$.

*Answer:* 28 percent

**4.9** Show that the percentage of error arising from using the classical definition of momentum ($p = m_0v$) instead of the relativistic momentum is 20 percent for a particle traveling at $3c/5$.

*Solution:*
With $v = 3c/5 \rightarrow \Gamma = 5/4$, we have

$$\frac{p - m_0v}{p} = 1 - \frac{m_0v}{p}$$
$$= 1 - \frac{m_0v}{\Gamma m_0v}$$
$$= 1 - \frac{1}{\Gamma} = 1 - \frac{4}{5} = 0.20.$$

**4.10** Derive Equation 4.46 by starting with Equation 4.25.

*Answer:* $E^2 = E_0^2 + p^2c^2$

**4.11** What is the momentum for a particle of rest energy 0.513 MeV and total energy 0.855 MeV?

*Solution:*
Linear momentum can be expressed in terms of $E_0 = 0.513$ MeV and $E = 0.855$ MeV using Equation 4.46 in the form

$$p = \frac{\sqrt{E^2 - E_0^2}}{c}.$$

Thus, direct substitution of $E_0$ and $E$ gives

$$p = \sqrt{0.855^2 - 0.513^2}\,\frac{\text{MeV}}{c} = 0.684\,\frac{\text{MeV}}{c}.$$

**4.12** Find the speed of the particle described in Problem 4.11.

*Answer:* $v = 0.800c$

**4.13** A particle of rest energy 3 MeV has a total energy of 5 MeV. Find the particle's speed $v$ and momentum $p$.

*Solution:*

We need an expression for $v$ in terms of $E$ and $E_0$. Squaring both sides of the Equation

$$\frac{E}{E_0} = \Gamma = \frac{1}{\sqrt{1 - \frac{v^2}{c^2}}}$$

and solving for $v$ gives

$$\begin{aligned} v &= c\sqrt{1 - \left(\frac{E_0}{E}\right)^2} \\ &= c\sqrt{1 - \left(\frac{3}{5}\right)^2} \\ &= c\sqrt{1 - \frac{9}{25}} = \frac{4}{5}c, \end{aligned}$$

where $E_0 = 3$ MeV and $E = 5$ MeV have been substituted. The momentum $p$ is now easily obtained by realizing that

$$\begin{aligned} p = mv &= \frac{Ev}{c^2} \\ &= \frac{(5\ \text{MeV})(4c/5)}{c^2} = 4\,\frac{\text{MeV}}{c}. \end{aligned}$$

This approach offers an alternative to that used in Problems 4.11 and 4.12.

**4.14** How much energy in terms of $E_0$ would be required to accelerate a particle of mass $m_0$ from rest to a speed of $0.8c$?

*Answer:* $T = \frac{2}{3}E_0$

**4.15** Two particles separated by a massless spring are forced closer together by a compressive force doing 18 J of work on the system. What is the change in mass of the system in units of kilograms?

*Solution:*
We know $\Delta E = 18$ J and need to find $\Delta m$. From Equation 4.42 we have

$$\Delta m = \frac{\Delta E}{c^2} = \frac{18 \text{ J}}{9 \times 10^{16} \frac{\text{m}^2}{\text{s}^2}} = 2 \times 10^{-16} \text{ kg}.$$

**4.16** If a particle of rest energy $E_0$ is traveling at a speed of $0.6c$, how much energy in terms of $E_0$ is needed to increase its speed to $0.8c$?

*Answer:* $E_f - E_i = \frac{5}{12}E_0$

**4.17** At what fraction of the speed of light must a particle travel so that its total energy is just double its rest energy?

*Solution:*
Under the condition

$$E = 2\text{E}_0,$$

we substitute $\Gamma E_0$ for $E$ and obtain

$$\Gamma = 2.$$

Substituting from Equation 4.24 for $\Gamma$ and squaring gives

$$\frac{1}{1 - \frac{v^2}{c^2}} = 4,$$

which is easily solved for

$$\frac{v^2}{c^2} = \frac{3}{4} \rightarrow v = 0.866c.$$

**4.18** At what fraction of the speed of light must a particle travel to have a kinetic energy that is exactly double its rest energy?

*Answer:* $v = 0.9428c$

**4.19** Observers in system S′ measure the speed of a $1.60 \times 10^{-29}$ kg particle traveling parallel to their $X'$-axis to be $0.6c$. If the relative speed between S and S′ is $0.8c$, what do observers in S measure for the momentum of the particle?

*Solution:*
We know $m_0 = 1.60 \times 10^{-29}$ kg, $v_x' = v' = 0.6c \rightarrow \Gamma' = 5/4$, and $u = 0.8c \rightarrow \gamma = 5/3$ and need to find $p_x$. From Equation 4.66 we have

$$p_x = \gamma\left(p_x' + \frac{E'u}{c^2}\right),$$

which is expressible in terms of the given information as

$$\begin{aligned} p_x &= \gamma\left(\Gamma' m_0 v' + \Gamma' \frac{E_0 u}{c^2}\right) \\ &= \gamma\,(\Gamma' m_0 v' + \Gamma' m_0 u) \\ &= \gamma \Gamma' m_0 (v' + u). \end{aligned}$$

Now, direct substitution yields

$$\begin{aligned} p_x &= \left(\frac{5}{3}\right)\left(\frac{5}{4}\right) m_0(0.6c + 0.8c) \\ &= \left(\frac{5}{3}\right)\left(\frac{5}{4}\right)(16.0 \times 10^{-30}\ \text{kg})\left(\frac{14}{10}\right) c \\ &= \left(\frac{5}{3}\right)\left(\frac{7}{4}\right)(16.0 \times 10^{-30}\ \text{kg})(3 \times 10^{8}\ \text{m/s}) \\ &= (5)(7)\left(4 \times 10^{-22}\ \text{kg} \cdot \frac{\text{m}}{\text{s}}\right) \\ &= 1.4 \times 10^{-20}\ \text{kg} \cdot \frac{\text{m}}{\text{s}}. \end{aligned}$$

Actually, this problem could have been solved more directly by using Equation 4.66, since we had already calculated its momentum and energy for an inertial system in Problems 4.3 and 4.5, respectively.

**4.20** In Problem 4.19, what do observers in system S measure for the particle's total energy?

*Answer:* $E = 4.44 \times 10^{-12}$ J

CHAPTER 5

# Quantization of Matter

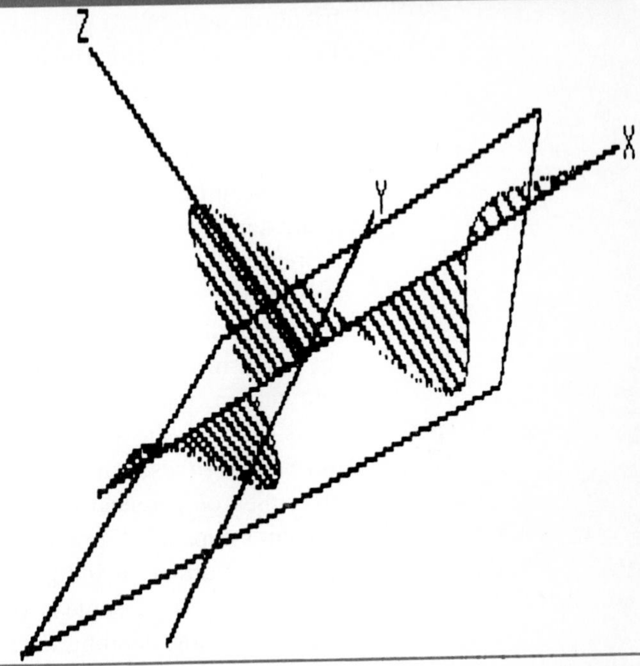


> Atoms of the different chemical elements are different aggregations of atoms [particles] of the same kind. . . . Thus on this view we have in the cathode rays matter in a new state, a state in which the subdivision of matter is carried very much further than in the ordinary gaseous state: a state in which all matter—that is, matter derived from different sources such as hydrogen, oxygen, etc.—is of one and the same kind; this matter being the substance from which all the chemical elements are built up.
>
> J. J. THOMSON, *Philosophical Magazine* 44, 293 (1897)

## Introduction

The study of Einstein's *special theory of relativity* has expanded and completely altered our fundamental view of nature from that suggested by *classical* mechanics. It is important to realize that our new perception of the concepts of length, mass, time, and energy resulted from essentially one new basic postulate of nature (the invariance of the speed of light) and an application of *classical mechanics* to fundamental physical considerations of *macroscopic* phenomena. Additional deviations from classical physics and insights of *microscopic* phenomena will be detailed in this and the next few chapters, as we consider other theoretical and experimental contributions to modern physics. The method of inquiry is similar to that utilized in the study of Einsteinian relativity, in that a few *new* fundamental postulates of nature are combined with well known principles of classical mechanics *and* electromagnetic theory to produce a new nonclassical view of nature on the *microscopic* level.

The immediate objective of this chapter is to study the **quantization** of matter, a concept that suggests matter is composed of basic constituents or minute particles. After a brief review of the evolution and scientific acceptance of this *atomic view*, the qualitative physical properties of an electron will be investigated. This is immediately followed by a study of the early measurements and estimates of the specific charge ($e/m_e$), absolute charge, mass, and size of an electron. The emphasis of these discussions is *not* on the actual experiments and analyses performed by physicists in obtaining these early estimates. Instead, the logical application of basic principles of classical physics is emphasized in the development of relationships capable of predicting these fundamental physical properties of an electron. Further, a limited discussion of the *modern model* of the atom and nucleus is presented, followed by theoretical considerations for the mass, size, and binding energy of an atom. As a number of fundamental relationships of classical electromagnetic theory will be utilized in this chapter, a review of the basic equations, SI units, defined units, and conventional symbols presented in a general physics textbook might prove beneficial.

## 5.1 Historical Perspective

The concept of matter being *quantized* (i.e., discrete) was suggested as early as the fifth century B.C. by the Greek philosopher Democritus. This view, however, was mostly disregarded for nearly two thousand years by scientists in favor of the Aristotelian philosophy that considered space and matter as being *continuous*. Serious theoretical support for the *atomic view* of matter by Pierre Gassendi, Robert Hooke, and Isaac Newton appeared in the middle and latter part of the seventeenth century. These efforts, however, were essentially ignored for nearly another hundred years, before initial exprimental evidence from quantitative chemistry was available in support of the *quantization of matter*.

Of the many scientists involved in the development of quantitative chemistry at the turn of the nineteenth century, the more noteworthy include chemists Antoine Lavoisier, J. L. Proust, John Dalton, J. L. Gay-Lussac, and the Italian physicist Amedeo Avogadro. The work of these individuals clearly established that basic substances participate in chemical reactions in discrete or quantized entities. Their efforts led to the definition of chemical elements and the concept of *atomic masses* (originally called atomic *weights*). In fact Dalton suggested each element was composed of physically and chemically identically *atoms* and that these atoms were different from the atoms of any other element. He also introduced the concept of atomic masses; however, it was Avogadro who provided the best rationale for finding atomic masses by way of his hypothesis that *at the same temperature and pressure equal volumes of gases contain the same number of particles*. He was also the first to recognize that two or more atoms could combine to form what he called a *molecule*, a concept that was not fully understood until the development of *quantum mechanics* in the twentieth century. His hypothesis is of fundamental importance to physics and physical chemistry in that it predicts the number of atoms or molecules in one *mole* of a substance (any element or compound) as being exactly equal to a number $N_o$, called Avogadro's constant. Although the absolute magnitude of $N_o$ was not known for more than fifty years *after* Avogadro's hypothesis, knowledge of its *existence* was sufficient and of primary importance in the development of relative atomic masses for the chemical elements. In the last section of this chapter Avogadro's hypothesis and the value of $N_o$ will be utilized in calculating the *absolute* mass and size of an atom.

An enormous amount of evidence for the quantization of matter was provided by the advent and development of *kinetic theory* in the nineteenth century, which was complementary to and independent of the view suggested by quantitative chemistry. Kinetic theory arises from the application of Newtonian mechanics to a gas considered as a system consisting of a very large number of identical particles. These particles are imagined to exist in a state of random motion and have elastic collisions with one another and the gas container. This large and very elegant subject was the first *microscopic* model of matter describing the physical properties of a gas. It was initially developed in part by Daniel Bernoulli in 1738; however, the major contributions and development occurred in the nineteenth century and were brought about notably by J. P. Joule, R. J. Clausius, J. C. Maxwell, L. Boltzmann, and J. W. Gibbs. Although kinetic theory per se is not germane to our immediate objectives, it is appropriate to acknowledge its contribution to the atomic view of nature. Many of the results of kinetic theory will be independently developed and discussed later in this textbook, when we consider the fundamental principles and physical applications of *statistical mechanics*.

One other contribution supporting the atomic view of matter came from the law of electrolysis developed by Michael Faraday in 1833. By allowing electricity to flow through electrolytic solutions and observing the components of the solution being liberated at the electrodes, Faraday was able to predict the existence of a discrete unit of electrical charge. His work supported not only the quantization of matter, but also the *quantization of electrical charge*. This discreteness in nature was later confirmed by experimental investigations of *cathode* and *canal* rays, which led to measurements of the elemental electrical charge in nature and measurements of atomic masses, respectively. The qualitative physical properties of *cathode* rays is the topic of discussion in the next section, while *canal* rays will be considered in some detail in Section 5.6.

## 5.2 Cathode Rays

During the second half of the nineteenth century considerable scientific effort was devoted to the investigation of electrical discharge through rarefied gases. In 1853 a Frenchman by the name of Masson discharged an electrical spark through a rarefied gas and found that the glass tube containing the gas was filled with a bright glow, instead of the normal spark as observed in air. A few years later the German glass blower Heinrich Geissler manufactured a number of these gaseous discharge tubes and sold them to scientists around the world. The *Geissler tube*, as illustrated in Figure 5.1, essentially contained an *anode* and *cathode* electrode embedded in a partially evacuated glass tube. As the internal pressure of the tube is further decreased, the electrical discharge through the rarefied gas undergoes a number of different phases, as was reported by W. Crookes, Faraday, and others. At a pressure of roughly 0.01 mm of Hg a glow discharge is produced, as the entire tube tends to glow with a faint greenish light. The initial explanation was that invisible rays, called *cathode rays*, emanating from the cathode electrode would strike the walls of the tube and cause a fluorescence of the glass. The existence of these invisible cathode rays caused considerable investigatory excitement in the scientific community during the remainder of the nineteenth century.

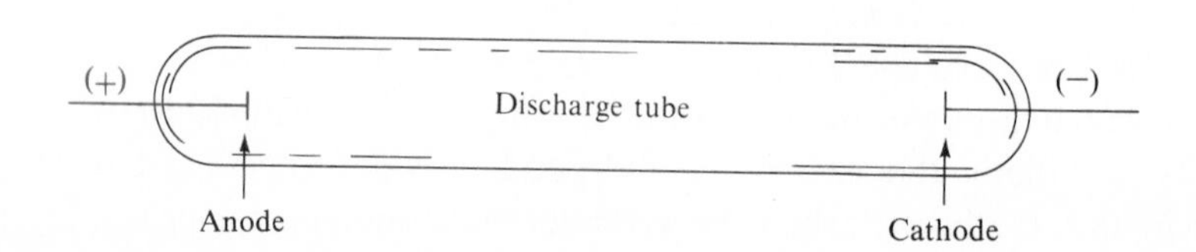


Figure 5.1
A simple Geissler discharge tube containing two electrodes.

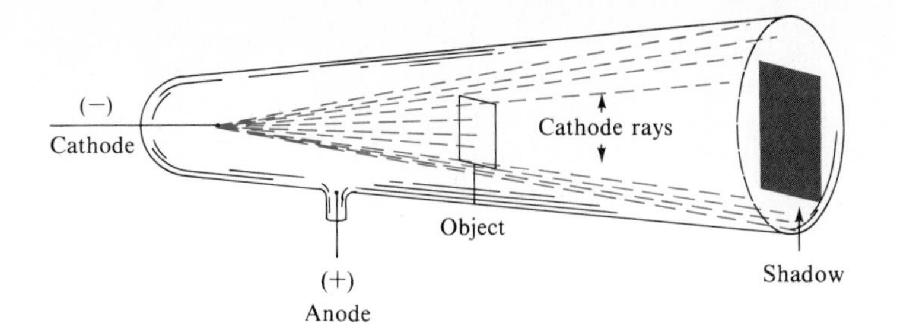


Figure 5.2
A Crookes demonstration tube illustrating the rectilinear propagation of cathode rays.

It can be easily demonstrated that the rays travel from the cathode to the anode in a straight line by using a discharge tube similar to that of Figure 5.2, where the dashed lines represent the rays emanating from a *point source*. A greenish fluorescence is observed where the rays strike the glass, while the glass in the shadow of the object remains dark. Since the shadow is distinctive and always on the side opposite the cathode, the rays must be traveling in straight lines and emanating from the cathode electrode. This 1869 discovery of the rectilinear propagation of cathode rays is credited to Johann W. Hittorf. One year later William Crookes demonstrated that cathode rays have energy and momentum by using a modified discharge tube similar to the one depicted in Figure 5.3. Here, the rays strike a frictionless pinwheel causing it to rotate in a counterclockwise fashion. That the rays are emanating from the cathode is also verifiable, since a reversal of the electrical polarity on the electrodes results in a clockwise rotation of the pinwheel. Because of the motion of the pinwheel Crookes concluded that cathode rays consisted of *invisible particles* possessing both mass and velocity and, consequently, momentum $mv$ and kinetic energy $\frac{1}{2}mv^2$.

Cathode rays were found to be *negatively charged particles* by Jean Perrin in 1895. A simple demonstration of this is illustrated in Figure 5.4, where a beam of cathode rays is created by a *pinhole* placed close to the cathode electrode. With the magnetic field *on* and directed *into* the plane of the page, the beam of rays is observed to cause fluorescence around region $B$. Without the magnetic field the region of fluorescence is around point $A$, while fluorescence is observed to occur at $B'$, when the direction of the field is reversed. With these results noted, an application of the *left-*

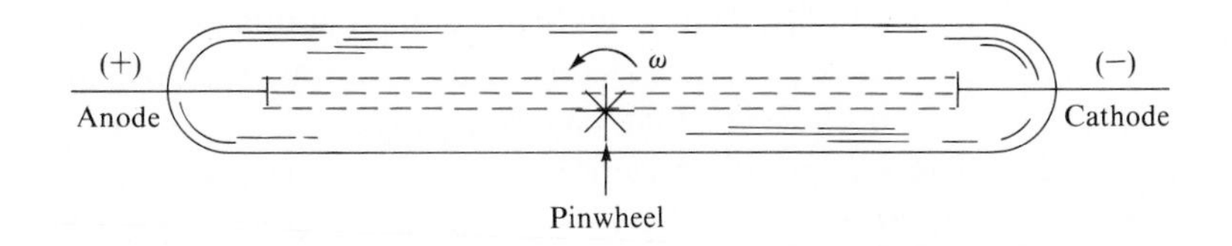


Figure 5.3
Demonstration of energy and momentum possessed by cathode rays.

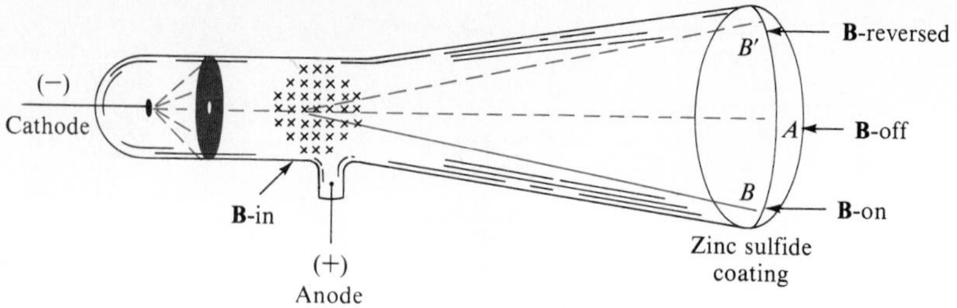

Figure 5.4
Demonstration of the negative charge associated with cathode rays.

*hand rule* shows cathode rays to be negatively charged particles. Recall that the *left-hand rule* is based on the famous **Lorentz force equation** of general physics,

Lorentz Equation

$$\mathbf{F}_B = q\mathbf{v} \times \mathbf{B}, \tag{5.1}$$

where $\mathbf{F}_B$ is the force experienced by a body having an electrical charge $q$ traveling with a velocity $\mathbf{v}$ through an external $\mathbf{B}$ field. For *negatively* charged particles the vector form of this equation suggests that the thumb, first, and second finger of the *left hand* can represent the directions of $\mathbf{F}_B$, $\mathbf{v}$, and $\mathbf{B}$, respectively. Thus, with the magnetic field directed as depicted in Figure 5.4, *negatively* charged particles will experience an acceleration due to the force $\mathbf{F}_B$ as they traverse the magnetic field and, subsequently, be deviated from their initial rectilinear path to a point like $B$ at the end of the tube.

## 5.3 Measurement of the Specific Charge $e/m_e$ of Electrons

At this time cathode rays were understood to consist of particles of some unknown mass and negative electrical charge. In 1897 J. J. Thomson successfully determined the charge-to-mass ratio of cathode particles by using a highly evacuated discharge tube. Although he used different gases in the discharge tube and different cathode metals, he always obtained the same value for the charge-to-mass ratio of the cathode particles. Calling these particles *cathode corpuscles*, Thomson properly concluded that they were common to all metals and different from the chemical atoms. He suggested a revolutionary new model for electrically neutral atoms as consisting of negatively electrified corpuscles that can be liberated from an atom by electrical forces. These *corpuscles* were later called **electrons** (a term first introduced by G. J. Stoney in 1874 to describe the charge carried by an *ion*) and recognized as possessing a *quantized charge* and as being fundamental constituents of all atoms.

Thomson's insight on the nature of electricity resulted from experiments using a highly evacuated discharge tube similar to the one depicted in Figure 5.5. The electrons emanating from the cathode electrode $C$ are strongly affected by the potential difference between the cathode and anode electrodes. This potential difference is *not* uniformly distributed between the electrodes, however, as roughly 0.95 of the potential drop is concentrated very close to and in front (within approximately 1 cm) of the cathode. Consequently, assuming the cathode metal $C$ to be small and approximating a point, the emanating electrons are radially accelerated and travel in straight lines away from the cathode. Some of these electrons will pass through the *apertures* $A_1$ and $A_2$ of Figure 5.5 and become a highly collimated beam of particles, which travel rectilinearly at nearly a constant speed $v_x$ along the axis of the tube. In this manner the apparatus creates a thin beam of electrons, which can pass through a region where a uniform electric field $\mathbf{E}$ (created by a parallel plate capacitor) coexists and is directionally perpendicular to a uniform magnetic field $\mathbf{B}$ (created by Helmholtz coils). In the absence of the electric and magnetic fields, the rectilinearly propagating electrons will strike the end of the tube at point $R$, as illustrated in Figure 5.5. The existence of the magnetic $\mathbf{B}$ field alone causes the beam of electrons to be deflected to position $B$ on the fluorescent end of the tube, while the electric $\mathbf{E}$ field existing alone results in a deflection of the beam to point $E$. Using the apparatus of Figure 5.5, a number of different methods and analyses will be described below for measuring the *specific charge* $e/m_e$ of electrons, where $e$ and $m_e$ are the conventional symbols used to represent the electrical charge and rest mass, respectively, of electrons. In all considerations the electric and magnetic fields are assumed to be uniform within a rather well-defined geometric region and zero outside this region. Further, we ignore as insignificant the gravitational force acting on electrons and the interaction of *their* electric fields, as they pass through the discharge tube.

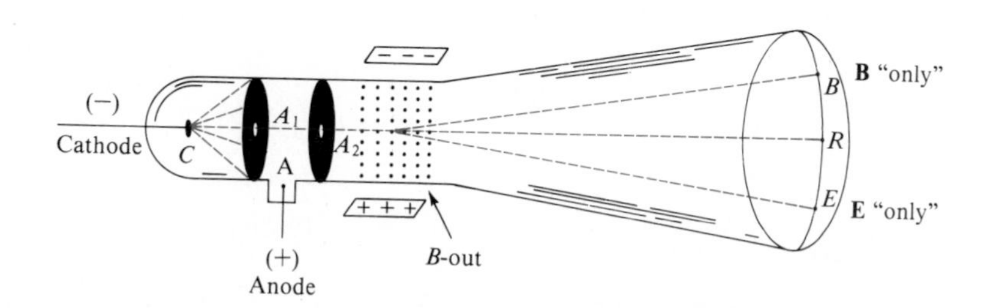


Figure 5.5
Experimental discharge tube for measuring $e/m_e$ for electrons.

## Speed of Electrons

Thomson knew the cathode electrons had a nearly uniform speed $v_x$ before entering the coexisting **E** and **B** fields, since in the absence of these fields the beam of electrons produced a well-defined fluorescent spot on the end of the tube. A value of $v_x$ can be determined by considering the electric **E** and magnetic **B** fields of the Thomson apparatus to be activated. Further, the magnitudes and directions of the **E** and **B** fields are adjusted such that the beam of particles is *undeflected* upon passing through the geometrical region where the fields coexist. With this adjustment of the apparatus, the particles will not experience any *net* external accelerating force, as they pass through the coexisting **E** and **B** fields. Consequently, in that region of the tube the upward magnetic force $\mathbf{F}_B$ on the particles must be *equal in magnitude* to the downward electric force $\mathbf{F}_E$,

$$F_B = F_E. \tag{5.2}$$

An equality for $F_B$ is directly obtained from Equation 5.1 as

$$F_B = ev_xB, \tag{5.3}$$

with $q$ and $v$ being replaced by $e$ and $v_x$, respectively. The *cross product* $\mathbf{v} \times \mathbf{B}$ in Equation 5.1 reduces to that given in Equation 5.3, since the velocity of the electrons is everywhere perpendicular to the magnetic field in Figure 5.5. An expression for $\mathbf{F}_E$ of Equation 5.2 is also easily obtained by recalling the defining equation for **electric field intensity,**

Electric Field Intensity

$$\mathbf{E} \equiv \frac{\mathbf{F}}{q}. \tag{5.4}$$

Replacing $q$ with $e$, this definition gives

$$F_E = eE, \tag{5.5}$$

which when substituted along with Equation 5.3 into Equation 5.2 gives

Speed of Electrons

$$v_x = \frac{E}{B}. \tag{5.6}$$

The values for $E$ and $B$ in this expression are easily determined by knowing the geometry of the capacitor and Helmholtz coils and by taking readings of the voltmeter and ammeter associated with each. For example, the parallel plate capacitor has a uniform electric field intensity given by

$$E = \frac{V_c}{d}, \qquad (5.7)$$

Parallel Plate Capacitor

where $V_c$ is the potential drop (read from a voltmeter) across the capacitor plates and $d$ is the plate separation distance. Although the acceleration of the cathode electrons from rest to the speed $v_x$ is accomplished by the nonuniform electric field between the electrodes, Equation 5.6 allows the determination of $v_x$ from knowledge of well-defined and uniform electric and magnetic fields. The approximate value of $v_x$ is also important to know, as we must decide whether classical physics or Einsteinian relativity is more appropriate in our derivations for the $e/m_e$ ratio of electrons. Thomson found $v_x$ to be on the order of 1/10 the speed of light, which means classical physics can be safely employed in the analyses (e.g., see example of kinetic energy following Equation 4.38).

## Analysis of $e/m_e$ Using the **B**-field Deflection of Electrons

Up to this point in our deliberation of Thomson's experiment, a beam of cathode electrons has been allowed to pass *undeflected* through a region of coexisting **E** and **B** fields. Now, if the electric field **E** is deactivated, the path of cathode electrons is depicted in Figure 5.6 as being uniformly

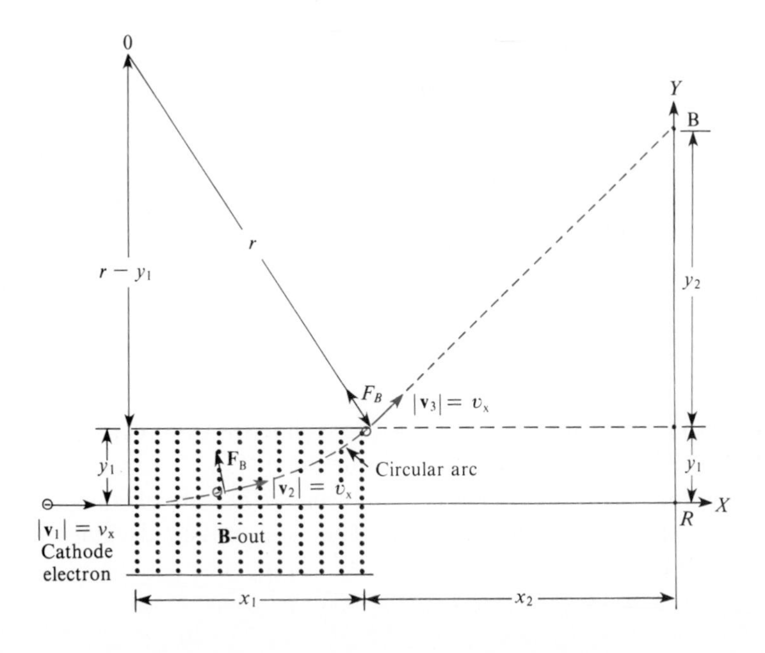


Figure 5.6
The effect of a uniform **B** field on cathode electrons.

deflected by the magnetic **B** field and as being rectilinear beyond the field. The uniform deflection of an electron in this magnetic field alone results from it experiencing an accelerating force $\mathbf{F}_B$, which is everywhere perpendicular to the electron's velocity **v** (e.g. $\mathbf{v}_1$, $\mathbf{v}_2$, and $\mathbf{v}_3$ of Figure 5.6) and the magnetic induction **B.** In this situation the electron will traverse the magnetic field in a *circular* path of radius $r$ in a plane perpendicular to the **B** field. Although the velocity of the electron undergoes a *directional change* due to the accelerating force $\mathbf{F}_B$, the electron's *speed* is *constant* as it traverses the magnetic field. Consequently, the accelerating force is of *constant magnitude* $F_B$, as given by Equation 5.3, and *changing direction*. Further, under these conditions $F_B$ is recognized as being a **centripetal force** $F_c$, which is given by

Centripetal Force

$$F_c = \frac{mv^2}{r}. \tag{5.8}$$

From this centripetal force expression and Equation 5.3 we obtain

$$ev_xB = \frac{m_e v_x^2}{r},$$

where $m$ and $v$ in Equation 5.8 have been replaced by $m_e$ and $v_x$, respectively. Thus,

$$\frac{e}{m_e} = \frac{v_x}{rB}, \tag{5.9}$$

where $m_e$ is the rest mass of an electron and $r$ is the radius of the arc depicted in Figure 5.6. This equation can be further modified by substitution from Equation 5.6 to obtain

Charge to Mass Ratio

$$\frac{e}{m_e} = \frac{E}{rB^2}, \tag{5.10}$$

which gives the *specific charge* of electrons in terms of directly measurable quantities $B$, $E$, and $r$. Although this equation differs (because of our analysis) somewhat from that used by Thomson, it has an advantage of simplicity in derivational steps and form. Thomson measured values for $e/m_e$ in the range $0.7 \times 10^{11}$ C/kg to $2 \times 10^{11}$ C/kg, whereas the more recent accepted value is

$$\frac{e}{m_e} = 1.758806 \times 10^{11} \frac{\text{C}}{\text{kg}}. \quad (5.11)$$

Since the value of $e$ in *coulombs* (C) is necessarily *micro* in size, the value given for $e/m_e$ portends the rest mass of an electron in *kilograms* (kg) must be extremely small.

## Analysis of $e/m_e$ Using the Cathode-Anode Potential

Equation 5.10 is somewhat inhibiting to use for the determination of the charge-to-mass ratio of electrons, since the radius $r$ of the circular arc of Figure 5.6 is usually difficult to accurately measure. An equation involving $e/m_e$ for the electrons can be obtained in terms of easily measurable variables by considering the work done on the cathode electrons by the impressed electric field between the cathode and anode electrodes. Assuming the *liberation energy* required to free the electrons from the cathode metal is negligibly small, then the work done by the electric field on the electrons goes into kinetic energy. From the definition of **electrical potential,**

$$V \equiv \frac{W}{q}, \quad (5.12)$$

Electrical Potential

it follows that the work done on the electrons of charge $e$ is related to the potential drop $V$ between the electrodes as

$$W = eV, \quad (5.13)$$

where $V$ is read directly from the apparatus voltmeter. Since the electrons of mass $m_e$ have a zero initial velocity after being liberated from the cathode, the apparatus collimates and the electric field of the electrodes accelerate the electrons to a final horizontal velocity of $v_x$. Thus, the work done by the accelerating potential is just

$$W = \tfrac{1}{2} m_e v_x^2. \quad (5.14)$$

which upon substitution into Equation 5.13 gives

$$\frac{e}{m_e} = \frac{v_x^2}{2V}. \quad (5.15)$$

With the coexisting **E** and **B** fields of the apparatus adjusted such that the cathode electrons are undeviated from their rectilinear path, $v_x$ is given by Equation 5.6 and Equation 5.15 becomes

Charge to Mass Ratio

$$\frac{e}{m_e} = \frac{E^2}{2VB^2} \tag{5.16}$$

for the charge to mass ratio of the electrons. The advantage of Equation 5.16 over Equation 5.10 is the ease and reliability of accurately measuring the values of the parameters $B$, $E$, and $V$. Further, by using this equation to determine the specific charge of electrons, there is no need to consider a magnetic deflection of the cathode rays.

## Analysis of $e/m_e$ Using the E-field Deflection of Electrons

It was fortunate that the electrons were of equal mass, charge, and nearly equal velocities $v_x$ before passing through Thomson's coexisting electric and magnetic fields. The uniformity of $v_x$ for the electrons can be verified by comparing the value of $e/m_e$ obtained from another analysis with that predicted by Equation 5.16. In this instance the **B** field is deactivated so the path of the electrons is dependent on only the uniform **E** field. The affect of the electric field alone on the beam of cathode electrons is depicted in Figure 5.7, where the direction of the **E** field has been reversed from that of Figure 5.5 for illustration purposes. The electrons of identical

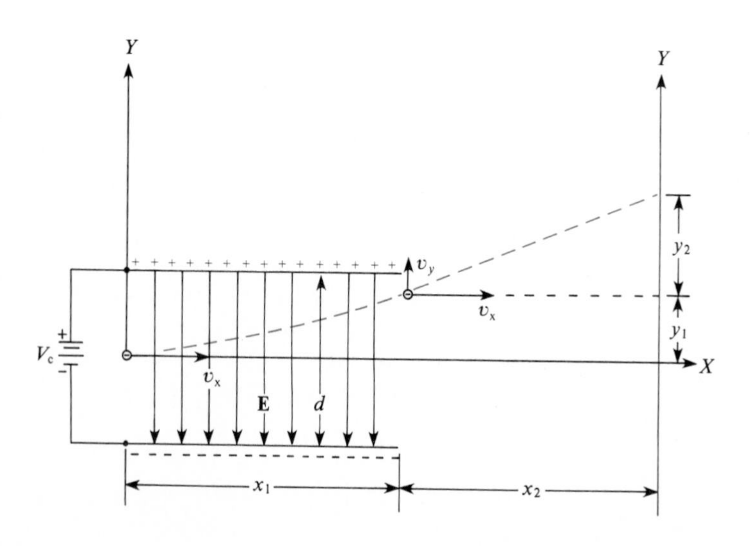


Figure 5.7
The affect of a uniform **E** field on cathode electrons.

mass $m_e$ enter the **E** field with a zero $y$-component of velocity. Because of their electrically charged state they are accelerated in the positive $y$-direction by the capacitor's uniform **E** field, such that they emerge from the capacitor with independent and uniform components of velocity $v_x$ and $v_y$. Unlike the accelerating force $\mathbf{F}_B$ due to the magnetic field alone, the cathode electrons now experience an accelerating force $\mathbf{F}_E$ that is *constant* in both magnitude *and* direction. Consequently, the electrons experience a *uniform* vertical acceleration $a_y$ due to the electric field, while their horizontal component of velocity $v_x$ that is *perpendicular* to **E** remains totally *unchanged.* For a uniform vertical acceleration $a_y$, the electrons vertical displacement due to the electric field only is given by classical kinematics as

$$y_1 = \overset{0}{\cancel{v_{0y}}} t_1 + \tfrac{1}{2} a_y t_1^2 = \tfrac{1}{2} a_y t_1^2, \tag{5.17}$$

where $t_1$ is the time required for the electrons to traverse the **E** field as given by

$$t_1 = \frac{x_1}{v_x}. \tag{5.18}$$

The acceleration in the $y$-direction due to the **E** field is simply expressed from Newton's second law of motion (see Equation 1.17) for $m \neq m(t)$ by

$$a_y = \frac{F_E}{m_e} = \frac{eE}{m_e}, \tag{5.19}$$

where Equation 5.5 has been utilized in obtaining the second equality. Clearly, substitution of Equations 5.18 and 5.19 into Equation 5.17 yields

$$y_1 = \frac{1}{2} \frac{eE}{m_e} \frac{x_1^2}{v_x^2} \tag{5.20}$$

for the electric field deflection of the cathode electrons. As the electrons emerge from the **E** field of the parallel plate capacitor, they have a constant speed of $v_y$ in the $y$-direction, which portends their vertical displacement $y_2$ as they traverse the horizontal distance $x_2$ at the constant speed $v_x$. In this case the vertical displacement is given by

$$y_2 = v_y t_2, \tag{5.21}$$

where the time involved is simply

$$t_2 = \frac{x_2}{v_x}. \tag{5.22}$$

Since the defining equation for average acceleration allows

$$v_y = \overset{0}{\cancel{v_{0y}}} + a_y t_1,$$

then from Equations 5.18 and 5.19 we have

$$v_y = \frac{eE}{m_e} \frac{x_1}{v_x}. \tag{5.23}$$

Now, substitution of Equations 5.22 and 5.23 into Equation 5.21 yields

$$y_2 = \frac{eEx_1x_2}{m_e v_x^2}. \tag{5.24}$$

Consequently, from Equations 5.20 and 5.24 the total deflection of the electrons is just

$$y = y_1 + y_2 = \frac{eEx_1}{m_e v_x^2} (\tfrac{1}{2}x_1 + x_2), \tag{5.25}$$

which is easily solved for

$$\frac{e}{m_e} = \frac{yv_x^2}{x_1E\ (\frac{1}{2}x_1 + x_2)}$$

or more simply

$$\frac{e}{m_e} = \frac{yE}{x_1B^2\ (\frac{1}{2}x_1 + x_2)} \tag{5.26}$$

by using Equation 5.6 for $v_x$. This equation could be reduced further in terms of easily measurable physical parameters by using Equation 5.7 ($E = V_c/d$). The point is that Equation 5.26 is but another analysis for the electron's charge-to-mass ratio using the Thomson apparatus. The results obtained by this equation should compare favorably with those predicted by Equation 5.16, if the electrons enter the **E** field with very nearly equal velocities $v_x$. It is also interesting to note that a *deflection analysis* very similar to the one just presented could be made for the situation where the **E** field is turned off instead of the **B** field. It is left to the reader to verify that when the deflection due to the **E** field alone is equated to the deflection due to the **B** field alone, Equation 5.6 is directly obtained.

The physical principles and analyses associated with the Thomson-like discharge tube are very important to students of physics and engineering, as a number of current electronic instruments utilize cathode ray tubes. For example, modern *oscilloscopes* use electric fields to deflect the cathode electrons, while *television tubes* utilize magnetic field deflection of electrons. Also, a diverging electron beam can be focused by a magnetic field applied along the axis of the beam using a solenoid, which is of considerable importance in the design and construction of *electron microscopes*.

## 5.4 Measurement of the Charge of an Electron

Although Thomson's investigation of cathode rays did not establish all electrons as having identical charges and rest masses, he is attributed with the discovery of the electron. Thomson realized it was possible for the electrons to differ slightly in mass and electrical charge in such a way as to preserve their charge-to-mass ratio. Consequently, it was necessary to measure either the mass or the charge of electrons to determine if either was quantized. Because of the suspected extremely small mass of the electron and the difficulty anticipated in determining it, researchers opted for measurements of the electron's charge. Experiments initiated by J. J. Thomson and J. S. Townsend and later modified by J. J. Thomson and H. A. Wilson were eventually refined by Robert A. Millikan, who in 1909 made the first successful determination of the electronic charge $e$. Millikan's research provided entirely independent evidence for the *quantization* of electrical charge and allowed for the accurate determination of the electron's rest mass $m_e$ (utilizing Thomson's $e/m_e$ result), Avogadro's number $N_o$ (see Section 5.7), and atomic masses.

Basic to Millikan's experimental apparatus was an *air filled* parallel plate capacitor, wherein minute oil drops were illuminated and viewed with a microscope. The oil droplets are normally produced by an *atomizer*, which will result in some droplets being electrically charged by *nozzle friction* of the atomizer, or they could be charged by external *irradiation* by x-rays or a radioactive material. Because of the retarding force of *fluid friction* on an oil droplet moving in air, a droplet quickly attains a uniform velocity called its **terminal velocity,** when acted upon by an accelerating force due to gravity or an external electric field. The terminal velocity $v_g$

of a negatively charged droplet *falling* in the *gravitational* field and the droplet's terminal velocity $v_E$ attained when the *electric* field of the capacitor is activated are determined by means of a scale in the eyepiece of the microscope. With measured values for the terminal velocities and the value of the uniform electric field, the total electrical charge of the droplet can be determined.

The physical fundamentals of the Millikan oil drop experiment are depicted in Figure 5.8. Since the charge of a droplet results from an *excess* or a *deficient* number of electrons, it is desirable to observe a droplet having the smallest electrical charge. Such a droplet is easily selected by observing the response of all droplets to the external electric field of the capacitor. With the electric field applied as in Figure 5.8b, uncharged droplets will be observed to *fall* under the influence of gravity, positively charged droplets (those deficient in electrons) will fall due to the electric and gravitational fields, and negatively charged droplets will *rise* under the accelerating force of the electric field. Since *terminal velocities* are attained very rapidly by the droplets, the slowest *rising* droplet would have the smallest number of *excess* electrons. Selecting this droplet and deactivating the electric field, the droplet would be observed to *fall* under the influence of gravity, as depicted in Figure 5.8a. When its uniform terminal velocity $v_g$ is attained, the net external force acting on the droplet is zero. Thus, we have from Figure 5.8a

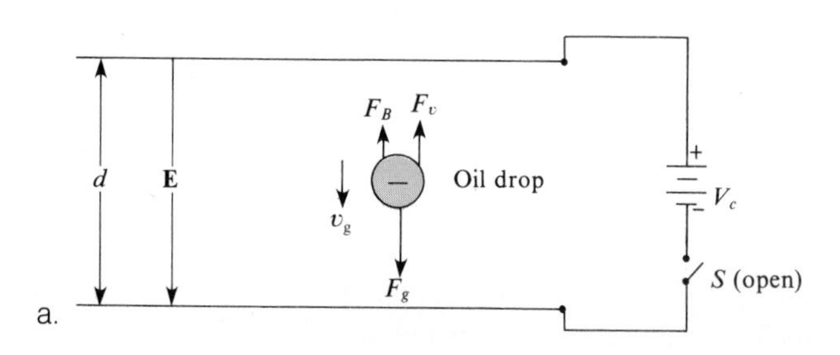


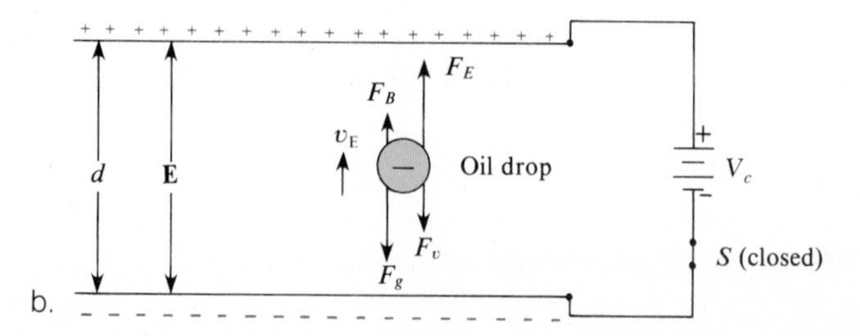


Figure 5.8
The dynamics of oil drop motion between *(a)* uncharged and *(b)* charged capacitor plates.

$$F_g = F_B + F_v, \quad (5.27)$$

where $F_g$ is the downward gravitational force (weight of the droplet), $F_B$ is the **buoyant** force of the air, and $F_v$ is the retarding force of *fluid friction*. Assuming the droplet to be a small *sphere*, $F_v$ is given by **Stokes' law** as

$$F_v = 6\pi r\eta v_g \quad (5.28)$$

Stokes' Law

for a spherical droplet of radius $r$ moving through a homogeneous resisting medium (air) of *viscosity coefficient* $\eta$. Combining Equations 5.27 and 5.28, we obtain

$$F_g - F_B = 6\pi r\eta v_g, \quad (5.29)$$

E Field OFF

where $v_g$ has already been defined as the terminal velocity of the droplet due to the gravitational field only, as determined from measurements of displacement and time.

When the electric field of the parallel plate capacitor is activated, we have the situation depicted in Figure 5.8b. The negatively charged oil droplet will *rise* due to the accelerating force $F_E$ given by Equation 5.5. Again, when the terminal velocity $v_E$ is attained, the net external force acting on the droplet is zero, and from Figure 5.8b we have

$$F_E + F_B = F_g + F_v. \quad (5.30)$$

In this equation the force due to the electric field $F_E$ is given by combining Equations 5.5 and 5.7,

$$F_E = qE = q\frac{V_c}{d},$$

and $F_v$ is given by Stokes' law as

$$F_v = 6\pi r\eta v_E,$$

where $v_E$ is the terminal velocity of the droplet under the influence of the uniform **E** field. Substituting these two equalities into Equation 5.30 and solving for the charge on the oil droplet gives

$$q = \frac{d}{V_c}(F_g - F_B + 6\pi r\eta v_E),$$

E Field ON

which can be further reduced by Equation 5.29 to the form

Droplet Charge

$$q = \frac{6\pi r\eta d}{V_c}(v_g + v_E). \tag{5.31}$$

Once the radius $r$ of the oil drop is determined, the electrical charge of the droplet is easily calculable by using Equation 5.31.

An expression for the radius of the droplet is obtainable from Equation 5.29, by realizing that $F_g$ is the weight of the droplet and $F_B$ is the weight of the volume of air displaced by the droplet. Since *weight*, as given by Equation 1.18, is the product of a mass and the acceleration of gravity $g$, then with $m_o$ being the mass of the *oil* droplet and $m_a$ being the mass of the *air* displaced by the droplet, we have

$$F_g = m_o g \tag{5.32a}$$

and

$$F_B = m_a g. \tag{5.32b}$$

Further, from the definition of **mass density,**

Mass Density

$$\rho \equiv \frac{M}{V}, \tag{5.33}$$

and the equation for the *volume of a sphere* of radius $r$,

Volume of a Sphere

$$V = \frac{4}{3}\pi r^3, \tag{5.34}$$

we obtain

$$m_o = \frac{4}{3}\pi r^3 \rho_o \tag{5.35a}$$

and

$$m_a = \frac{4}{3}\pi r^3 \rho_a. \tag{5.35b}$$

Now, substitution of Equations 5.32 and 5.35 into 5.29 yields

$$\frac{4}{3}\pi r^3 g(\rho_o - \rho_a) = 6\pi r\eta v_g,$$

which is easily solved for the radius of the droplet in the form

$$r = 3\left[\frac{\eta v_g}{2g\,(\rho_o - \rho_a)}\right]^{1/2}. \tag{5.36}$$

Droplet Radius

In this equation $\rho_o$ and $\rho_a$ represent the mass density of the oil and air, respectively, which are normally known or easily measured quantities. As a point of interest, Millikan obtained an experimental correction to Stokes' law, which effectively results in a correction factor to Equation 5.36. Thus, the best value for the charge of an oil droplet is obtained by calculating the radius of the droplet using Equation 5.36 and employing Millikan's *correction factor* before using Equation 5.31. The point of interest, however, is that Millikan was able to directly calculate the minute charge on an oil droplet from basic experimental data, which is easily visualized by combining Equations 5.36 and 5.31 to obtain

$$q = \frac{18\pi\eta d}{V_c}\left[\frac{\eta v_g}{2g\,(\rho_o - \rho_a)}\right]^{1/2}(v_g + v_E). \tag{5.37}$$

Droplet Charge

With the relationship given by Equation 5.37, an experimenter can measure the value of $v_g$ for a particular oil droplet in the absence of the **E** field, then a number of values for $v_E$ can be determined for the same droplet with the **E** field activated. Since the charge on the droplet will change over time, due to a loss or gain in electrons, the different values measured for $v_E$ will result in a set of values for $q$ when they are separately substituted into Equation 5.37. Now, if the electronic charge is always unique and discrete, the difference $q_1 - q_2$ between any two *different* negative changes of the set will always be an *integral multiple* of the charge of an electron $e$. Although Millikan personally conducted or supervised measurements on hundreds of droplets, he *always* found the electrical charge on a droplet to be an integral multiple of one electrical charge, which he proposed as the fundamental unit of electric charge. Thus, electron charge is *quantized*, having a currently accepted value of

$$e = 1.60219 \times 10^{-19}\ \text{C} \tag{5.38}$$

Electron Charge

to six significant figures. Clearly, any one electron is just like every other electron, having a definite rest mass $m_e$ and a quantized negative charge $e$. The rest mass of an electron is now immediately calculable by combining the results of Thomson and Millikan. That is,

$$m_e = \frac{e}{e/m_e} = \frac{1.60219 \times 10^{-19}\ \text{C}}{1.758806 \times 10^{11}\ \text{C/kg}}$$

from Equations 5.11 and 5.38, which will give the rest mass of an electron as

Electron Rest Mass

$$m_e = 9.10953 \times 10^{-31} \text{ kg} \tag{5.39}$$

to six significant figures. The values for $e$ and $m_e$ have been verified many times by numerous experimentalists, with more recent measurements to eight significant figures.

A new unit of energy commonly used in modern physics is now definable in terms of the electron charge $e$. The work done in accelerating a particle of charge $e$ through a potential difference $V$ is given by Equation 5.13. The work done on the particle goes into kinetic energy and this energy is independent of the *mass* of the particle, according to Equation 5.13. Since many calculations in modern physics involve electrons and other elementary particles being accelerated through a potential difference, it is convenient to compute kinetic energy in terms of a new unit of energy, called the **electron volt** and abbreviated eV. One eV is defined as *the kinetic energy received by any particle of charge e that is accelerated through a potential difference of one volt*. Thus, in accordance with this definition and Equation 5.13, we have

Electron Volt

$$1\text{eV} \equiv 1.60219 \times 10^{-19} \text{ J}, \tag{5.40}$$

where the abbreviation J for Joule represents the defined unit of energy in the SI system.

## 5.5 Determination of the Size of an Electron

Just as no direct method of measuring the electron's mass or charge exists, none are available for determining its size. A rough idea of the electron's physical volume can be approximated by considering its *mass* as being *electromagnetic in nature*. Since Einsteinian relativity gives the proportionality between mass and energy as

$$E_0 = m_0 c^2, \tag{4.39}$$

then the electron's *mass* may be considered as a manifestation of the *energy* associated with its *electrostatic charge*. These considerations suggest that the work done in assembling the charge of an electron may be thought

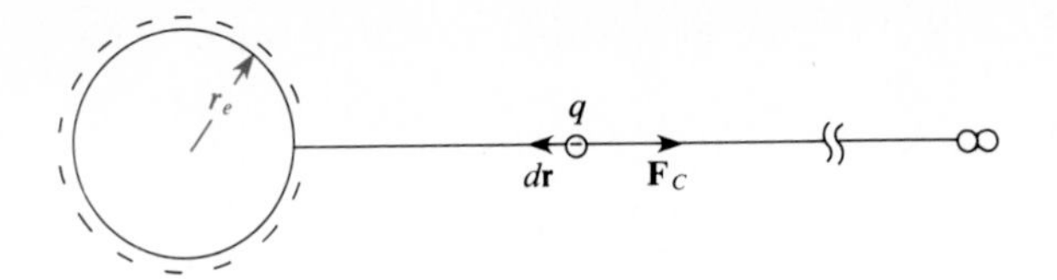


Figure 5.9
The assemblage of an electron from $N$ negative charges $q$.

of as representative of its Einsteinian rest energy. It must be emphasized that these ideas in the construction of a theoretical model are at best only *approximate*, and should not be taken literally. None the less, assuming the electron to be a sphere of radius $r_e$, as illustrated in Figure 5.9, then its assemblage of total charge $e$ may be thought of as consisting of a *large* number $N$ of *minute* negative charges $q$ that have been brought from infinity up to the electron's sphere. Clearly,

$$N = \frac{e}{q} \tag{5.41}$$

and the work done in bringing the first *imaginary* and negligibly small charge $q$ from infinity up to the electron's sphere is

$$W_1 = \int_\infty^{r_e} \mathbf{F}_C \cdot \mathbf{dr} = 0. \tag{5.42}$$

The work $W_1$ is zero since the Coulombic force defined by

$$F_C \equiv k\frac{Qq}{r^2} \tag{5.43}$$

Coulomb's Law

is necessarily zero (i.e., $\mathbf{F}_C = k(0)q/r^2 = 0$). In bringing up the second charge $q$ the charge on the electron is $q$, thus the Coulombic force is $F_C = kqq/r^2$ and the resulting work is

$$W_2 = \int_\infty^{r_e} \mathbf{F}_C \cdot \mathbf{dr} = -kqq\int_\infty^{r_e}\frac{dr}{r^2} = \frac{kqq}{r_e}. \tag{5.44}$$

The first negative sign in Equation 5.44 is necessary since $\mathbf{F}_C$ is oppositely directed to **dr,** which results in cos 180° = −1 from the *scalar product* of the two vectors. In a similar fashion it is easily verified that

$$W_3 = 2W_2,\ W_4 = 3W_2,\ \cdot\cdot\cdot,\ W_n = (N - 1)W_2. \tag{5.45}$$

Thus, the total work done in assembling the $N$ minute charges $q$ on the electron's sphere of radius $r_e$ is

$$\begin{aligned} W_{\text{TOTAL}} &= W_1 + W_2 + W_3 + \cdots + W_N \\ &= (0 + 1 + 2 + 3 + \cdots + N - 1)W_2 \\ &= \frac{(N-1)Nkq^2}{2r_e}, \end{aligned} \tag{5.46}$$

with the last equality coming from Equation 5.44 and the obvious *series* identity. This result may be further reduced to

$$W_{\text{TOTAL}} = \frac{N^2kq^2}{2r_e} = \frac{ke^2}{2r_e}, \tag{5.47}$$

by considering $N \gg 1$ and using Equation 5.41. Now, equating Equations 4.39 and 5.47,

$$m_ec^2 = \frac{ke^2}{2r_e},$$

and solving for the radius of an electron gives

Electron's Radius

$$r_e = \frac{ke^2}{2m_ec^2}, \tag{5.48}$$

where $m_e$ has been substituted for $m_0$ in Equation 4.39. With $k$, defined in terms of the **permittivity of free space** $\epsilon_0$ by

$$k \equiv \frac{1}{4\pi\epsilon_0}, \tag{5.49}$$

and the speed of light $c$ having values of

$$k = 8.98755 \times 10^9 \frac{\text{N} \cdot \text{m}^2}{\text{C}^2}, \tag{5.50}$$

$$c = 2.99792 \times 10^8 \frac{\text{m}}{\text{s}} \tag{5.51}$$

substituted along with the values for $e$ and $m_e$ (Equations 5.38 and 5.39, respectively) into Equation 5.48, we obtain

$$r_e = 1.40898 \times 10^{-15} \text{ m} \quad (5.52)$$

Electron's Radius

for the radius of an electron.

It should be emphasized that the value given by Equation 5.52 for the radius of an electron is only correct within its *order of magnitude*. Our value differs somewhat from other determinations for the size of an electron (e.g., magnetic field calculations for the energy of an accelerated electron, x-ray scattering experiments, etc.), but it is a reasonable one to use until a more precise value is obtained. Further, even though Equation 5.48 represents only an approximation to the size of an electron, it clearly suggests a surprising *inverse proportionality* between the radius of an electron and its mass. This implies that any attempt at reducing the size of the electron, by close packing of the electrostatic charge, will result in an increase in the electron's mass, because of the extra work required against the repulsive Coulombic forces arising from the spatial distribution of the electron's charge.

## 5.6 Canal Rays and Thomson's Mass Spectrograph

During the experimental investigations of cathode rays, E. Goldstein observed in 1886 rays propagating in the opposite direction *toward* the cathode electrode. He designed a special discharge tube (schematically illustrated in Figure 5.10) to isolate these rays, which were originally called **canal rays.** Shortly after J. J. Thomson's determination of the specific charge of electrons in 1897, W. Wien deflected a beam of *canal rays* by a magnetic field and concluded that they consisted of *positively* charged particles. Since that time, they have been found to be positively charged *atoms* of different masses, having a much smaller charge-to-mass ratio than electrons.

The processes taking place in the Goldstein discharge tube that result in the origin of canal rays are best explained by using the *modern model*

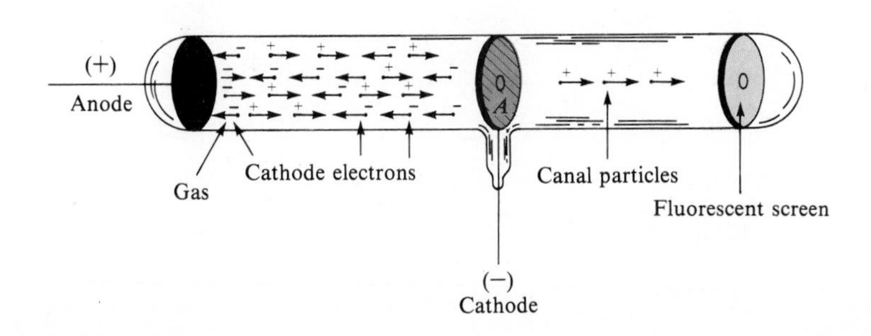


Figure 5.10
A discharge tube illustrating the existence of canal rays.

*of an atom*, which is briefly presented in the next section. Referring to Figure 5.10, as the cathode electrons move toward the anode, they occasionally have an inelastic collision with the atoms and molecules of the residual gas in the tube. In this manner some atoms and molecules are *ionized* (lose an electron) and are thus attracted (accelerated) toward the cathode electrode. Between the cathode and anode, there exist both electrons and positively charged atoms moving in opposite directions toward the anode and cathode, respectively. Of the many positively charged particles striking the cathode, those passing through the small aperture $A$ represent the observed canal rays. When these canal ray particles strike the fluorescent screen at the end of the tube, tiny flashes of light, called *scintillations*, are produced.

In 1911 J. J. Thomson took advantage of the properties of canal rays in developing the *mass spectrograph*, which is depicted schematically in Figure 5.11. A small amount of gas is injected between the cathode and anode of the apparatus and the inelastic collisions between the gaseous atoms and electrons result in the observed *canal rays*. After being accelerated to the cathode electrode, these positively charged particles pass through a region where a **B** field and an **E** field coexist parallel to one another. If the gas in the apparatus contains only one type of atom, then a single parabolic curve will be observed on the fluorescent screen or photographic plate. Particles ionized close to the anode will be greatly accelerated while traveling toward the cathode and, being under the deflecting fields' influence for a short time, their rectilinear paths will be only slightly bent by the external **E** and **B** fields to a point like $A$ on the screen. On the other hand, particles ionized fairly close to the cathode are only slightly accelerated by the electric field between the anode and cathode. These particles clearly remain longer in the deflecting **E** and **B** fields, and thus their rectilinear paths are bent considerably to a point like $C$ on the screen.

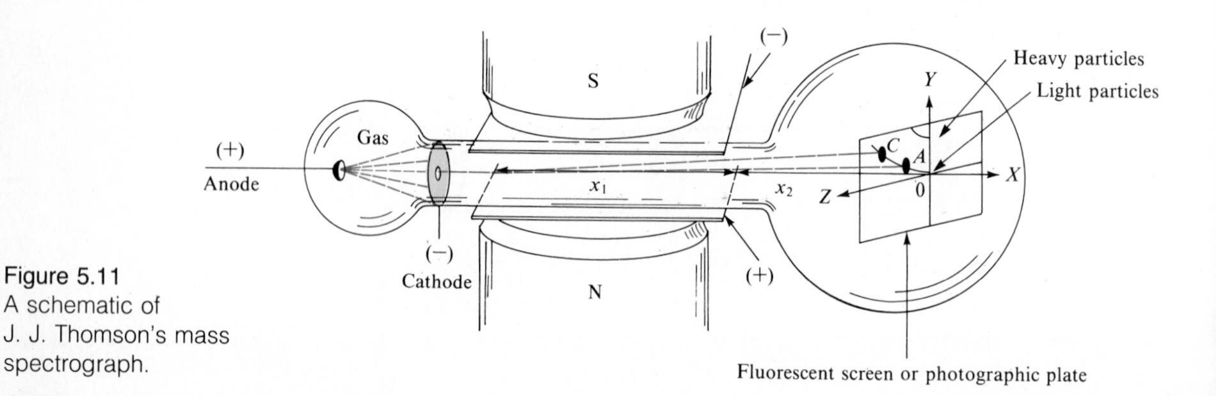


Figure 5.11
A schematic of J. J. Thomson's mass spectrograph.

If there are different types of atoms in the gas, different curves will be recorded on the screen or photographic plate, each parabola corresponding to one particular type of atom or molecule. From knowledge of the values for the **E** and **B** fields, and the assumption that each canal particle possesses a unit positive charge because of being singly ionized, it is a relatively simple matter, as detailed below, to calculate the mass of the atoms producing each parabola.

Unlike Thomson's charge-to-mass ratio analysis presented in Section 5.3, where electrons were accelerated only along the $y$-axis, the positively charged atoms comprising canal rays are deflected in the positive $y$-direction due to the **E** field, while simultaneously being accelerated in the positive $z$-direction by the coexisting **B** field. Assuming the fields to be uniform and of length $x_1$, as depicted in Figure 5.11, the analysis here is similar to the one detailed in Section 5.3.

The total deflection in the $y$-direction due to the **E** field is given by the derivation

$$\begin{aligned} y &= y_1 + y_2 = (\overset{0}{\cancel{v_{0y}}} t_1 + \tfrac{1}{2} a_y t_1^2) + v_y t_2 \\ &= \tfrac{1}{2} a_y t_1^2 + a_y t_1 t_2 \\ &= \tfrac{1}{2} a_y \frac{x_1^2}{v_x^2} + a_y \frac{x_1 x_2}{v_x^2} \\ &= \frac{a_y x_1}{v_x^2} (\tfrac{1}{2} x_1 + x_2), \end{aligned} \qquad (5.53)$$

which is in essence the same derivation presented previously. Taking into account that the acceleration $a_y$ is due to the **E** field (see Equation 5.19), then Equation 5.53 becomes

$$y = \frac{qEx_1}{mv_x^2} (\tfrac{1}{2} x_1 + x_2). \qquad (5.54)$$

**E Field Deflection**

This result is identical to that of Equation 5.25, except for the presence of $q$ and $m$ instead of $e$ and $m_e$. In this consideration for a *singly ionized* atom, the magnitude of $q$ is identical to $e$ and $m$ is the mass of the atom in kilograms.

Although the deflection of the ionized atoms in the positive $z$-direction is a bit more complicated than the $y$-direction deflection, it may be handled in a similar fashion. From Section 5.3 we know the magnetic field exerts an accelerating force $\mathbf{F}_B$ on a charged particle that is *uniform in magnitude* and *changing in direction*. These properties of $\mathbf{F}_B$ are easily observed in Figure 5.6, where the displacement of the particle due to the

**B** field is indicated as $y_1$. Imagining $y_1 \equiv z_1$, $y_2 \equiv z_2$, and $\mathbf{B} = -\mathbf{B}$ in Figure 5.6, then the displacement $z_1$ is expressible in terms of $x_1$ and $r$. This is easily accomplished by using the Pythagorean theorem on the right triangle of sides $r - z_1$ and $x_1$ and hypothenuse $r$ in Figure 5.6. Accordingly,

$$r^2 = (r - z_1)^2 + x_1^2,$$

which when solved for $r$ gives

$$r = \frac{x_1^2 + z_1^2}{2z_1} \approx \frac{x_1^2}{2z_1}, \qquad z_1 \ll x_1. \tag{5.55}$$

The approximation in Equation 5.55 is good if the displacement $z_1$ is *very small* compared with the length of the **B** field (i.e., $z_1 \ll x_1$). Another expression for the radius of the circular arc traversed by the particle in the **B** field is obtained by realizing the accelerating force $\mathbf{F}_B$ is a centripetal force. Thus, from Equations 5.3 and 5.8 we have

$$qv_xB = \frac{mv_x^2}{r},$$

which when solved for $r$ gives

$$r = \frac{mv_x}{qB}. \tag{5.56}$$

As the radius $r$ is difficult to measure, we eliminate $r$ from Equations 5.55 and 5.56 to obtain

$$z_1 = \frac{qBx_1^2}{2mv_x}. \tag{5.57}$$

Consequently, for small **B** field deflections of the particle, the circular path approximates the parabolic path given by Equation 5.57. Interestingly, the *small deflection approximation* is equivalent to approximating the accelerating force $\mathbf{F}_B$ by a *constant* force in both *magnitude* and *direction*. This is easily realized by applying kinematics for *uniform acceleration* to the problem. That is,

$$z_1 = \overset{0}{\cancel{v_{0y}}}t_1 + \tfrac{1}{2}a_zt_1^2$$

$$= \frac{a_z x_1^2}{2v_x^2}$$

$$= \frac{F_B x_1^2}{2mv_x^2}$$

$$= \frac{qv_x B x_1^2}{2mv_x^2}$$

$$= \frac{qBx_1^2}{2mv_x},$$

which is an identical result to that given by Equation 5.57.

From the above discussion it is clear that assuming the **B** field deflection of the canal ray particle to be *very small* compared with the length $x_1$ of the field, the force $\mathbf{F}_B$ can be considered as constant in both magnitude and direction. Thus, the accelerating force $\mathbf{F}_B$, which is normal to the plane of Figure 5.11, can be incorporated in our analysis by the same method as that used above for $\mathbf{F}_E$. Clearly, the total displacement in the $z$-direction is just

$$z = z_1 + z_2 = \frac{a_z x_1}{v_x^2}\left(\tfrac{1}{2}x_1 + x_2\right), \tag{5.58}$$

which is identical to Equation 5.53 except for the presence of $a_z$ instead of $a_y$. Now, however, the acceleration due to the **B** field is given by

$$a_z = \frac{F_B}{m} = \frac{qv_x \mathrm{B} \sin\theta}{m}, \tag{5.59}$$

where the Lorentz force equation (Equation 5.1) has been used in obtaining the second equality. Under the assumption of small **B** field deflections, the angle $\theta$ between $\mathbf{v}_x$ and **B** is always very nearly 90°. Consequently, Equation 5.59 reduces to

$$a_z = \frac{qv_x B}{m}, \tag{5.60}$$

which when substituted into Equation 5.58 yields

$$z = \frac{qBx_1}{mv_x}\left(\tfrac{1}{2}x_1 + x_2\right). \tag{5.61}$$

**B** Field Deflection

Now, solving Equation 5.61 for $v_x$, substituting into Equation 5.54, and solving the resultant equation for the mass $m$ of the canal particles yields

Mass of Atom

$$m = \frac{qyx_1B^2}{z^2E}\left(\tfrac{1}{2}x_1 + x_2\right). \tag{5.62}$$

This equation allows for the determination of the mass or $q/m$ ratio of an atom or molecule of the residual gas in the Thomson apparatus in terms of easily measurable physical quantities. The equation is clearly that of a *parabola*, since $z^2$ is proportional to $y$. Taking $q$ to be the *absolute magnitude* of the electronic charge, the smallest value of $m$ would be for the *hydrogen ion* (or **proton**). The hydrogen ion mass $m_p$ was found to be approximately $1836m_e$, which means electrons contribute very little to the mass of atoms.

## 5.7 Modern Model of the Atom

The first information concerning the existence of the atomic *nucleus* resulted from the discovery of *radioactive* atoms. *Radioactivity* simply refers to the *disintegration* or *decay* of one atom into another. It was originally discovered by H. Becquerel in 1896 when he observed radiation emitting from a uranium salt. It was later found that radioactive rays subjected to a transverse magnetic field split into three rays, classified by E. Rutherford as $\alpha$-, $\beta$-, and $\gamma$-rays. The physical properties of these rays are quite different: $\alpha$-rays are helium nuclei, $\beta$-rays consist of high speed electrons, and $\gamma$-rays are very short wavelength electromagnetic radiation. It should be mentioned that $\gamma$-rays are very similar to x-rays, which were discovered in the year 1895 by W. K. Roentgen, but with greater penetrating power. The scattering of radioactive rays and x-rays, when used to bombard nuclei, has resulted in a wealth of information about the atom.

It is beneficial at this point to introduce some new terminology and the basic model of an atom and its constituents. An atom of any chemical element can be thought of as containing *nucleons* in the nucleus, with electrons *encircling* the nucleus in some naturally fixed energy levels. **Nucleons** are defined as being either positively charged particles, called **protons,** or electrically neutral particles, called **neutrons.** Although the *proton* was not named until 1920 by E. Rutherford, it was easily observed in the Thomson parabola apparatus as a hydrogen ion. The *proton* has a rest mass of

Proton Rest Mass

$$m_p = 1.67265 \times 10^{-27} \text{ kg} \tag{5.63}$$

and an electrical charge of

$$q_p = 1.60219 \times 10^{-19} \text{ C}. \tag{5.64}$$

Proton Charge

It should be noted that the electrical charge of a proton is identical in magnitude to the charge of an electron, except it is electrically *positive*. Its mass, however, is mysteriously larger than the mass of an electron by a factor of 1836,

$$\frac{m_p}{m_e} = \frac{1.67265 \times 10^{-27} \text{ kg}}{9.10953 \times 10^{-31} \text{ kg}} = 1836.$$

The *neutron*, on the other hand, has a rest mass of

$$m_n = 1.67495 \times 10^{-27} \text{ kg} \tag{5.65}$$

Neutron Rest Mass

which is larger than the combined masses of the *proton* and the *electron*. The existence of the *neutron* was postulated as early as 1920 by Rutherford; however, it was not identified until 1932 by J. Chadwick.

Although there are a number of other *subatomic* particles that are of interest in nuclear physics (e.g., the antielectron or *positron*, antiproton, neutrino, antineutrino, etc.), our purposes will be completely served by considering electrons, protons, and neutrons as the basic constituents of an atom. A *normal* atom of any chemical element will be taken as one that has the same number of electrons as protons and is thus electrically neutral. Any process by which an atom *loses* an electron is called **ionization.** An atom can be singly ionized, doubly ionized, and so forth by losing one, two, and so forth electrons, respectively. Some atoms have an *affinity* for more than their *normal* number of electrons. For our purposes such atoms will be referred to as being singly or doubly **countervailed,** if they *gain* one or two additional electrons, respectively. Further, a **molecule** is simply taken to be a combination of two or more elemental chemical atoms.

It is convenient at this point to define a few other terms that are commonly referenced in the study of atomic structure. The **atomic number** $Z$ is the number ascribed to an element that specifies its position in a *periodic table* by defining the *number of protons* in that normal atom. The *atomic* **mass number** $A$ specifies the combined number of *neutrons and protons* in a nucleus and is often referred to as the *nucleon number*. Consequently, the **neutron number** $N$ may be defined by $N \equiv A - Z$. To summarize the above,

$$
\begin{aligned}
Z &\equiv \textit{Atomic Number}\\
&= \text{number of protons in the nucleus}\\
&= \text{number of electrons in the atom;}\\
A &\equiv \textit{Mass Number} \text{ or } \textit{Nucleon Number}\\
&= \text{number of nucleons in the nucleus;}\\
N &\equiv \textit{Neutron Number}\\
&= \text{number of neutrons in the nucleus}\\
&= A - Z.
\end{aligned}
$$

The **nuclide** for a species of atom is characterized by the constitution of its nucleus and hence by the values of the $A$ and $Z$ numbers. It is normally denoted by ${}^{A}_{Z}S$, where $S$ represents the *chemical symbol* for the particular element. Measuring $q/m$ for atoms by the parabola method, Thomson realized as early as 1912 that atoms of different mass could belong to the same chemical element. *Atoms having identical electronic configurations but differing in the number of neutrons in the nucleus* were later named **isotopes** by F. Soddy and are recognized as nuclides of identical $Z$ but different $N$ numbers. Thus, for $Z = 1$ the *isotopes of hydrogen* are denoted by the *nuclides* ${}^{1}_{1}H$ for hydrogen, ${}^{2}_{1}H$ (or ${}^{2}_{1}D$) for deuterium, and ${}^{3}_{1}H$ (or ${}^{3}_{1}T$) for tritium. In the year 1933 K. T. Bainbridge developed a high precision mass spectrograph and discovered what are now commonly called **isobars.** These are *atoms having essentially the same mass but differing in their electronic configuration* and thus belonging to different chemical elements. That is, *isobars* are nuclides of *identical A* but *different* $Z$ and, consequently, $N$ numbers (e.g., ${}^{3}_{1}H$ and ${}^{3}_{2}He$). Further, *nuclides* having *identical N* but *different* $Z$ numbers, such as ${}^{2}_{1}H$ and ${}^{3}_{2}He$, are classified as **isotones.** The common terms defined above are restated for emphasis as follows:

Nuclide = Nuclear configuration characterized by $A$ and $Z$,
Isotopes = Nuclides of identical $Z$ but different $A$,
Isobars = Nuclides of identical $A$ but different $Z$,
Isotones = Nuclides of identical $N$ but different $Z$.

The atomic number $Z$, the mass number $A$, the number of isotopes, and the relative abundance of isotopes in nature for the chemical elements are listed in Appendix C. This constitutes only a partial list of isotopes, as well over 1000 nuclides have been identified as either stable or radioactive.

## 5.8 Specific and Molal Atomic Masses

In most textbooks the table of Appendix C includes a listing of either *atomic masses* or *atomic weights*. These two quantities are different *by definition* and need to be carefully considered. Originally, the mass of an individual atom (or molecule) was completely unknown. However, Avogadro's hypothesis provided the rationale for the comparison of the masses of equal numbers of different kinds of atoms (or molecules). As the hydrogen atom was the least massive of all chemical atoms, its mass was arbitrarily taken as one. Then, the mass of a given volume of hydrogen gas could be compared with the mass of an equal volume of another gas, both at the same temperature and pressure, to obtain *relative masses* for other atoms (and molecules). The basis for assigning *relative masses* changed from hydrogen to oxygen and more recently to carbon. Currently, the basic unit for *relative masses* is called the **unified atomic mass unit** u, or amu. It is defined to be exactly 1/12 the mass of the most common isotope of carbon, ${}^{12}_{6}C$, and has a value to six significant figures of

$$u \equiv 1.66057 \times 10^{-27} \text{ kg}. \tag{5.66}$$

Atomic Mass Unit

This equation should be regarded as nothing more than a *conversion factor* between the mass unit kg and the new mass unit u. Thus, in atomic mass units the electron has a mass of

$$\begin{aligned} m_e &= \frac{9.10953 \times 10^{-31} \text{ kg}}{1.66057 \times 10^{-27} \text{ kg/u}} \\ &= 5.48579 \times 10^{-4} \text{ u}, \end{aligned}$$

while the rest masses of the proton and neutron are

$$\begin{aligned} m_p &= 1.00727 \text{ u}, \\ m_n &= 1.00866 \text{ u}. \end{aligned}$$

Frequently, we will employ the notation $m_u$ to denote the *mass* ($1.66057 \times 10^{-27}$ kg) of the *atomic mass unit*. When compared to the mass $m_e$ of an electron, $m_u$ is nearly 1823 times larger, which means that the hydrogen atom (being essentially 1837 times larger than $m_e$) is *slightly* larger than $m_u$. It should also be obvious from the above considerations that $m_u$ is very nearly equal to the mass of a proton $m_p$ and to the mass of a neutron $m_n$. These comparisons are significant, as the mass of an atom is *essentially* dependent on its constitution of nucleons. Consequently, with u (or $m_u$) taken as the basis, *relative* atomic (and molecular) masses will be very nearly equal to integers, with that of the hydrogen atom being close to unity.

It is important to emphasize that relative atomic (and molecular) masses are *dimensionless* quantities. With this in mind, let us now make the distinction between *atomic weight* and *atomic mass*. The *chemical* **atomic weight** can be defined as *the average mass of all the isotopes of an element, weighted according to their relative abundance in nature, in atomic mass units*. Although this *average* relative mass is useful in chemistry, the study of physics requires knowledge of the *absolute* mass of atoms and their nuclei. With $m_a$ representing the absolute mass of an atom, including its $Z$ electrons, and $m_\mathrm{u}$ being the mass of the unified atomic mass unit, then we can define the **relative atomic mass** of an atom, denoted by $(\mathrm{AM})_a$, as the ratio of $m_a$ to $m_\mathrm{u}$,

Relative Atomic Mass

$$(\mathrm{AM})_a \equiv \frac{m_a}{m_\mathrm{u}}. \tag{5.67}$$

Clearly, this definition of *atomic mass* compares the mass of an electrically *neutral* atom with $m_\mathrm{u}$ and is, consequently, a *relative* and *dimensionless* quantity. It is sometimes loosely referred to as the *specific* atomic mass, by analogy with the definitions of specific heat, specific thermal capacity, specific internal energy, and so forth in thermal physics.

A listing of *relative atomic masses* for neutral atoms of all stable and many radioactive nuclides is given in the table of Appendix C. It should be noted that the atomic mass listed for each isotope is nearly equal to the corresponding atomic mass number $A$. The reason for this is easily understood by considering an atom of any isotope as consisting of a number of electrons $N_e$, a number of protons $N_p$, and a number of neutrons $N_n$. Accordingly, Equation 5.67 could be interpreted as

Relative Atomic Mass

$$\begin{aligned}(\mathrm{AM})_a &= \frac{m_a}{m_\mathrm{u}}\\ &\approx \frac{N_e m_e + N_p m_p + N_n m_n}{m_\mathrm{u}}\\ &\approx \frac{N_p m_p + N_n m_n}{m_\mathrm{u}}\\ &\approx N_p + N_n\\ &= Z + (A - Z) = A,\end{aligned} \tag{5.68}$$

where $m_p$, $m_n$, and $m_\mathrm{u}$ have been considered as *nearly* equal in magnitude and much greater than $m_e$. The second equality is indicated as an approximation because the nuclear *binding energy* (discussed in Section 5.9) has been ignored. Further, it should be noted that *chemical atomic weights* have been given for each element in the table of Appendix B, which are easily calculated from the table of Appendix C using the *relative atomic mass* and the *relative abundance* datum for the stable isotopes of each chemical element.

We have already discussed the basic importance of Avogadro's hypothesis in the assignment of atomic masses. The Avogadro constant $N_o$ is also integrally related to atomic mass and the value of $m_u$. To see this relationship consider the definition of a **mole** as being *the amount of a substance (gas, liquid, or solid) whose actual number of particles (atoms or molecules) is exactly equivalent to* $N_o$. Accordingly, the defining equation for the *number of moles n* of a substance is

$$n \equiv \frac{N}{N_o}, \qquad (5.69)$$

Number of Moles

where $N$ is the total number of particles (atoms or molecules) and $N_o$ is the Avogadro constant given by

$$N_o = 6.022045 \times 10^{23}. \qquad (5.70)$$

Avogadro Constant

If the total mass $M$ of a substance is known to be

$$M = m_a N, \qquad (5.71)$$

Total Mass

then Equation 5.69 can be expressed as

$$n = \frac{M}{m_a N_o}. \qquad (5.72)$$

Since $m_a$ is the absolute mass of a particular atom, then the denominator $m_a N_o$ of this equation must represent the *total mass of one mole* of such atoms. Denoting the product $m_a N_o$ by the symbol $\mathcal{M}$,

$$\mathcal{M} \equiv m_a N_o, \qquad (5.73)$$

Molal Atomic Mass

and using Equation 5.67 we obtain

$$\begin{aligned} \mathcal{M} &= (\mathrm{AM})_a m_u N_o \\ &= (\mathrm{AM})_a (1.00000 \times 10^{-3}\ \mathrm{kg}) \\ &= (\mathrm{AM})_a \cdot \mathrm{grams}, \end{aligned} \qquad (5.74)$$

Molal Atomic Mass

where values for $m_u$ (Equation 5.66) and $N_o$ (Equation 5.70) have been used in obtaining the second equality. Clearly, the absolute mass of one mole of a substance $\mathcal{M}$ is equivalent to the relative *atomic mass* in units of *grams*. The quantity $\mathcal{M}$, defined by Equation 5.73, could be called the *gram atomic mass*, but we shall call it the **molal atomic mass.** From the first and last equality in Equation 5.74, it is clear that

$$N_o = \frac{\text{grams}}{m_u}. \tag{5.75}$$

Thus, in the sense of this equation Avogadro's number $N_o$ is the reciprocal of the unified atomic mass unit u. If $m_a$ in Equation 5.73 referred to the mass of a molecule, then $\mathcal{M}$ would be interpreted as the *molal molecular mass*. The point, however, is that the *molal atomic* (or *molecular*) *mass* is the *absolute mass* of one mole of atoms (or molecules), as given by the relative atomic (or molecular) mass in grams. Because the atomic masses are very nearly equal to the atomic mass numbers in the table of Appendix C, essentially 2 g of hydrogen represents a mole of $H_2$, 32 g of oxygen constitutes a mole of $O_2$, and 18 g of water represents a mole of $H_2O$.

One of the most useful relationships for solving problems is obtained by combining Equations 5.69, 5.72, and 5.73 to obtain

$$n = \frac{N}{N_o} = \frac{M}{\mathcal{M}}. \tag{5.76}$$

This allows for the determination of any one of the three quantities $n$, $N$, or $M$ by knowing either one of the other two. For example the mass of a hydrogen atom is easily found by realizing $N = 1$ and $M = m_H$ and using the second equality of this equation,

$$\begin{aligned} m_H &= \frac{\mathcal{M}_H}{N_o} = \frac{1.007825 \text{ g}}{6.022045 \times 10^{23}} \\ &= 1.673559 \times 10^{-24} \text{ g}. \end{aligned}$$

Alternatively, Equation 5.67 could be used with identical results. That is,

$$\begin{aligned} m_H &= (\text{AM})_H m_u \\ &= (1.007825)(1.6605655 \times 10^{-27} \text{ kg}) \\ &= 1.673559 \times 10^{-24} \text{ g}. \end{aligned}$$

The number of moles $n$ or the number of particles $N$ (atoms or molecules) in a known mass of a substance is also easily computed using Equation 5.76. For example, 1 g (exactly) of hydrogen gas contains

$$\begin{aligned} N_{\mathrm{H}} &= \frac{M_{\mathrm{H}} N_o}{\mathcal{M}_{\mathrm{H}}} \\ &= \frac{(1\ \mathrm{g})(6.022045 \times 10^{23})}{1.007825\ \mathrm{g}} \\ &= 5.975288 \times 10^{23} \end{aligned}$$

atoms, which represents

$$\begin{aligned} n_{\mathrm{H}} &= \frac{N_{\mathrm{H}}}{N_o} \\ &= \frac{5.975288 \times 10^{23}}{6.022045 \times 10^{23}} \\ &= 0.9922357 \end{aligned}$$

or

$$\begin{aligned} n_{\mathrm{H}} &= \frac{M_{\mathrm{H}}}{\mathcal{M}_{\mathrm{H}}} \\ &= \frac{1\ \mathrm{g}}{1.007825\ \mathrm{g}} \\ &= 0.9922358 \end{aligned}$$

moles. The last two answers differ by rounding-off errors.

## 5.9 Size and Binding Energy of an Atom

The size of a particular atom can be estimated from knowing its molal atomic mass and a few other fundamental physical relations. To be more specific, the volume of space occupied by an atom $V_a$ is simply the volume of one mole of such atoms $V_o$ divided by $N_o$

$$V_a = \frac{V_o}{N_o}. \tag{5.77}$$

From the definition of mass density we obtain

$$V_a = \frac{M_o}{\rho_a N_o} = \frac{\mathcal{M}_a}{\rho_a N_o}, \tag{5.78}$$

where the mass of one mole $M_o$ has been identified as equivalent to $\mathcal{M}_a$ (see Equation 5.76) in the second equality. If, further, $V_a$ is imagined to be a sphere of radius $r_a$

$$V_a = \frac{4}{3}\pi r_a^3,$$

then from Equation 5.78 we obtain

Radius of an Atom

$$r_a = \left(\frac{3\mathcal{M}_a}{4\pi\rho_a N_o}\right)^{1/3} \tag{5.79}$$

for the radius of an atom. Using carbon $^{12}_{6}C$ as an example, with $\mathcal{M}_C$ = 12.0 g and $\rho_C$ = 2.25 g/cm$^3$ from Appendix B, Equation 5.79 gives

$$r_C = \left[\frac{(3)(12.0 \text{ g})}{4\pi(2.25 \text{ g/cm}^2)\,(6.02 \times 10^{23})}\right]^{1/3} = 1.28 \times 10^{-10} \text{ m}.$$

Consequently, the radius of a carbon atom is approximately one angstrom (1 Å = $10^{-10}$ m), which is enormous compared with the radius of the electron ($r_e \approx 10^{-15}$ m) computed previously in Section 5.5. Further, the proton radius is estimated to be roughly $10^{-15}$ m and recent estimates of the nuclear radius, $r_N$, place an upward limit of about $10^{-14}$ m for the more massive nuclear radius. Our picture of the atom from these estimates reveals it to be largely *empty*, with the volume of the atom to the volume of the nucleus being

$$\frac{V_a}{V_N} \approx \left(\frac{r_a}{r_N}\right)^3 \approx \left(\frac{1.28 \times 10^{-10} \text{ m}}{10^{-14} \text{ m}}\right)^3 \approx 2.10 \times 10^{12} \tag{5.80}$$

for carbon. Thus, a collapse of atomic structure would result in an increase in the mass density of matter by a factor of roughly $10^{12}$. Such a collapse of atomic structure is postulated for *white dwarf* and *neutron* stars, where 1 $cm^3$ of such matter would weigh several *million tons* on the surface of the Earth. For example, a collapse of carbon atoms would result in a mass density $\rho'_C$ given roughly by

$$\rho'_C = \frac{m_a}{V_N} = \frac{m_a}{V_a}\frac{V_a}{V_N} = \rho_a \frac{V_a}{V_N}$$
$$\approx (2.25 \text{ g/cm}^3)(2.10 \times 10^{12}) \approx 4.73 \times 10^{12} \text{ g/cm}^3.$$

Consequently, one cubic centimeter of such atoms would have a weight on Earth of

$$F_g \equiv mg = \rho'_C V_C g$$
$$= \left(4.73 \times 10^{12} \frac{\text{g}}{\text{cm}^3}\right)(1 \text{ cm}^3)\left(980 \frac{\text{cm}}{\text{s}^2}\right)$$
$$\approx 4.64 \times 10^{15} \text{ dy} = (4.64 \times 10^{15} \text{ dy})\left(2.248 \times 10^{-6} \frac{\text{lb}}{\text{dy}}\right)$$
$$\approx 1.04 \times 10^{10} \text{ lb} = (1.04 \times 10^{10} \text{ lb})\left(\frac{1 \text{ ton}}{2 \times 10^3 \text{ lb}}\right)$$
$$= 5.20 \times 10^6 \text{ tons}$$

As a last consideration of the data in the table of Appendix C and using the accuracy of other constants from the inside cover, note that the mass of the hydrogen atom $^1_1H$ given by

$$\begin{aligned} m_H &= (AM)_H m_u \\ &= (1.007825)(1.6605655 \times 10^{-27} \text{ kg}) \\ &= 1.673559 \times 10^{-27} \text{ kg} \end{aligned} \tag{5.81}$$

is exactly the sum of the proton and electron masses

$$\begin{aligned} m_p + m_e &= (1.6726485 \times 10^{-27} \text{ kg}) + (9.109534 \times 10^{-31} \text{ kg}) \\ &= 1.673559 \times 10^{-27} \text{ kg}, \end{aligned} \tag{5.82}$$

within the degree of accuracy assumed. For deuterium $^2_1D$, however, the atomic mass,

$$m_{\mathrm{D}} = (2.014102)(1.6605655 \times 10^{-27}\ \mathrm{kg}) = 3.344548 \times 10^{-27}\ \mathrm{kg},$$

is not the same as the sum of the proton mass $m_p$, the neutron mass $m_n$, and the electron mass $m_e$:

$$m_p + m_n + m_e = 3.348513 \times 10^{-27}\ \mathrm{kg}.$$

This difference or *loss in mass*

$$\Delta M = 3.965000 \times 10^{-30}\ \mathrm{kg}$$

between the *free* particles and the *bound* particles goes into the **binding energy** of the atom, as given by Einstein's formula (Equation 4.40),

$$\begin{aligned} E_B &= \Delta M c^2 \\ &= (3.965000 \times 10^{-30}\ \mathrm{kg})(2.997925 \times 10^8\ \mathrm{m/s})^2 \\ &= \frac{3.563565 \times 10^{-13}\ \mathrm{J}}{1.602189 \times 10^{-19}\ \mathrm{J/eV}} \\ &= 2.224185 \times 10^6\ \mathrm{eV} \\ &= 2.224185\ \mathrm{MeV}. \end{aligned} \tag{5.83}$$

As we shall see in Chapter 7, the *binding energy* of the electron is on the order of 10 eV, thus the result given by Equation 5.83 is essentially the *binding energy* of the nucleons, often called the *nuclear binding energy* and denoted as $B_N$. Using the same symbolic notation as in the derivation of Equation 5.68, $B_N$ can be expressed as

$$\begin{aligned} B_N &\approx (N_p m_p + N_n m_n + N_e m_e - m_a)c^2 \\ &= (N_p m_p + N_e m_e)c^2 + N_n m_n c^2 - m_a c^2 \\ &= Z(m_p + m_e)c^2 + (A - Z)m_n c^2 - m_a c^2, \end{aligned} \tag{5.84}$$

where for a normal atom $N_p = N_e \equiv Z$. But from comparing the results of Equations 5.81 and 5.82, we can replace $m_p + m_e$ with *essentially* the mass of the hydrogen atom $m_{\mathrm{H}}$ and obtain

Nuclear Binding Energy

$$B_N \approx Z m_{\mathrm{H}} c^2 + (A - Z)m_n c^2 - m_a c^2 \tag{5.85}$$

for the *nuclear binding energy* of *any* atom of mass $m_a$ having $Z$ protons. In this equation $m_H$ and $m_a$ would be calculated using either Equation 5.67 or Equation 5.73. That is,

$$m_H = (AM)_H m_u = \frac{\mathcal{M}_H}{N_o}, \tag{5.86}$$

where $\mathcal{M}_H$ is simply the *molal atomic mass* of *hydrogen*.

## Review of Fundamental and Derived Equations

A listing of the fundamental and derived equations of this chapter is presented below, along with *new* defined units, terms, and symbols. Not included are the well-known definitions and derived equations of kinematics.

### FUNDAMENTAL EQUATIONS—CLASSICAL PHYSICS

| | | |
|---|---|---|
| $V = \frac{4}{3}\pi r^3$ | | Volume of a Sphere |
| $\rho \equiv \frac{M}{V}$ | | Mass Density |
| $\mathbf{F} = m\mathbf{a}$, | $m \neq m(t)$ | Newton's Second Law |
| $F_g \equiv mg$ | | Weight |
| $F_c = \frac{mv^2}{r}$ | | Centripetal Force |
| $F_v = 6\pi r\eta v$ | | Stokes' Law |
| $W \equiv \int \mathbf{F} \cdot d\mathbf{r}$ | | Work |
| $T \equiv \frac{1}{2}mv^2$ | | Kinetic Energy |
| $F_C \equiv \frac{kqQ}{r^2}$ | | Coulomb's Law |
| $\mathbf{E} \equiv \frac{\mathbf{F}_E}{q}$ | | Electric Field Intensity |

$$V \equiv \frac{W}{q}$$ Electric Potential

$$E = \frac{V_c}{d}$$ Parallel Plate Capacitor

$$\mathbf{F}_B = q\mathbf{v} \times \mathbf{B}$$ Lorentz Force Equation

$$n \equiv \frac{N}{N_o}$$ Number of Moles

## FUNDAMENTAL EQUATIONS—MODERN PHYSICS

$$E \equiv mc^2$$ Energy-Mass Equivalence

$$(\text{AM})_a \equiv \frac{m_a}{m_\text{u}}$$ Relative Atomic Mass

$$\mathcal{M} \equiv m_a N_o$$ Molal Atomic Mass

$$n \equiv \frac{N}{N_o} = \frac{M}{\mathcal{M}}$$ Number of Moles

## DEFINED UNITS

$\text{eV} \equiv 1.60219 \times 10^{-19}\ \text{J}$ Electron Volt

$\text{u} \equiv 1.66057 \times 10^{-27}\ \text{kg}$ Unified Atomic Mass Unit

## MODERN PHYSICS SYMBOLS

$Z \equiv$ Atomic Number

$A \equiv$ Mass Number

$N \equiv$ Neutron Number

${}^{A}_{Z}S \equiv$ Nuclide

## DERIVED EQUATIONS

### *Thomson's $e/m_e$ Apparatus*

$$v_x = \frac{V_c}{dB}$$ Speed of Electrons $\leftarrow$ **E** and **B** Fields

$$\frac{e}{m_e} = \frac{E}{rB^2}$$ Specific Charge ← **B** Field Only

$$\frac{e}{m_e} = \frac{E^2}{2VB^2}$$ Specific Charge ← Cathode Potential

$$y_1 = \frac{eEx_1^2}{2m_e v_x^2}$$ **E** Field Displacement

$$y_2 = \frac{eEx_1x_2}{m_e v_x^2}$$ Displacement Beyond **E** Field

$$\frac{e}{m_e} = \frac{yE}{x_1B^2\,(\frac{1}{2}x_1 + x_2)}$$ Specific Charge ← Displacement

### *Millikan's Oil Drop Experiment*

$$F_g - F_B = 6\pi r\eta v_g$$ **E** Field OFF

$$F_E = F_g - F_B + 6\pi r\eta v_E$$ **E** Field ON

$$q = \frac{6\pi r\eta d}{V_c}(v_g + v_E)$$ Droplet Charge

$$F_g = \frac{4}{3}\pi r^3\rho_o g$$ Droplet Weight

$$F_B = \frac{4}{3}\pi r^3\rho_a g$$ Droplet Buoyant Force

$$r = 3\left[\frac{\eta v_g}{2g(\rho_o - \rho_a)}\right]^{1/2}$$ Droplet Radius

### *Size of an Electron*

$$r_e = \frac{ke^2}{2m_e c^2}$$ Electron's Radius

### *Thomson's Mass Spectrograph*

$$y = \frac{qEx_1}{mv_x^2}(\tfrac{1}{2}x_1 + x_2)$$ **E** Field Deflection

$$z = \frac{qBx_1}{mv_x}\left(\tfrac{1}{2}x_1 + x_2\right) \qquad \textbf{B} \text{ Field Deflection}$$

$$m = \frac{qyx_1B^2}{z^2E}\left(\tfrac{1}{2}x_1 + x_2\right) \qquad \text{Mass of Ionized Atom}$$

### Relative and Molal Atomic Mass

$$(\mathrm{AM})_a = A \qquad \text{Relative Atomic Mass}$$

$$\mathcal{M} = (\mathrm{AM})_a \cdot \text{grams} \qquad \text{Molal Atomic Mass}$$

### Size of Atom

$$r_a = \left(\frac{3\mathcal{M}}{4\pi\rho N_o}\right)^{1/3} \qquad \text{Radius of an Atom}$$

### Nuclear Binding Energy

$$B_N \approx Zm_{\mathrm{H}}c^2 + (A - Z)m_nc^2 - m_ac^2$$

## Problems

**5.1** If the accelerating potential between the cathode and anode of Thomson's $e/m_e$ apparatus is 182.2 V, what uniform velocity $v_x$ will the electrons acquire before entering the coexisting **E** and **B** fields? Assume accuracy to three significant figures and derive the appropriate equation.

*Solution:*
With knowledge of $V = 182.2$ V, $v_x$ is easily obtained by realizing the work done on the electron by the electric field, $W = \mathrm{eV}$, goes into kinetic energy. That is,

$$\mathrm{eV} = \tfrac{1}{2}m_ev_x^2,$$

which is identical to Equation 5.15. Solving this equation for $v_x$ and substituting the known quantities yields

$$v_x = \left(\frac{2\mathrm{eV}}{m_e}\right)^{1/2}$$

$$= \left[\frac{2(1.60 \times 10^{-19}\ \mathrm{C})\ (182.2\ \mathrm{V})}{91.1 \times 10^{-32}\ \mathrm{kg}}\right]^{1/2}$$

$$= 8.00 \times 10^6\ \mathrm{m/s}.$$

**5.2** An electron is accelerated from rest by an electrical potential $V$. If the velocity squared of the electron is $32 \times 10^{12}$ m$^2$/s$^2$, derive the equation for $V$ and find its value.

*Answer:* $V = 91.1$ V

**5.3** Electrons are directed through a region where uniform **B** and **E** fields coexist such that the path of the electrons is not altered. The uniform **E** field is established by a parallel plate capacitor having a 5 cm plate separation distance and the capacitor is connected to a 50 V battery. If $B = 2 \times 10^{-3}$ Wb/m$^2$ *(tesla),* derive the equation and calculate the value for $v_x$.

*Solution:*
With $V_c = 50$ V, $d = 5 \times 10^{-2}$ m, $B = 2 \times 10^{-3}$ Wb/m$^2$, and $F_E = eE$ equated to $F_B = ev_xB \sin 90°$ we obtain

$$\begin{aligned} v_x &= \frac{E}{B} = \frac{V_c}{dB} \\ &= \frac{50 \text{ V}}{(5 \times 10^{-2} \text{ m})(2 \times 10^{-3} \text{ Wb/m}^2)} \\ &= 5 \times 10^5 \text{ m/s.} \end{aligned}$$

**5.4** Electrons with a speed of $1.60 \times 10^7$ m/s enter a uniform **B** field at right angles to the induction lines. If $B = 4.555 \times 10^{-3}$ Wb/m$^2$, derive the equation for the radius of the electrons circular path and calculate its value.

*Answer:* $r = 2 \times 10^{-2}$ m

**5.5** If a beam of electrons, moving with a speed of $2 \times 10^7$ m/s, enters a uniform **B** field at right angles to the lines of force and describes a circular path with a 30 cm radius, what is the magnetic induction? Derive the appropriate equation for $B$ before substituting the physical data.

*Solution:*
Given $v_x = 2 \times 10^7$ m/s, $\theta = 90°$, and $r = 3 \times 10^{-1}$ m, how do we find $B$? Since $F_B = F_c$, then

$$ev_x\text{B} \sin \theta = \frac{m_e v_x^2}{r},$$

which yields

$$B = \frac{m_e v_x}{er \sin \theta}.$$

As $\sin 90° = 1$, substitution of the physical data yields

$$B = \frac{(9.11 \times 10^{-31}\ \text{kg})\ (2 \times 10^{7}\ \text{m/s})}{(1.60 \times 10^{-19}\ \text{C})(0.3\ \text{m})}$$

$$= 3.80 \times 10^{-4}\ \text{Wb/m}^2.$$

**5.6** Electrons with $6.396404 \times 10^{4}$ eV kinetic energy enter a uniform magnetic field at 65.6378° with respect to the induction lines. If the magnetic induction is $6.24146 \times 10^{-3}$ T, derive the equation and find the value for the radius of the electrons circular arc?

*Answer:* $r = 0.15$ m

**5.7** A parallel plate capacitor 25 cm long with a 5 cm separation between the plates is connected to a 91.1 V battery. If an electron enters this field with a velocity of $2 \times 10^{9}$ cm/s at an angle of 90° to the **E** field, how far will the electron be deviated from its original rectilinear path immediately after passing through the electric field?

*Solution:*
The physical data given is $x = 0.25$ m, $d = 0.05$ m, $V_c = 91.1$ V, $v_x = 2 \times 10^{7}$ m/s, and $\theta = 90°$, and we want to find the displacement $y_1$ of the electron due to only the **E** field. The derivation starting with Equation 5.17 and ending with Equation 5.20 is appropriate for this problem. That is, with $v_{0y} = 0$, we have

$$y = \tfrac{1}{2}a_y t_1^2 = \tfrac{1}{2}\frac{F_E}{m_e}t_1^2$$

$$= \frac{eEt_1^2}{2m_e} = \frac{e(V_c/d)t_1^2}{2m_e}$$

$$= \frac{e(V_c/d)(x_1/v_x)^2}{2m_e}$$

$$= \frac{eV_c x_1^2}{2m_e d v_x^2}$$

$$= \frac{(1.60 \times 10^{-19}\ \text{C})(91.1\ \text{V})(0.25\ \text{m})^2}{2(9.11 \times 10^{-31}\ \text{kg})(0.05\ \text{m})(2 \times 10^{7}\ \text{m/s})^2}$$

$$= 2.5 \times 10^{-2}\ \text{m} = 2.5\ \text{cm}.$$

**5.8** Let the electron of Problem 5.7 travel a horizontal distance of 80 cm after exiting the **E** field. Derive the equation and calculate its additional vertical deflection $y_2$.

*Answer:* $y_2 = 16$ cm

**5.9** A parallel plate capacitor 25 cm long with a 5 cm separation between the plates is connected to 182.2 V. If electrons enter this capacitor at right angles to the **E** field and are deviated by $y_1 = 5$ cm from their original rectilinear path after passing through the capacitor, what is their original horizontal speed $v_x$? Further, if these electrons travel a horizontal distance of 80 cm after exiting the **E** field, what additional vertical deflection $y_2$ will they experience?

*Solution:*
For the first part of this problem we know $x_1 = 0.25$ m, $d = 0.05$ m, $V_c = 182.2$ V, $\theta = 90°$, and $y_1 = 0.05$ m, and we want to find $v_x$. From Problem 5.7 we have

$$y_1 = \frac{eV_c x_1^2}{2m_e d v_x^2},$$

which when solved for $v_x$ yields

$$\begin{aligned} v_x &= \left(\frac{eV_c x_1^2}{2m_e d y_1}\right)^{1/2} \\ &= \frac{(1.60 \times 10^{-19}\text{ C})(182.2\text{ V})(0.25\text{ m})^2}{2(91.1 \times 10^{-32}\text{ kg})(0.05\text{ m})(0.05\text{ m})} \\ &= [(16)(25)(10^{12})\text{m}^2/\text{s}^2]^{1/2} \\ &= 2 \times 10^7\text{ m/s}. \end{aligned}$$

For the second part of this problem we have additional knowledge of $x_2 = 0.80$ m and we want to find $y_2$. From Problem 5.8 or Equations 5.7 and 5.24 we have

$$\begin{aligned} y_2 &= \frac{eV_c x_1 x_2}{m_e d v_x^2} \\ &= \frac{(16 \times 10^{-20}\text{ C})(182.2\text{ V})(25 \times 10^{-2}\text{ m})(80 \times 10^{-2}\text{ m})}{(91.1 \times 10^{-32}\text{ kg})(5 \times 10^{-2}\text{ m})(4 \times 10^{14}\text{ m}^2/\text{s}^2)} \\ &= (16)(2)(10^{-2})\text{m} = 0.32\text{ m}. \end{aligned}$$

**5.10** Verify that Equation 5.6 is directly obtained for an undeflected electron passing through the Thomson $e/m_e$ apparatus, by equating the deflection of the electron due to the **E** field alone to the deflection due to the **B** field alone.

*Answer:* $v_x = E/B$

**5.11** An electron traveling at $8 \times 10^6$ m/s enters that region of the Thomson $e/m_e$ apparatus where the **E** and **B** fields coexist and are adjusted to be counter balancing. The **E** field is created by a parallel plate capacitor connected to a 91.1 V battery and having a 6.4 cm plate separation. If the **E** field is deactivated, what is the radius of the electron's circular arc through the counter balancing magnetic field?

*Solution:*
In this problem we know $v_x = 8 \times 10^6$ m/s, $V_c = 91.1$ V, $d = 0.064$ m, and $\theta = 90°$, and we need to find $r$. By equating $F_B$ and $F_c$ (i.e., $ev_xB \sin \theta = m_ev_x^2/r$), we obtain

$$r = \frac{m_ev_x}{eB},$$

since $\sin 90° = 1$. Also, with $F_B = F_E$ giving

$$ev_xB = eE,$$

we have

$$B = \frac{E}{v_x} = \frac{V_c}{dv_x},$$

where Equation 5.7 has been used for $E$. Now, substitution of this expression for $B$ into our radius equation yields

$$\begin{aligned} r &= \frac{m_edv_x^2}{eV_c} \\ &= \frac{(91.1 \times 10^{-32} \text{ kg})(0.064 \text{ m})(64 \times 10^{12} \text{ m}^2/\text{s}^2)}{(1.60 \times 10^{-19} \text{ C})(91.1 \text{ V})} \\ &= 0.256 \text{ m} = 25.6 \text{ cm}. \end{aligned}$$

**5.12** Consider the situation described in Problem 5.11 only now allow the counter balancing **B** field to be deactivated instead of the **E** field. If the electron is deflected vertically by 5 cm while traversing the **E** field of the capacitor, how long is the capacitor and what is the vertical speed acquired by the electron?

*Answer:* $x_1 = 16$ cm, $v_y = 5 \times 10^6$ m/s

**5.13** In the Millikan oil-drop experiment consider a droplet having a terminal velocity to fall 0.240 cm in 18 s with the **E** field deactivated. Find the radius of the droplet for $\rho_o = 891$ kg/m$^3$, and $\eta = 1.80 \times 10^{-5}$ kg/m · s.

*Solution:*
With $\Delta y = 2.40 \times 10^{-3}$ m and $\Delta t = 18$ s, $v_g$ is found to be

$$v_g = \frac{\Delta y}{\Delta t} = \frac{24.0 \times 10^{-4} \text{ m}}{18 \text{ s}} = (4/3) \times 10^{-4} \text{ m/s}.$$

Now, suppressing the units and substituting the physical data into Equation 5.36 (which should be derived from first principles) gives

$$r = 3\left[\frac{\eta v_g}{2g(\rho_o - \rho_a)}\right]^{1/2}$$

$$= 3\left[\frac{(1.80 \times 10^{-5})(4/3) \times 10^{-4}}{2(9.80)(890)}\right]^{1/2}$$

$$= 1.11 \times 10^{-6} \text{ m}.$$

**5.14** If the droplet in Problem 5.13 experiences terminal velocity of $1.11 \times 10^{-5}$ m/s when the **E** field is activated, what is the charge on the droplet? Allow the **E** field to be established by a parallel plate capacitor having a 1.5 cm plate separation being connected to a 169.56 V battery.

*Answer:* $q = 30e$

**5.15** The atomic mass of cobalt (Co) is 58.9332. Find the mass in grams of one Co atom using the definition of *atomic mass* and the definition of a *mole*.

*Solution:*
From Equation 5.67 we have

$$m_{\text{Co}} = (\text{AM})_{\text{Co}} m_{\text{u}}$$
$$= (58.9332)(1.6605655 \times 10^{-24} \text{ g})$$
$$= 9.78624 \times 10^{-23} \text{ g},$$

while from Equation 5.76 ($n = N/N_o = M/\mathcal{M}$), we obtain the same result for $N = 1$ and $M \equiv m_{\text{Co}}$. That is,

$$m_{\text{Co}} = \frac{\mathcal{M}_{\text{Co}}}{N_o}$$
$$= \frac{58.9332 \text{ g}}{6.022045 \times 10^{23}}$$
$$= 9.78624 \times 10^{-23} \text{ g}.$$

**5.16** The atomic mass of the most abundant isotope of copper (Cu) is 62.9296. How many atoms are there in exactly one gram of Cu and how many moles are represented by this mass?

*Answer:* $N_{\text{Cu}} = 9.56950 \times 10^{21}$, $n_{\text{Cu}} = 1.58908 \times 10^{-2}$

**5.17** What is the mass in grams of exactly 3.5 moles of carbon (C) and how many atoms does this amount represent?

*Solution:*
We know $n_C = 3.5$ moles and $\mathcal{M}_C = 12.0000$ g, and we want to find $M_C$ and $N_C$. Since $n = M/\mathcal{M}$, then

$$\begin{aligned} M_C &= n_C \mathcal{M}_C \\ &= (3.50000)(12.0000 \text{ g}) \\ &= 42.0000 \text{ g}. \end{aligned}$$

Further, since $n = N/N_o$ we have

$$\begin{aligned} N_C &= n_C N_o \\ &= (3.500000)(6.022045 \times 10^{23}) \\ &= 2.107716 \times 10^{24}. \end{aligned}$$

**5.18** How many atoms are there in a 15 kg bar consisting of 70 percent Cu and 30 percent Zn by mass?

*Answer:* $N = 1.42869 \times 10^{26}$

**5.19** A massive bar of $10^{26}$ atoms is composed of 70 percent Cu (AM = 62.93) atoms and 30 percent Fe (AM = 55.94) atoms. What is the mass of the bar?

*Solution:*
With $N = 10^{26}$, $N_{Cu} = 7 \times 10^{25}$, and $N_{Fe} = 3 \times 10^{25}$, the mass $M$ of the bar is given by

$$\begin{aligned} M &= M_{Cu} + M_{Fe} \\ &= \frac{N_{Cu}\mathcal{M}_{Cu}}{N_o} + \frac{N_{Fe}\mathcal{M}_{Fe}}{N_o} \\ &= (0.7\mathcal{M}_{Cu} + 0.3\mathcal{M}_{Fe}) \frac{N}{N_o} \\ &= [0.7(62.93 \text{ g}) + 0.3(55.94 \text{ g})] \frac{N}{N_o} \\ &= (44.05 \text{ g} + 16.78 \text{ g}) \frac{N}{N_o} \\ &= \frac{(60.83 \text{ g})(10^{26})}{6.022045 \times 10^{23}} \\ &= 1.010 \times 10^4 \text{ g} = 10.10 \text{ kg}. \end{aligned}$$

**5.20** A beam of doubly ionized Zn atoms (AM = 63.9) enter the electric field of a 16.6 m long parallel plate capacitor, which has a separation distance between the plates of 8 cm. The Zn atoms enter the capacitor with a horizontal speed of $2 \times 10^7$ m/s at right angles to the existing **E** field.

If the capacitor is connected to a 63.9 V battery, how far vertically will the Zn atoms be deviated from their original rectilinear path after passing through the **E** field?

*Answer:* $y = 8.30 \times 10^{-4}$ m

**5.21** A beam of triply ionized Zn (AM = 63.9) atoms, moving with a speed of $1.60 \times 10^7$ m/s, enters a uniform field of $4.98 \times 10^{-14}$ Wb/m$^2$ magnitude at an angle of 30° with respect to the magnetic flux lines. What is the radius of the circular arc described by the beam?

*Solution:*
The given information includes $(\text{AM})_{\text{Zn}} = 63.9$, $v = 1.60 \times 10^7$ m/s, $B = 4.98 \times 10^{-14}$ Wb/m$^2$, $\theta = 30°$, and $q = 3e$, and we need to find the radius $r$ described by the beam of ionized Zn atoms as it traverses the **B** field. Since $F_c = F_B$ we have

$$\frac{m_{\text{Zn}}v^2}{r} = qvB \sin\theta,$$

which allows $r$ to be described by

$$\begin{aligned} r &= \frac{m_{\text{Zn}}v}{qB \sin\theta} \\ &= \frac{(\text{AM})_{\text{Zn}}m_{\text{u}}v}{3eB \sin\theta} \\ &= \frac{(63.9)(1.66 \times 10^{-27})(1.60 \times 10^7)}{3(1.60 \times 10^{-19})(4.98 \times 10^{-14})(0.500)} \\ &= 1.42 \times 10^{14} \text{ m.} \end{aligned}$$

**5.22** Derive the equation and find the nuclear binding energy $B_N$ of a carbon atom in MeV?

*Answer:* $B_N = 92.16484$ MeV

**5.23** Derive the equation and find the radius of a copper atom, using the data of Appendix B.

*Solution:*
Assuming an atom to be a sphere of radius $r_a$, then we approximate its volume as $V_a = (4/3)\,\pi r_a^3$, from which

$$\begin{aligned} r_a^3 &= \frac{3}{4\pi} V_a \\ &= \frac{3}{4\pi}\frac{m_a}{\rho_a} \end{aligned}$$

$$= \frac{3}{4\pi}\frac{m_a N_o}{\rho_a N_o}$$

$$= \frac{3\mathcal{M}_a}{4\pi\,\rho_a N_o}.$$

Now, using $\mathcal{M}_{Cu} = 63.546$ g and $\rho_{Cu} = 8.96$ g/cm$^3$ from Appendix B, direct substitution yields

$$r_{Cu} = \left(\frac{3\mathcal{M}_{Cu}}{4\pi\rho_{Cu}N_o}\right)^{1/3}$$

$$= \left[\frac{3(63.546\text{ g})}{4(3.1417)(8.96\text{ g/cm}^3)(6.02 \times 10^{23})}\right]^{1/3}$$

$$= (2.81 \times 10^{-24}\text{ cm}^3)^{1/3}$$

$$= 1.41 \times 10^{-8}\text{ cm} = 1.41 \times 10^{-10}\text{ m}.$$

**5.24** Consider Thomson's mass spectrograph where a **B** field of $4.15 \times 10^{-3}$ Wb/m$^2$ is antiparallel to a coexisting **E** field. Assume doubly ionized atoms are accelerated through the distance $x_1 = 24$ cm and then travel $x_2 = 88$ cm farther at a uniform speed before striking a fluorescent screen. If the $y$-deflection data associated with the **E** field yields $v_x = 2 \times 10^5$ m/s and the total $z$-displacement is 24 cm, what is the mass of the ions?

*Answer:* $m = 6.64 \times 10^{-27}$ kg

CHAPTER 6

# Quantization of Electromagnetic Radiation

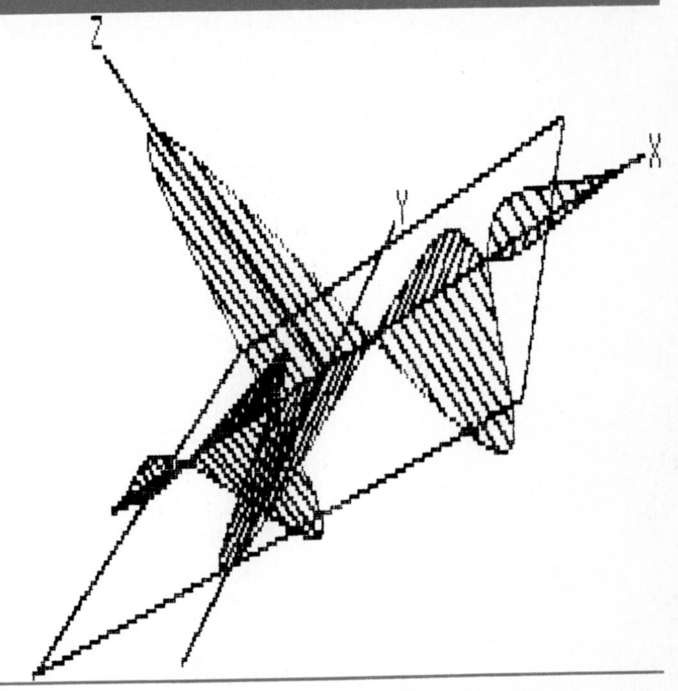


Are not gross Bodies and Light convertible into one another, and may not Bodies receive much of their activity from the Particles of Light which enter their Composition?

I. NEWTON, *Opticks* (1730)

## Introduction

In the seventeenth century there were two conflicting views concerning the nature of electromagnetic radiation (often referred to as simply light). Newton and his followers believed light consisted of very small and fast moving elastic particles called *corpuscles*. This view satisfactorily accounted for the **law of reflection** in geometrical optics, as *the angle of incidence is equal to the angle of reflection* for perfectly elastic bodies and light rays being reflected from a plane surface. The theory also predicted the *law of refraction*, allowing corpuscles of light to be attracted toward a transparent material medium (e.g., air, water, glass, etc.), with a resulting increase in their component of velocity that was perpendicular to the medium's surface. Accordingly, Newton's corpuscular theory predicted the speed of light to be *greater* in a transparent material medium than in free space and its direction of propagation to be bent toward the normal. The other view by Christian Huygens regarded light as being composed of *waves*, which also explained the reflection and refraction of light. According to this theory, light waves would also be bent toward the normal upon entering a transparent material medium, but the speed of wave propagation in the medium would be *less* than its speed in free space. The debate surrounding these two different theories continued until the middle of the nineteenth century, when the French physicists A. H. Fizeau in 1849 and J. B. Foucault in 1850 measured the speed of light in air and water, respectively. Their results of the speed of light in air (Fizeau) being greater than the speed of light in water (Foucault) confirmed Huygens' wave theory, completely negating Newton's corpuscular view.

The wave nature of electromagnetic radiation was well established and almost universally accepted by the end of the nineteenth century. However, the particle view was once again to gain support, as the result of a fundamentally new interpretation of electromagnetic radiation initiated by Max Planck in 1900 and later modified by Einstein in 1905. Planck assumed atoms to be capable of absorbing and emitting *quanta* of electromagnetic energy, by considering atoms as tiny electromagnetic oscillators having allowed energy states that are quantized in nature. Planck's quantization of energy for atoms was *generalized* by Einstein to be a fundamental property of electromagnetic radiation and *not* just a special property of atoms. In 1905 Einstein explained the photoelectric effect by assuming electromagnetic radiation to behave as if its energy was concentrated into discrete bundles or packets, called **quanta** or more commonly **photons.** Later in 1923 A. H. Compton provided evidence that *photons* undergo particle-like collisions with atoms, by considering the energy *and* linear momentum of a beam of x-rays to be concentrated in *photons*. In general, physicists were most reluctant to accept the quantum explanations of the photoelectric and Compton effects, because of the apparent contradiction to the successful wave theory. In fact, for many years after Einstein's successful explanation of the photoelectric effect, Planck considered light as propagating through space as an electromagnetic wave and Einstein's *photon concept* as being wholly untenable.

Although the major objective of this chapter is to illuminate the particle-like behavior of electromagnetic radiation, we begin with a review of the classical properties and generation of electromagnetic waves. The wave properties of electromagnetic radiation are further illustrated by energy considerations and Bragg reflection (actually diffraction) of x-rays. We then emphasize two experiments where the quantum or particle-like nature of light dominates its wave nature, by a discussion of the photoelectric effect and the Compton effect. The failure of classical wave theory to explain the former phenomenon and the success of Einstein's *photon concept* are fully detailed in one section. Finally, an alternative derivation for the relativistic Doppler effect is presented, which demonstrates the consistency between the *photon quantization hypothesis* and Einstein's *special theory of relativity*. This chapter illustrates that electromagnetic radiation appears to possess a dual personality, behaving at times like waves and at other times like particles. This dual-like behavior of radiation, later recognized to be a general characteristic of all physical entities, is *not* explainable by classical physics nor by the *old quantum theory* being presented in Chapters 5 through 7. It is, however, satisfactorily reconciled with the aid of the theory of quantum mechanics and will be discussed in considerable detail in Chapter 8.

## 6.1 Properties and Origin of Electromagnetic Waves

The wavelength spectrum of electromagnetic radiation, illustrated in Table 6.1, consists of radiation ranging from $\gamma$-rays of wavelength $10^{-14}$ m to long waves of wavelength $10^5$ m. The ranges indicated for the differently named *bands* of radiation are only *approximate*, as there is considerable overlapping presented in Table 6.1. It is of interest to note that visible

TABLE 6.1
The approximate wavelength spectrum of electromagnetic waves.

| *Name of Radiation* | *Wavelength Range* (m) |
|---|---|
| $\gamma$-Rays | $10^{-14} - 10^{-10}$ |
| x-Rays | $10^{-11} - 10^{-8}$ |
| Ultraviolet | $10^{-8} - 10^{-7}$ |
| Visible | $10^{-7} - 10^{-6}$ |
| Infrared | $10^{-6} - 10^{-4}$ |
| Heat | $10^{-5} - 10^{-1}$ |
| Microwaves | $10^{-2} - 10$ |
| Radio Waves | $10 - 10^3$ |
| Long Waves | $10^3 - 10^5$ |

light waves represent only a very small slice of the total spectrum. The wavelength in **angstrom units,**

Angstrom Unit

$$\text{Å} \equiv 10^{-10}\ \text{m},$$

of each color of visible light corresponding to the approximate center of each color band is

| | |
|---|---|
| red | $\lambda_r = 6600\ \text{Å},$ |
| orange | $\lambda_o = 6100\ \text{Å},$ |
| yellow | $\lambda_y = 5800\ \text{Å},$ |
| green | $\lambda_g = 5500\ \text{Å},$ |
| blue | $\lambda_b = 4700\ \text{Å},$ |
| violet | $\lambda_v = 4100\ \text{Å}.$ |

As pointed out previously in Chapter 3, all electromagnetic radiation propagates in free space (vacuum) at the speed of light

$$c = 2.99792 \times 10^8\ \frac{\text{m}}{\text{s}} \approx 3 \times 10^8\ \frac{\text{m}}{\text{s}}, \tag{5.51}$$

and obeys the wave equation

$$c = \lambda\nu, \tag{3.30}$$

where $\lambda$ is the wavelength and $\nu$ is the frequency of a particular radiation. From this wave equation it is obvious that a short wavelength corresponds to a high frequency and a long wavelength to a low frequency. Further, the speed of light in air is very nearly the same as its speed in free space, but its speed in other optically dense media is slower. Consequently, the ratio of the speed of light $c$ in free space to the speed of light $v$ in a transparent material medium is always greater than one (e.g., approximately 4/3 for water and 3/2 for glass) and is defined as the **index of refraction,**

Index of Refraction

$$n \equiv \frac{c}{v}. \tag{6.1}$$

The classical concept of an electromagnetic wave is, as its name implies, a combination of a varying electric field and a varying magnetic field propagating through space at the speed $c$. To better understand the properties of the electric and magnetic fields associated with an electromagnetic

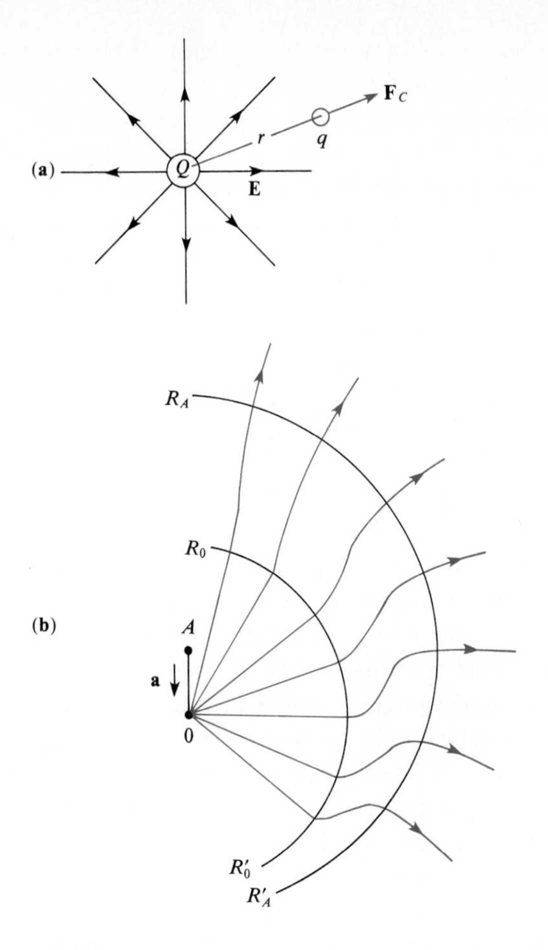


Figure 6.1
The electric field lines of force associated with *a*, a stationary and *b*, an accelerating positive charge *Q*.

wave, we will discuss, qualitatively, how the origin of radiation is ultimately an accelerated electric charge. First, however, consider the small positive electric charge at rest in Figure 6.1a, where the electric field is depicted by *imaginary* lines of force extending radially from the charge. Each line of force gives the direction of the electric field **E** and the Coulombic force $\mathbf{F}_C$ on a very small positive test charge $q$ placed at any point along the line. This is in total agreement with the definition of electric field intensity given by

$$\mathbf{E} \equiv \frac{\mathbf{F}_C}{q}, \qquad (5.4)$$

Electric Field Intensity

where the magnitude of $\mathbf{F}_C$ is defined by Coulomb's law as

$$\mathbf{F}_C \equiv k\frac{Qq}{r^2} \qquad (5.43)$$

Coulomb's Law

in terms of the distance $r$ between the charges. It should be emphasized that although we restrict our discussion to lines of force in a plane, they extend radially in all directions of real space. Further, if the charge $Q$ were moving with a *small* uniform velocity in the plane of the figure, the associated lines of force would be the same as depicted in Figure 6.1a. Now, however, in accordance with Ampere's law the *electric current* is surrounded by a concentric magnetic field **B**, whose lines of force are normal (perpendicular) to the plane of Figure 6.1a. For these imaginary electric and magnetic fields, a *tangent* constructed at any point on an electric or magnetic *line of force* would give the direction of the electric **E** field or magnetic **B** field, respectively, at that point.

A major point of the above discussion is that *steady* electric and magnetic fields are associated with *steady* electric currents. Clearly, an electromagnetic *wave* cannot be produced by any steady electric current. It is suggested, however, that an electric and a magnetic *wave* can be produced by a *varying* or *alternating* electric current. Such a current can be thought of as resulting from an *oscillating* electric charge, which necessarily requires a *periodic* acceleration of the charge. Before discussing the more general case of an electric charge undergoing periodic motion, we first consider a charge undergoing a single acceleration. A positive electric charge $Q$ is depicted in Figure 6.1b as experiencing a rapid acceleration from point $A$ to point $O$. Let the acceleration from $A$ to $O$ require a time $\Delta t_a$ and a time $\Delta t_o$ elapse after the charge reaches position $O$. After the time $\Delta t_a + \Delta t_o$ has elapsed, the *original* electric field lines of force about $Q$ at position $A$ are depicted in Figure 6.1b beyond the arc $R_A R'_A$, which is drawn about point $A$ with the radius $c(\Delta t_a + \Delta t_o)$. The uniform lines of force about $Q$ between point $O$ and the arc $R_O R'_O$, which is drawn with the radius $c\Delta t_o$, represent the uniform lines of force of $Q$ at position $O$ during the time $\Delta t_o$. The lines of force during the acceleration time $\Delta t_a$ of $Q$ are, consequently, represented by the connecting *wavy lines* between the arcs $R_O R'_O$ and $R_A R'_A$. The form of these *wavy* lines of force will depend upon the exact kind of acceleration experienced by $Q$ between points $A$ and $O$.

The acceleration of an electric charge is accompanied by *changes* in its uniform lines of force (e.g., the wavy lines of force in Figure 6.1b) and these *changes* propagate away from the accelerated charge at the speed of light $c$. In this manner an accelerated electric charge produces a *pulse* of electromagnetic radiation. This pulse of radiation can be better understood by considering Figure 6.2a, where a *wavy* line of force normal to $AO$ in Figure 6.1b is enlarged. At point $P_1$ on this line of force a tangent is constructed, which gives the actual direction of the electric **E** field at point $P_1$. The vector **E** can be regarded as the resultant of a *transverse* field $E_t$ and a field $E_o$ that would be associated with the charge at rest. If tangents at a number of points along the *wavy* line of force were constructed, we

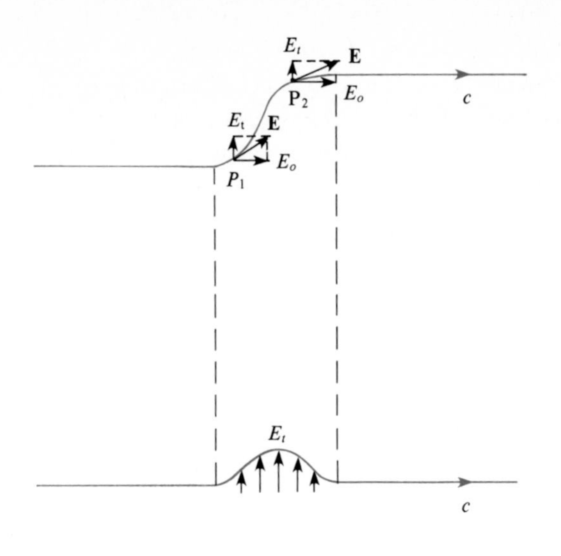


(a) Electric field line distorted by linear acceleration.

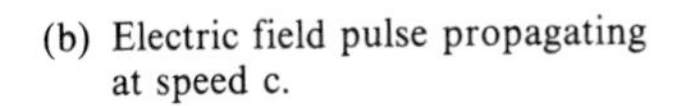
(b) Electric field pulse propagating at speed c.

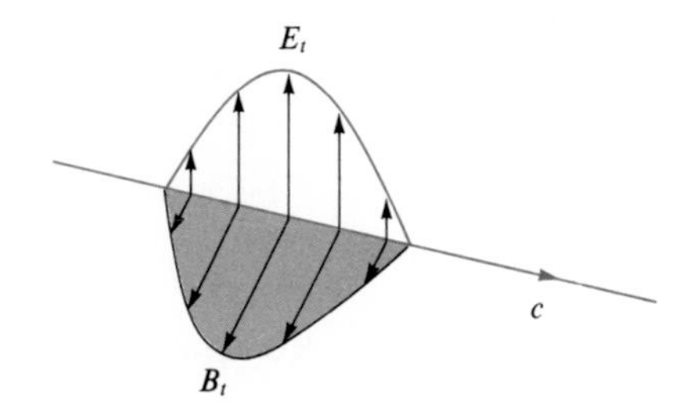


(c) Electromagnetic pulse consisting of transverse and magnetic field lines.

Figure 6.2
The construction of an electromagnetic pulse from an accelerated electric charge line of force.

would obtain the various *transverse* components depicted in Figure 6.2b. This is clearly not a wave but merely a *pulse* consisting of transverse electric field vectors. A similar analysis of the magnetic **B** field associated with the accelerated charge $Q$ of Figure 6.1 would yield a *magnetic field pulse* that is in phase and perpendicular to the electric field pulse. Thus, as illustrated in Figure 6.2c, an electric charge undergoing a linear acceleration produces a pulse of electromagnetic radiation having electric and magnetic field components that are perpendicular to one another and their direction of propagation.

If the electric charge in Figure 6.1 is forced to oscillate with simple periodic motion, electromagnetic waves are produced like the one illustrated in Figure 6.3. This wave results from the electric field line of force that is perpendicular to and in the same plane as the oscillating electric charge and is depicted at an *instant* in time. It is recognized as a *transverse wave,* since the alternating electric and magnetic field vectors are at right angles to the direction of propagation. Because the fields are acknowledged to consist of only *transverse* vectors, the $t$-subscript has been dropped from $E$ and $B$ in this figure. Since the alternating electric field vectors at all points in the wave are parallel, the wave is said to be *polarized* or more specifically, *plane polarized*. The *plane of vibration*, which is commonly called the *plane of polarization*, is defined by the direction of polarization (the $Y$-axis) and the direction of propagation (the $X$-axis). It is important to realize, however, that the wavy lines of Figure 6.3 simply depict the strengths of the electric and magnetic field vectors and that *nothing is vibrating* in the electromagnetic wave. The direction of the alternating magnetic field is interrelated with the electric field in that it must be normal to the plane of polarization. Consequently, the three vectors **E**, **B**, and **c** constitute a set of mutually orthogonal vectors, where the direction of propagation is given by $\mathbf{E} \times \mathbf{B}$.

A quantitative description of the electric and magnetic field components of a plane-polarized electromagnetic wave propagating in the $x$-direction should now be rather obvious. In Figure 6.3, $E_x = E_z = 0$ and the sinusoidal form of $E_y$ is dependent on only $x$ and $t$. Thus, we postulate the electric field vector **E** by the equation

E-Field Vector

$$\mathbf{E} = E_m \sin(kx - \omega t)\mathbf{j}, \tag{6.2}$$

where $E_m$ is the maximum amplitude and $k$ is the **wave number** defined by

Wave Number

$$k \equiv \frac{2\pi}{\lambda}. \tag{6.3}$$

In a similar manner, since $B_x = B_y = 0$ and $B_z = B_z(x, t)$ in Figure 6.3, we postulate the magnetic field vector **B** to be of the form

B-Field Vector

$$\mathbf{B} = B_m \sin(kx - \omega t)\mathbf{k}, \tag{6.4}$$

with the symbol $B_m$ representing the maximum amplitude. The symbol $\omega$ in Equations 6.2 and 6.4 represents the **angular speed,** which is often called the *angular* (or *circular*) *frequency* because of the relationship

Angular Frequency

$$\omega = 2\pi\nu. \tag{6.5}$$

It is interesting to note that from Equations 3.30, 6.3, and 6.5 the speed of propagation $c$ is equal to the ratio of $\omega$ and $k$,

$$c = \lambda\nu = \frac{\lambda}{2\pi} 2\pi\nu = \frac{\omega}{k}. \tag{6.6}$$

There are a couple of points that are important to realize about Equations 6.2 and 6.4 in relationship to Figure 6.3. The first point is that the **E** and **B** field components of the electromagnetic wave are *in phase* with each other in space and time. This can be visualized by realizing that as time goes on the entire field structure of Figure 6.3 moves along as a unit at the speed $c$. If the wave moves past a point in space, however, the electric and magnetic fields at that point *change in phase every instant*, with both **E** and **B** attaining their maximum or minimum at the same point in space and at the same instant in time. The second point to be realized is that Figure 6.3 represents a plot of **E** versus the position coordinate $x$ at a constant value of time, say $t = t_0 \equiv 0$. Thus $\mathbf{E}(x, t_0)$ has a sinusoidal dependence on $x$ with a *wavelength* $\lambda = 2\pi/k$. Likewise, if $x$ is held constant, say $x = x_0 \equiv 0$, a plot of $\mathbf{E}(x_0, t)$ versus $t$ would look like Figure 6.3, with the $X$-axis being replaced by a $t$-axis. In this case, the *period of oscillation* (instead of wavelength) would be given by $T = 1/\nu = 2\pi/\omega$.

Before leaving this section, it should be emphasized that the qualitative discussion of the origin of electromagnetic waves has been concerned with only those waves produced by *linear* acceleration of an electric charge. However, electromagnetic radiation occurs whenever an electric charge is accelerated, irrespectively of the manner in which it is is accelerated. For example, a charge in uniform circular motion experiences centripetal acceleration that produces a *circularly polarized* electromagnetic wave. Such waves are commonly produced by a *synchrotron*, which im-

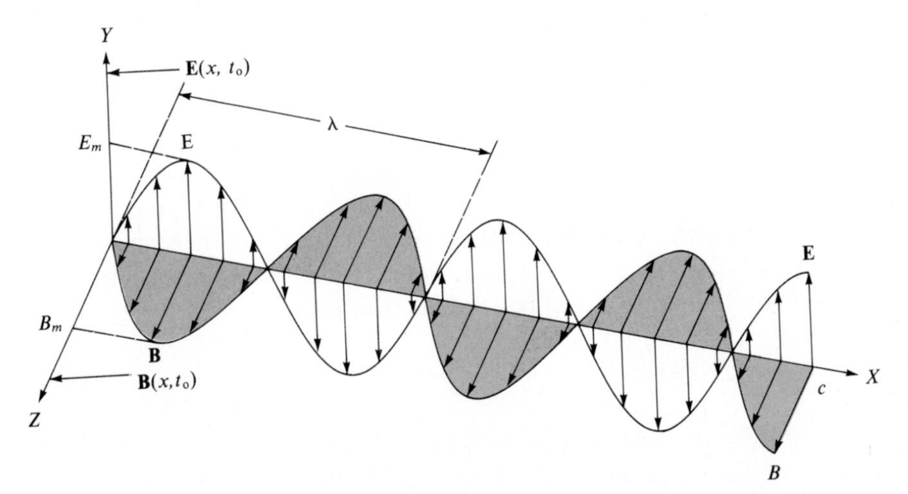


Figure 6.3
An electromagnetic plane polarized monochromatic traveling wave, with the transverse **E**-field vectors in the *XY* plane.

parts very high speeds to charged particles by a high-frequency electric field combined with a low-frequency magnetic field. As a last point of interest, we can infer from this section that the frequency of an electromagnetic wave produced by an accelerating charge depends on the frequency of oscillation of that charge. Conversely, an electric charge, say the electrons in a receiving antenna, will be accelerated by the forces they encounter from passing electromagnetic waves. The frequency of the resulting alternating current will then depend on the frequency of the incident electromagnetic waves.

## 6.2 Intensity, Pressure, and Power of Electromagnetic Waves

The laws of electricity, magnetism, optics, and the propagation of electromagnetic waves were well understood by 1864 and completely contained in a set of four partial differential equations—known as Maxwell's equations. Although it would not serve our objectives to develop James Clark Maxwell's electrodynamics, one of his equations in differential form, namely

Maxwell's Equation

$$\nabla \times \mathbf{E} = -\frac{\partial \mathbf{B}}{\partial t}, \tag{6.7}$$

will be briefly utilized. This equation is normally derived in general physics starting with *Faraday's law of induction* and invoking the calculus of *Stokes' formula*. The *inverted delta* symbol $\nabla$, as given in Appendix A Section A.9, is called the **del operator** and defined by

Del Operator

$$\nabla \equiv \frac{\partial}{\partial x}\mathbf{i} + \frac{\partial}{\partial y}\mathbf{j} + \frac{\partial}{\partial z}\mathbf{k}. \tag{6.8}$$

The *curly dees* in the expression $\partial/\partial x$ are, as you may well know, simply interpreted as the **partial derivative** *with respect to x*. We need only consider the operational nature of Equations 6.7 and 6.8 and their application to the plane-polarized electromagnetic wave of the previous section. That is, direct substitution of Equations 6.2 and 6.4 into Equation 6.7 gives

$$E_m k \cos(kx - \omega t)\mathbf{k} = -B_m(-\omega)\cos(kx - \omega t)\mathbf{k}. \tag{6.9}$$

The left-hand side of this equation is directly obtained from the *curl* of $\mathbf{E}$ (i.e., $\nabla \times \mathbf{E}$), when it is realized that for $\mathbf{E} = \mathbf{E}(x, t)$ the partials with respect to $y$ and $z$ become identically zero, and the remaining term, $(\partial E/\partial x)$ $(\mathbf{i} \times \mathbf{j})$, gives a vector pointing in the **k**-direction (i.e., $\mathbf{i} \times \mathbf{j} = \mathbf{k}$). Equation 6.9 immediately reduces to

$$E_m = cB_m \tag{6.10}$$

by substitution from Equation 6.6. Further, considering only the magnitudes of **E** and **B** given in Equations 6.2 and 6.4, the ratio of these two equations gives

$$\frac{E}{B} = \frac{E_m}{B_m},$$

which combines with Equation 6.10 to give

$$E = cB \tag{6.11}$$

for the instantaneous magnitudes of **E** and **B.** Thus taking into account the vector properties of **E, B,** and **c** we obtain

$$\mathbf{E} = \mathbf{B} \times \mathbf{c}. \tag{6.12}$$

Plane Waves

The results given by Equations 6.10 to 6.12 show the *interdependence* of the electric and magnetic field vectors, and they will prove most useful in developing equations for the energy transported by an electromagnetic wave. Usually, the energy transmitted in a radiation field by an electromagnetic wave is specified in terms of the **intensity**, which can be simply thought of as energy per unit area per unit of time. More specifically, *intensity is the energy per unit time transmitted across a unit area that is normal to the direction of propagation of a wave*. It can be calculated using the **Poynting vector S,** which is defined in general physics by the equation

$$\mathbf{S} \equiv \frac{1}{\mu_0} \mathbf{E} \times \mathbf{B} \tag{6.13}$$

Poynting Vector

The quantity $\mu_0$ in this equation is called the **permeability constant,** which has the defined value (exactly) of

$$\mu_0 \equiv 4\pi \times 10^{-7} \frac{\text{N}}{\text{A}^2}. \tag{6.14}$$

Permeability Constant

The units of **S** are easily obtained by realizing that **E** has units of N/C = N/A · s, since it is defined as a force per unit charge, and **B** has the defined unit of a *tesla* (T). Thinking of **B** as being defined as a force $\mathbf{F}_B$ per unit *magnetic pole m′* of units A · m,

Magnetic Induction

$$\mathbf{B} \equiv \frac{\mathbf{F}_B}{m'}, \tag{6.15}$$

then T = N/A · m. Consequently, the units of **S** are given by

$$\begin{aligned} \mathbf{S} \rightarrow \frac{\mathrm{A}^2}{\mathrm{N}} \frac{\mathrm{N}}{\mathrm{A \cdot s}} \frac{\mathrm{N}}{\mathrm{A \cdot m}} &= \frac{\mathrm{N}}{\mathrm{m \cdot s}} \\ &= \frac{\mathrm{N}}{\mathrm{m \cdot s}} \frac{\mathrm{m}}{\mathrm{m}} \\ &= \frac{\mathrm{J}}{\mathrm{m^2 s}} = \frac{\mathrm{W}}{\mathrm{m^2}}, \end{aligned}$$

as expected for a quantity representing energy per unit time per unit area. It should also be realized that the magnitude of **S** divided by the speed of light $c$ would give *the amount of radiant energy per unit volume of space,* which is called the **energy density.** That is, using the symbol $\epsilon$ to represent electromagnetic wave energy to distinguish it from particle energy, $S$ could be thought of as

Instantaneous Intensity

$$S = \frac{1}{A}\frac{d\epsilon}{dt}, \tag{6.16}$$

where $A$ represents the unit surface area. Now, since the electromagnetic wave is propagating in the $x$-direction with a speed $c = dx/dt$, then

Energy Density

$$\frac{S}{c} = \frac{1}{A}\frac{d\epsilon}{dt}\left(\frac{1}{dx/dt}\right) = \frac{1}{A}\frac{d\epsilon}{dx} = \frac{d\epsilon}{dV}. \tag{6.17a}$$

Frequently, the ratio of $S$ to $c$ is called the **radiation pressure,** since

Radiation Pressure

$$\begin{aligned} \frac{S}{c} &= \frac{1}{A}\frac{d\epsilon}{dx} \\ &= \frac{1}{A}\frac{F_\epsilon dx}{dx} \\ &= \frac{F_\epsilon}{A} \end{aligned} \tag{6.17b}$$

and pressure is recognized as force per unit area. Equation 6.17b gives the radiation pressure for a totally *absorbed* wave; whereas a totally *reflected* wave undergoes a change of momentum that is *twice* as great, and consequently, the resulting pressure is $2S/c$.

It should be emphasized that the Poynting vector, as defined by Equation 6.13, is perfectly applicable to any kind of electromagnetic radiation. For the plane-polarized monochromatic traveling wave given by Equations 6.2 and 6.4, the instantaneous value of **S** is given by

$$\mathbf{S} = \frac{1}{\mu_0} E_m B_m \sin^2(kx - \omega t)\mathbf{i}, \tag{6.18}$$

which is seen to point in the direction of wave propagation as expected. In terms of the magnitudes of **S, E,** and **B** we have

$$S = \frac{EB}{\mu_0}, \tag{6.19}$$

which from Equation 6.12 can be rewritten as

$$S = \frac{E^2}{\mu_0 c}. \tag{6.20}$$

Instantaneous Intensity

This equation (actually, all of the equations involving **S**) represents the *instantaneous* rate of energy flow per unit area, which is characterized at a point in space at a particular *instant* in time. Normally in optics, the **intensity** of radiation is taken as the *time-averaged value* of $S$ at a particular point. With $I_\epsilon$ representing the intensity of electromagnetic radiation and $\langle\ \rangle$ representing a *time average*, we have

$$I_\epsilon = \langle S \rangle. \tag{6.21}$$

Time Average Intensity

Thus, from the last two equations, the intensity of a plane polarized electromagnetic wave is

$$I_\epsilon = \frac{\langle E^2 \rangle}{\mu_0 c}. \tag{6.22}$$

As a final expression for the intensity of our electromagnetic wave, Equation 6.22 is less than satisfying. We need to establish a definition for *time-averaging* and then evaluate $<E^2>$. The **time average** of any function of time $f(t)$ over one complete cycle can be defined as

Time Average

$$<f(t)> \equiv \frac{1}{T}\int_0^T f(t)dt, \tag{6.23}$$

where $T$ represents the *period* of one complete cycle or oscillation. Since $T = 2\pi/\omega$ and $E = E_m \sin(kx - \omega t)$ for the plane polarized electromagnetic wave, then at the point $x = x_0 \equiv 0$ we have

$$<E^2> = \frac{\omega}{2\pi}E_m^2\int_0^{2\pi/\omega} \sin^2 \omega t\, dt. \tag{6.24}$$

The integral of Equation 6.24 can be easily handled by changing the variable of integration. That is, with $\theta \equiv \omega t$, $dt = d\theta/\omega$, and the limits of integration are given by $t = 0 \rightarrow \theta = 0$ and $t = 2\pi/\omega \rightarrow \theta = 2\pi$. Thus, Equation 6.24 becomes

$$\begin{aligned} <E^2> &= \frac{E_m^2}{2\pi}\int_0^{2\pi} \sin^2\theta\, d\theta \\ &= \frac{E_m^2}{2\pi}\int_0^{2\pi} (\tfrac{1}{2} - \tfrac{1}{2}\cos 2\theta)d\theta \\ &= \frac{E_m^2}{4\pi}\left(\int_0^{2\pi} d\theta - \int_0^{2\pi} \cos 2\theta\, d\theta\right) \\ &= \frac{E_m^2}{4\pi}(2\pi - 0) \\ &= \frac{E_m^2}{2}, \end{aligned} \tag{6.25}$$

where a few math identities of Appendix A have been utilized. Finally, substitution of Equation 6.25 into Equation 6.22 yields

Time Average Intensity

$$I_\epsilon = \frac{E_m^2}{2\mu_0 c} \tag{6.26}$$

for the wave *intensity* over one cycle for plane polarized electromagnetic radiation.

There are two major points of Equation 6.26 that should be fully realized. The first is that *intensity is directly proportional to the square of the amplitude,* which is a general property of all waves. The second point is that the time-average *power* dissipated perpendicularly to a particular unit area $A$ is given by $I_e A$ or

$$<\mathbf{P}> = \frac{E_m^2 A}{2\mu_0 c}. \qquad (6.27)$$

Time Average Power

## 6.3 Diffraction of Electromagnetic Waves

When two waves *collide* in a region of space, the *collision* is quite dissimilar from one involving two *particles* in that the two waves combine according to the *principle of linear superposition* (see Chapter 8, Section 8.5), then each wave emerges from the *collision* with its original physical characteristics unchanged. This particular property of waves produces the phenomenon known as **interference,** which is commonly demonstrated in general physics by a *resonance* experiment using sound waves or by double-slit and diffraction grating experiments using visible light. We generally observe *interference* when two or more waves of the same type and similar physical properties (i.e., amplitude, frequency, and phase) enter the same region of space at the same time. Conversely, if *interference* is observed, like in a diffraction experiment, it indicates a *wave-like* phenomenon. This is an important point to emphasize, as later (in Chapter 8, Section 8.4) we will see how *electrons* exhibit *wave-like* behavior in a *diffraction experiment*. For now, however, we will concentrate on how x-rays were *first* demonstrated to consist of electromagnetic waves by an experiment involving *diffraction*.

The *interference* of electromagnetic radiation is easily demonstrated for visible light using a mechanically constructed *diffraction grating*. This is because the *line-spacing* $d(\approx 3\ \mu\text{m})$ of the grating is only a few times larger than the wavelength of visible light, $\lambda_{\text{visible}} \approx 0.5\ \mu\text{m}$. For x-rays, however, it is impossible to mechanically produce a grating having $d \approx \lambda_{\text{x-rays}}$, as the wavelength of x-rays (Table 6.1) is on the order of $10^{-10}$ m. It was suggested by Max von Laue in 1912 that the atoms in a crystal might serve as a three-dimensional grating for x-rays, since the atomic spacing was known to be on the order of an angstrom. Atomic spacing is easily determined from knowledge of atomic masses and the mass density for a particular crystal. For example, consider the periodic array of atoms illustrated in Figure 6.4 for common salt (NaCl), which has a simple structure called **face-centered cubic.** In the **primitive cell,** illustrated by the shaded region in the figure, there are eight **lattice points,** one at each corner of the cube of edge $a$. These *lattice points* locate the positions of the Na ions and Cl ions, which are indicated by either shaded or unshaded

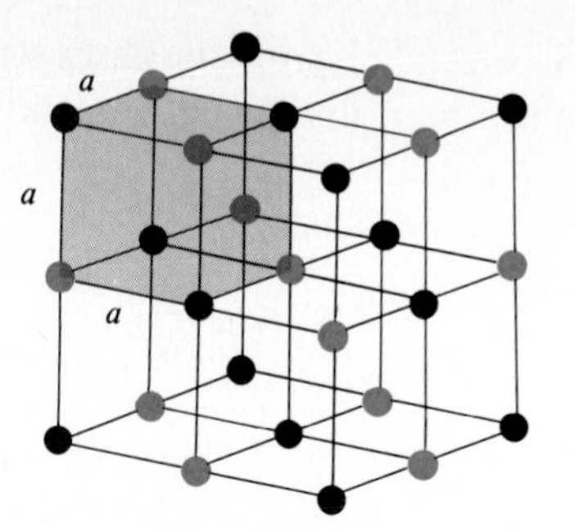


Figure 6.4
The space lattice of a face-centered cubic crystal, where the primitive cell is illustrated by a shaded cube.

circles. Looking at the *lattice point* in the geometrical center of the figure, we note that it is *shared* by the eight adjoining cells. Thus, there is only *one* lattice point per primitive cell in the face-centered cubic structure. This also means that the mass of the eight atoms (four Na and four Cl) in a *primitive cell* of NaCl is the sum of the individual masses divided by *eight*. Thus, the *mass density* of NaCl can be expressed as

$$\begin{aligned}\rho_{\text{NaCl}} &\equiv \frac{M}{V} \\ &= \frac{(4m_{\text{Na}} + 4m_{\text{Cl}})/8}{a^3} \\ &= \frac{m_{\text{Na}} + m_{\text{Cl}}}{2a^3},\end{aligned} \tag{6.28}$$

where $a^3$ is just the *volume* of the *primitive cell*. The quantities $m_{\text{Na}}$ and $m_{\text{Cl}}$ are easily determined by using Equation 5.67 and the *atomic masses* of Na and Cl given in the table of Appendix C. That is, for the one common isotope of Na we have

$$\begin{aligned}m_{\text{Na}} &= (\text{AM})_{\text{Na}} m_{\text{u}} \\ &= (22.9898)(1.6605655 \times 10^{-27}\ \text{kg}) \\ &= 3.81761 \times 10^{-26}\ \text{kg},\end{aligned} \tag{6.29a}$$

while for the two naturally occurring isotopes of Cl

$$\begin{aligned}m_{\text{Cl}} &= [(34.9689)(0.7577) + (36.9659)(0.2423)]\ m_{\text{u}} \\ &= (35.4528)(1.6605655 \times 10^{-27}\ \text{kg}) \\ &= 5.8872 \times 10^{-26}\ \text{kg}.\end{aligned} \tag{6.29b}$$

Now, with the mass density of NaCl being $2.18 \times 10^3$ kg/m$^3$, substitution into Equation 6.28 gives

$$\begin{aligned} a &= \left(\frac{m_{\text{Na}} + m_{\text{Cl}}}{2\rho_{\text{NaCl}}}\right)^{1/3} \\ &= \left[\frac{9.70 \times 10^{-26}\ \text{kg}}{2(2.18 \times 10^3\ \text{kg/m}^3)}\right]^{1/3} \\ &= (2.23 \times 10^{-29}\ \text{m}^3)^{1/3} \\ &= 2.81 \times 10^{-10}\ \text{m}. \end{aligned} \tag{6.30}$$

Clearly, this value for the atomic spacing of NaCl is on the order of an x-ray wavelength, so Max von Laue's suggestion to use crystals as diffraction gratings appears to be reasonable and applicable.

In 1913 William L. Bragg presented a simplistic analysis of the diffraction of x-rays by crystalline solids. Bragg considered the *constructive interference* of x-rays could result from the *scattering* of waves from two adjacent atoms lying in separate but parallel planes, which are now referred to as **Bragg planes.** Figure 6.5a illustrates two sets of *Bragg planes*

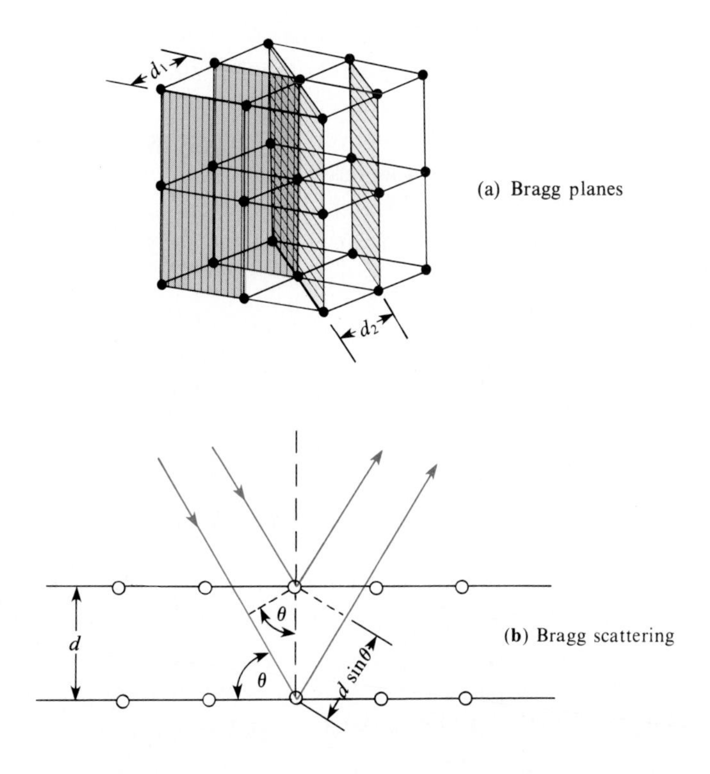


Figure 6.5
A face-centered cubic crystal illustrating *a*, two sets of Bragg planes and *b*, Bragg scattering from successive planes.

for a face-centered cubic crystal, although there are many other possible sets that could be drawn. Figure 6.5b depicts the Bragg scattering of a beam of x-rays from two successive planes of atomic spacing $d$. Although each plane scatters part of the incident beam in random directions, a small fraction of the beam, depicted by two solid rays, is **specularly** *(angle of incidence is equal to angle of reflection)* reflected from the Bragg planes at an angle $\theta$. The two parallel scattered rays will *interfere constructively*, if their paths differ by an integral number of wavelengths. That is, the path difference between the two rays must be $n\lambda$ for monochromatic waves, where $n = 1, 2, 3$, and so forth. Since the *bottom* ray travels a distance of $2d \sin\theta$ further than the *top* ray, the condition for constructive interference of *specularly* scattered waves is satisfied if

Bragg's Law

$$2d \sin\theta = n\lambda. \tag{6.31}$$

This equation is known as **Bragg's law** for x-ray *diffraction*. Because of the condition of *specular scattering*, it is often, though incorrectly, referred to as *Bragg reflection*. The x-ray diffraction experiments by Laue and Bragg confirmed two presumptions made by turn-of-the-century physicists: *that x-rays consist of electromagnetic waves, and that crystals contain atoms in a periodic array*.

The theory of waves and its applications in *physical optics* is a broad and interesting subject that we have barely touched in these three sections. Our discussions have been totally concerned with the wave-like behavior of electromagnetic radiation. We now turn our attention to the theory and experiments that suggest radiation as having additional properties in nature that are *particle-like*. Our understanding of wave properties and the associated theory will, however, be most useful in the following sections and chapters.

## 6.4 Energy and Momentum of Electromagnetic Radiation

Although the transmission of energy by electromagnetic waves is well-known and common to everyday experiences (e.g., the energy transmitted to a closed car on a sunny day), less known is that electromagnetic waves transport *momentum*. In this section we will derive a relationship between the energy $\epsilon$ and momentum $p_\epsilon$ of an electromagnetic wave and see how, in a sense, it suggests a particle-like behavior for radiation. This relationship will be developed by capitalizing on previously established wave properties and fundamental physical relationships. In particular, consider a plane polarized electromagnetic wave traveling in the $x$-direction to be incident on a positively charged particle at rest. The charged particle will initially experience a force

$$\mathbf{F}_E = qE\mathbf{j},$$

due to the electric field component of the incident wave, and undergo an acceleration in the positive $y$-direction. Once the particle starts moving, however, it experiences an additional force

$$\mathbf{F}_B = q\mathbf{v} \times \mathbf{B}$$

due to the magnetic field component of the wave. Since *initially* $\mathbf{v}$ in this expression is in the positive $y$-direction and $\mathbf{B}$ is in the positive $z$-direction (i.e., $\mathbf{B} = B\mathbf{k}$), then according to the right-hand rule the particle will undergo an additional acceleration in the positive $x$-direction. Consequently, a positively charged particle at rest will have a contribution to its velocity of $v_y\mathbf{j}$ from the electric field and $v_x\mathbf{i}$ from the magnetic field components of the incident electromagnetic wave. Realizing that $v_x$ and $v_y$ are rapidly varying with time, due to the alternating electric and magnetic field vectors of the wave, we can say that *very quickly* after the particle is exposed to the radiation it has an instantaneous velocity given by

$$\mathbf{v} = v_x\mathbf{i} + v_y\mathbf{j} = \frac{dx}{dt}\mathbf{i} + \frac{dy}{dt}\mathbf{j}. \tag{6.32}$$

The acceleration of the particle is now governed by the total force

$$\mathbf{F} = q\mathbf{E} + q\mathbf{v} \times \mathbf{B}, \tag{6.33}$$

where $\mathbf{v}$ is given by Equation 6.32. Combining these two equations and realizing that $\mathbf{E} = E\mathbf{j}$ and $\mathbf{B} = B\mathbf{k}$ we immediately obtain

$$\begin{aligned}\mathbf{F} &= qE\mathbf{j} + q(v_x\mathbf{i} + v_y\mathbf{j}) \times B\mathbf{k} \\ &= qE\mathbf{j} + qv_xB(-\mathbf{j}) + qv_yB\mathbf{i},\end{aligned}$$

which can be expressed more simply as

$$\mathbf{F} = qv_yB\mathbf{i} + q(E - v_xB)\mathbf{j}. \tag{6.34}$$

In this form, the components of force are easily recognized as

$$F_x = qv_yB, \tag{6.35}$$

and

$$F_y = q(E - v_xB). \tag{6.36}$$

The components of force given by the last two equations can now be utilized to obtain an expression for the momentum transferred to the charged particle by invoking Newton's second law,

Newton's Second Law

$$\mathbf{F} \equiv \frac{d\mathbf{p}}{dt}, \tag{1.16}$$

in terms of $x$- and $y$-components,

$$F_x\mathbf{i} + F_y\mathbf{j} = \frac{dp_x}{dt}\mathbf{i} + \frac{dp_y}{dt}\mathbf{j}. \tag{6.37}$$

From Equations 6.36 and 6.37 the infinitesimal $y$-component of momentum is just

$$\begin{aligned} dp_y &= F_y dt \\ &= q(E - v_xB)dt \\ &= qEdt - qB\frac{dx}{dt}\,dt \\ &= qEdt - qBdx. \end{aligned} \tag{6.38}$$

Integrating over one complete cycle after substitution from Equations 6.2 and 6.4 gives

$$\begin{aligned} p_y &= \int_0^{p_y} dp_y \\ &= qE_m\int_0^{2\pi/\omega} \sin(kx - \omega t)dt - qB_m\int_0^{2\pi/k} \sin(kx - \omega t)dx \\ &= -qE_m\int_0^{2\pi/\omega} \sin(\omega t)dt - qB_m\int_0^{2\pi/k} \sin(kx)dx = 0. \end{aligned} \tag{6.39}$$

The two integrals in the second equality can be evaluated by using $\sin(kx - \omega t) = \sin kx \cos \omega t - \sin \omega t \cos kx$ or, as illustrated, by letting $x = x_0 \equiv 0$ in the first integral and $t = t_0 \equiv 0$ in the second integral. With the latter, they obviously go to zero, since $\cos 2\pi - \cos 0 = 1 - 1 = 0$ is obtained in both cases. In a similar manner Equations 6.35 and 6.37 give

$$\int_0^{p_x} dp_x = \int F_x dt$$

$$= \int qv_y \mathrm{B} \mathrm{dt}$$

$$= qB \int_0^y dy,$$

where the last integral is just over $dy$, as $B = B(x, t)$. Thus, we obtain

$$p_x = qBy \tag{6.40}$$

for the $x$-component of momentum. Since $\mathbf{p} = p_x\mathbf{i} + p_y\mathbf{j}$, then

$$\mathbf{p} = qBy\mathbf{i} \tag{6.41}$$

Time Average Momentum

for the total momentum transferred to the charged particle by one complete cycle of the electromagnetic wave. Actually, this result is valid for any number of complete cycles of the wave, as the two integrals involving the sine function still vanish. Further, even though **E** and **B** reverse direction during every cycle of the wave, with $F_y$ averaging to zero, the net force $F_x$ on the particle in the $x$-direction does not average to zero over one cycle and its direction remains constant.

By itself, Equation 6.41 is not particularly important, unless we choose a value for $q$ and estimate values for $B$ and $y$. However, it can be combined with an expression for energy to obtain a significant result. From the definitions of instantaneous power,

$$P \equiv \frac{dW}{dt}, \tag{6.42}$$

Instantaneous Power

and work (Equation 1.20), the energy $W$ of the electromagnetic radiation is given by

$$\frac{dW}{dt} = \frac{\mathbf{F} \cdot d\mathbf{r}}{dt} = \mathbf{F} \cdot \mathbf{v}, \tag{6.43}$$

where **F** and **v** are given by Equations 6.34 and 6.32. Thus,

$$\begin{aligned}\frac{dW}{dt} &= [qv_y B\mathbf{i} + q(E - v_x B)\mathbf{j}] \cdot [v_x\mathbf{i} + v_y\mathbf{j}] \\ &= qv_x v_y B + q(E - v_x B)v_y \\ &= qEv_y.\end{aligned} \tag{6.44}$$

Now, treating the differentials algebraically we obtain

$$\begin{aligned}dW &= qE\frac{dy}{dt}\,dt \\ &= qE\,dy,\end{aligned} \tag{6.45}$$

which when integrated over one cycle,

$$\int_0^{\epsilon} dW = qE\int_0^{y} dy,$$

gives

Time Average Energy

$$\epsilon = qEy. \tag{6.46}$$

Since $E = cB$ (Equation 6.11), substitution into Equation 6.46 along with the magnitude of Equation 6.41 yields

Energy-Momentum of an Electromagnetic Wave

$$\epsilon = pc. \tag{6.47}$$

This result is most significant in its interpretation that *as a charged particle absorbs radiant energy* $\epsilon$ *in the time* $2\pi/\omega$, *the linear momentum* $p$ *transferred to the particle in the same time is* $\epsilon/c$. We have in essence described a *perfectly inelastic collision* between the incident electromagnetic radiation and the charged particle. This is very analogous to an inelastic collision between two particles that are very dissimilar in mass. If one particle of mass $M$ is stationary and the other of mass $m << M$ is incident with kinetic energy $T = \frac{1}{2}mv^2 = \frac{1}{2}pv$, then effectively the momentum $p = 2T/v$ is transferred to the mass $M + m \approx M$ during a perfectly inelastic collision. In this sense, the electromagnetic wave is exhibiting a particle-like behavior. Aside from this analogy, our result is totally consistent with Einsteinian relativity (Chapter 4, Section 4.5), since the energy of a particle with *zero rest mass* traveling at the speed $c$ is given by $E = pc$. Interestingly, Maxwell knew of this energy-momentum relationship (Equation 6.47) for electromagnetic waves for well over thirty years before the development of the special theory of relativity. However, he was so entrenched with his differential equations of wave theory that he totally overlooked any particle-like behavior of electromagnetic radiation.

## 6.5 Photoelectric Effect

Although the wave nature of light characterized by interference, diffraction, and polarization was supported by overwhelming evidence prior to the twentieth century, classical physics recognized that a *quantum* of electromagnetic radiation possessed momentum. Einstein's special theory of relativity (Chapter 4, Section 4.4) also acknowledged a *quanta of light-energy* now called a *photon*, as possessing energy and momentum; however, the theory predicted that a photon, necessarily, has a zero rest mass and cannot be accelerated. These latter characteristics are not normally associated with a particle, such as an electron, since it can be accelerated and we can determine its mass, size, charge, and kinetic energy.

The classical wave nature of light is displaced by a *quantum* or particle behavior in a phenomenon known as the **photoelectric effect,** where a satisfactory explanation assumes a single photon to interact directly with an electron. A basic description of the *photoelectric effect* is *the ejection of electrons from a metal surface that is irradiated by electromagnetic radiation*. In general photoelectrons are produced by most metals when exposed to ultraviolet light. If visible light is incident on the alkali metals (lithium, sodium, potassium, rubidium, and cesium), the production of photoelectrons is observed. The photoelectron phenomenon was first observed by Heinrich Hertz in 1887 and later by W. Hallwachs in 1888. Hallwachs observed that ultraviolet light neutralized a negatively charged metal, whereas a positively charged body was unaffected by irradiation.

In 1898 J. J. Thomson and Philipp Lenard observed the photoelectric phenomenon by experimental apparatus similar to that shown schematically in Figure 6.6. Electromagnetic radiation from the light source S causes electrically charged particles to be liberated at the cathode metal C. The deflection of these particles by a magnetic field and the determination of their *specific charge* by the methods described in Chapter 5, Section 5.3, identified the particles as electrons. Such electrons are called **photoelectrons** in reference to their source of excitation. If an electrical potential is impressed across the cathode and anode electrodes, then any photoelectrons produced at C will migrate to A as the result of Coulombic forces of attraction by A and repulsion by C. Any photoelectron current produced is measured by a micro-ammeter $\mu$-A. Further, as indicated in the schematic, the amount and polarity of the impressed voltage is controlled by a vari-

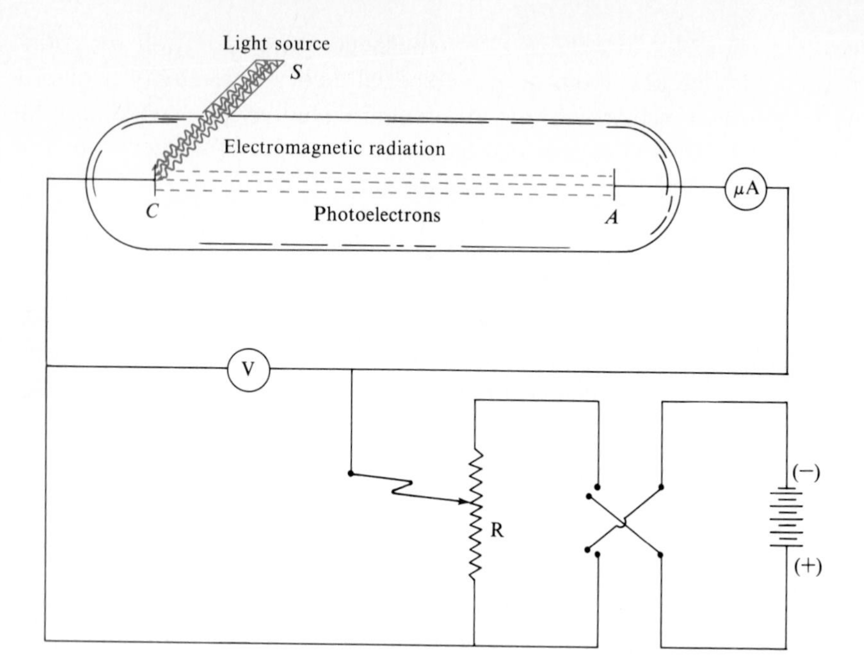


Figure 6.6
An illustrative schematic of an experimental system for measurements of the photoelectric effect.

able resistor R and a switching arrangement, respectively, and measured by a voltmeter V. As will be presently discussed, various types of measurements can be made using this apparatus. First, however, we will adopt the following symbolic notation that will be utilized throughout the remainder of our discussion of this phenomenon:

$I_e$ = photoelectron current,
$V$ = impressed electrical potential,
$I_\epsilon$ = intensity of the electromagnetic radiation,
$\nu$ = frequency of the electromagnetic radiation,
C = cathode metal material.

A plot of $I_e$ versus $V$ is illustrated in Figure 6.7a, where the physical measurements involved varying $V$ and measuring $I_e$ while $I_\epsilon$, $\nu$, and C were maintained constant. Since an impressed negative voltage tends to keep photoelectrons from reaching A, then $V = -V_s$, the so called **stopping potential,** is the electrical potential required to stop the most *energetic* photoelectrons. Clearly, the most energetic photoelectrons would normally be the surface electrons of C, since once liberated these electrons would not loose kinetic energy by way of atomic collisions within the metal before escaping from the cathode surface. It follows that the maximum kinetic energy of a photoelectron is given by

$$T_{\max} = eV_s, \tag{6.48}$$

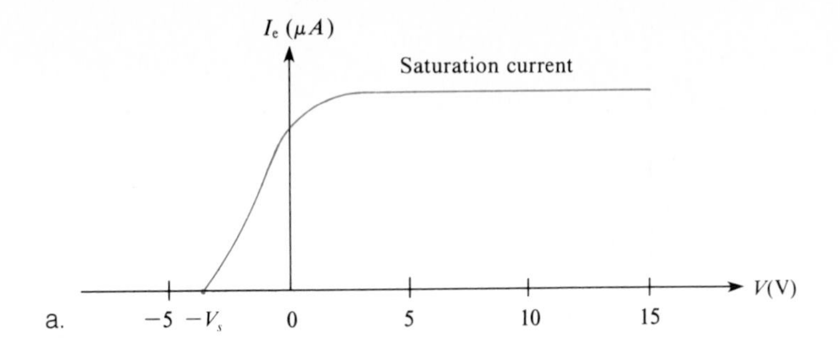


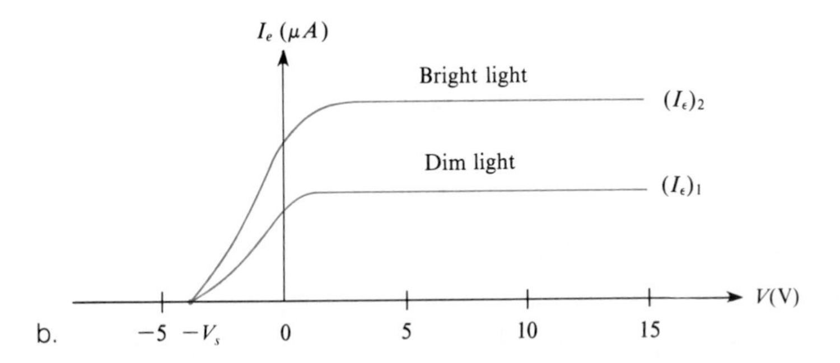


Figure 6.7
*a*, A plot of photocurrent $I_e$ versus $V$ for constant $I_\epsilon$, $\nu$, and C. *b*, A plot of $I_e$ versus $V$ for two values of $I_\epsilon$ with $\nu$ and $C$ constant.

where $e$ is the magnitude of an electron charge. It should also be observed in Figure 6.7a that for a small positive voltage impressed across C and A, the *saturation current* is attained, as all the photoelectrons produced reach A. Allowing the light intensity $I_\epsilon$ to vary results in the graph illustrated in Figure 6.7b. The surprising result illustrated here is that $V_s$ and consequently $T_{max}$ is *independent* of the *intensity (brightness)* of the incident electromagnetic radiation. If the frequency of the incident light is altered, the value of $V_s$ would be affected but $T_{max}$ would still be independent of the light intensity. A similar statement could also be made for the data if C was changed.

If $V$, $\nu$, and C are held constant and $I_\epsilon$ is allowed to vary, it is possible to realize more directly how the photocurrent $I_e$ is dependent on the intensity of the incident light. A representative plot of this data is illustrated in Figure 6.8a. Obviously, the rate at which photoelectrons are emitted from the cathode metal is directly proportional to the intensity of light incident on it, other variables being held constant. A change in $\nu$ or C results in the slope of the graph being changed. Another interesting result is observed when the photocurrent $I_e$ is measured while allowing the electromagnetic wave frequency $\nu$ to vary, as depicted in Figure 6.8b. Here, two different cathode materials are used with similar results. There appears to exist in nature a *minimum frequency*, $\nu_0$, for incident electromagnetic

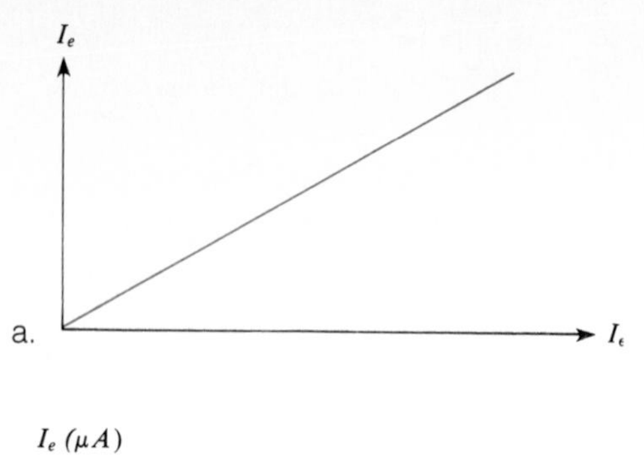


Figure 6.8
*a*, A plot of photocurrent $I_e$ versus $I_\epsilon$ for constant $V$, $\nu$, and $C$. *b*, A plot of photocurrent $I_e$ versus $\nu$ for constant $V$ and $I_\epsilon$.

waves on a particular cathode material, below which no photoelectrons will be produced. This explains why blue light on zinc does not produce photoelectrons while ultraviolet light does. Apparently, for every metal there is a **threshold frequency,** $\nu_0$, necessary to produce a photoelectric effect.

In Figure 6.9a we have a plot of $V_s$ versus $\nu$ and an equally surprising observation. Photoelectrons are produced with their *minimum* (zero) kinetic energy at the *threshold frequency* of light for a particular cathode. For higher frequencies of the incident light the photoelectrons are produced with a correspondingly greater kinetic energy. A plot of $V_s$ versus $I_\epsilon$ is illustrated in Figure 6.9b for two different cathode materials. This is not a surprising result, since it is rather obvious from Figure 6.7b that the *stopping potential* $V_s$ is independent of the wave intensity $I_\epsilon$ of the incident electromagnetic radiation.

## 6.6 Classical and Quantum Explanations of the Photoelectric Effect

Physicists were perplexed by the experimental findings of the photoelectric phenomenon, as they were not able to explain the experimental data by treating the incident electromagnetic radiation as waves. Classical physics considers the *valence* electrons of a cathode metal to exist as *conduction*

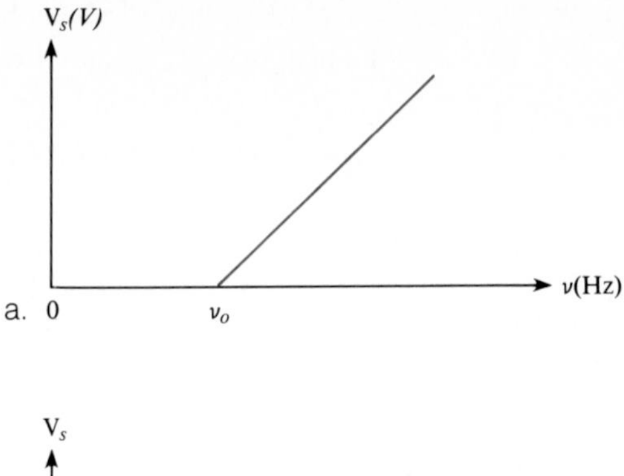


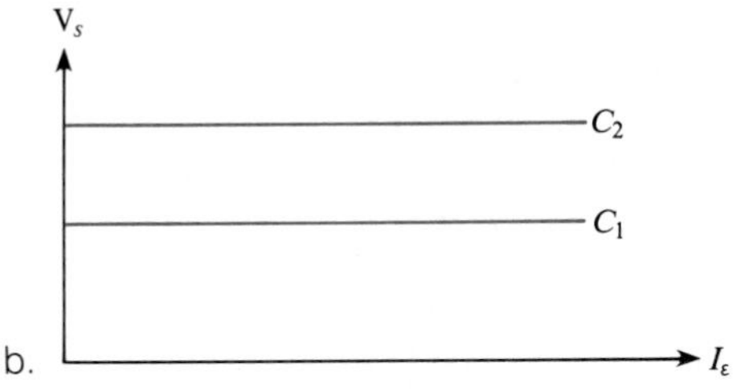


Figure 6.9
*a*, A plot of $V_s$ versus $\nu$ for constant $I_\epsilon$ and *c*. *b*, A plot of stopping potential $V_s$ versus $I_\epsilon$ for constant $\nu$.

*electrons* or, essentially, *free* electrons. These electrons move about the metal in a fairly unrestricted manner and freely respond to any externally imposed electric field. Conduction electrons are weakly *bound* to the metal, however, because of the Coulombic force of attraction existing between the positively charged *ionized* atoms of the metal and each conduction electron. It seems reasonable that electrons (conduction electrons or valence electrons of surface atoms) could be liberated from the cathode metal by absorbing enough energy from incident electromagnetic waves. We would expect an increase in the number of electrons liberated to occur with an increase in the light *intensity*, in agreement with Figure 6.8a. However, the absorption of electromagnetic wave energy by electrons should occur at *any* frequency of incident light, so the existence of a *threshold frequency* in Figure 6.8b is completely contradictory to classical wave theory.

A closer analysis of the photoelectric phenomenon by *classical wave theory* suggests that cathode electrons should oscillate in response to the alternating electric field of the incident electromagnetic waves. The waves should cause the electrons to vibrate with an *amplitude* $A_e$ that is directly proportional to the *maximum amplitude* $E_m$ (see Equation 6.2) of the incident wave,

$$A_e \propto E_m.$$

Further, classical physics predicts the average kinetic energy of the photoelectrons to be directly proportional to the square of their *vibrational amplitude*, that is,

$$T_{\text{avg}} \propto A_e^2.$$

Thus, from these two proportionalities, the average kinetic energy of the photoelectrons is proportional to the square of the maximum amplitude of the incident electromagnetic waves. That is,

$$T_{\text{avg}} \propto E_m^2,$$

but from Equation 6.26

$$I_\epsilon \propto E_m^2,$$

so we have $T_{\text{avg}}$ directly proportional to the *intensity* of the electromagnetic wave,

$$T_{\text{avg}} \propto I_\epsilon.$$

This suggests that for electrons at the surface of the cathode metal

$$T_{\text{avg}} = T_{\max} \propto I_\epsilon$$

and from Equation 6.48 we obtain the relation

Classical Theory

$$V_s \propto I_\epsilon.$$

This resulting proportionality is clearly contradictory to the experimental data illustrated in Figure 6.9b, since the *stopping potential* for the most energetic photoelectrons is *independent* of the incident light *intensity*.

Another failure of classical physics is the prediction of a *lag time* between the activation of the light source and the emission of a photoelectron. We can estimate this lag time by allowing the light source to be a common helium-neon laser, which has a maximum power of

$$P_L = 10^{-3}\ \text{W}$$

and a beam area of approximately

$$A_L = 10^{-4}\ \text{m}^2.$$

The laser intensity is thus

$$I_\epsilon = \frac{P_L}{A_L} = 10\,\frac{\text{W}}{\text{m}^2},$$

which falls on cathode atoms having an approximate radius (see Chapter 5, Section 5.9) of

$$r_a \approx 10^{-10}\ \text{m}.$$

Thus, the wave energy per unit time available to a cathode atom would be

$$\begin{aligned}
P_a &= I_\epsilon A_a \\
&= I_\epsilon(\pi r_a^2) \\
&= \left(10\,\frac{\text{W}}{\text{m}^2}\right)(\pi \cdot 10^{-20}\ \text{m}^2) = \pi \cdot 10^{-19}\,\frac{\text{J}}{\text{s}} \\
&= \frac{\pi \cdot 10^{-19}\ \text{J/s}}{1.6 \times 10^{-19}\ \text{J/eV}} \\
&\approx 2\,\frac{\text{eV}}{\text{s}}.
\end{aligned} \tag{6.49}$$

Consequently, it should take approximately a *second* for a valence electron to absorb enough radiant energy (~ 1 to 2 eV) from our laser for the production of a photoelectron. This result is not consistent with the experimental findings of E. O. Lawrence and J. W. Beams in 1927, which placed an upper limit of $10^{-9}$ s on the *liberation time* of photoelectrons. Our example is indeed generous, as the photoelectric effect for a sodium surface is detectable for violet light intensities on the order of $I_\epsilon \approx 10^{-6}$ W/m$^2$. In this case Equation 6.49 predicts

$$\begin{aligned}
P_a &= \pi \cdot 10^{-26}\ \text{J/s} \\
&\approx 2 \times 10^{-7}\ \text{eV/s},
\end{aligned}$$

or a time for the absorption of 1 eV of energy per atom of

$$\begin{aligned}
\Delta t &= \frac{1\text{eV}}{\text{P}_a} \\
&= (0.5 \times 10^7\ \text{s})\,\frac{1\ \text{hr}}{3600\ \text{s}}\,\frac{1\ \text{day}}{24\ \text{hrs}} \\
&\approx 58\ \text{days}.
\end{aligned}$$

Classical Theory

These rough classical estimates for the photoelectron *liberation time* are clearly inconsistent by many orders of magnitude with the phenomenon observed.

Realizing that the experimental data illustrated in Figures 6.7 through 6.9 could not be understood on the basis of the classical electromagnetic *wave theory*, Albert Einstein proposed an explanation in 1905 for the photoelectric phenomenon that was based on a *quantum hypothesis* suggested earlier by Max Planck in 1900. Planck had been concerned with an explanation of *blackbody radiation* and developed an empirical formula that predicted the intensity of radiation emitted by hot (luminous) bodies as a function of wavelength and the temperature of the body. He sought a theoretical basis for his formula (see Chapter 12, Section 12.6) in terms of an atomic model that considered the atoms of the blackbody to behave like small electromagnetic oscillators. He made a radical departure from classical physics by assuming an atomic oscillator (see Chapter 7, Section 7.6 for a derivation) to have an energy given by

Planck's Quantum Hypothesis

$$E = nh\nu, \tag{6.50}$$

where $\nu$ is the oscillator frequency, $h$ is a constant, and $n$ (now called the **principal quantum number**) is any *one* number of the positive integers (i.e., 1, 2, 3, $\cdots$ ). The constant $h$, originally estimated by Planck by fitting his formula to experimental data, is now called **Planck's constant** and has the value

Planck's Constant

$$h = 6.626176 \times 10^{-34}\ \mathrm{J \cdot s} \tag{6.51}$$

Planck's hypothesis (Equation 6.50) asserts that the energy of an atomic oscillator does not represent a continuum of energy states in accordance with classical physics but, rather, a discrete or *quantized* set of values. He further assumed that the oscillators could not radiate energy continuously but only in *quanta*, as given by

Emission Quanta

$$\Delta E = \Delta n h\nu, \tag{6.52}$$

when the oscillator changes from one *allowed* energy state to another. So long as an oscillator remains in a *quantized* or *stationary* state, energy is neither emitted nor absorbed.

Although Planck recognized the energy emitted by atomic oscillators to be quantized by Equation 6.52, he considered the radiant energy to propagate through space as a continuum of electromagnetic waves. This reasoning is entirely consistent with classical physics and the discussions presented in Section 6.1. Einstein, however, reasoned that if atomic oscillators could neither emit *nor absorb* light energy except in quantized amounts, then this suggests that electromagnetic radiation consists of *quanta of energy*. He postulated that the energy in a light beam propagates through space in concentrated bundles or *quanta*, which are now called **photons,** each having an energy given by

$$\epsilon = h\nu, \tag{6.53}$$

Einstein's Photon Postulate

Here, the Greek letter *epsilon* ($\epsilon$) is used to denote *photon energy* to distinguish it from the oscillator energy $E$. Einstein was able to immediately test his hypothesis by applying the *photon concept* to an explanation of the photoelectric effect. He considered an incident photon to have a perfectly *inelastic* collision with a *bound electron* in the cathode metal, thereby annihilating itself and giving up its energy to the electron. Part of this energy, called the **work function** $W_0$, is consumed in liberating the electron from the metal surface, with the remainder being transformed into the electron's kinetic energy $T_{max}$. By conservation of energy, Einstein obtained

$$\epsilon = W_0 + T_{max}, \tag{6.54}$$

Energy Conservation

where $\epsilon$ is given by Equation 6.53 and $T_{max} = \frac{1}{2}m_e v^2$. For an electron at the metal surface that receives *just* enough energy from a photon to be liberated $T_{max} = 0$, and Equation 6.54 combined with the requirements of Figure 6.8b or Figure 6.9a suggest the *work function* be defined by

$$W_0 \equiv h\nu_0. \tag{6.55}$$

Work Function

Thus, Einstein's conservation of energy equation can be rewritten in the form

$$h\nu = h\nu_0 + T_{max}, \tag{6.56}$$

Photoelectric Equation

which is his famous **photoelectric equation.** Since a *photon* travels at the speed of light $c = \lambda\nu$, this equation can be further modified by substituting $c/\lambda$ for $\nu$ or $c/\lambda_0$ for $\nu_0$, allowing for the solution of several different types of photoelectric problems given at the end of the chapter.

Einstein's equation is consistent with the experimental features of the photoelectric effect. Substitution of Equation 6.48 into Equation 6.56 and solving for $V_s$ gives

$$V_s = \frac{h}{e}(\nu - \nu_0), \tag{6.57}$$

which is in *perfect* agreement with the experimental data of Figure 6.9a. It is also obvious from the equation that $V_s$ is independent of the light intensity $I_\epsilon$ but dependent on the cathode metal *work function* $h\nu_0$, in agreement with Figure 6.9b. Furthermore, the *lag time* between incident light and photoelectron emission is expected to be quite small, due to the annihilation of a photon by an electron. Shortly after Einstein's published work, Robert A. Millikan confirmed the photoelectric equation and measured the value of Planck's constant $h$ from experimental data of the photoelectric phenomenon. His graph was consistent with the equation

$$h = \frac{\frac{1}{2}m_e v^2}{\nu - \nu_0}, \tag{6.58}$$

which is directly obtainable from Equation 6.56 by substitution of $T_{\max} = \frac{1}{2}m_e v^2$.

As an instructive illustration of the photoelectric equation, we will derive an equation for the *change in stopping potential* of photoelectrons emitted from a surface for a *change in* the *wavelength* of the incident light. With

$$e(V_s)_1 = \frac{hc}{\lambda_1} - h\nu_0$$

and

$$e(V_s)_2 = \frac{hc}{\lambda_2} - h\nu_0,$$

where $c = \lambda\nu$ has been used with Equation 6.57, then

$$\Delta V_s \equiv (V_s)_2 - (V_s)_1 = \frac{hc}{e}\left(\frac{1}{\lambda_2} - \frac{1}{\lambda_1}\right). \tag{6.59}$$

Hence, the stopping potential depends on the wavelength (or frequency) of the incident photons and not on the intensity of light, in agreement with our discussion of Figure 6.7b.

Although Einstein's *photon concept* was strikingly successful in explaining photoelectric phenomena, it was in direct conflict with the classical wave theory of electromagnetic radiation. His quantum theory of light is fundamentally different from the wave theory of light in that neither can be approximated nor derived from the other. This case of contradictory theories is even more profound than the case of Einsteinian versus classical relativity, where the latter was seen to be an approximation of the former.

Here, we see that light has a dual property, behaving as a *wave* in interference and diffraction phenomena and as a *particle*, or *photon*, in the photoelectric phenomenon. Our modern view of the nature of light considers both theories as *complementary* to each other and as necessary for a complete description of electromagnetic radiation. This *wave-particle duality* in nature is discussed at length in Chapter 8; however, for now we continue our discussion with another experiment on which the *photon concept* is firmly based.

## 6.7 Quantum Explanation of the Compton Effect

The analysis of the photoelectric effect by Einstein did not tell us that a photon behaves like a *particle* with *localized* properties of mass and momentum. Compelling evidence of these properties was provided by Arthur H. Compton in 1923 when he analyzed the scattering of well defined incident x-rays from a metallic foil. In his investigation Compton found that incident x-rays of wavelength around $7.10 \times 10^{-11}$ m became x-rays of a slightly greater wavelength (about $7.34 \times 10^{-11}$ m) when scattered by the electrons in a foil. In particular, Compton found the scattered wavelengh $\lambda'$ to be greater than and independent of the incident wavelength $\lambda$ of the x-ray photon, but the scattered photon's wavelength $\lambda'$ was strongly dependent on the angle $\theta$ through which it was scattered.

To understand this phenomenon consider the situation depicted in Figure 6.10. Compton considered that the incident x-ray photon has a speed $c$ in accordance with Einstein's second postulate of special relativity and a

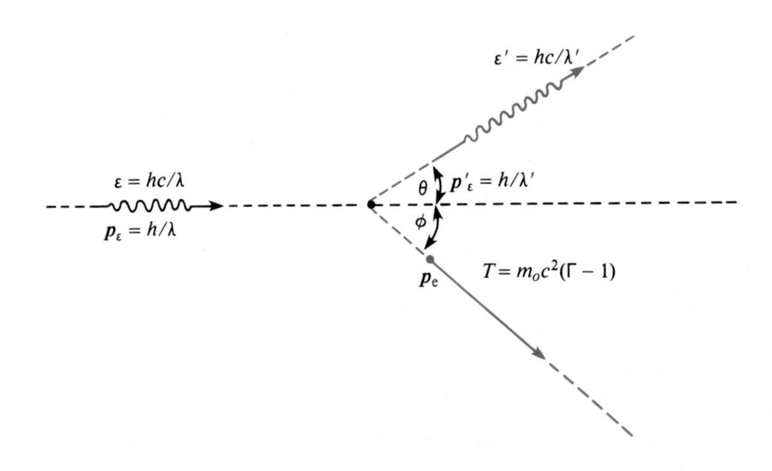


Figure 6.10
The scattering of an x-ray photon by a "bound" electron.

finite energy $\epsilon$ as given by Einstein's photon postulate. He further reasoned the relativistic mass of a photon to be given by

$$m_\epsilon = \Gamma(m_0)_\epsilon, \tag{6.60}$$

where

$$\Gamma \equiv \frac{1}{\sqrt{1 - \frac{v^2}{c^2}}} = \frac{1}{0} \tag{4.23}$$

because $v = c$ for a photon. These equations suggest that the photon's rest mass $(m_0)_\epsilon$ must be zero,

Photon Rest Mass

$$(m_0)_\epsilon = 0, \tag{6.61}$$

and consequently Equation 6.60 is an *indeterminate* expression (i.e., $m_\epsilon = 0/0$). Although previously discussed in Chapter 4, Section 4.4 by an alternative derivation, in no way does this result imply the photon's momentum to be zero. Compton reasoned that if the relativisitic energy of a photon be given by

$$\epsilon^2 = p_\epsilon^2 c^2 + (m_0)_\epsilon^2 c^4, \tag{6.62}$$

then for a zero rest mass its energy is simply

Photon Energy

$$\epsilon = p_\epsilon c, \tag{6.63}$$

which is consistent with Einsteinian relativity (Equation 4.45) and classical electrodynamics (Equation 6.47). Substituting Einstein's photon postulate (Equation 6.53) and $c = \lambda\nu$, Compton obtained

Photon Momentum

$$p_\epsilon = \frac{h}{\lambda} \tag{6.64}$$

for the photon's momentum.

To obtain an expression involving the wavelengths of the incident and scattered photons, $\lambda$ and $\lambda'$ respectively, Compton employed conservation principles. From Figure 6.10 it is easy to write down

$$p_\epsilon - p'_\epsilon \cos\theta = p_e \cos\theta \tag{6.65}$$

for the conservation of momentum along the $X$-axis. Likewise, the conservation of momentum along the $Y$-axis is simply

$$p'_\epsilon \sin\theta = p_e \sin\theta. \tag{6.66}$$

Squaring Equations 6.65 and 6.66 and adding the results gives

$$p_\epsilon^2 + p_\epsilon'^2 - 2p_\epsilon p'_\epsilon \cos\theta = p_e^2, \tag{6.67}$$

Momentum Conservation

which involves two unknowns $p_e$ and $p'_\epsilon$. The momentum of the electron $p_e$ is expressible in terms of the electron's relativistic energy because

$$E^2 = p_e^2c^2 + E_0^2, \tag{6.68}$$

where, of course,

$$E_0 \equiv m_0c^2 \tag{4.39}$$

and $m_0$ is the *rest mass* of the scattered electron. Solving Equation 6.68 for $p_e^2$ yields

$$p_e^2 = \frac{(E - E_0)(E + E_0)}{c^2} \tag{6.69}$$

and using the relativistic equation of kinetic energy,

$$T = E - E_0, \tag{4.41a}$$

Relativistic Kinetic Energy

we obtain

$$p_e^2 = \frac{T(T + 2E_0)}{c^2}. \tag{6.70}$$

Employing Equation 4.39 with Equation 6.70 gives

$$p_e^2 = \frac{T^2}{c^2} + 2m_0T. \tag{6.71}$$

This result could be substituted into Equation 6.67; however, at this point the kinetic energy of the electron is unknown. It can be conveniently expressed in terms of $p_\epsilon$ and $p'_\epsilon$ by employing the conservation of relativistic energy principle. That is,

$$\epsilon + (m_0)_e c^2 = \epsilon' + (m_0)_e c^2 + T, \tag{6.72}$$

which immediately reduces to

$$T = \epsilon - \epsilon' = c(p_\epsilon - p'_\epsilon) \tag{6.73}$$

by utilization of Equation 6.63. At this point, substitution of Equation 6.73 into Equation 6.71 yields another expression for the electron's momentum in terms of the photon momenta $p_\epsilon$ and $p'_\epsilon$, that is,

Energy Conservation

$$p_e^2 = p_\epsilon^2 + p_\epsilon'^2 - 2p_\epsilon p'_\epsilon + 2m_0c(p_\epsilon - p'_\epsilon). \tag{6.74}$$

Now, $p_e^2$ in this equation can be substituted into Equation 6.67 to obtain

$$\frac{p_\epsilon - p'_\epsilon}{p_\epsilon p'_\epsilon} = \frac{1}{m_0 c}(1 - \cos\theta). \tag{6.75}$$

This result can be written in a more amenable form, since

$$\frac{p_\epsilon - p'_1}{p_\epsilon p'_\epsilon} = \frac{1}{p'_\epsilon} - \frac{1}{p_\epsilon} = \frac{\lambda' - \lambda}{h}, \tag{6.76}$$

where Equation 6.64 has been used in obtaining the last equality. The well-known **Compton equation** is now easily obtained by substitution of Equation 6.76 into Equation 6.75. That is,

Compton Equation

$$\lambda' - \lambda = \lambda_C (1 - \cos\theta), \tag{6.77}$$

where the so-called **Compton wavelength,** $\lambda_C$, is defined by

Compton Wavelength

$$\lambda_C \equiv \frac{h}{m_0 c}. \tag{6.78}$$

Equation 6.77 shows that the *increase* in the scattered x-ray wavelength $(\lambda' - \lambda)$ is dependent on only the *scattering angle* $\theta$ and the constant

Compton Wavelength

$$\lambda_C = 2.42631 \times 10^{-12}\ \text{m}. \tag{6.79}$$

This rather surprising theoretical prediction was later verified by many experiments. Compton's theory provides a particularly strong case for the existence of electromagnetic quanta; stronger even than Einstein's analysis of the photoelectric effect.

As an instructive example of the Compton analysis, we will consider whether or not a free electron at rest can annihilate a photon and thereby obtain a nonzero kinetic energy. Assuming the annihilation of the photon to occur, then the conservation of momentum requires that the momentum of the photon *before* the perfectly inelastic collision be equal to the momentum of the electron *after* the collision. That is,

$$p_\epsilon = p_e,$$

which upon multiplication by the speed of light yields

$$\begin{aligned}\epsilon &= p_\epsilon c = p_e c \\ &= \sqrt{E^2 - E_0^2} \\ &= \sqrt{(E - E_0)(E + E_0)} \\ &= \sqrt{T(T + 2E_0)}, \end{aligned} \tag{6.80}$$

where Equations 6.63, 6.68, and 4.41a have been employed. Conservation of relativistic energy requires the energy of the photon plus the *rest energy* of the electron before the collision be equal to the *energy of motion* of the electron after the collision. That is, $\epsilon + E_0 = E$, which immediately becomes

$$\epsilon = T \tag{6.81}$$

from Equation 4.41a. The obvious contradiction between these last two numbered equations indicates that such a process (the annihilation of a photon by a free electron) could never occur in nature.

As a more practical example of the Compton equation, consider an incident photon of wavelength $7.10 \times 10^{-11}$ m colliding with a bound electron in such a manner that a scattered photon occurs at an 88° angle. Clearly, the scattered wavelength is given by

$$\begin{aligned}\lambda' &= \lambda + \lambda_C (1 - \cos\theta) \\ &= 7.10 \times 10^{-11}\ \text{m} + (2.43 \times 10^{-12}\ \text{m})(1 - 0.035) \\ &= 7.34 \times 10^{-11}\ \text{m}.\end{aligned}$$

The energy of the scattered electron is obtainable from Equation 6.72 with the aid of Einstein's photon postulate $\epsilon = h\nu$ and $c = \lambda\nu$, that is

$$\frac{hc}{\lambda} = \frac{hc}{\lambda'} + T. \tag{6.82}$$

With the value of $h$ taken from Equation 6.51 we have

$$hc \approx \frac{(6.63 \times 10^{-34}\ \mathrm{J \cdot s})\ (3.00 \times 10^{8}\ \mathrm{m/s})}{1.60 \times 10^{-19}\ \mathrm{J/eV}},$$

which reduces nicely to

$$hc = 12.4 \times 10^{-7}\ \mathrm{m \cdot eV}. \tag{6.83}$$

Now, Equation 6.82 yields

$$\begin{aligned} T &= hc\left(\frac{1}{\lambda} - \frac{1}{\lambda'}\right) \\ &= (12.4 \times 10^{-7}\ \mathrm{m \cdot eV})\left(\frac{1}{7.10 \times 10^{-11}\ \mathrm{m}} - \frac{1}{7.34 \times 10^{-11}\ \mathrm{m}}\right) \\ &= 571\ \mathrm{eV} \end{aligned}$$

for the *kinetic energy* of the scattered electron. Of course this is rather small when compared to the *electron's rest energy* of

$$E_0 \equiv m_0c^2 = 5.12 \times 10^5\ \mathrm{eV}. \tag{6.84}$$

## 6.8 Relativistic Doppler Effect Revisited

As a last example of the quantization of electromagnetic radiation, we consider the consistency between Einstein's *photon postulate* (Equation 6.53) and his *special theory of relativity*. It seems reasonable that since a photon has energy related to its frequency, then the relativistic frequency transformations (Chapter 3, Section 3.6) should be derivable from the relativistic energy transformations (Chapter 4, Section 4.6). As a particular example, we imagine a source (emitter) of *monochromatic* photons to be at the origin of coordinates of system S′ and an observer (receiver) to be at the origin of S. The photons in S′ have energy denoted by $\epsilon'$, while observers in S denote their energy as $\epsilon$. These two energy quantities should transform according to Equations 4.68 and 4.69, with $E$ and $E'$ being replaced accordingly by $\epsilon$ and $\epsilon'$. That is, Equation 4.68 becomes

$$\epsilon = \gamma(\epsilon' + p_x' u), \tag{6.85}$$

where $p_x'$ is interpreted as the momentum of the photon in the $x$-direction. Since the emitter of photons in S′ must be allowed to approach S from the left, pass S, and then recede from S to the right, then the photons *emission angle* $\theta'$ relative to the $X'$-axis must vary from 0 to $\pi$. Accordingly, $p_x'$ in Equation 6.85 is replaced by

$$p_x' = p_\epsilon' \cos\theta', \tag{6.86}$$

which can be rewritten as

$$p_x' = \frac{\epsilon' \cos\theta'}{c} \tag{6.87}$$

from Equation 6.63 (or Equation 6.47). Thus, Equation 6.85 becomes

$$\begin{aligned} \epsilon &= \gamma\left[\epsilon' + \epsilon' \cos\theta' \left(\frac{u}{c}\right)\right] \\ &= \gamma\epsilon'(1 + \beta\cos\theta'), \end{aligned} \tag{6.88}$$

where Equation 2.5 for $\beta$ has been utilized in the second equality. Now, using Einstein's *photon postulate* $\epsilon = h\nu$ and $\epsilon' = h\nu'$, our last equation immediately becomes

$$\nu = \gamma\nu'(1 + \beta\cos\theta'). \tag{6.89}$$

Doppler Effect

This equation is a general transformation formula for the *relativistic Doppler effect* for emitted photons of *proper* frequency $\nu'$ and their frequency $\nu$ measured by an inertial observer. The inverse of this equation, given by

$$\nu' = \gamma\nu(1 - \beta\cos\theta), \tag{6.90}$$

Inverse Doppler Effect

can be derived by analogous arguments starting with Equation 4.69 for the situation where photons are emitted from system S and detected by an inertial observer in S′.

Because of the angle ($\theta'$ or $\theta$) variable in the Doppler effect equations, three different phenomena are easily predicted. First consider Equation 6.89 and the case where the emitter of photons S′ is approaching the observer in S from the left. In this instance the photons are emitted at the angle $\theta' = 0$, with respect to the positive $X'$-axis, and the frequency detected in S is given by

$$\nu = \gamma\nu'(1 + \beta), \tag{6.91}$$

since $\cos 0° = 1$. Further, since

$$\gamma(1 \pm \beta) = \frac{1}{\gamma(1 \mp \beta)} \tag{6.92}$$

(see Chapter 3, Section 3.6 and Problem 3.18), then Equation 6.91 becomes

Approaching Case

$$\nu = \frac{\nu'}{\gamma(1 - \beta)}, \tag{6.93}$$

which is identical to our previously derived result given by Equation 3.47. When S′ passes S, $\theta' = \pi/2$ and Equation 6.89 gives

Transverse Case

$$\nu = \gamma\nu', \tag{6.94}$$

which is a new *transverse* Doppler effect not previously considered in Chapter 3, Section 3.6. When S′ is receding from S, $\theta' = \pi$ and Equation 6.89 yields

$$\nu = \gamma\nu'(1 - \beta),$$

which immediately reduces to

Receding Case

$$\nu = \frac{\nu'}{\gamma(1 + \beta)} \tag{6.95}$$

by using the identity given in Equation 6.92. Thus, the general Doppler effect equation for photons emitted from S′ to S (Equation 6.89) yields identical results for the *longitudinal* phenomena previously derived in Chapter 3, Section 3.6, and it predicts a *transverse* phenomenon not previously considered. Crucial to the derivation was the assumption that electromagnetic radiation propagates as *quanta of energy*, as defined quantitatively by Einstein's *photon postulate*. Consequently, the *concept of photons* propagating at the speed of light $c$ with energy given by $\epsilon = h\nu$ and momentum $p_\epsilon = \epsilon/c$ is entirely consistent with Einsteinian relativity.

The *inverse* situation, where light is emitted from S and detected in S′, is directly obtained from Equation 6.90. For the three cases we obtain

$$\theta = 0 \rightarrow \nu' = \frac{\nu}{\gamma(1+\beta)}, \quad (6.96)$$ S Approaching S′

$$\theta = \frac{\pi}{2} \rightarrow \nu' = \gamma\nu, \quad (6.97)$$ S Passing S′

$$\theta = \pi \rightarrow \nu' = \frac{\nu}{\gamma(1-\beta)}. \quad (6.98)$$ S Receding S′

Although the first and last results are directly obtained from Equations 6.93 and 6.95 by algebra, the *proper frequency* here is denoted by $\nu$ and not $\nu'$. Further, the *transverse effect* is clearly an example of the *time dilation* phenomenon.

## Review of Fundamental and Derived Equations

A listing of the fundamental equations of classical and relativistic dynamics used in this chapter is presented below, along with newly introduced physical constants and fundamental postulates. Further, equations derived from the wave theory and quantum theory of electromagnetic radiation are separated in logical listings that parallel their development in each section of the chapter.

### FUNDAMENTAL EQUATIONS—CLASSICAL PHYSICS

$\mathbf{F} \equiv \frac{d\mathbf{p}}{dt}$ Newton's Second Law

$dW \equiv \mathbf{F} \cdot d\mathbf{r}$ Infinitesimal Work

$T = \frac{1}{2}mv^2$ Kinetic Energy

$P \equiv \frac{dW}{dt}$ Instantaneous Power

$F_C \equiv k\frac{qQ}{r^2}$ Coulomb's Law

$\mathbf{E} \equiv \frac{\mathbf{F}}{q}$ Electric Field Intensity

| | |
|---|---|
| $V \equiv \frac{W}{q}$ | Electric Potential |
| $\mathbf{B} \equiv \frac{\mathbf{F}}{m'}$ | Magnetic Induction |
| $\mathbf{F}_B = q\mathbf{v} \times \mathbf{B}$ | Lorentz Force Equation |
| $c = \lambda\nu$ | Speed of Light in a Vacuum |
| $n \equiv \frac{c}{v}$ | Index of Refraction |
| $\omega = 2\pi\nu$ | Angular Speed |
| $k \equiv \frac{2\pi}{\lambda}$ | Wave Number |
| $\mathbf{E} = E_m \sin(kx - \omega t)\mathbf{j}$ | Electric Field Vector |
| $\mathbf{B} = B_m \sin(kx - \omega t)\ \mathbf{k}$ | Magnetic Field Vector |
| $\nabla \times \mathbf{E} = -\frac{\partial \mathbf{B}}{\partial t}$ | Maxwell's Equation |
| $\mathbf{S} \equiv \frac{1}{\mu_0}\mathbf{E} \times \mathbf{B}$ | Poynting Vector |
| $S = \frac{1}{A}\frac{d\epsilon}{dt}$ | Instantaneous Intensity |
| $\langle f(t) \rangle \equiv \frac{1}{T}\int_0^T f(t)\, dt$ | Time Average |
| $I_\epsilon = \langle S \rangle$ | Time Average Intensity |

## FUNDAMENTAL EQUATIONS—EINSTEINIAN RELATIVITY

| | |
|---|---|
| $m = \Gamma m_0$ | Relativistic Mass |
| $E_0 \equiv m_0 c^2$ | Rest Energy |
| $E \equiv mc^2$ | Total Energy |
| $T = E - E_0$ | Kinetic Energy |
| $E^2 = E_0^2 + p^2c^2$ | Energy-Momentum Invariant |
| $E = \gamma(E' + p_x' u)$ | Energy Transformation |

## NEW PHYSICAL CONSTANTS

$\mu_0 \equiv 4\pi \times 10^{-7} \dfrac{\mathrm{N}}{\mathrm{A}^2}$ Permeability Constant

$h = 6.62617 \times 10^{-34}\ \mathrm{J \cdot s}$ Planck's Constant

$\lambda_C \equiv \dfrac{h}{m_e c} = 2.42631 \times 10^{-12}\ \mathrm{m}$ Compton Wavelength

## FUNDAMENTAL POSTULATES

### *Atomic Oscillators*

$E = nh\nu$ Planck's Quantum Hypothesis

$\Delta E = \Delta n h\nu$ Emission Quanta

### *Electromagnetic Radiation*

$\epsilon = h\nu$ Einstein's Photon Postulate

## DERIVED EQUATIONS—ELECTROMAGNETIC WAVES

### *Intensity, Pressure, and Power*

$c = \dfrac{\omega}{k}$ Speed of Light versus Wave Number

$\mathbf{E} = \mathbf{B} \times \mathbf{c}$ Plane Electromagnetic Waves

$\dfrac{S}{c} = \dfrac{d\epsilon}{dV}$ Energy Density

$= \dfrac{F_\epsilon}{A}$ Radiation Pressure

$$S = \frac{E^2}{\mu_0 c}$$ Instantaneous Intensity

$$I_\epsilon = \langle S \rangle$$

$$= \frac{E_m^2}{2\mu_0 c}$$ Time Average Intensity

$$\langle P \rangle = I_\epsilon A$$

$$= \frac{E_m^2 A}{2\mu_0 c}$$ Time Average Power

### *Diffraction*

$$2d \sin \theta = n\lambda$$ Bragg's Law

### *Energy and Momentum*

$$\mathbf{F} = q\mathbf{E} + q\mathbf{v} \times \mathbf{B}$$

$$= qv_y B\mathbf{i} + q(E - v_x B)\,\mathbf{j}$$

$$\mathbf{p} = \mathbf{i} \int F_x \, dt + \mathbf{j} \int F_y \, dt = qBy\mathbf{i}$$ Time Average Momentum

$$\epsilon = \int_0^\epsilon dW = qEy$$ Time Average Energy

$$\epsilon = pc$$ Energy-Momentum Relation

## DERIVED EQUATIONS—ELECTROMAGNETIC QUANTA

### *Photoelectric Effect*

$$\epsilon = W_0 + T_{max}$$ Conservation of Energy

$$W_0 \equiv h\nu_0$$ Work Function

$$V_s = \frac{T_{max}}{e}$$ Stopping Potential

$$h\nu = h\nu_0 + T_{max}$$ Photoelectric Equation

### Compton Effect

|  |  |
|---|---|
| $m_\epsilon = \Gamma(m_0)_\epsilon = 0$ | Photon Rest Mass |
| $\epsilon = p_\epsilon c$ | Photon Total Energy |
| $p_\epsilon = \dfrac{h}{\lambda}$ | Photon Momentum |
| $E_0 = m_0 c^2$ | Electron Rest Energy |
| $E^2 = p_e^2 c^2 + E_0^2$ | Electron Total Energy |
| $T = E - E_0$ | Electron Kinetic Energy |
| $p_\epsilon^2 + p_\epsilon'^2 - 2p_\epsilon p_\epsilon' \cos\theta = p_e^2$ | Momentum Conservation |
| $\epsilon + (m_0)_e c^2 = \epsilon' + (m_0)_e c^2 + T$ | Energy Conservation |
| $p_\epsilon^2 + p_\epsilon'^2 - 2p_\epsilon p_\epsilon' + 2m_0 c(p_\epsilon - p_\epsilon') = p_e^2$ | Energy Conservation |
| $\lambda' - \lambda = \lambda_C(1 - \cos\theta)$ | Compton Equation |

### Doppler Effect $S' \rightarrow S$

|  |  |
|---|---|
| $\epsilon = \gamma(\epsilon' + p_x' u)$ | |
| $= \gamma(\epsilon' + \beta\epsilon' \cos\theta)$ | Energy Transformation |
| $\nu = \gamma\nu'(1 + \beta\cos\theta')$ | Frequency Transformation |
| $\nu = \dfrac{\nu'}{\gamma(1 - \beta)}$ | Approaching Case |
| $\nu = \gamma\nu'$ | Transverse Case |
| $\nu = \dfrac{\nu'}{\gamma(1 + \beta)}$ | Receding Case |

## Problems

**6.1** An observer is 1.0 m from a point source of light whose power output is 1.5 kW. What are the maximum amplitudes of the electric and magnetic field vectors at that point, if the source radiates monochromatic plane waves uniformly in all directions?

*Solution:*
With $r = 1.0$ m and $P = 1.5 \times 10^3$ W, $E_m$ is obtainable from Equation 6.27,

$$\langle P \rangle = \frac{E_m^2 A}{2\mu_0 c},$$

with $A$ being replaced by the *area of a sphere* of radius $r$,

$$A = 4\pi r^2.$$

Thus, we find the maximum amplitude of the electric field vector to be

$$E_m = \frac{1}{r}\left(\frac{\mu_0 c P}{2\pi}\right)^{1/2}$$

$$= \left(\frac{1}{1.0\text{m}}\right)\left[\frac{4\pi \times 10^{-7}\ \text{N/A}^2)(3 \times 10^8\ \text{m/s})(1.5 \times 10^3\ \text{W})}{2\pi}\right]^{1/2}$$

$$= \left(\frac{1}{1.0\text{m}}\right)\left(3 \times 10^2 \frac{\text{N}\cdot\text{m}}{\text{A}\cdot\text{s}}\right)$$

$$= 3 \times 10^2 \frac{\text{V}}{\text{m}}.$$

From knowledge of $E_m$, the maximum amplitude of the magnetic field vector is easily obtained by using the relation $E_m = cB_m$ (Equation 6.10). That is,

$$B_m = \frac{E_m}{c} = \frac{3 \times 10^2\ \text{V/m}}{3 \times 10^8\ \text{m/s}} = 10^{-6}\ \text{T}.$$

**6.2** If a NaCl crystal is irradiated by x-rays of 0.250 nm wavelength and the first Bragg *reflection* is observed at 26.4°, what is the atomic spacing of the crystal?

*Answer:* $d = 2.81 \times 10^{-10}$ m

**6.3** A laser beam of energy flux 60 W/m$^2$ falls on a square metal surface of edge 10 mm for one hour. Assuming total absorption of the beam by the metal, find the momentum delivered to the metal surface during the irradiation time.

*Solution:*
Knowing $S = 60\ \text{W/m}^2$, $A = (10\ \text{mm})^2$, and $\Delta t = 3.6 \times 10^3$ s, the total momentum $p_t$ of the laser beam transferred to the metal surface is similar to Equation 6.47,

$$p = \frac{\epsilon}{c}.$$

The total radiation energy $W_\epsilon$ is calculated using

$$W_\epsilon = SA\Delta t,$$

thus we have for the total momentum

$$\begin{aligned} p_t &= \frac{SA\,\Delta t}{c} \\ &= \frac{(60\ \text{W/m}^2)(10\ \text{mm})^2(3.6 \times 10^3\ \text{s})}{3 \times 10^8\ \text{m/s}} \\ &= 7.2 \times 10^{-8}\ \frac{\text{kg} \cdot \text{m}}{\text{s}}. \end{aligned}$$

**6.4** Photons of $2 \times 10^{-27}$ kg · m/s momentum are incident normally to a 10 cm$^2$ surface. If the intensity of the photons is $30 \times 10^{-2}$ W/m$^2$, how many photons strike the surface per second?

*Answer:* $\dfrac{N}{t} = 5 \times 10^{14}\ \dfrac{\text{photons}}{\text{second}}$

**6.5** If a radio station operates at 110 MHz with a power output of 300 kW, what is the *rate of emission* of photons from the station?

*Solution:*
With $P = 3 \times 10^5$ W and $\nu = 1.10 \times 10^8$ Hz, a *dimensional analysis* gives

$$\begin{aligned} \frac{N}{t} &= \frac{N\epsilon}{t\epsilon} = \frac{W_\epsilon}{t\epsilon} = \frac{P}{\epsilon} = \frac{P}{h\nu} \\ &= \frac{3 \times 10^5\ \text{J/s}}{(6.63 \times 10^{-34}\ \text{J} \cdot \text{s})(1.10 \times 10^8/\text{s})} \\ &= 4.11 \times 10^{30}\ \frac{\text{photons}}{\text{second}}, \end{aligned}$$

where Einstein's photon postulate has been used.

**6.6** Find the wavelength of the photon that will just liberate an electron of 20 eV *binding energy* in a cathode metal.

*Answer:* $\lambda = 6.20 \times 10^{-8}$ m

**6.7** What is the energy in eV and momentum in kg · m/s of ultraviolet photons of wavelength 310 nm?

*Solution:*
With $\lambda = 3.1 \times 10^{-7}$ m and Einstein's photon postulate,

$$\epsilon = h\nu = \frac{hc}{\lambda},$$

we obtain

$$\epsilon = \frac{12.4 \times 10^{-7}\ \text{m} \cdot \text{eV}}{3.1 \times 10^{-7}\ \text{m}}$$
$$= 4\ \text{eV},$$

where $hc$ given by Equation 6.83 has been used. From this result and the relation $\epsilon = pc$, the momentum of the ultraviolet photon is

$$p_\epsilon = \frac{\epsilon}{c}$$
$$= \frac{4\ \text{eV}}{3 \times 10^8\ \text{m/s}}\left(1.6 \times 10^{-19}\ \frac{\text{J}}{\text{eV}}\right)$$
$$= 2.13 \times 10^{-27}\ \frac{\text{kg} \cdot \text{m}}{\text{s}}.$$

Alternatively, since

$$\epsilon = p_\epsilon c = \frac{hc}{\lambda},$$

we could have used (see Equation 6.64)

$$p_\epsilon = \frac{h}{\lambda}$$
$$= \frac{6.63 \times 10^{-34}\ \text{J} \cdot \text{s}}{3.10 \times 10^{-7}\ \text{m}}$$
$$= 2.14 \times 10^{-27}\ \frac{\text{kg} \cdot \text{m}}{\text{s}}.$$

The difference in these two answers results from rounding-off errors in our conversion from electron-volts to Joules.

**6.8** Assuming the kinetic energy of photoelectrons to be negligibly small, find the threshold frequency for the production of photoelectrons emitted by incident light of $6 \times 10^{-7}$ m wavelength.

*Answer:* $\nu_0 = 5 \times 10^{14}$ Hz

**6.9** Assuming the work function of sodium to be negligibly small, what is the velocity of photoelectrons resulting from incident light of $3 \times 10^{-8}$ m wavelength?

*Solution:*
We know $W_0 = 0$ and $\lambda = 3 \times 10^{-8}$ m, and we need to find $v_e$. Since

$$\epsilon = W_0 + T_{\text{max}},$$

then with $\epsilon = hc/\lambda$ and $T_{\text{max}} = \frac{1}{2}mv_e^2$, we have

$$\frac{hc}{\lambda} = 0 + \tfrac{1}{2}mv_e^2,$$

which can be solved for $v_e$ to obtain

$$\begin{aligned} v_e &= \left(\frac{2hc}{m\lambda}\right)^{1/2} \\ &= \left[\frac{2(6.63 \times 10^{-34})(3 \times 10^8)}{(9.11 \times 10^{-31})(3 \times 10^{-8})}\right]^{1/2} \\ &= 3.82 \times 10^6 \frac{\text{m}}{\text{s}}. \end{aligned}$$

**6.10** Incident light on a cathode metal has a wavelength of $2.00 \times 10^{-7}$ m. If the kinetic energy of the photoelectrons produced range from zero to $6 \times 10^{-19}$ J, find the stopping potential for the incident light and the threshold wavelength for the metal.

*Answer:* $V_s = 3.75$ V, $\lambda_0 = 5.06 \times 10^{-7}$ m

**6.11** If the largest wavelength for photoelectron emission from potassium is 5000 Å, what is the maximum kinetic energy of photoelectrons produced by illumination of 2000 Å light?

*Solution:*
In this problem we know $\lambda_0 = 5 \times 10^{-7}$ m, $\lambda = 2 \times 10^{-7}$ m, and need to find $T_{max}$. Using Einstein's photoelectric equation we have

$$\begin{aligned} T_{max} &= h\nu - h\nu_0 \\ &= hc\left(\frac{1}{\lambda} - \frac{1}{\lambda_0}\right) \\ &= (12.4 \times 10^{-7}\ \text{m} \cdot \text{eV})(5 \times 10^6\ \text{m}^{-1} - 2 \times 10^6\ \text{m}^{-1}) \\ &= 3.72\ \text{eV}. \end{aligned}$$

**6.12** The threshold frequency of beryllium is $9.4 \times 10^{14}$ Hz. Assume light of wavelength $\tfrac{1}{2}\lambda_0$ illuminates beryllium, what is the maximum kinetic energy in electron volts of emitted photoelectrons?

*Answer:* $T_{max} = 3.90$ eV

**6.13** What is the threshold wavelength for a cathode material of 3.75 V stopping potential, if the incident light has a momentum of $3.315 \times 10^{-27}$ kg · m/s?

*Solution:*
Knowing $V_s = 3.75$ V and $p_\epsilon = 3.315 \times 10^{-27}$ kg · m/s, then an equation for $\lambda_0$ in terms of $V_s$ and $p_\epsilon$ must be derived. From Einstein's photoelectric equation

$$\epsilon = W_0 + T_{max}$$

we have

$$p_\epsilon c = \frac{hc}{\lambda_0} + eV_s,$$

which immediately yields

$$\lambda_0 = \frac{hc}{p_\epsilon c - eV_s}$$

$$= \frac{(6.63 \times 10^{-34}\ \text{J} \cdot \text{s})(3 \times 10^{8}\text{m/s})}{(3.315 \times 10^{-27}\ \text{kg} \cdot \text{m/s})(3 \times 10^{8}\ \text{m/s}) - (1.6 \times 10^{-19}\ \text{C})(3.75\ \text{V})}$$

**6.14** What is the maximum speed of a photoelectron resulting from an incident photon of momentum $3.31 \times 10^{-27}$ kg · m/s, if the threshold wavelength is $5.06 \times 10^{-7}$ m?

*Answer:* $v_e = 1.15 \times 10^6 \dfrac{\text{m}}{\text{s}}$

**6.15** If the scattering angle is 90°, what is the *increase* in the scattered photon's wavelength in a Compton experiment?

*Solution:*
The wavelength $\lambda'$ of the scattered photon is *greater* than that of the incident photon, as given by the Compton equation. That is,

$$\begin{aligned} \lambda' - \lambda &= \lambda_C(1 - \cos\theta) \\ &= (2.43 \times 10^{-12}\ \text{m})(1 - 0) \\ &= 2.43 \times 10^{-12}\ \text{m}. \end{aligned}$$

**6.16** If a 0.2 Å x-ray photon in a Compton experiment is scattered through an angle of 60°, what is the *fractional change* $(\lambda' - \lambda)/\lambda$ in the wavelength?

*Answer:* $\dfrac{\lambda' - \lambda}{\lambda} = 0.0608$

**6.17** If a Compton electron attains a kinetic energy of 0.024 MeV when an incident x-ray photon of energy 0.124 MeV strikes it, then what is the wavelength of the scattered photon?

*Solution:*
With $T = 2.4 \times 10^4$ eV and $\epsilon = 1.24 \times 10^5$ eV, $\lambda'$ can be obtained from the conservation of energy requirement for the Compton experiment. That is, from Equation 6.72,

$$\epsilon = \epsilon' + T,$$

with the substitution

$$\epsilon' = p'c = \frac{hc}{\lambda'},$$

we obtain

$$\begin{aligned}\lambda' &= \frac{hc}{\epsilon - T} \\ &= \frac{12.4 \times 10^{-7}\ \text{m} \cdot \text{eV}}{10^5\ \text{eV}} \\ &= 1.24 \times 10^{-11}\ \text{m}.\end{aligned}$$

**6.18** What is the increase in the scattered wavelength of an x-ray photon, if the Compton electron attains a kinetic energy of 0.024 MeV from an incident photon of 0.124 MeV energy?

*Answer:* $\lambda' - \lambda = 2.4 \times 10^{-12}$ m

**6.19** What angle does the scattered photon of Problem 6.17 make with respect to the direction of the incident photon?

*Solution:*
Knowing $\epsilon = 1.24 \times 10^5$ eV and $\lambda' = 1.24 \times 10^{-11}$ m, then $\theta$ can be obtained from the Compton equation. That is,

$$\cos\theta = 1 - \frac{\lambda' - \lambda}{\lambda_C},$$

where $\lambda$ is given by

$$\begin{aligned}\lambda &= \frac{hc}{\epsilon} \\ &= \frac{12.4 \times 10^{-7}\ \text{m} \cdot \text{eV}}{1.24 \times 10^5\ \text{eV}} \\ &= 10^{-11}\ \text{m}.\end{aligned}$$

Direct substitution gives

$$\cos\theta = 1 - \frac{0.24 \times 10^{-11}\ \text{m}}{2.43 \times 10^{-12}\ \text{m}}$$

$$= 1 - 0.988 = 0.012,$$

thus $\theta \approx 89.3°$.

**6.20** In a Compton scattering experiment let $T = 0.04$ MeV and $\epsilon = 0.2$ MeV. Find the angle that the scattered photon makes with respect to the direction of the incident photon.

*Answer:* $\theta = 68.8°$

**6.21** Prove that a free electron moving at a relativistic speed $v$ cannot emit a photon of energy $\epsilon$ and continue at a slower speed $v'$.

*Solution:*

From the conservation of momentum principle we have

$$p_e = p'_e + p_\epsilon,$$

which can be written as

$$mv = m'v' + \frac{\epsilon}{c}.$$

Using Einstein's relativistic mass equation, this equation becomes

$$\Gamma m_0 v = \Gamma' m_0 v' + \frac{\epsilon}{c},$$

which can be solved for the energy of the emitted photon in the form

$$\epsilon = (\Gamma v - \Gamma' v') m_0 c$$

or

$$\epsilon = \frac{E_0}{c}(\Gamma v - \Gamma' v').$$

However, the conservation of energy requires that

$$E = E' + \epsilon,$$

which can be rewritten as

$$\Gamma E_0 = \Gamma' E_0 + \epsilon$$

and solved for $\epsilon$ to obtain

$$\epsilon = E_0(\Gamma - \Gamma').$$

Because of the difference between these two equations for $\epsilon$, the process will never occur in nature.

**6.22** Consider an electron moving at a speed $v$ to annihilate a photon of energy $\epsilon$ and continue moving with an increased speed of $v'$. Show that conservation of energy predicts $\epsilon = E_0(\Gamma' - \Gamma)$, while conservation of momentum gives $\epsilon = (E_0/c)(\Gamma' v' - \Gamma v)$.

CHAPTER 7

# Quantization of One-Electron Atoms

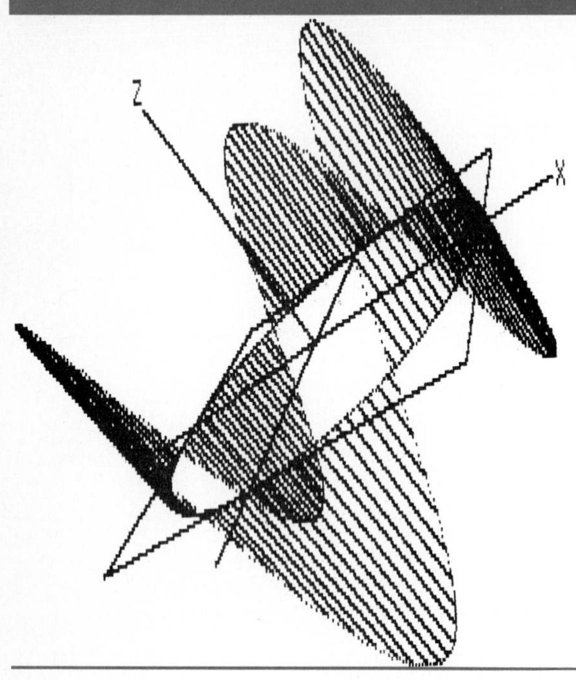


In any molecular system consisting of positive nuclei and electrons in which the nuclei are at rest relative to each other and the electrons move in circular orbits, the angular momentum of every electron round the centre of its orbit will in the permanent state of the system be equal to $h/2\pi$, where $h$ is Planck's constant.

N. BOHR, "On the Constitution of Atoms and Molecules," *Philosophical Magazine* 26, 1 (1913)

## Introduction

Early in the twentieth century much was known about the interrelationship between, and the quantization of, matter and electromagnetic radiation. There was the need, however, for a descriptive quantitative model of the atom that would properly account for the many physical properties (e.g., electrical neutrality, energy quantization, etc.) that had been experimentally determined. At this time, the chemical elements were widely acknowledged as consisting of electrically neutral atoms, but very little was known about the actual structure of atoms. As we have discussed, the first insight into atomic structure was provided by J. J. Thomson's discovery of the electron in 1897 and the subsequent determination of its quantized properties of electrical charge and rest mass by R. A. Millikan in 1909. Further, the discovery of canal rays and the development of the mass spectrograph by J. J. Thomson in 1911 clearly established atoms as possessing a positively charged constituent, later called the proton, that was 1836 times more massive than an electron. Since the proton was found to possess an electrically positive charge that is equal in magnitude to the charge of an electron, it was logical to assume electrons and protons were fundamental constituents of all electrically neutral atoms comprising the chemical elements. This assumption was a reasonable inference from a number of different experiments, including those discussed in the last chapter pertaining to the photoelectric and Compton effects. There, however, we discussed the nonclassical property of atoms to exist in quantized energy states that allow for the emission and absorption of quanta of electromagnetic radiation. Thus, it appears that quantization principles and electrical neutrality must be incorporated in any complete physical description of the structure of the atom.

The modern model of the atom presented in Chapter 5, Section 5.7 was not immediately obvious to scientists of the early twentieth century. In fact, considerable difficulty was encountered when attempts were made to theoretically describe the existence of both electrons and protons in a stable atom by purely classical arguments. J. J. Thomson proposed a *plum pudding* atomic model as early as 1898, where the atom was regarded as a heavy sphere of uniformly distributed positive charge (the pudding) with enough electrons (the plums) embedded to make it electrically neutral. This model was found to be inconsistent with scattering experiments conducted by E. Rutherford in 1911, where a beam of high energy alpha particles (helium nuclei) were used to bombard a thin gold foil. Expecting most of the alpha particles to pass directly through the foil and a few passing near or through a Thomson atom to be only slightly deflected from their original rectilinear path, Rutherford was amazed to find particles deflected through large angles and a few to actually be *reflected*. From the experimental results, Rutherford proposed a *planetary model* of the atom, where he considered the rather massive and positively charged protons to be concentrated in a very small nucleus at the center of the atom. To account for electrical neutrality, he further considered the nucleus to be surrounded, at a relatively large distance away, by a cloud of the appropiate number of electrons. This model satisfactorily explained the scattering of alpha particles, since the atoms of the foil, consisting primarily of empty space, would allow the majority of particles to pass through undeflected. A close encounter of an alpha particle with a nucleus, however, would result in the particle experiencing a large repulsive Coulombic force and being deflected through a large angle. Obviously, a collision with an electron would result in the electron being appreciably deflected, owing to its comparatively very small mass.

Although the Rutherford model of the atom met with initial success in the explanation of experimental data, it was quickly demonstrated to be incapable of explaining the long term stability of the atom's constituents. Further, a number of classical theories for atomic structure were also developed around the turn of the century, but they could not adequately predict the observed spectrum of electromagnetic radiation emitted by atoms. Accordingly, we begin this chapter with a review of the well-known energy spectrum of the hydrogen atom and the empirical equations used to predict its *line emission spectra*. Then an analysis of the *planetary model* for a one-electron atom (e.g., hydrogen, singly ionized helium, doubly ionized lithium, etc.) using arguments of classical physics is considered. The success of the model within Newtonian mechanics is fully discussed, along with its failure in connection with electromagnetic theory. We will find that an electron orbiting a nucleus in an assumed circular orbit is classically expected to have an energy *spectrum* consisting of a continuum of frequencies, as it spirals inward toward the nucleus. This theoretical prediction of electrodynamics is contrary to the spectrum of discrete frequencies observed for atoms. Next, Bohr's postulates for the one-electron atom are introduced and quantitatively developed, followed by a discussion of their successful explanation of the radius, orbital frequency, and energy of the hydrogen electron and the observed energy spectrum of the hydrogen atom. Since our initial treatment of the Bohr model assumes a stationary nucleus (i.e., an infinite nuclear mass as compared to the mass of the electron) with the electron moving in a *stable* circular orbit, the effect of a *finite nuclear mass* on the Bohr model is taken into account. We then consider the *Wilson-Sommerfeld quantization rule* and its generalized applicability to systems exhibiting periodic motion, including the Bohr electron and the classical linear harmonic oscillator. These quantization calculations are followed by a discussion of the principal and orbital quantum numbers associated with each electron in an atom, the Bohr-Sommerfeld scheme for denoting the electron configuration of atoms in the periodic table, the magnetic and spin quantum numbers, and the *Pauli exclusion principle*.

## 7.1 Atomic Spectra

A successful atomic model should be capable of explaining not only the long term stability of an electrically neutral atom consisting of electrons and protons, but also the observed spectrum of emitted electromagnetic radiation. To better understand the latter requirement, we will now consider some of the more salient features of atomic spectra in general and the hydrogen spectrum in particular. A spectrum is simply an orderly array of the wavelengths of light, described by either a continuum or a discrete set of sequential wavelengths. As illustrated in Figure 7.1, a spectrum may be produced by collimating rays of light from a *source* by a slit (or a lens) and then allowing the collimated light to pass through a prism (or diffraction grating), where it is broken up into its spectrum, and finally recorded on a photographic plate. Spectra may be classified as either **emission** or **absorption,** depending on whether they are created by the *emission* of photons from a system of atoms constituting the *source*, or by the *absorption* of incident photons on a *system of atoms* resulting in the *unabsorbed photons* constituting the *source*. An *emission* or *absorption spectrum* can be further subdivided into **continuous** or **line spectra,** depending on whether the light recorded on the photographic plate appears as a *continuum* of wavelengths or as a *discrete set* of wavelengths characterized normally by lines. **Continuous emission spectra** arise from the photons emitted by *hot solids* serving as the *source*, while **line emission spectra** arise from the photons emitted by a *hot rarefied gas* serving as the *source*.

Because of its simplicity, the *line emission spectrum* of hydrogen gas was carefully studied by physicists before the turn of the century. At that time, the wavelength of the first nine spectral lines were known accurately from spectroscopic measurements. The six spectral lines in the visible region of the electromagnetic spectrum are given in Table 7.1.

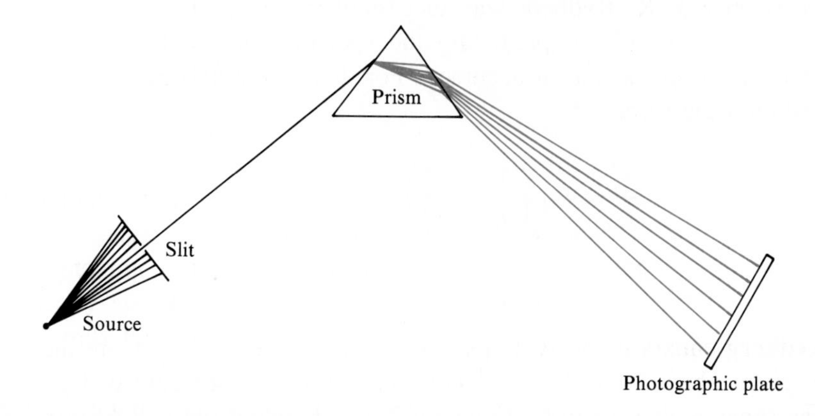


Figure 7.1
The basic components of a prism spectrograph.

TABLE 7.1
The visible spectr
lines of hydrogen.

alize the model slightly, we also assume the nucleus to have a positive charge of

Nuclear Charge

$$q_N = Ze, \tag{7.4}$$

where $Z$ is the *atomic number* and $e$ is the absolute magnitude of the charge of an electron. This allows our model to describe the hydrogen atom ($Z = 1$), the singly ionized helium atom ($Z = 2$), the doubly ionized lithium atom ($Z = 3$), and so forth for *one-electron* atoms. Since the mass of a proton is considerably greater than the mass of an electron (i.e., $m_p = 1836m_e$), then to a first order approximation we can assume the nuclear mass $M$ ($M \gg m$) to be stationary in space, as illustrated by Figure 7.2. Our model is thus one of a single electron of mass $m$ and charge $-e$ traversing a *stationary* nucleus of mass $M$ and charge $Ze$ in a circular orbit.

In the one-electron model, the accelerating force of the electron is provided by its Coulombic attraction to the nucleus and given by

Balmer's F

$$\begin{aligned} F_C &= \frac{kq_Nq_e}{r^2} \\ &= \frac{Zke^2}{r^2}, \end{aligned} \tag{7.5}$$

where $r$ is the radius of the electron's circular path. Since this *inward directed* force is perpendicular to the electron's velocity vector at every point in its path, the force is recognized as being a centripetal force

Centripetal Force

$$F_c = \frac{mv^2}{r}, \tag{5.8}$$

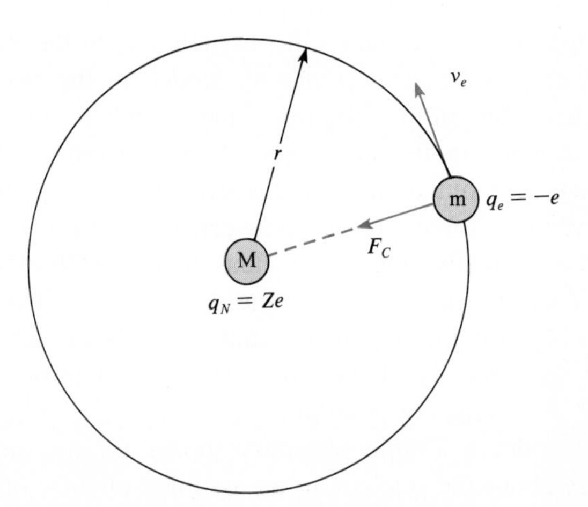


Figure 7.2
The Rutherford "planetary model" of the one-electron atom.

Rydberg F

Rydberg C

where $v$ is the uniform speed of the electron. The *mechanical* or *orbital* stability of the electron is given by equating the centripetal and Coulombic forces,

$$\frac{mv^2}{r} = \frac{Zke^2}{r^2}, \tag{7.6}$$

Orbital Stability

which immediately yelds

$$v = \left(\frac{Zke^2}{mr}\right)^{1/2} \tag{7.7}$$

Electron Velocity

for the velocity of the electron in terms of its orbital radius. With the electron's translational speed related to its angular speed by

$$v = r\omega, \tag{7.8}$$

then from Equation 6.5 ($\omega = 2\pi\nu$) and Equation 7.7 we obtain

$$\nu = \frac{v}{2\pi r}$$

$$= \frac{1}{2\pi}\left(\frac{Zke^2}{mr^3}\right)^{1/2} \tag{7.9}$$

for the classical *orbital frequency* of the electron. Using $Z = 1$ for hydrogen and the value $r = 5.29 \times 10^{-11}$ m obtained below, along with the known values of $k$, $e$, and $m$, Equation 7.9 yields $\nu = 6.58 \times 10^{15}$ Hz. This result for the number of revolutions per second made by a hydrogen electron in orbit about a single proton agrees well with the orbital frequency determined by other methods.

The value used above for the atomic radius of the electron's circular orbit can be estimated from energy considerations. Since the nucleus of the one-electron model is considered to be stationary, then the total energy $E_t$ of the two-body system is given by

$$E_t = E_k + E_p, \tag{7.10}$$

which consists of the electron's kinetic energy $E_k$

$$E_k = \tfrac{1}{2}mv^2, \tag{7.11}$$

and electrostatic potential energy $E_p$. The potential energy of the electron in the electrostatic field of the nucleus is obtained by calculating the work done on the system in removing the electron from position $r$, relative to the nucleus, to infinity. From the definition of work (Equation 1.20) and Equation 7.5 we have

$$E_p = \int_r^\infty \mathbf{F}_C \cdot d\mathbf{r}$$
$$= -Zke^2 \int_r^\infty r^{-2}\, dr, \tag{7.12}$$

where the negative sign indicates an *attractive* Coulombic force between the electron and nucleus. Integration of Equation 7.12 yields

$$E_p = -\frac{Zke^2}{r}, \tag{7.13}$$

which can be substituted along with Equation 7.11 into Equation 7.10 to obtain

$$E_t = \tfrac{1}{2}mv^2 - \frac{Zke^2}{r} \tag{7.14}$$

for the total energy of the electron. With

$$mv^2 = \frac{Zke^2}{r}$$

from Equation 7.6, our expression for the total energy can be reduced in form and solved for $r$ to obtain

$$r = -\frac{Zke^2}{2E_t}. \tag{7.15}$$

From experimental findings the energy required to *ionize* a hydrogen atom is 13.6 eV. Thus, the electron's *binding energy* must be $E_t = -13.6$ eV, since $E_t$ in Equation 7.15 is *negative* valued. It should be noted that if the electron had zero or positive valued energy, it would not exist in a bound stable orbit about the nucleus. Substitution of $Z = 1$ and $E_t = (-13.6 \text{ eV})(1.60 \times 10^{-19} \text{ J/eV}) = -12.18 \times 10^{-18}$ J, along with the values for $k$ and $e^2$, into Equation 7.15 yields $r = 5.29 \times 10^{-11}$ m. This value for the radius of the electron's circular orbit is in good agreement with estimates made by other experimental techniques.

In spite of the success of the Rutherford model in explaining alpha particle scattering and in predicting the orbital radius and frequency of the hydrogen electron, it was found to be in conflict with the predictions of classical electromagnetic theory. Radiation theory of classical physics predicts the energy radiated per unit time, by a charge $e$ experiencing an acceleration $a$, to be given by

$$P = \frac{2}{3}\frac{ke^2a^2}{c^3}. \tag{7.16}$$

Electron Radiation

The electron of the Rutherford model undergoes a centripetal acceleration given by

$$a = \frac{v^2}{r}, \tag{7.17}$$

Centripetal Acceleration

which takes the form

$$a = \frac{Zke^2}{mr^2} \tag{7.18}$$

by substitution from Equation 7.6 for $v^2$. Thus, substitution of this result into Equation 7.16 gives

$$P = \frac{2}{3}\frac{Z^2k^3e^6}{m^2c^3r^4}, \tag{7.19}$$

which for a hydrogen electron in a circular orbit of radius $r = 5.29 \times 10^{-11}$ m yields $P = 4.63 \times 10^{-8}$ J/s $= 2.89 \times 10^{11}$ eV/s. Since the orbital frequency of the hydrogen electron given by Equation 7.9 is $\nu = 6.58 \times 10^{15}\ \text{s}^{-1}$, the *period* for one complete orbit is just $T = 1/\nu = 1.52 \times 10^{-16}$ s. Now, multiplication of $P$ in Equation 7.19 by the electron's period gives $PT = 4.39 \times 10^{-5}$ eV, which is the amount of energy radiated by the electron in just one complete orbit about the hydrogen proton. This means that $E_t$ in Equation 7.15 becomes more negative and $r$ necessarily becomes smaller. As $r$ becomes smaller, the time rate at which the electron radiates energy increases markedly, since $P$ in Equation 7.19 is inversely proportional to $r^4$. Thus, according to classical electromagnetic theory applied to the Rutherford model, the electron cannot exist in a stable circular orbit, but rather spirals in toward the nucleus as it rapidly radiates more and more energy. In the creation of a hydrogen atom, for example, an electron of zero free energy would spiral within a distance of $r = 5.29 \times 10^{-11}$ m in a time roughly given by

$$\begin{aligned} t_s &\approx \frac{E_t}{P} \\ &= \frac{13.6 \text{ eV}}{2.89 \times 10^{11} \text{ eV/s}} \\ &= 4.71 \times 10^{-11} \text{ s.} \end{aligned}$$

From this distance to the proton, the electron would spiral in even more rapidly, since $P$ very quickly increases for any decrease in the value of $r$. Classical estimates predict the electron to collapse on the proton within about $10^{-16}$ s after the formation of a hydrogen atom.

According to classical physics the Rutherford model had a fatal flaw in predicting the long term stability of atoms. All atoms of the chemical elements should collapse in a *very* short time after their formation, and the energy spectra of all atoms should be a continuum, owing to the continuous radiation of energy by spiralling electrons. Both conclusions, however, were contradicted by experimental data that negated the classical planetary model as a viable explanation of atomic structure. It was revived, however, two years later when Niels Bohr combined the essentials of the model, a very small and massive nucleus surrounded at some distance by electrons, with Planck's *quantum hypothesis* for a simple yet brilliant description of atomic structure. Bohr's postulates, his break with the classical theory of radiation, and the amazing success of his model will be the next subject of inquiry.

## 7.3 Bohr Model of the One-Electron Atom

Immediately after obtaining his doctorate degree in Copenhagen in 1911, Niels Bohr was involved in post-graduate research under J. J. Thomson, which was followed by additional studies under Ernest Rutherford. He returned to Copenhagen in 1913 to develop and publish his famous theory on the atomic structure of the hydrogen atom. Being aware of Planck's quantum hypothesis, the predictability of the observed hydrogen spectrum by empirical equations, and the limited success of the Rutherford model of the atom, Bohr recognized that a successful theoretical model of the hydrogen atom had to depart, somewhat, from classical physics. He conceived a remarkable set of postulates for atomic structure that retained the laws of classical mechanics and abandoned the classical theory of radiation. Although the success of the Bohr model was immediate and most impressive, it will be seen to be limited in applicability and seriously inadequate as a generalized model for atomic structure. Its subsequent displacement within a decade by a more accurate *quantum mechanics* model does not detract, however, from its mathematical and pictorial simplicity, elegance, and usefulness. In fact, there are several important ideas of the Bohr model (i.e., stationary states, quantum jumps, conservation of energy, and the correspondence principle) that have been retained as essential aspects of modern physics. Furthermore, Bohr's model of the hydrogen atom was a significant and most important contribution to the acceptance of the *quantum concept* and the development of *quantum mechanics*.

Bohr's model of the hydrogen atom considers classical physics to be *limited* in its applicability for the description of the motion of the electron about the proton. He assumed the Rutherford planetary model, illustrated in Figure 7.2, where a single electron traverses the nucleus in a circular orbit with a constant speed. It should be emphasized that our generalization of the model to be descriptive of any *one-electron* atom of nuclear charge $Ze$ does not alter the considerations and results obtained by Bohr for the hydrogen atom. Recognizing the inconsistency between the planetary model and electromagnetic radiation theory, Bohr found it necessary to make several assumptions concerning the structure of atoms. Bohr's work on the binding energy of electrons by positive nuclei was first published in the July 1913 issue of the *Philosophical Magazine* under the title "On the Constitution of Atoms and Molecules." This was the first of a trilogy of papers published by Bohr in that scholarly journal in 1913, and the primary assumptions of his model for atomic structure were most clearly stated in his third publication appearing in the November 1913 issue. Rearranged somewhat in order, his assumptions on pages 874 to 875 were as follows:

> That the dynamical equilibrium of the systems in the stationary states is governed by the ordinary laws of mechanics, while these laws do not hold for the passing of the systems between the different stationary states.
>
> That the different stationary states of a simple system consisting of an electron rotating round a positive nucleus are determined by the condition that the ratio between the total energy, emitted during the formation of the configuration, and the frequency of revolution of the electron is an entire multiple of $h/2$. Assuming that the orbit of the electron is circular, this assumption is equivalent with the assumption that the angular momentum of the electron round the nucleus is equal to an entire multiple of $h/2\pi$.
>
> That energy radiation is not emitted (or observed) in the continuous way assumed in the ordinary electrodynamics, but only during the passing of the system between different "stationary" states.

That the radiation emitted during the transition of a system between two stationary states is homogeneous, and that the relation between the frequency $\nu$ and the total amount of energy emitted $E$ is given by $E = h\nu$, where $h$ is Planck's constant.

To capitalize on our previous discussion and maximize the information content, we restate Bohr's postulates as follows:

1. An electron obeys the laws of classical mechanics while encircling the nucleus of an atom in a *stable orbit* with a uniform speed under the influence of a Coulombic force of attraction.
2. Only highly restricted orbits are allowed by nature for the electron, the selection of which is specified by the *quantization* of the electron's angular momentum to the value $nh/2\pi$, where the *principal quantum number* $n$ takes on the values $n = 1, 2, 3, \cdots$.
3. An electron is prohibited from emitting any electromagnetic quanta while encircling the nucleus along a *permitted* orbit, allowing the electron's total energy to remain constant.
4. An electron may make a direct transition from one *permitted* orbit to another by the emission of a single Planck photon having an energy equal to the energy difference of the two electron states of motion.

Bohr's first postulate allowed the *orbital stability* of the electron to be given by classical mechanics, where the attractive Coulombic force is equated to the centripetal force, as given by Equation 7.6. That is,

Bohr's 1st Postulate

$$mv^2 = \frac{Zke^2}{r} \tag{7.20}$$

is a simplified quantitative expression for Bohr's first postulate.

Bohr's second postulate *quantized* the angular momentum of the electron (see also Equations 8.56 to 8.57 of Chapter 8, Section 8.4)

$$L = I\omega = (mr^2)\omega = mr^2\frac{v}{r} = mvr, \tag{7.21}$$

to an integer value

Principal Quantum Number

$$n = 1, 2, 3, \cdots \tag{7.22}$$

(now called the **principal quantum number**) times $h/2\pi$. With the symbolic definition

$$\hbar \equiv \frac{h}{2\pi}, \tag{7.23}$$

the *quantization of angular momentum* for the Bohr electron is obtained by equating Equation 7.21 to the product of $n$ times Equation 7.23. That is,

$$mvr = n\hbar \tag{7.24}$$

Bohr's 2nd Postulate

is a quantitative representation of Bohr's famous *quantization postulate*.

The allowed *radii* of the Bohr electron is now easily obtained by combining Bohr's first and second postulates. That is, solving Equation 7.24 for the translational speed $v$ of the electron and squaring gives

$$v^2 = \frac{n^2\hbar^2}{m^2r^2},$$

which can be substituted into Equation 7.20 to yield

$$\frac{n^2\hbar^2}{mr} = Zke^2.$$

Because the value of $r$ in this expression varies as the *principal quantum number* $n$ takes on the values 1, 2, 3, $\cdots$, we adopt $n$ as a subscript for $r$. Now, solving the equation for the radius and using the new subscript notation, we obtain

$$\begin{aligned} r_n &= \frac{n^2}{Z}\left(\frac{\hbar^2}{mke^2}\right) \\ &\equiv \frac{n^2}{Z} r_1 \end{aligned} \tag{7.25}$$

Bohr Radii

for the radius of the electron in the $n$th allowed orbit. The use of the principal quantum number $n$ as a subscript on the radius of the Bohr electron allows for the identification of the *permitted* stable electron orbits, $r_n = r_1, r_2, r_3, \cdots$. The value of $r_1$, which is often referred to as the **Bohr Radius,** is directly calculated by using the values of the known constants in the defining equation

$$r_1 \equiv \frac{\hbar^2}{mke^2} \tag{7.26}$$

Bohr Radius

to obtain

$$r_1 = 0.529178 \text{ Å}. \tag{7.27}$$

Clearly, the value obtained for $r_1$ corresponds to the *innermost* radius ($n = 1$) allowed for a hydrogen ($Z = 1$) atom. Further, Equation 7.25 indicates that for a large value of $n$ the radius of the permitted electron orbit is considerably larger than the radius of the innermost or **ground state** orbit $r_1$ (e.g., $r_2 = 4r_1$, $r_3 = 9r_1$, and $r_4 = 16r_1$ for a hydrogen electron).

A generalized equation for the *translational speed* $v_n$ of an electron in any of its *n allowed* orbits is also obtainable from Bohr's first and second postulates. That is, solving Equation 7.20 (Bohr's first postulate) for $v = Zke^2/mvr$ and substituting from Equation 7.24 (Bohr's second postulate) immediately gives

Bohr Velocities

$$\begin{aligned} v_n &= \frac{Z}{n}\left(\frac{ke^2}{\hbar}\right) \\ &\equiv \frac{Z}{n} v_1, \end{aligned} \tag{7.28}$$

where the *subscript notation* has again been employed. The *ground state* translational speed, defined by

Bohr Velocity

$$v_1 \equiv \frac{ke^2}{\hbar}, \tag{7.29}$$

has the value

$$v_1 = 2.18769 \times 10^6 \text{ m/s}, \tag{7.30}$$

which is the maximum allowed speed of the electron in a hydrogen atom. To a good approximation (four significant numbers) the value of $v_1$ can be expressed by

$$v_1 = \frac{c}{137} \equiv c\alpha, \tag{7.31}$$

where $c$ is the speed of light in a vacuum and $\alpha$ is the so-called **fine structure constant.** From Equations 7.29 and 7.31 we obtain

Fine Structure Constant

$$\alpha = \frac{ke^2}{\hbar c}, \tag{7.32}$$

and, occasionally, the allowed Bohr velocities, Equation 7.28, are expressed in terms of $\alpha$ as

$$v_n = \frac{Z}{n}\alpha c. \tag{7.33}$$

Further, Equation 7.28 predicts that the orbital speed of the Bohr electron decreases with increasing values of the principal quantum number (e.g., $v_2 = v_1/2$, $v_3 = v_1/3$, and $v_4 = v_1/4$ for a hydrogen electron).

Knowing the generalized equation for the velocity of the Bohr electron in any allowed orbit, it now becomes an easy task to obtain an equation for the electron's *orbital frequencies*. Since $v = r\omega$ and $\omega = 2\pi\nu$, the orbital frequency is given by

$$\nu = \frac{v}{2\pi r}. \tag{7.34}$$

Substitution from Equations 7.25 and 7.28 gives

$$\begin{aligned} \nu_n &= \frac{Z^2}{n^3}\left(\frac{mk^2e^4}{h\hbar^2}\right) \\ &\equiv \frac{Z^2}{n^3}\nu_1, \end{aligned} \tag{7.35}$$

Bohr Frequencies

where the subscript notation has been employed. The *Bohr frequency* for the hydrogen electron in the *ground state* is defined by

$$\nu_1 = \frac{mk^2e^4}{h\hbar^2}, \tag{7.36}$$

Bohr Frequency

which, upon substitution of the values for $k$, $e$, and $h$, has the value

$$\nu_1 = 6.57912 \times 10^{15}\ \text{Hz}. \tag{7.37}$$

From Equation 7.35 it is obvious that the *orbital frequency* of the Bohr electron decreases *very rapidly* with any increase in the value of the principal quantum number (e.g., $\nu_2 = \nu_1/8$, $\nu_3 = \nu_1/27$, and $\nu_4 = \nu_1/64$ for a hydrogen electron), because of its inverse dependence on $n^3$.

Our derivations of the generalized equations for the allowed radii, velocities, and orbital frequencies of an atomic electron constrained to move in one of the permitted circular orbits has been based on the laws of classical mechanics and *Bohr's quantization postulate*. Although the classical theory of radiation predicts that the accelerated electron must radiate energy continuously and, thus, should spiral inward to the nucleus, Bohr's third postulate recognized the validity of electromagnetic theory was restricted to macroscopic phenomena and not applicable to the microscopic atom. Classical mechanics is still valid, however, according to his first postulate, so the *total energy* of the one-electron atom is given by Equation 7.15. Using the *subscript notation* involving the principal quantum number, Equation 7.15 becomes

Bohr's 3rd Postulate

$$E_n = -\frac{Zke^2}{2r_n}, \tag{7.38}$$

which is a quantitative representation of Bohr's third postulate. Substitution from Equation 7.25 for $r_n$ gives

Bohr Energies

$$\begin{aligned} E_n &= -\frac{Z^2}{n^2}\left(\frac{mk^2e^4}{2\hbar^2}\right) \\ &\equiv -\frac{Z^2}{n^2}|E_1| \end{aligned} \tag{7.39}$$

for the energy of the Bohr electron in any one of the permitted orbits defined by $n$. The *absolute magnitude* of the Bohr electron's *ground state* energy, defined by

$$|E_1| \equiv \frac{mk^2e^4}{2\hbar^2}, \tag{7.40}$$

has the value

$$|E_1| = 2.17972 \times 10^{-18}\ \text{J} = 13.6046\ \text{eV}. \tag{7.41}$$

By using this value for $|E_1|$ in Equation 7.39, it is easily shown that values of $E_n$ become less negative rather rapidly for increasing values of the principal quantum number (e.g., $E_1 = -13.6046$ eV, $E_2 = -3.40115$ eV, $E_3 = -1.51162$ eV, $E_4 = -0.850288$ eV, and $E_5 = -0.544184$ eV for a hydrogen electron), with $E_n$ approaching zero as $n$ increases to infinity. It should be emphasized that Equation 7.39 yields *negative energies*, because the electron is *bound* to the nucleus in a *stationary state*. If the electron was *free*, it could have either *zero* energy or *positive* valued kinetic energy. Further, a *bound* electron is in its *most stable state* when it is in the state of *lowest* total energy, that state characterized by $n = 1$.

An interesting relationship between the energy of a Bohr electron in

| $n$ | $r_n$ (Å) | $v_n$ ($10^6$ m/s) | $\nu_n$ ($10^{15}$ Hz) | $E_n$ (eV) |
|---|---|---|---|---|
| 1 | 0.529178 | 2.18769 | 6.57912 | −13.6046 |
| 2 | 2.11671 | 1.09385 | 0.822390 | −3.40115 |
| 3 | 4.76260 | 0.729230 | 0.243671 | −1.51162 |
| 4 | 8.46685 | 0.546923 | 0.102799 | −0.850288 |
| 5 | 13.2295 | 0.437538 | 0.0526330 | −0.544184 |
| 6 | 19.0504 | 0.364615 | 0.0304589 | −0.377906 |
| 7 | 25.9297 | 0.312527 | 0.0191811 | −0.277645 |
| 8 | 33.8674 | 0.273461 | 0.0128498 | −0.212572 |
| 9 | 42.8634 | 0.243077 | 0.0090249 | −0.167958 |

TABLE 7.3
A few quantized states (defined by the value of $n$) of the hydrogen atom, characterized by the electron's radius, speed, orbital frequency, and energy predicted by the Bohr model.

the $n$th stationary state and its orbital frequency can be obtained from Equations 7.35 and 7.39. That is, Equation 7.35 can be expressed as

$$\begin{aligned} \nu_n &= \frac{1}{n}\frac{Z^2}{n^2}\frac{mk^2e^4}{\hbar^2}\frac{1}{h} \\ &= \frac{2}{nh}|E_n|, \end{aligned}$$

which when solved for the absolute magnitude of $E_n$ gives

$$|E_n| = \tfrac{1}{2}nh\nu_n. \tag{7.42}$$

Energy Quantization

This relationship is strikingly similar to Planck's quantization hypothesis for atomic oscillators and is, indeed, recognized as the *correct* expression for **energy quantization,** instead of that given by Equation 6.50. This result and Bohr's fourth postulate, which will be quantitatively developed and discussed in the next section, provide the *link* between atomic structure and Planck's quantum theory of radiation. Although Equation 7.42 emphasized the *quantization of energy* of the Bohr atom, we should realize that *Bohr's quantization postulate* for the electron's angular momentum has led to the *quantization* of the electron's orbital radius, velocity, frequency, and energy (i.e., Equations 7.25, 7.28, 7.35, and 7.39, respectively), as illustrated for the hydrogen atom in Table 7.3.

## 7.4 Emission Spectra and the Bohr Model

The known visible spectral lines of the hydrogen atom are fairly well predicted by the Bohr model, as illustrated in the fourth column of Table 7.1. To understand how the values presented in Table 7.1 were obtained, we will consider Bohr's fourth postulate. Accordingly, the energy of the emitted photon representing any one of the illustrated spectral lines of Figure 7.3 results from the Bohr electron making a downward *quantum transition* from an initial energy level $E_i$ to a *lower* final energy level $E_f$. For example, the third Lyman emission corresponds to the Bohr electron making a direct transition from the $n = 4$ level to the $n = 1$ level, where the energy of the emitted photon associated with this *quantum jump* is just equal to the difference between $E_4$ and $E_1$ of the electron. Thus, the emitted photon would have a positive valued energy given by (see Table 7.3) $\epsilon = E_4 - E_1 = -0.850288 \text{ eV} - (-13.6046 \text{ eV})$ or $\epsilon = 12.754312$ eV. In general the energy $\epsilon$ of any *emitted* photon is expressible in terms of the *initial* $n_i$ and *final* $n_f$ *quantum levels* of the electron. From Bohr's fourth postulate

Bohr's 4th Postulate

$$\epsilon = E_i - E_f, \tag{7.43}$$

and Equation 7.39 we have

$$\epsilon = -\frac{Z^2|E_1|}{n_i^2} - -\frac{Z^2|E_1|}{n_f^2}, \tag{7.44}$$

which is more conveniently written in the form

Photon Energy

$$\epsilon = Z^2|E_1|\left(\frac{n_i^2 - n_f^2}{n_i^2 n_f^2}\right). \tag{7.45}$$

Now, using Einstein's photon postulate as expressed by Equation 6.53, the frequency of an *emitted photon* is given by

$$\nu = Z^2\frac{|E_1|}{h}\left(\frac{n_i^2 - n_f^2}{n_i^2 n_f^2}\right). \tag{7.46}$$

The ratio of constants in this expression can be solved for

$$\frac{|E_1|}{h} = \frac{mk^2e^4}{2h\hbar^2} = 3.28956 \times 10^{15} \text{ Hz}, \tag{7.47}$$

allowing Equation 7.46 to become

$$\nu = Z^2(3.28956 \times 10^{15} \text{ Hz})\left(\frac{n_i^2 - n_f^2}{n_i^2 n_f^2}\right). \tag{7.48}$$

Of course if a photon is absorbed, the electron makes an *upward* transition and $n_f > n_i$. In this case knowledge of $\nu$ and $n_i$ is sufficient for a determination of $n_f$ by using Equation 7.48. Since electromagnetic radiation propagates at the speed of light $c$, the wavelength of an *emitted* or *absorbed* photon is directly obtainable from Equation 7.46. That is, with $\nu = c/\lambda$ substituted into Equation 7.46, we obtain

$$\lambda = \frac{hc}{Z^2|E_1|}\left(\frac{n_i^2 n_f^2}{n_i^2 - n_f^2}\right), \tag{7.49}$$

which can be reduced to

$$\lambda = \frac{911.346\ \text{Å}}{Z^2}\left(\frac{n_i^2 n_f^2}{n_i^2 - n_f^2}\right) \tag{7.50}$$

by substitution for the physical constants $h$, $c$, and $E_1$. Equation 7.50 predicts, with reasonable accuracy, the experimentally reported emission lines of the hydrogen atom, listed in Table 7.1, as it agrees rather closely with Balmer's formula given by Equation 7.1. Very often you will see Equation 7.50 written in terms of the Rydberg constant $R_H$. According to the Bohr model this constant can be evaluated from

Rydberg Constant

$$R_H = \frac{|E_1|}{hc}, \tag{7.51}$$

which upon substitution of constants gives

$$R_H = 1.09714 \times 10^7\ \text{m}^{-1}. \tag{7.52}$$

Thus, Equation 7.49 can take the form

$$\frac{1}{\lambda} = Z^2 R_H\left(\frac{n_i^2 - n_f^2}{n_i^2 n_f^2}\right), \tag{7.53}$$

which agrees reasonably well with Rydberg's formula given by Equation 7.2. Actually, with the correction to the Bohr model obtained in the next section, the model predicted value of $R_H$ agrees almost perfectly with the experimental value given by Equation 7.3.

The predictions of the Bohr model contained in Equations 7.45, 7.46, and 7.49 are most important in that they predict the energy, frequency, and wavelength of observed spectral lines for hydrogen, denoted as the Lyman, Balmer, Paschen, Brackett, and Pfund series in Figure 7.3. Normally, an atom of hydrogen exists in its *ground state* corrresponding to

$n = 1$, where the electron has its lowest energy. If the atom absorbs energy through collisions, irradiation, and so forth, the electron makes a direct transition to an *excited state* for which $n > 1$. As is the tendency for all physical systems, the atom in the *excited state* will return to its *ground state* by the emission of its excess energy. This can be accomplished by the electron making a direct transition or *quantum jump* from the excited state $n_i > 1$ to the ground state $n_f = 1$, with the emission of a *photon* of energy $\epsilon$ given by Equation 7.45. Alternatively, the electron might *cascade* from its initial excited state to successively lower energy states, until the ground state is attained. In this instance each quantum jump to a lower energy state is accomplished by the emission of a single photon. For example, an electron excited into the $n = 5$ state, could cascade successively through the energy states $n = 4$, $n = 3$, $n = 2$, and $n = 1$. Four spectral lines of the hydrogen atom would be emitted with wavelengths given by Equation 7.49 for the quantum jump $n_i = 5$ to $n_f = 4$, $n_i = 4$ to $n_f = 3$, $n_i = 3$ to $n_f = 2$, and $n_i = 2$ to $n_f = 1$. These spectral lines represent the first line of the Brackett, Paschen, Balmer, and Lyman series, respec-

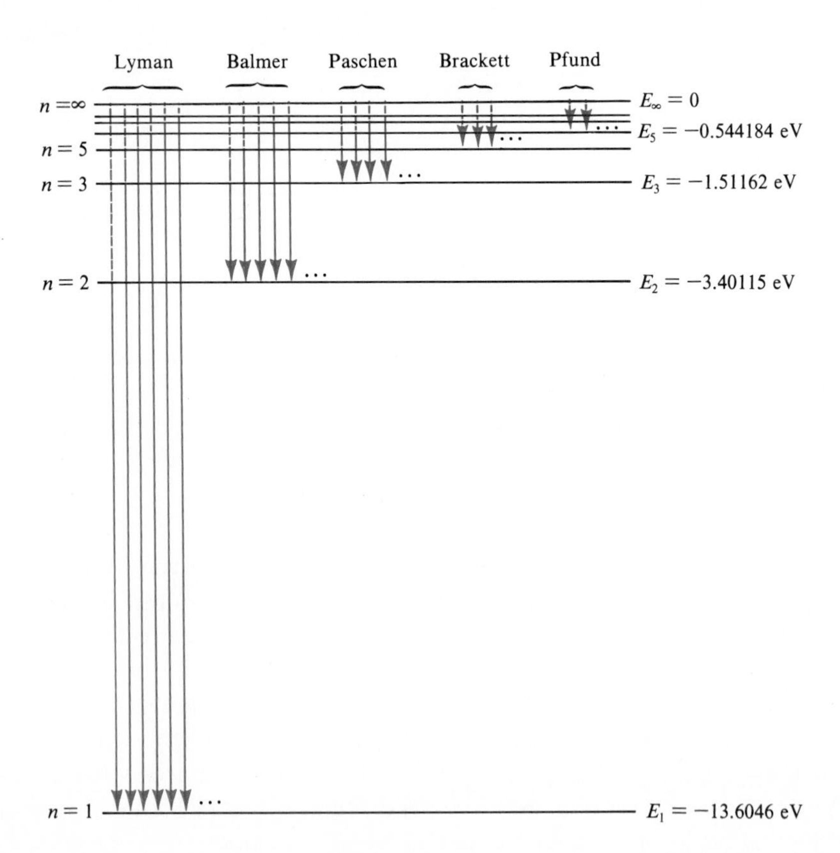


Figure 7.3
An energy level diagram for the hydrogen atom, illustrating the five well-known spectral series occurring in nature.

tively, and are illustrated in Figure 7.3 by an arrow going from the initial quantum state $n_i$ to the final quantum state $n_f$.

The success of the Bohr theory for the one-electron atom is most impressive, as the spectral lines of the Lyman, Balmer, Paschen, Brackett, and Pfund series, illustrated in Figure 7.3, are all accurately predicted by Equation 7.50. It should be emphasized that the Lyman, Brackett, and Pfund series had not been experimentally observed prior to the publication of Bohr's theory in 1913. They were sought after and soon discovered, however, because of the general acceptance of the Bohr model. In addition to the success of the Bohr theory in predicting the hydrogen spectrum, the model worked equally well for other one-electron atoms, like the singly ionized helium atom.

## 7.5 Correction to the Bohr Model for a Finite Nuclear Mass

Implicit in our derivations involving the Bohr model of the one-electron atom is the assumption of an infinitely large nucleus that remains fixed in space. Actually, this was a reasonable first approximation for even the lightest atom, since the nucleus of a hydrogen atom is 1836 times more massive than the bound electron. Even though the mass of a nucleus is considerably greater than the mass of an electron, its mass is *finite* and should not be considered as *fixed* in space. To enhance the accuracy of the Bohr model for the one-electron atom, we need to take into account the finite mass of the nucleus and attribute some degree of motion to both the electron and the nucleus.

Consider the more realistic picture of an atom as that depicted in Figure 7.4, where the electron of mass $m$ and nucleus of mass $M$ each revolve about their *center of mass* (C.M.) such that it remains fixed in space. The center-to-center distance between the nucleus and the electron is taken to be $r$, while $x$ is the distance from the nucleus and $r - x$ is the distance from the electron to the center of mass. By definition of the center of mass, we have

$$Mx = m(r - x), \tag{7.54}$$

which can be solved for $x$ in the form

$$x = \frac{mr}{m + M}. \tag{7.55}$$

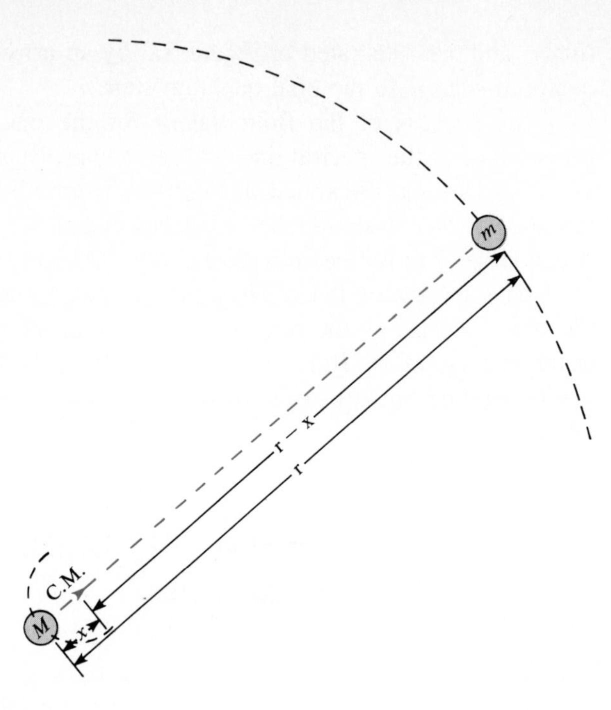


Figure 7.4
A nucleus of mass $M$ and an electron of mass $m$ in circular motion about their common center of mass.

Also, from Equation 7.54

$$r - x = \frac{Mx}{m},$$

and substitution from Equation 7.55 gives

$$r - x = \frac{Mr}{m + M}. \tag{7.56}$$

These last two numbered equations are important and will be frequently utilized in our derivational considerations. The former equation gives the radius of the circular orbit made by the nucleus about the center of mass, while the latter gives the radius of the electron's circular orbit.

As a first consideration, let us derive an expression for the angular momentum of the electron-nucleus system. This is easily accomplished by adding the angular momentum of the electron $L_e$ and the angular momentum of the nucleus $L_N$, that is

$$\begin{aligned} L &= L_e + L_N \\ &= I_e\omega_e + I_N\omega_N \\ &= m(r - x)^2\omega_e + Mx^2\omega_N. \end{aligned} \tag{7.57}$$

Substituting from Equations 7.55 and 7.56 for the quantities $r - x$ and $x$ gives

$$L = m\left(\frac{Mr}{m + M}\right)(r - x)\omega_e + M\left(\frac{mr}{m + M}\right)x\omega_N$$
$$= \left(\frac{mM}{m + M}\right)r[(r - x)\omega_e + x\omega_N].$$

Realizing that

$$\omega_e = \omega_N \equiv \omega, \tag{7.58}$$

since the distance of separation $r$ between the nucleus and the electron must be constant at all times for the center of mass to be fixed in space, then our expression for the angular momentum reduces to

$$L = \left(\frac{mM}{m + M}\right)r^2\omega. \tag{7.59}$$

At this point it is convenient to introduce the so-called **reduced mass,**

$$\mu \equiv \frac{mM}{m + M}, \tag{7.60}$$

Reduced mass

which allows Equation 7.59 to be written as

$$L = \mu r^2\omega. \tag{7.61}$$

Total Angular Momentum

There is a tendency, at this point, to let the electron's speed $v$ be given by $v = r\omega$ and write Equation 7.61 as $L = \mu vr$. This result for the *total angular momentum* of the electron-nucleus system is identical to Equation 7.21, except for the presence of the *reduced mass* $\mu$ instead of the mass of the electron $m$. The result $L = \mu vr$, however, is *not correct*, as the translational speed of the electron $v$ is related to its angular speed $\omega_e$ by the equation

$$v = (r - x)\omega_e. \tag{7.62}$$

Speed of Electron

Since $\omega$ in Equation 7.61 is equal to $\omega_e$ by Equation 7.58, Equation 7.61 can be recast in the form

$$L = \mu r^2\omega_e$$

$$= \mu r^2\left(\frac{v}{r - x}\right) \qquad \text{(using Eq. 7.61)}$$

$$= \mu r^2\left[\frac{v}{Mr/(m + M)}\right] \qquad \text{(using Eq. 7.56)}$$

$$= \mu \frac{mvr}{mM/(m + M)}$$

Total Angular Momentum

$$= mvr, \qquad (7.63)$$

where the defining equation for the reduced mass $\mu$ (Equation 7.60) has been used in obtaining the last equality. Clearly, the total angular momentum of the electron-nucleus system is *identical* to the angular momentum of the electron (Equation 7.21) in the *fixed nucleus system*.

The Bohr theory for the electron-nucleus system of Figure 7.4 is now slightly different than that developed in the previous two sections. The electron moves uniformly in a circular orbit of radius $r - x$ about the center of mass of the two-body system. The Coulombic force of attraction between the electron and nucleus, given by

$$F_C = \frac{Zke^2}{r^2}, \qquad (7.5)$$

causes the electron to experience a centripetal acceleration. The centripetal force on the electron in this case, however, is given by

$$F_c = \frac{mv^2}{r - x}$$

$$= m(r - x)\omega^2 \qquad \text{(using Eq. 7.62)}$$

$$= m\left(\frac{Mr}{m + M}\right)\omega^2 \qquad \text{(using Eq. 7.56)}$$

$$= \mu r\omega^2, \qquad (7.64)$$

where the defining equation (Equation 7.60) for the reduced mass $\mu$ has been used in obtaining the last equality. Consequently, the *orbital stability* of the electron in this reduced mass model is given by

Bohr's 1st Postulate

$$\mu r\omega^2 = \frac{Zke^2}{r^2}, \qquad (7.65)$$

which is a quantitative expression for Bohr's first postulate. Because of the result expressed in Equation 7.63, Bohr's second postulate has the quantitative form

$$mvr = n\hbar, \tag{7.24}$$

which is identical to that obtained for the *fixed nucleus model*. In our present case, however, it must be interpreted as *the quantization* of the **total angular momentum** of the *two-body system*. With $mvr = \mu r^2\omega$ from Equations 7.61 and 7.63, then Equation 7.24 becomes

$$\mu r^2\omega = n\hbar, \tag{7.66}$$

Bohr's 2nd Postulate

which is a more amenable form of Bohr's second postulate for the present considerations. Solving this last equation for $\omega$ and substituting into Equation 7.65 immediately yields

$$r_n = \frac{n^2}{Z}\left(\frac{\hbar^2}{\mu ke^2}\right) \tag{7.67}$$

Atomic Radii

for the quantized *stationary states* of the atom. This result is identical to that given by Equation 7.25, except the mass of the electron has been replaced by the reduced mass $\mu$. The interpretation of $r_n$ for a particular value of $n$, however, is a bit different in this case. It represents the *mean distance* of *separation between* the *electron* and *nucleus*, not the radius of the electron's circular orbit in the $n$th allowed stationary state of the atom.

The total energy of the electron-nucleus system is still given by Equation 7.10,

$$E_t = E_k + E_p, \tag{7.10}$$

now, however, the kinetic energy has contributions from both the electron and the nucleus. That is, with $v_e$ and $v_N$ representing the translational speed of the electron and nucleus, respectively, about their common center of mass, we have

$$\begin{aligned} E_k &= \tfrac{1}{2}mv_e^2 + \tfrac{1}{2}Mv_N^2 \\ &= \tfrac{1}{2}m(r - x)^2\omega^2 + \tfrac{1}{2}Mx^2\omega^2 \\ &= \tfrac{1}{2}\omega^2[m(r - x)^2 + Mx^2] \\ &= \tfrac{1}{2}\omega^2[\mu r(r - x) + \mu rx] \\ &= \tfrac{1}{2}\mu r^2\omega^2, \end{aligned} \tag{7.68}$$

where we have used

Speed of Nucleus

$$v_N = x\omega_N \tag{7.69}$$

for the translational speed of the nucleus in the second equality. Because of Equation 7.65, the kinetic energy is also expressible as

$$E_k = \frac{Zke^2}{2r}. \tag{7.70}$$

Now, with the potential energy $E_p$ given by Equation 7.13,

$$E_p = -\frac{Zke^2}{r}, \tag{7.13}$$

the total energy for this case (Equation 7.10) is identical to that obtained previously, that is

$$E_t = -\frac{Zke^2}{2r}. \tag{7.15}$$

Consequently, substitution from Equation 7.67 and introducing the $n$-subscript notation, we obtain

Quantized Energy States

$$E_n = -\frac{Z^2}{n^2}\left(\frac{\mu k^2 e^4}{2\hbar^2}\right) \tag{7.71}$$

for the permitted quantized energy states of the one-electron atom having a *finite* nuclear mass.

It is instructive to reproduce all of the derivations of the Bohr theory, with the inclusion of the *finite nuclear mass* model of Figure 7.4. The net effect to the theory, however, is that all of the equations are identical to those previously derived, except the electron mass $m$ is replaced by the reduced mass $\mu$. Further, by comparing the value of the electron's rest mass $m$ to the value of $\mu$ for hydrogen, we find

$$\begin{aligned} \frac{m}{\mu} &= \frac{m + M}{M} \\ &= \frac{m_e + m_p}{m_p} \\ &= 1.00055. \end{aligned} \tag{7.72}$$

Thus, the ground state energy of the *hydrogen atom* is now given by

$$\begin{aligned} E_1 &= \frac{\mu}{m}(-13.6046 \text{ eV}) \\ &= -13.5971 \text{ eV} \end{aligned} \tag{7.73}$$

and it is separated from the proton by a distance of

$$\begin{aligned} r_1 &= \frac{m}{\mu}(0.529178 \text{ Å}) \\ &= 0.529469 \text{ Å}. \end{aligned} \tag{7.74}$$

As a last example, the theoretical value of the Rydberg constant for hydrogen (Equation 7.51) now becomes

$$\begin{aligned} R_H &= (\mu/m)\frac{|E_1|}{hc} \\ &= 1.09654 \times 10^7 \text{ m}^{-1}, \end{aligned} \tag{7.75}$$

which varies from the experimental value given in Equation 7.3 by only 24 parts in 100,000. Actually, a more careful calculation, using eight significant figure accuracy for the physical constants, gives $R_H = 1.0967758 \times 10^7 \text{ m}^{-1}$.

Before leaving this section, it needs to be emphasized that most references give the total angular momentum of the electron-nucleus system as $L = \mu vr$ instead of our result given in Equation 7.63. It is, however, easy to show that the angular moment of only the electron of Figure 7.4 is given by $L_e = \mu vr$. That is,

$$\begin{aligned} L_e &= I_e\omega_e \\ &= m(r - x)^2\omega_e \\ &= \frac{m(r - x)^2 v}{r - x} \qquad \text{(using Figure 7.4)} \\ &= m(r - x)v \qquad \text{(using Equation 7.62)} \\ &= m\left(\frac{Mr}{m + M}\right)v \quad \text{(using Equation 7.56)} \\ &= \mu vr, \end{aligned} \tag{7.76}$$

Electron's Angular Momentum

where Equation 7.60 has been used in obtaining the last equality. Further, by starting with Equation 7.57, it is straight forward to derive

$$L = \mu(v_e + v_N)r \tag{7.77}$$

for the total angular momentum of the electron and nucleus. This result reduces to $L = \mu v_e r$ under the assumption $v_e \gg v_N$, but this is equivalent to assuming the nuclear mass to be fixed in space, which is clearly contradictory to our finite nuclear mass model.

## 7.6 Wilson-Sommerfeld Quantization Rule

Planck's quantization of atomic oscillators and Bohr's quantization of the one-electron atom raised questions as to the existence of a fundamental relationship between these two quantization conditions. In 1916 William Wilson and Arnold Sommerfeld enunciated a general rule for the quantization of any physical system having coordinates that are periodic functions of time. Their rule is

Wilson-Sommerfeld Quantization Rule

$$\oint p_q dq = n_q h, \tag{7.78}$$

where $dq$ is an *infinitesimal generalized coordinate*, $p_q$ is the *generalized momentum* associated with the coordinate, $n_q$ is a *quantum number* which takes on integral values, and the integration is over *one complete cycle* of the generalized coordinate. The importance of this quantization rule is its utilization in expanding the range of applicability of the *old* quantum theory to all systems exhibiting a periodic dependence on time. The rule is perfectly general and, as will be illustrated, capable of predicting both the Bohr and the Planck quantization conditions by straight forward analyses of simple models.

### Quantization of Angular Momentum for the Bohr Electron

As a specific example of the application of the Wilson-Sommerfeld quantization rule, we will consider the Bohr model of the electron moving with periodic motion in a circular orbit about a nucleus fixed in space. The best coordinates to use for the Bohr model of a one-electron atom are the polar coordinates $r$ and $\theta$. According to Equation 7.78 the *radial* quantization of the electron is given by

$$\oint m \frac{dr}{dt} dr = n_r h, \tag{7.79}$$

where $r$ is the radius of the electron's circular orbit. Since the radius is constant, $r \neq r(t)$, then $dr/dt = 0$ and the left-hand side of Equation 7.79 becomes exactly zero. Thus, there is no quantization condition arising from the $r$-coordinate.

The quantization of the *angular* motion is obtained by realizing

$$p_\theta = I_\theta \frac{d\theta}{dt}, \tag{7.80}$$

where the *moment of inertia* is given by

$$I_\theta = mr^2. \tag{7.81}$$

Since the angular speed defined by

$$\omega \equiv \frac{d\theta}{dt} \tag{7.82}$$

Angular Speed

is a constant, by substitution of Equations 7.81 and 7.82 into Equation 7.80 we have

$$p_\theta = mr^2\omega. \tag{7.83}$$

The Wilson-Sommerfeld quantization formula now becomes

$$\oint p_q dq = \oint p_\theta d\theta = n_\theta h,$$

which upon substitution from Equation 7.83 becomese

$$mr^2\omega \oint d\theta = n_\theta h. \tag{7.84}$$

One complete cycle of the angular coordinate $\theta$ is accomplished by allowing $\theta$ to take on the values from 0 to $2\pi$. As such, Equation 7.84 becomes

$$mr^2\omega \int_0^{2\pi} d\theta = n_\theta h, \tag{7.85}$$

which upon integration and substitution from Equation 7.8 yields

$$mvr = n_\theta \hbar.$$

This result is in perfect agreement with the Bohr quantization of the electron's angular momentum, as given by Equation 7.24.

The Wilson-Sommerfeld quantization rule is also applicable to the *reduced mass model* of the one-electron atom. As before, the *radial quantization* of the atom is exactly zero, since the radii of the circular orbits described by the electron and the nucleus are constant in time, that is, $d(r - x)/dt = 0$ and $dx/dt = 0$. Consequently, there is only the need to consider the quantization of the *angular* motion of both the electron and nucleus. In this case the total angular momentum is given by Equation 7.61, so instead of Equation 7.83 we have

$$p_\theta = \mu r^2 \omega.$$

The Wilson-Sommerfeld quantization rule now gives

$$\begin{aligned} n_\theta h &= \oint p_q dq \\ &= \oint p_\theta d\theta \\ &= \mu r^2 \omega \oint d\theta \\ &= \mu r^2 \omega \int_0^{2\pi} d\theta \\ &= \mu r^2 \omega 2\pi, \end{aligned}$$

which reduces exactly to the quantization of angular momentum that is expressed in Equation 7.66, that is

$$\mu r^2 \omega = n_\theta \hbar.$$

## Quantization of a Linear Harmonic Oscillator

As a second example of the applicability of the Wilson-Sommerfeld quantization rule, consider a classical *harmonic oscillator* oriented along the $x$-axis, as illustrated in Figure 7.5. By way of a review, the particle of mass $m$ executes *periodic motion* from $A$ to $-A$ and back to $A$, after it is ini-

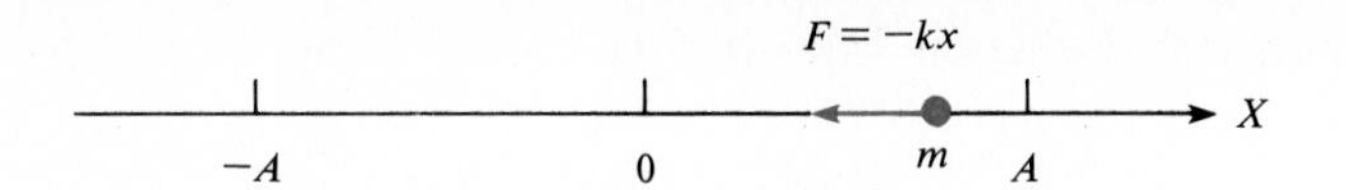


Figure 7.5
The classical one-dimensional harmonic oscillator.

tially displaced to the right through the distance $A$. We can take the *restoring force* on the particle to be that defined by the well known Hooke's law

$$F = -kx, \tag{7.86}$$

Hooke's Law

where $k$ is the *coefficient of elasticity*. In the absence of any frictional forces, Newton's second law of motion gives

$$m\ddot{x} + kx = 0, \tag{7.87}$$

Equation of Motion

where the *double dot* above the spatial coordinate denotes a second order differentiation with respect to time. The total energy of the system is just the sum of the kinetic energy $T$ and potential energy $V$, as given by

$$E = T + V. \tag{7.88}$$

For a conservative field with $V = V(x)$, general physics predicts the restoring force to be given by

$$F = -\frac{dV}{dx}, \tag{7.89}$$

Conservative Force

which can be rewritten as

$$\int_0^V dV = -\int_0^x F dx. \tag{7.90}$$

After substitution from Equation 7.86, integration yields

$$V = \tfrac{1}{2}kx^2 \tag{7.91}$$

for the potential energy of the linear oscillator. With the kinetic energy defined by the usual equation,

$$T = \tfrac{1}{2}mv^2 = \tfrac{1}{2}m\dot{x}^2, \tag{7.92}$$

the total energy (Equation 7.88) becomes

Equation of Motion

$$\tfrac{1}{2}m\dot{x}^2 + \tfrac{1}{2}kx^2 = E. \tag{7.93}$$

Actually, this last *equation of motion* can be obtained direrctly from Equation 7.87 by integration. That is,

$$\begin{aligned} m\ddot{x} + kx &= m\dot{v} + kx \\ &= m\frac{dv}{dx}\frac{dx}{dt} + kx \\ &= mv\frac{dv}{dx} + kx = 0. \end{aligned}$$

Integration at this point is accomplished by using

$$\int mv dv + \int kx dx = 0,$$

which gives

$$\tfrac{1}{2}mv^2 + \tfrac{1}{2}kx^2 = E,$$

where $E$ is the integration constant.

Since the harmonic oscillator is periodic in time with a maximum amplitude $A$, we assume the coordinate solution of Equation 7.87 and Equation 7.93 to be of the form

Oscillator Coordinate

$$x = A \sin \omega t, \tag{7.94}$$

which has a dependence on time as that illustrated in Figure 7.6. Now, direct substitution of this assumed solution into Newton's second law (Equation 7.87) gives

$$-m\omega^2 x + kx = 0,$$

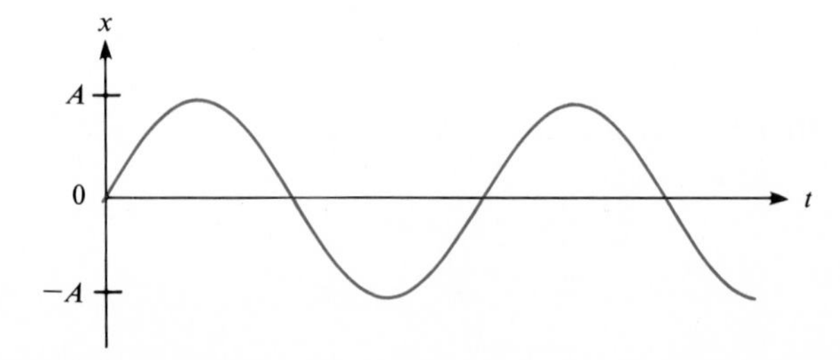


Figure 7.6
The coordinate dependence on time of the linear harmonic oscillator.

which simplifies to

$$\omega = \sqrt{\frac{k}{m}} \tag{7.95}$$

Oscillator Speed

for the oscillator's angular speed or

$$\nu = \frac{1}{2\pi}\sqrt{\frac{k}{m}} \tag{7.96}$$

Oscillator Frequency

for the *frequency* of the oscillator. Also, the total energy of the oscillator is now directly obtained from the assumed solution (Equation 7.94) and Equation 7.93 as

$$\tfrac{1}{2}m(\omega A \cos \omega t)^2 + \tfrac{1}{2}k(A \sin \omega t)^2 = E,$$

which is easily reduced by Equation 7.95 to

$$E = \tfrac{1}{2}kA^2. \tag{7.97}$$

Oscillator Energy

In this form the oscillator has only potential energy; whereas, when the particle is at the origin 0, it has only kinetic energy. Since the total energy is always equal to the sum of the kinetic and potential energies, it is always equal to the *maximum* of either. That is, in general we have

$$\begin{aligned} E &= \tfrac{1}{2}kA^2 \\ &= \tfrac{1}{2}mv^2 \end{aligned} \tag{7.98}$$

for the *maximum* total energy of the oscillator.

Having reviewed the classical theory of the linear harmonic oscillator, we can now proceed to apply the Wilson-Sommerfeld quantization rule. In this case the rule (Equation 7.78) becomes

$$\oint p_x dx = nh,$$

where the momentum is given by

$$\begin{aligned} p_x &= mv \\ &= m\frac{dx}{dt}. \end{aligned}$$

With the assumed coordinate solution given in Equation 7.94, we can write the translational velocity as

$$\frac{dx}{dt} = \omega A \cos \omega t$$

from whence we obtain

$$dx = \omega A \cos \omega t \, dt.$$

Thus, the Wilson-Sommerfeld rule gives

$$\begin{aligned} nh &= \oint p_x dx \\ &= \oint m \frac{dx}{dt} dx \\ &= \oint m(\omega A \cos \omega t)(\omega A \cos \omega t \, dt) \\ &= m\omega^2 A^2 \oint \cos^2 \omega t \, dt \\ &= m \frac{k}{m} A^2 \oint \cos^2 \omega t \, dt \\ &= 2E \oint \cos^2 \omega t \, dt, \end{aligned} \tag{7.99}$$

where Equations 7.95 and 7.97 have been used, respectively, in obtaining the last two equalities. The integral in this last expression can be simplified by letting

$$\theta = \omega t$$

and using Equation 7.82 in the form

$$dt = \frac{d\theta}{\omega}.$$

Realizing that over one complete cycle $t$ goes from 0 to $2\pi/\omega$, then our new variable of integration goes from 0 to $2\pi$. With these substitutions, Equation 7.99 gives

$$nh = \frac{2E}{\omega}\int_0^{2\pi} \cos^2\theta \, d\theta$$

$$= \frac{2E}{\omega}\int_0^{2\pi} \tfrac{1}{2}(\cos 2\theta + 1)\, d\theta$$

$$= \frac{E}{\omega}\left[\int_0^{2\pi} \cos 2\theta \, d\theta + \int_0^{2\pi} d\theta\right], \tag{7.100}$$

where the trigonometric identity

$$\cos^2\theta = \tfrac{1}{2}\cos 2\theta + \tfrac{1}{2}$$

from Appendix A, Section A.6 has been employed. As the first integral of Equation 7.100 goes to zero, we obtain

$$nh = \frac{2\pi E}{\omega},$$

which immediately reduces to

$$E = nh\nu \tag{7.101}$$

Oscillator Quantization

by using $\omega = 2\pi\nu$. Consequently, the total energy of the classical linear harmonic oscillator is quantized to an integral number $n$ times the energy $h\nu$, which is identical with *Planck's quantum hypothesis* given by Equation 6.50.

## 7.7 Quantum Numbers and Electron Configurations

Although the *circular orbit* theory of the Bohr model was successful in explaining the line spectrum of hydrogen-like atoms and the small wavelength shift arising from the relative motion of the nucleus, very precise measurements on hydrogen reveal that the energy levels have *fine structure*. Bohr considered the more general case of *elliptical orbits* to explain the fine structure splitting of spectral lines; however, his results lead to exactly the same spectral lines as that predicted by the *circular orbits* theory. In 1915 Sommerfeld was successful in generalizing the Bohr model to include elliptical orbits *and* relativistic effects, which explained the fine structure of the hydrogen spectrum. Although the translational velocities of bound electrons are considerably smaller than the velocity of light (e.g., $2.18769 \times 10^6/2.99792 \times 10^8 \approx 7.3 \times 10^{-3}$), the very small relativistic corrections to the electron mass account for the fine structure splitting of the hydrogen spectral lines.

Sommerfeld's results for the quantization of elliptical orbits revealed that the single energy state of the Bohr model actually consisted of several energy states. He came up with an additional quantum number and some special selection rules governing the *allowable* transitions for the hydrogen atom. However, the selection rules did not always agree with observed transitions and it was realized that the *mechanics of quanta*, as perpetuated by Planck, Einstein, Bohr, Sommerfeld, and others, was in itself limited. This started a new era in physical reasoning and the development of the new *quantum mechanics*. In as much as the Bohr model and Sommerfeld's extensions allow the physicist of today to quickly approximate physical reality, we will consider one aspect of Sommerfeld's generalizations before discussing the fundamental concepts of *quantum mechanics*.

An interesting result of Sommerfeld's work is that for any of the allowed energy states, the electron can move in any one of a number of elliptical orbits. More specifically, for each energy level denoted by the principal quantum number $n$ there exists $n$ possible orbits for the electron, as given by

Orbital QN

$$l = 0, 1, 2, 3, \cdots, n - 1, \tag{7.102}$$

where $l$ is called the **orbital quantum number.** Further, all orbits for a particular value of $n$ have the *same* energy as that given by Bohr's equation (Equation 7.39) for the assumed circular orbits. Thus, for $n = 1$ the electron moves in a circular orbit characterized by $l = 0$, whereas for $n = 2$ the electron can move in a circular orbit ($l = 0$) or an elliptical orbit ($l = 1$) of the same total energy. The *orbital quantum number* was actually advanced independently by Wilson and Sommerfeld in 1915, and their arguments supported the generalized Wilson-Sommerfeld quantization rule discussed in the previous section. The quantity $l$ represents the *quantum number* of the *orbital angular momentum* of the electron in both the original Wilson-Sommerfeld theory and the later developed quantum mechanics. It specifies the quantization of *angular momentum* in units of $\hbar$ that are associated with an electron in any of the allowed states that correspond to classical elliptical orbits having equal energy but different shapes or *eccentricities*. For example, there are three allowed angular momentum states ($l = 0, 1, 2$) corresponding to three *degenerate* states of motion for $n = 3$ in hydrogen.

Using the results of the Bohr theory and the Sommerfeld generalizations, we can develop the *Bohr-Sommerfeld scheme* for the building up of the chemical atoms. Electrons are imagined to exist in energy states called **shells,** which are specified by the letter $K$ corresponding to $n = 1$, the letter $L$ to $n = 2$, and so on according to the following scheme:

Energy States

Electron Shells

$$\begin{array}{cccccccc} n = & 1 & 2 & 3 & 4 & 5 & \cdots \\ & \downarrow & \downarrow & \downarrow & \downarrow & \downarrow & \\ & K & L & M & N & O & \cdots. \end{array}$$

Because of the *orbital quantum number* $l$, each *shell* is further imagined to be divided into *degenerate* energy states called **subshells,** which are specified by a letter according to the following scheme:

Angular Momentum States

Electron Subshells

$$\begin{array}{cccccccc} l = & 0 & 1 & 2 & 3 & 4 & \cdots \\ & \downarrow & \downarrow & \downarrow & \downarrow & \downarrow & \\ & s & p & d & f & g & \cdots. \end{array}$$

This particular convention for the names of the subshells originated from the empirical classification of the spectra of alkali metals (lithium, sodium, and potassium) into series called *sharp*, *principal*, *diffuse*, and *fundamental*. Later these *series* were recognized as resulting from electron transitions to the $l = 0$, 1, 2, and 3 states, respectively, so electrons in the $l = 0, 1, 2, 3, 4, 5$, and so forth states have conventionally been described as being in the $s$, $p$, $d$, $f$, $g$, $h$, and so forth states. The resulting **atomic notation** specifies the state of an electron by indicating the value of $n$ before the letter denoting $l$. For example, an electron in the $2s$ state corresponds to one having quantum numbers $n = 2$ and $l = 0$, while a $4f$ electron has $n = 4$ and $l = 3$. The atomic states for hydrogen are illustrated in Table 7.4 through quantum numbers $n = 6$ and $l = 5$.

The *atomic notation* illustrated in Table 7.4 is further extended to

TABLE 7.4
The atomic notation for a few electron states in hydrogen.

| SUBSHELLS | SHELLS $n = 1$ | $n = 2$ | $n = 3$ | $n = 4$ | $n = 5$ | $n = 6$ | ··· |
|---|---|---|---|---|---|---|---|
| $l = 0 \rightarrow s$ | $1s$ | $2s$ | $3s$ | $4s$ | $5s$ | $6s$ | ··· |
| $l = 1 \rightarrow p$ | | $2p$ | $3p$ | $4p$ | $5p$ | $6p$ | ··· |
| $l = 2 \rightarrow d$ | | | $3d$ | $4d$ | $5d$ | $6d$ | ··· |
| $l = 3 \rightarrow f$ | | | | $4f$ | $5f$ | $6f$ | ··· |
| $l = 4 \rightarrow g$ | | | | | $5g$ | $6g$ | ··· |
| $l = 5 \rightarrow h$ | | | | | | $6h$ | ··· |
| ⋮ | | | | | | | |

include an atom containing more than one electron with identical values of $n$ and $l$. In this case the number of electrons having identical $n$ and $l$ values is written as a superscript to the letter denoting $l$, that is

$$\text{ATOMIC NOTATION} \equiv nl^{(\text{No. of Electrons})}. \tag{7.103}$$

Consequently, an atom having 6 electrons for which $n = 3$ and $l = 1$ is said to have $3p^6$ electrons. The maximum number of electrons allowed in a *shell* is given by

Maximum Shell Electrons

$$N_n = 2n^2, \tag{7.104}$$

while the maximum number allowed in a *subshell* can be obtained from the equation

Maximum Subshell Electrons

$$N_l = 2(2l + 1). \tag{7.105}$$

These equations are easily justified in *quantum mechanics* by the inclusion of two additional quantum numbers, which will be briefly discussed later. For now, however, we accept Equations 7.104 and 7.105 as empirical equations in the Bohr-Sommerfeld scheme. As an example of this scheme, consider the $M$-shell ($n = 3$), where we have the existence of the $s$, $p$, and $d$ subshells (i.e., $l = 0$, 1, and 2). From Equation 7.104 we find $N_n = 2(3)^2 = 18$ for the maximum allowable number of electrons in the $M$-shell, and these electrons are imagined to fill up the $s$, $p$, and $d$ subshells in that order. For an $s$-subshell Equation 7.105 gives $N_s = 2(2 \cdot 0 + 1) = 2$ electrons, whereas for the $p$ and $d$ subshells we obtain a maximum of 6 ($N_p = 2(2 \cdot 1 + 1) = 6$) and 10 ($N_d = 2(2 \cdot 2 + 1) = 10$) allowed electrons, respectively. Consequently, since the $s$, $p$, and $d$ subshells are associated with the $M$-shell, we have a maximum number of electrons $N_l = N_s + N_p + N_d = 2 + 6 + 10 = 18$, which is in agreement with the $N_n = 2n^2$ calculation. Table 7.5 illustrates the maximum number of electrons in each shell and subshell and the atomic scheme for $n = 1$ through $n = 4$. This model suggests the shells and subshells of atoms are to be filled by electrons in the order presented. Generalizing Table 7.5 and employing the *atomic notation* given in Equation 7.103, we obtain the following **Bohr-Sommerfeld scheme** for the building up of chemical atoms:

Bohr-Sommerfeld Scheme

$$1s^2 2s^2 2p^6 3s^2 3p^6 3d^{10} \cdots. \tag{7.106}$$

| Values of n | Name of Shell | Maximum no. of Electrons | Values of l | Name of Subshell | Maximum no. of Electrons |
|---|---|---|---|---|---|
| 1 | K | 2 | 0 | s | 2 |
| 2 | L | 8 | 0 | s | 2 |
| | | | 1 | p | 6 |
| 3 | M | 18 | 0 | s | 2 |
| | | | 1 | p | 6 |
| | | | 2 | d | 10 |
| 4 | N | 32 | 0 | s | 2 |
| | | | 1 | p | 6 |
| | | | 2 | d | 10 |
| | | | 3 | f | 14 |
| . | . | . | . | . | . |
| . | . | . | . | . | . |
| . | . | . | . | . | . |

TABLE 7.5
Bohr-Sommerfeld scheme for electron configurations.

The Bohr-Sommerfeld scheme of Equation 7.106 allows us to write down the *electron configuration* for a normal unexcited atom. For example, in *atomic notation* the electron configuration for a silicon atom ($Z = 14$) is given by

$$_{14}\text{Si}: 1s^2 2s^2 2p^6 3s^2 3p^2,$$

where the *atomic number* $Z = 14$ indicates the number of protons, and hence electrons, in a neutral atom of silicon. This scheme works well for chemical atoms through $Z = 18$ (argon); however, a departure occurs for $Z = 19$ (potassium) and the heavier elements. For potassium ($Z = 19$) the electronic configuration is given by

$$_{19}\text{K}: 1s^2 2s^2 2p^6 3s^2 3p^6 4s^1,$$

where the nineteenth electron goes into the $4s$ state *instead* of the expected $3d$ state suggested by the Bohr-Sommerfeld scheme (Equation 7.106). This type of departure occurs frequently for the rest of the elements in the chemical periodic table. However, a scheme known as **Paschen's triangle** appears to accommodate most of these departures, except for a few heavy chemical atoms. This scheme can be obtained from the electron states illustrated in Table 7.4, by imagining diagonal lines of positive slope to be drawn through the states. This is illustrated in Table 7.6, with the resulting **Paschen scheme** given by

Paschen Scheme

$$1s^2 2s^2 2p^6 3s^2 3p^6 4s^2 3d^{10} 4p^6 \cdots. \quad (7.107)$$

Using this scheme, the electron configuration for cobalt,

$$_{27}\text{Co: } 1s^2 2s^2 2p^6 3s^2 3p^6 4s^2 3d^7,$$

indicates an unfilled $3d$ subshell; whereas, zinc ($_{30}$Zn) has a completely filled $3d$ subshell. Irregularities from the *Paschen scheme* are indicated by an asterisk in Appendix B.

The departure of electronic states from the Bohr-Sommerfeld scheme and the justification of the Paschen scheme can not be understood from the *old quantum theory*. The correct model that is consistent with experimental spectroscopic data can only be understood with the theory of *quantum mechanics* and the inclusion of two additional quantum numbers. Although we are not theoretically prepared to develop these additional quantum numbers from first principles, we can consider them empirically and somewhat superficially. Historically, these additional quantum numbers were introduced to explain the observed fine structure of some spectral lines. The third quantum number is associated with the so called *Zeeman effect*, where spectral lines are observed to split up into components, when the source of radiation is placed in an external **B**-field. The *orbital* **magnetic**

TABLE 7.6
Paschen's triangle of subshells illustrating an atomic scheme for electron configurations.

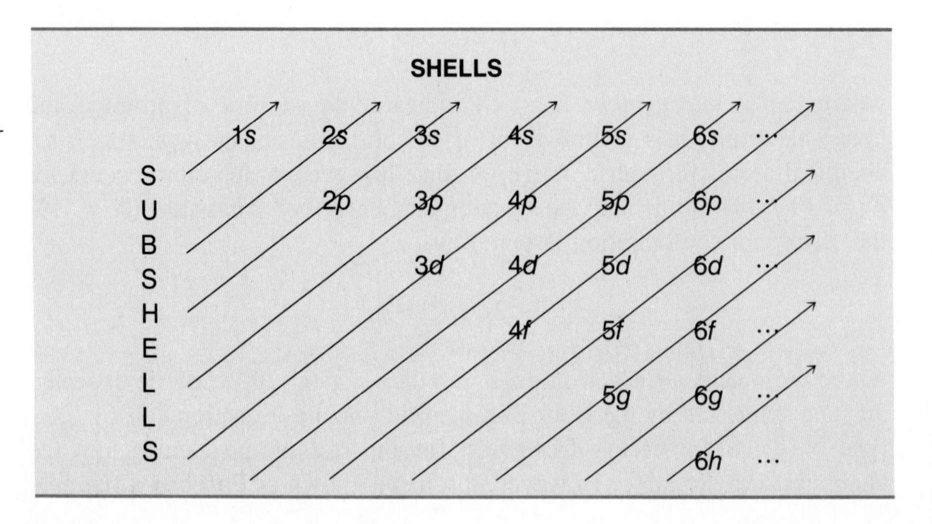

**quantum number,** $m$ or $m_l$, was introduced to explain the *magnetic moment* associated with an electron in each of its allowed orbits. It takes on the values

$$m_l = -l, \cdots, 0, \cdots, l \qquad (7.108)$$

Magnetic QN

and determines the components of the orbital angular momentum of an electron in an external magnetic field. Because of the restrictions, due to the orbital quantum number $l$, on the orientation of electron orbits, the orbits are said to be *space quantized* by the values of $m_l$. O. Stern and W. Gerlach verified *space quantization* of atoms in 1921. Later, in 1925 G. E. Uhlenbeck and S. A. Goudsmit postulated a fourth quantum number, called the *electron spin magnetic quantum number*, to account for additional fine structure in the hydrogen spectrum and space quantization of the atoms. The **spin quantum number,** $s$ or $m_s$, accounts for the intrinsic spin of an electron as it orbits the nucleus, which gives rise to not only an additional *magnetic moment*, but also an *angular momentum* that is independent of any orbital angular momentum. The quantity $m_s$ specifies the quantized spin angular momentum of an electron and has only the allowed values

$$m_s = +\tfrac{1}{2}, -\tfrac{1}{2}. \qquad (7.109)$$

Spin Magnetic QN

These values for $m_s$ are often referred to as *spin up* and *spin down*, respectively.

With the inclusion of magnetic and spin quantizations, we have the *state of motion* (or quantum state) of an electron in an atom characterized by the *set*

$$\text{Electron State} \equiv (n, l, m_l, m_s) \qquad (7.110)$$

Electron Quantum State

of quantum numbers that have restricted values that are summarized in Table 7.7. This set of quantum numbers along with the **Pauli exclusion principle,** which states that *no two electrons in an atom can have identical sets of quantum numbers*, allows us to specify a scheme for the *electronic configuration* of the chemical atoms. The model illustrated in Table 7.8 is only exact for the hydrogen atom; however, all of the chemical atoms are found to follow the same scheme. In its lowest energy state, the hydrogen electron is characterized by $n = 1$, $l = 0$, $m_l = 0$, and $m_s = -\frac{1}{2}$ or $m_s = +\frac{1}{2}$. This means that there are only two allowed and distinct states of motion for the hydrogen electron in the $K$-shell, which differ only in spin orientation. Thus, a maximum of 2 electrons can be accommodated in the $K$-shell, having *quantum states* defined by $(1, 0, 0, -\frac{1}{2})$ and (1, 0,

0, $+\frac{1}{2}$). In the $L$-shell Table 7.8 gives 8 distinct orbital states allowed for the hydrogen electron, which means that 8 electrons can be accommodated with different quantum numbers in this shell. Similarly, for $n = 3$ the $M$-shell is seen to be filled by 18 electrons, 2 in the $s$-subshell ($l = 0$), six in the $p$-subshell ($l = 1$), and 10 in the $d$-subshell ($l = 2$). Thus, the maximum number of electrons in the shells and subshells are correctly predicted by Equations 7.104 and 7.105, respectively. Further, the order in which subshells are filled by electrons is given by spectroscopic data to be

TABLE 7.7
The restricted values of the four quantum numbers $n$, $l$, $m_l$, and $m_s$.

| | |
|---|---|
| Principal | $n = 1, 2, 3, 4, 5, \cdots$ |
| Orbital | $l = 0, 1, 2, \cdots, n - 1$ |
| Magnetic | $m_l = -l, -l + 1, \cdots, 0, \cdots, l - 1, l$ |
| Spin | $m_s = -\frac{1}{2}, +\frac{1}{2}$ |

TABLE 7.8
The allowed values of the four quantum numbers for the lowest three states in the H-atom.

| $n$ | $l$ | $m_l$ | $m_s$ |
|---|---|---|---|
| 1 | 0 | 0 | $\mp\frac{1}{2}$ |
| 2 | 0 | 0 | $\mp\frac{1}{2}$ |
| | | −1 | $\mp\frac{1}{2}$ |
| | 1 | 0 | $\mp\frac{1}{2}$ |
| | | 1 | $\mp\frac{1}{2}$ |
| 3 | 0 | 0 | $\mp\frac{1}{2}$ |
| | 1 | −1 | $\mp\frac{1}{2}$ |
| | | 0 | $\mp\frac{1}{2}$ |
| | | 1 | $\mp\frac{1}{2}$ |
| | 2 | −2 | $\mp\frac{1}{2}$ |
| | | −1 | $\mp\frac{1}{2}$ |
| | | 0 | $\mp\frac{1}{2}$ |
| | | 1 | $\mp\frac{1}{2}$ |
| | | 2 | $\mp\frac{1}{2}$ |

$$1s2s2p3s3p4s3d4p5s4d5p6s4f5d6p7s6d$$

which is the same as that predicted by the *Paschen scheme* up to $Z = 88$.

## Review of Fundamental and Derived Equations

A listing of the fundamental and derived equations of this chapter is presented below, along with newly introduced physical constants and postulates. The derivations of modern physics are presented in a logical listing, which is similar to their development in each section of the chapter.

### FUNDAMENTAL EQUATIONS—CLASSICAL PHYSICS

|  |  |
|---|---|
| $\mu \equiv \dfrac{mM}{m + M}$ | Reduced Mass |
| $a_c = \dfrac{v^2}{r}$ | Centripetal Acceleration |
| $F = m\mathbf{a}, \qquad m \neq m(t)$ | Newton's Second Law |
| $F = -kx$ | Hooke's law |
| $F = -\dfrac{dV}{dx}$ | Conservative Force |
| $F_c = \dfrac{mv^2}{r}$ | Centripetal Force |
| $F_C = k\dfrac{qQ}{r^2}$ | Coulomb's Law |
| $W = \int \mathbf{F} \cdot d\mathbf{r}$ | Work |
| $T = \frac{1}{2}mv^2$ | Kinetic Energy |
| $E_t = E_k + E_p$ | Total Mechanical Energy |
| $P = \dfrac{2}{3}\dfrac{ke^2a^2}{c^3}$ | Power—Electron Radiation |
| $c = \lambda\nu$ | Speed of Light in a Vacuum |
| $\omega = \dfrac{d\theta}{dt}$ | Instantaneous Angular Velocity |

$v = r\omega$ Velocity Transformation

$\omega = 2\pi\nu$ Frequency

$I = mr^2$ Moment of Inertia

$L = I\omega$ Angular Momentum

$Q_N = Ze$ Nuclear Charge

$E = nh\nu$ Planck's Quantum Hypothesis

$\epsilon = h\nu$ Einstein's Photon Postulate

## EMPIRICAL EQUATIONS

$$\lambda = 911.4 \text{ Å} \left(\frac{n_i^2 n_f^2}{n_i^2 - n_f^2}\right)$$ Balmer's Formula

$$\frac{1}{\lambda} = R_{\text{H}} \left(\frac{1}{n_f^2} - \frac{1}{n_i^2}\right)$$ Rydberg Formula

$$\oint p_q dq = n_q h$$ Wilson-Sommerfeld Quantization Rule

## NEW PHYSICAL CONSTANTS

$$\hbar \equiv \frac{h}{2\pi}$$ Reduced Planck Constant

$$R_{\text{H}} = \frac{|E_1|}{hc} = \frac{\mu k^2 e^4}{2h\hbar^2 c} = 1.0967758 \times 10^7 \text{ m}^{-1}$$ Rydberg Constant

$$r_1 = \frac{\hbar^2}{mke^2} = 0.529178 \text{ Å}$$ Bohr Radius

$$v_1 = \frac{ke^2}{\hbar} = 2.18769 \times 10^6 \text{ m/s}$$ Bohr Velocity

$$\alpha = \frac{ke^2}{\hbar c} = 7.29735 \times 10^{-3}$$ Fine Structure Constant

$$\nu_1 = \frac{mk^2e^4}{h\hbar^2} = 6.57912 \times 10^{15} \text{ Hz}$$ Bohr Frequency

$$E_1 = -\frac{mk^2e^4}{2\hbar^2} = -13.6046 \text{ eV}$$ Bohr Energy

## FUNDAMENTAL POSTULATES

1. Bohr's Four Postulates for the One-Electron Atom
2. Pauli Exclusion Principle

## DERIVED EQUATIONS

### *Bohr Model—Infinite Nuclear Mass*

| | |
|---|---|
| $mv_n^2 = \dfrac{Zke^2}{r_n}$ | Bohr's 1st Postulate |
| $mv_nr_n = n\hbar$ | Bohr's 2nd Postulate |
| $r_n = \dfrac{n^2}{Z}\left(\dfrac{\hbar^2}{mke^2}\right)$ | Bohr Radii |
| $v_n = \dfrac{Z}{n}\left(\dfrac{ke^2}{\hbar}\right)$ | Bohr Velocities |
| $\nu_n = \dfrac{Z^2}{n^3}\left(\dfrac{mk^2e^4}{h\hbar^2}\right)$ | Bohr Frequencies |
| $E_n = -\dfrac{Zke^2}{2r_n}$ | Bohr's 3rd Postulate |
| $E_n = -\dfrac{Z^2}{n^2}\left(\dfrac{mk^2e^4}{2\hbar^2}\right)$ | Bohr Energies |
| $E_n = -\frac{1}{2}nh\nu_n$ | Energy Quantization |
| $\epsilon = E_i - E_f$ | Bohr's 4th Postulate |
| $\epsilon = Z^2\lvert E_1\rvert\left(\dfrac{n_i^2 - n_f^2}{n_i^2n_f^2}\right)$ | Photon Energy |

### *Bohr Model—Finite Nuclear Mass*

| | |
|---|---|
| $v_e = (r - x)\omega_e$ | Speed of Electron |
| $v_N = x\omega_N$ | Speed of Nucleus |
| $L = L_e + L_N = \mu r^2\omega = mvr$ | Total Angular Momentum |
| $L_e = \mu vr$ | Electron Angular Momentum |
| $\mu r\omega^2 = \dfrac{Zke^2}{r^2}$ | Bohr's 1st Postulate |

| | |
|---|---|
| $\mu r^2\omega = n\hbar$ | Bohr's 2nd Postulate |
| $r_n = \frac{n^2}{Z}\left(\frac{\hbar^2}{\mu k e^2}\right)$ | Atomic Radii |
| $E_n = -\frac{Zke^2}{2r_n}$ | Bohr's 3rd Postulate |
| $E_n = -\frac{Z^2}{n^2}\left(\frac{\mu k^2 e^4}{2\hbar^2}\right)$ | Atomic Energy States |

### Wilson-Sommerfeld Angular Momentum Quantization

| | |
|---|---|
| $mvr = n_\theta \hbar$ | Bohr Model—Infinite Nuclear Mass |
| $\mu r^2\omega = n_\theta \hbar$ | Bohr Model—Finite Nuclear Mass |

### Wilson-Sommerfeld Quantization—Linear Harmonic Oscillator

| | |
|---|---|
| $V = \frac{1}{2}kx^2$ | Potential Energy |
| $E = \frac{1}{2}mv^2 + \frac{1}{2}kx^2$ | Total Energy |
| $x = A \sin \omega t$ | Coordinate Solution |
| $\omega = \sqrt{\frac{k}{m}}$ | Angular Speed |
| $nh = \oint p_x \, dx$ | |
| $= 2E\oint \cos^2 \omega t \, dt = \frac{E}{\nu}$ | Energy Quantization |

## QUANTUM NUMBERS AND ELECTRON CONFIGURATIONS

| | |
|---|---|
| $n = 1, 2, 3, 4, 5, \cdots$ | Principal Quantum Number |
| $l = 0, 1, 2, \cdots, n - 1$ | Orbital Quantum Number |
| $m_l = -l, \cdots, 0, \cdots, l$ | Orbital Magnetic Quantum Number |
| $m_s = -\frac{1}{2}, +\frac{1}{2}$ | Spin Magnetic Quantum Number |
| $N_n = 2n^2$ | Maximum Number of Shell Electrons |

| | |
|---|---|
| $N_l = 2(2l + 1)$ | Maximum Number of Subshell Electrons |
| $n = 1\ 2\ 3\ 4\ 5 \cdots$ | Atomic Energy States |
| $\downarrow\downarrow\downarrow\downarrow\downarrow$ $K\ L\ M\ N\ O \cdots$ | Names of Electron Shells |
| $l = 0\ 1\ 2\ 3\ 4 \cdots$ | Angular Momentum States |
| $\downarrow\downarrow\downarrow\downarrow\downarrow$ $s\ p\ d\ f\ g \cdots$ | Names of Electron Subshells |
| $nl^{\text{No. of Electrons}}$ | Atomic Notation for Electron States |
| $1s,2s,2p,3s,3p,3d, \cdots$ | Bohr-Sommerfeld Scheme |
| $1s,2s,2p,3s,3p,4s,3d,4p,5s,4d,5p,6s,4f, \cdots$ | Paschen Triangle Scheme |
| $1s,2s,2p,3s,3p,4s,3d,4p,5s,4d,5p,6s,4f, \cdots$ | Quantum Mechanic's Scheme |

## Problems

**7.1** Classically, an electron in an orbit about a fixed proton obeys *Kepler's third law*, which has the form $T^2 = (\text{CONSTANT})\ r^3$. Here $T$ is the electron's orbital period and $r$ is the mean distance of separation between the electron and proton. Assuming a circular orbit for the electron, show that the constant in Kepler's law is $16\pi^3\epsilon_o m/e^2$ for the Rutherford model of the hydrogen atom.

*Solution:*

From fundamental relationships the period of the electron's orbit is given by (see Equation 7.9)

$$T = \frac{1}{\nu} = \frac{2\pi}{\omega} = \frac{2\pi r}{v},$$

so that the square of the period is

$$T^2 = \frac{4\pi^2 r^2}{v^2}.$$

Also, with orbital stability for ${}^1_1\text{H}$ given by Equation 7.6,

$$mv^2 = \frac{ke^2}{r},$$

the velocity squared can be solved for and substituted into the equation for $T^2$ to obtain

$$T^2 = \left(\frac{4\pi^2 m}{ke^2}\right) r^3.$$

Since $k \equiv 1/4\pi\epsilon_o$ (see Equation 5.49), our last equation gives the desired result.

**7.2** Like Problem 7.1, find the constant in Kepler's third law for the *reduced mass model* of the hydrogen atom.

*Answer:* $\dfrac{4\pi^2\mu}{ke^2}$

**7.3** After deriving the generalized equations for the Bohr radii (Equation 7.25) and Bohr velocities (Equation 7.28), find the principal quantum number and translational speed of an electron encircling a single fixed proton with a 1 m radius.

*Solution:*
Knowing $Z = 1$ and $r_n = 1$ m, we want to find $n$ and $v_n$. From Equation 7.25 we have

$$\begin{aligned} n &= \left[\frac{Zr_n}{r_1}\right]^{1/2} \\ &= \left[\frac{(1)\ (1\text{ m})}{(0.529178 \times 10^{-10}\text{ m})}\right]^{1/2} \\ &= (1.88972 \times 10^{10})^{1/2} \\ &= 137{,}467. \end{aligned}$$

The translational speed for such a Bohr electron is given by Equation 7.28,

$$\begin{aligned} v_n &= \frac{Z}{n} v_1 \\ &= \frac{(1)\ (2.18769 \times 10^6\text{ m/s}}{1.37467 \times 10^5} \\ &= 15.9143\text{ m/s}, \end{aligned}$$

where the value of $v_1$ from Equation 7.30 has been used.

**7.4** Show that the dimensions of the derived expressions for the Bohr radius (Equation 7.26) and the Bohr velocity (Equation 7.29) reduce to those of length and length divided by time, respectively.

*Answer:* m, m/s

**7.5** Show that the *fine structure constant* $\alpha$ is dimensionless. Find the ratio of the Bohr radii to the *reduced* Compton wavelength ($\hbar/mc$) and illustrate algebraically how it is related to $\alpha$.

*Solution:*
From Equation 7.32 we have

$$\alpha = \frac{ke^2}{\hbar c} \to \frac{(\text{Nm}^2/\text{C}^2)\text{C}^2}{(\text{Js})(\text{m/s})}$$
$$\to \frac{\text{Nm}}{\text{J}} = 1.$$

Also, with $\lambda_C' \equiv \hbar/mc$ and Equation 7.25 we obtain

$$\frac{r_n}{\lambda_C'} = \frac{(n^2/Z)\,(\hbar^2/mke^2)}{\hbar/mc}$$
$$= \frac{n^2}{Z}\left(\frac{\hbar c}{ke^2}\right)$$
$$= \frac{n^2}{Z}\frac{1}{\alpha}.$$

**7.6** After deriving the generalized equation for the Bohr frequencies (Equation 7.35), find the Bohr electron's radius, translational speed, and orbital frequency for $Z = 12$ and $n = 6$.

*Answer:* 1.58753 Å, $4.37538 \times 10^6$ m/s, $4.38608 \times 10^{15}$ Hz

**7.7** After deriving the generalized equation for the Bohr energies (Equation 7.39), find the translational speed and energy of a Bohr electron for $Z = 5$ and $n = 10$.

*Solution:*
For $n = 10$ and $Z = 5$, the speed of the electron is obtained from Equations 7.28 and 7.30, that is

$$v_{10} = \frac{Z}{n} v_1$$
$$= \frac{5}{10}\left(2.18769 \times 10^6 \frac{\text{m}}{\text{s}}\right)$$
$$= 1.09398 \times 10^6 \frac{\text{m}}{\text{s}},$$

while the energy is given by Equations 7.39 and 7.41,

$$E_{10} = -\frac{Z^2}{n^2}|E_1|$$
$$= -\frac{25}{100}(13.6046 \text{ eV})$$
$$= -3.40115 \text{ eV}.$$

**7.8** Show that the Bohr electron's binding energy in the ground state of a hydrogen atom is directly proportional to $\alpha^2$.

*Answer:* $E_1 = -\frac{1}{2}mc^2\alpha^2$.

**7.9** If the electron in a hydrogen atom is replaced by a negative muon ($m_\mu = 207m_e$), what changes occur in the allowable radii, velocities, and energies?

*Solution:*
Since only the mass of the negatively charged particle in the Bohr model has changed, the generalized equations for the radii and energies are the same as Equations 7.25 and 7.39, except $m$ is replaced with $m_\mu$. There is no change in the Bohr velocities, since Equation 7.28 is independent of the electron's mass. Accordingly, our equation for $r_n$ takes the form of

$$r_n = \frac{n^2}{Z}r_1',$$

where

$$r_1' = \frac{m_e}{m_\mu}r_1 = 2.55642 \times 10^{-13} \text{ m}.$$

Likewise, the equation for $E_n$ is of the form

$$E_n = -\frac{Z^2}{n^2}|E_1'|,$$

with $|E_1'|$ given by

$$|E_1'| = \frac{m_\mu}{m_e}|E_1| = 2816.15 \text{ eV}.$$

**7.10** A hydrogen electron is in its first excited state. Find the ionization energy of the atom and the frequency of a photon of this energy.

*Answer:* $E_2 = -3.40115$ eV, $\nu = 8.22388 \times 10^{14}$ Hz

**7.11** If the energy of a hydrogen electron is $-0.544$ eV, find the electron's principal quantum number $n$ and orbital frequency $\nu_n$.

*Solution:*
With $Z = 1$ and $E_n = -0.544$ eV, Equation 7.39 gives $n$ as

$$n = \left(\frac{-Z^2|E_1|}{E_n}\right)^{1/2}$$

$$= \left[-\frac{(1)(13.6 \text{ eV})}{(-0.544 \text{ eV})}\right]^{1/2}$$

$$= \sqrt{25} = 5.$$

The orbital frequency for the hydrogen electron in the $n = 5$ state is most easily obtained from Equation 7.42, that is

$$\begin{aligned} \nu_n &= \frac{-2E_n}{nh} \\ &= \frac{(-2)(-0.544\ \text{eV})(1.60 \times 10^{-19}\ \text{J/eV})}{5h} \\ &= \frac{3.48 \times 10^{-20}\ \text{J}}{6.63 \times 10^{-34}\ \text{Js}} \\ &= 5.25 \times 10^{13}\ \text{Hz}. \end{aligned}$$

Alternatively, Equation 7.35 could be used to directly calculate $\nu_n$ for $n = 5$ and $Z = 1$.

**7.12** If the radius and orbital speed of a Bohr electron is 4.232 Å and $1.10 \times 10^6$ m/s, respectively, what are the values for $n$ and $Z$?

*Answer:* $n = 4, Z = 2$

**7.13** Find the energy of a photon required to excite a hydrogen electron from the $K$-shell to the $M$-shell and from the $L$-shell to the $O$-shell.

*Solution:*

The absorbed or annihilated photon would have an energy given by the absolute magnitude of Equation 7.45, that is

$$\epsilon = Z^2|E_1| \frac{|n_i^2 - n_f^2|}{n_i^2 n_f^2},$$

For the $K$ to $M$ transition ($n_i = 1$ and $n_f = 3$) we obtain

$$\begin{aligned} \epsilon &= (13.6046\ \text{eV}) \frac{|1 - 9|}{(1)(9)} \\ &= 12.0930\ \text{eV}, \end{aligned}$$

while the $L$ to $O$ transition ($n_i = 2$ and $n_f = 5$) requires a photon of energy

$$\begin{aligned} \epsilon &= (13.6046\ \text{eV}) \frac{|4 - 25|}{(4)(25)} \\ &= 2.85697\ \text{eV}. \end{aligned}$$

**7.14** If a hydrogen electron makes a single transition from the $O$-shell to the $M$-shell, emitting the second spectral line of the Paschen series, what is the energy and wavelength of the emitted photon?

*Answer:* $\epsilon = 0.967438$ eV, $\lambda = 1.28158 \times 10^{-6}$ m

**7.15** Compute the wavelengths for the third, fourth, and fifth lines of the Lyman series of hydrogen.

*Solution:*
For the Lyman series $n_f = 1$ and the three spectral lines correspond to $n_i = 4, 5, 6$. Thus, direct substitution into Equation 7.50,

$$\lambda = \frac{911.346\ \text{Å}}{Z^2}\left(\frac{n_i^2 n_f^2}{n_i^2 - n_f^2}\right)$$

gives the following:

$$n_i = 4: \qquad \lambda_L = (911.346\ \text{Å})\,\frac{16}{15} = 972.102\ \text{Å},$$

$$n_i = 5: \qquad \lambda_L = (911.346\ \text{Å})\,\frac{25}{24} = 949.319\ \text{Å},$$

$$n_i = 6: \qquad \lambda_L = (911.346\ \text{Å})\,\frac{36}{35} = 937.384\ \text{Å}.$$

**7.16** Find the wavelengths of the first three spectral lines of the Paschen series.

*Answer:* 18747.7 Å, 12815.8 Å, 10936.2 Å

**7.17** Beginning at $n = 7$ an electron in a hydrogen atom makes six successive quantum jumps as it *cascades* downward through every lower energy level. Find the wavelength of each of the six photons emitted.

*Solution:*
With $Z = 1$ and $\lambda_B \equiv 911.346$ Å, Equation 7.50 immediately yields

$$n_i = 7,\ n_f = 6: \qquad \lambda = \lambda_B\,\frac{(49)(36)}{13} = 123663\ \text{Å},$$

$$n_i = 6,\ n_f = 5: \qquad \lambda = \lambda_B\,\frac{(36)(25)}{11} = 74564.7\ \text{Å},$$

$$n_i = 5,\ n_f = 4: \qquad \lambda = \lambda_B\,\frac{(25)(16)}{9} = 40504.3\ \text{Å},$$

$$n_i = 4,\ n_f = 3: \qquad \lambda = \lambda_B\,\frac{(16)(9)}{7} = 18747.7\ \text{Å},$$

$$n_i = 3,\ n_f = 2: \qquad \lambda = \lambda_B\,\frac{(9)(4)}{5} = 6561.69\ \text{Å},$$

$$n_i = 2,\ n_f = 1: \qquad \lambda = \lambda_B\,\frac{(4)(1)}{3} = 1215.13\ \text{Å}.$$

**7.18** A ground state helium electron annihilates a 243 Å photon and then emits a photon of $2.46 \times 10^{15}$ Hz. What is the electron's final state binding energy in eV?

*Answer:* $E_f = -13.6$ eV

**7.19** A ground state hydrogen atom absorbs 10.20345 eV of *excitation energy*. Afterward, the electron absorbs a 4860.512 Å photon. What is the binding energy of the electron's final state?

*Solution:*
With $n_i = 1$, $Z = 1$, $E_{\text{ABSORBED}} = 10.20345$ eV, and $\lambda_\epsilon = 4860.512$ Å (absorbed), we have the final state energy given by

$$\begin{aligned} E_f &= E_i + E_{\text{ABSORBED}} + \epsilon_{\text{ABSORBED}} \\ &= -\frac{Z^2}{n_i^2}|E_1| + E_{\text{ABSORBED}} + \epsilon_{\text{ABSORBED}} \\ &= -13.6046 + 10.20345 + \epsilon_{\text{ABSORBED}} \\ &= -3.40115 \text{ eV} + \epsilon_{\text{ABSORBED}}. \end{aligned}$$

Thus, after being excited to the $n = 2$ state (see Table 7.3) the electron annihilates a photon of energy

$$\epsilon_{\text{ABSORBED}} = \frac{hc}{\lambda_\epsilon} = 2.550862 \text{ eV}.$$

Now, substitution into the equation for $E_f$ *gives*

$$E_f = -0.850288,$$

which corresponds to $n_f = 4$.

**7.20** In the reduced mass model of the one-electron atom, the centripetal force on the nucleus must be equal in magnitude and oppositely directed to the centripetal force on the electron. Show that an application of Newton's third law of motion on this model of the atom reduces to the definition of the center of mass of the system, as given by Equation 7.54.

*Answer:* $Mx = m(r - x)$

**7.21** Verify that the dimension of the Rydberg constant is inverse length and show that the product $R_\text{H} r_1$ is dimensionless, as given by $\alpha/4\pi$.

*Solution:*
From Equation 7.51 we have

$$\begin{aligned} R_\text{H} &= \frac{|E_1|}{hc} \\ &= \frac{|E_1|}{h\lambda\nu} \\ &= \frac{|E_1|}{\epsilon}\frac{1}{\lambda}, \end{aligned}$$

so the units are obviously those of inverse wavelength or $\text{m}^{-1}$. From Equations 7.51, 7.26, and 7.40 we obtain

$$
\begin{aligned}
R_H r_1 &= \frac{|E_1|}{hc} r_1 \\
&= \frac{mk^2e^4/2\hbar^2}{hc} \frac{\hbar^2}{mke^2} \\
&= \frac{ke^2}{2hc} \\
&= \frac{1}{4\pi} \frac{ke^2}{\hbar c} \\
&= \frac{\alpha}{4\pi},
\end{aligned}
$$

where Equation 7.32 has been used in obtaining the last equality. This result is clearly the same whether we use the fixed nuclear mass model or the finite nuclear mass model.

**7.22** Name the first four *shells* of the Bohr-Sommerfeld atomic model and calculate the maximum number of electrons allowed in each shell.

*Answer:* $K \rightarrow 2,\ L \rightarrow 8,\ M \rightarrow 18,\ N \rightarrow 32$

**7.23** Using the *atomic notation* defined in Equation 7.103 and Paschen's triangle, write down the *electronic configuration* for argon ($Z = 18$), iron ($Z = 26$), and silver ($Z = 47$).

*Solution:*
For the three elements given we have

$$
\begin{aligned}
{}_{18}\text{Ar}&: 1s^2 2s^2 2p^6 3s^2 3p^6, \\
{}_{26}\text{Fe}&: 1s^2 2s^2 2p^6 3s^2 3p^6 4s^2 3d^6, \\
{}_{47}\text{Ag}&: 1s^2 2s^2 2p^6 3s^2 3p^6 4s^2 3d^{10} 4p^6 5s^2 4d^9.
\end{aligned}
$$

**7.24** Name each of the *subshells* associated with the $N$-shell and calculate the maximum number of electrons allowed in each subshell.

*Answer:* $s \rightarrow 2,\ p \rightarrow 6,\ d \rightarrow 10,\ f \rightarrow 14$

**7.25** Using the Pauli exclusion principle and Equation 7.110, write down the eight quantum states allowed for electrons in the $L$-shell.

*Solution:*
For the $L$-shell $n = 2$ and the quantum numbers $l$, $m_l$, and $m_s$ have allowed values as given in Table 7.8. Accordingly, the allowed quantum states are given by

$$
\begin{aligned}
&(2,0,0,\mp\tfrac{1}{2}), \\
&(2,1,-1,\mp\tfrac{1}{2}), \\
&(2,1,0,\mp\tfrac{1}{2}), \\
&(2,1,1,\mp\tfrac{1}{2}).
\end{aligned}
$$

CHAPTER 8

# Introduction to Quantum Mechanics

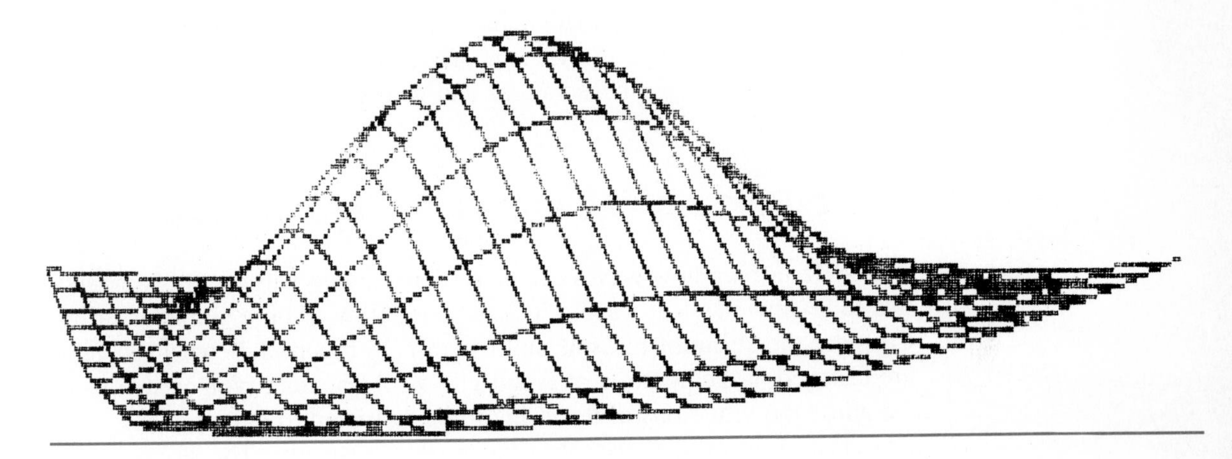


## Introduction

The preceding chapters were concerned with departures from classical views of relative motion, matter, and electromagnetic radiation in the conceptualization and theoretical description of natural phenomena. The classical wave model of light was found to have limited applicability and, consequently, a particle view of radiation as consisting of *quanta* was postulated by Einstein. The wave and particle models were demonstrated to be complementary in that together they provide a complete description of all observed radiation phenomena. In this chapter it is postulated that a dual wave-particle behavior is characteristic of not only electromagnetic radiation but also fundamental *particles*, such as electrons and protons. As will be discussed, this newly postulated *wave nature of matter* necessitates the description of a particle as a *matter wave*, which is *not localized* in space and time.

We begin with a discussion of the *classical wave equation* and its solutions for a vibrating string. This review of the classical wave equation will prove to be most instructive, as later in E. Schrödinger's quantum mechanics the *wave function* that characterizes a particle will be found to be a solution to an analogous partial differential equation. Before introducing de Broglie's postulates for the frequency and wavelength of particles, our review of classical waves is extended to include a generalized theoretical discussion of *traveling waves*. After detailing *de Broglie's hypothesis* for the wave nature of particles, we consider a confirmation of the hypothesis by a *diffraction* experiment involving electrons and its consistency with Bohr's quantization hypothesis and Einsteinian relativity. Our next subject of inquiry considers *matter waves* and the *principle of linear superposition*, followed by a theoretical discussion of *group*, *phase*, and *particle velocities*. Finally, the famous *Heisenberg uncertainty principle* is developed from general observations and insights on *matter waves*.

## 8.1 Equation of Motion for a Vibrating String

Before describing how waves can be associated with particles, it is instructive to review classical waves, such as those allowed on a vibrating string. Consider the string of mass $M$ and length $L$ in Figure 8.1, where its ends are fastened at the points $x = 0$ and $x = L$. As a simplifying assumption, we allow the vibrations of the string to be restricted to the $x$-$y$ plane, such that each point on the string can move only vertically. The string is also assumed to be of uniform linear density defined by

Linear Density

$$\mu \equiv \frac{M}{L}, \tag{8.1}$$

which allows the linear density of an elemental segment of length $dx$ and mass $dm$ to be expressed as

$$\mu = \frac{dm}{dx}. \tag{8.2}$$

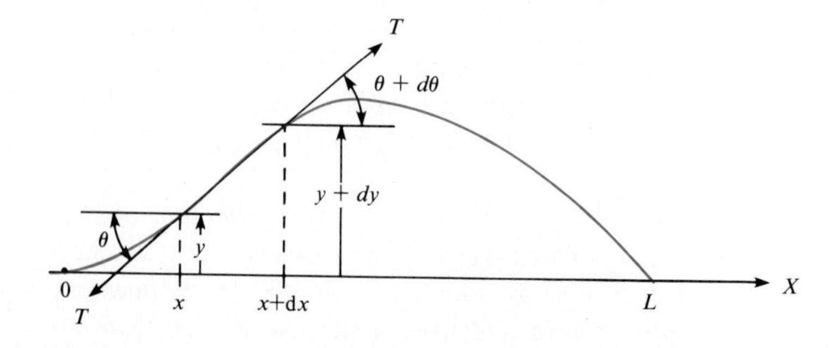


Figure 8.1
The vibrating string fastened at $x = 0$ and $x = L$.

Further, the amplitude of vibration is assumed to be sufficiently small such that the tension $T$ in the string can be considered as essentially constant. With these assumptions and Figure 8.1, the *net longitudinal* (horizontal) *force* acting on an elemental string segment of length $dx$ is given by

$$dF_x = T\cos(\theta + d\theta) - T\cos\theta \approx 0,$$

which is essentially zero for *very small* angular displacements $\theta$ (i.e., $\cos\theta \approx 1$ and $\cos(\theta + d\theta) \approx 1$). This result is obviously necessary for consistency with our initial *simplifying assumption*, which allows each elemental segment of the string to experience only transverse (vertical) motion. Since $\sin\theta \approx \theta$ and $\sin(\theta + d\theta) \approx \theta + d\theta$ for small angular displacements, the *net transverse force* acting on the segment $dx$ is given by

$$\begin{aligned} dF_y &= T\sin(\theta + d\theta) - T\sin\theta \\ &\approx T(\theta + d\theta) - T\theta \\ &= Td\theta. \end{aligned} \tag{8.3}$$

From Newton's second law of motion and Equation 8.2, this force is given by

$$\begin{aligned} dF_y &= dm\,a_y \\ &= \mu dx\,\frac{\partial^2 y}{\partial t^2}. \end{aligned} \tag{8.4}$$

In the last equality the transverse acceleration has been expressed by the second order partial derivative

$$a_y = \frac{\partial^2 y}{\partial t^2}, \tag{8.5}$$

since the transverse or vertical displacement of any point on the string is a function of both $x$ and $t$, that is, $y = y(x, t)$. Thus, by equating Equations 8.3 and 8.4 the equation representing the transverse motion of an elemental segment $dx$ of the string is obtained as

$$Td\theta = \mu dx\,\frac{\partial^2 y}{\partial t^2},$$

which is easily rearranged in the form

$$\frac{d\theta}{dx} = \frac{\mu}{T}\frac{\partial^2 y}{\partial t^2}. \tag{8.6}$$

The *equation of motion* represented by Equation 8.6 can be expressed in a more recognizable form by operating on

$$\tan\theta = \frac{\partial y}{\partial x}$$

with $d/dx$. That is,

$$\frac{d}{dx}\tan\theta = \sec^2\theta\,\frac{d\theta}{dx},$$

thus we obtain

$$\begin{aligned}\frac{d\theta}{dx} &= \frac{1}{\sec^2\theta}\frac{d}{dx}\tan\theta \\ &= \cos^2\theta\,\frac{\partial^2 y}{\partial x^2},\end{aligned}$$

which under the *small angle approximation* becomes

$$\frac{d\theta}{dx} \approx \frac{\partial^2 y}{\partial x^2}. \tag{8.7}$$

With this result for $d\theta/dx$, Equation 8.6 yields the more recognizable *wave equation*

$$\frac{\partial^2 y}{\partial x^2} = \frac{\mu}{T}\frac{\partial^2 y}{\partial t^2}. \tag{8.8}$$

Since the fundamental units of $\mu/T$ are inverse velocity squared, we take the liberty of defining

$$w = \sqrt{\frac{T}{\mu}}, \tag{8.9}$$

which will later be identified as the *phase velocity* with which traveling waves propagate on a long string. From Equations 8.8 and 8.9 we obtain the well known second order partial differential equation representing the *equation of motion* for the vibrating string

Classical Wave Equation

$$\frac{\partial^2 y}{\partial x^2} = \frac{1}{w^2}\frac{\partial^2 y}{\partial t^2}. \tag{8.10}$$

Although this equation was derived for the vibrating string, it should be recognized as the **classical wave equation,** whose solutions $y(x, t)$ can represent waves traveling along a string or wire, sound waves propagating through a material medium, or electromagnetic waves propagating in a vacuum. In all cases the *phase velocity* $w$ is dependent on only the physical properties of the medium and in the case of light waves in a vacuum $w = c$. The solution $y(x, t)$ to Equation 8.10 must be obtained for any given *initial position* $y(x)$ and *velocity* $v(x)$ of each point along the string in conjunction with the imposed *boundary conditions*. That is, at time $t = 0$, $y(x, t)$ must satisfy the **initial conditions**

$$y(x, 0) = y_0(x), \tag{8.11a}$$

Initial Conditions

$$\left(\frac{\partial y}{\partial t}\right)_{t=0} = v_0(x) \tag{8.11b}$$

and also the **boundary conditions**

$$y(0, t) = y(L, t) = 0. \tag{8.12}$$

Boundary Conditions

These *boundary conditions* simply express the fact that the string is tied at both ends and the displacements at these two points must always be zero. The *particular solution* of Equation 8.10 will be the usual *standing wave*, which will be obtained in the next section by the *linear superposition* of *traveling waves*.

## 8.2 Normal Modes of Vibration for the Stretched String

The solution to the *classical wave equation* will be presented in detail, since the mathematical procedures are useful in Schrödinger's quantum mechanics. One method for solving the second order partial differential equation is the method of **separation of variables,** which consists in looking for solutions of the form

$$y(x, t) = f(x)h(t). \tag{8.13}$$

Separation of Variables

Taking derivatives of Equation 8.13 with respect to $x$ and $t$ gives

$$\frac{\partial^2 y}{\partial x^2} = h\,\frac{d^2 f}{dx^2},$$

$$\frac{\partial^2 y}{\partial t^2} = f\frac{d^2 h}{dt^2},$$

which can be substituted into Equation 8.10 to obtain

$$\frac{w^2}{f}\frac{d^2 f}{dx^2} = \frac{1}{h}\frac{d^2 h}{dt^2}. \tag{8.14}$$

Since the left side of this equation is a function of $x$ only and the right side is a function of $t$ only, both sides may be set equal to a constant. Deciding what physical constant to use is easily facilitated by observing that the right side expresses acceleration divided by a displacement. Consequently, the constant will have units of $1/\text{s}^2$ and should be *negative valued*, as acceleration must be opposite the displacement at every value of $t$ for the string to return to its equilibrium position. Since angular velocity $\omega$ has fundamental units of $1/\text{s}$, then the constant chosen is $\omega^2$:

$$\frac{1}{h}\frac{d^2 h}{dt^2} = -\omega^2, \tag{8.15a}$$

$$\frac{w^2}{f}\frac{d^2 f}{dx^2} = -\omega^2. \tag{8.15b}$$

These equations can be rewritten as

$$\frac{d^2 h}{dt^2} + \omega^2 h = 0, \tag{8.16}$$

$$\frac{d^2 f}{dx^2} + \frac{\omega^2}{w^2} f = 0, \tag{8.17}$$

which are well known and relatively simple differential equations to solve.

Equations 8.16 and 8.17 are of the same form as the *equation of motion* (Equation 7.87) for the linear harmonic oscillator of Chapter 7. Although the sine function was used as a solution in that problem, the general solution to Equation 8.16 should be a combination of sine and cosine functions of the form

$$h(t) = A\cos\omega t + B\sin\omega t, \tag{8.18}$$

where $A$ and $B$ are arbitrary constants. The solution to Equation 8.17 is very similar and of the form

$$f(x) = C \cos \frac{\omega x}{w} + D \sin \frac{\omega x}{w}, \tag{8.19}$$

but it must satisfy the boundary conditions expressed by Equation 8.12. That is, at $x = 0$, $f(x)$ of Equation 8.19 must equal zero, which can only be satisfied if

$$C = 0. \tag{8.20}$$

Consequently, Equation 8.19 reduces to

$$f(x) = D \sin \frac{\omega x}{w}. \tag{8.21}$$

The second boundary condition requires $f(x)$ to be equal to zero at $x = L$. This condition is satisfied when

$$f(L) = D \sin \frac{\omega L}{w} = 0, \tag{8.22}$$

which requires that

$$\frac{\omega L}{w} = n\pi, \qquad n = 1, 2, 3, 4, \cdots. \tag{8.23}$$

Considering the vibrating string in its first few normal modes, as illustrated in Figure 8.2, we have the general equation

$$L = \frac{n\lambda}{2} \tag{8.24}$$

Standing Waves

suggested for the wavelength as a function of the **vibrating mode** $n$. From this equation the condition for standing waves is clearly that the length of

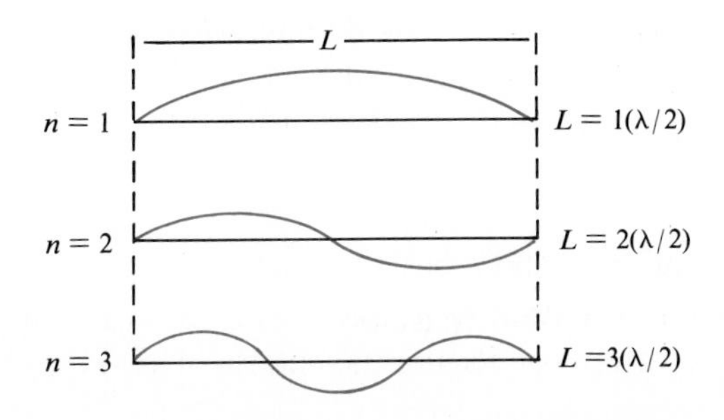


Figure 8.2
The normal modes of a vibrating string.

the string must be an integral number of half wavelengths. Substituting Equation 8.24 into Equation 8.23 gives

$$\frac{\omega}{w} = \frac{n\pi}{L} = \frac{n\pi}{n\lambda/2}$$
$$= \frac{2\pi}{\lambda} \equiv k, \tag{8.25}$$

where $k$ is the *wave number* or propagation constant originally defined in Equation 6.3.

Using the result expressed by Equation 8.25 in Equation 8.21, the solution to Equation 8.17 is

$$f(x) = D \sin kx, \tag{8.26}$$

while the solution to Equation 8.16 is given by Equation 8.18. Using Equations 8.26 and 8.18 for the assumed general solution (Equation 8.13) and then substituting into the *equation of motion* of the vibrating string (Equation 8.10) gives

$$y(x, t) = AD \sin kx \cos \omega t + BD \sin kx \sin \omega t. \tag{8.27}$$

Since the coefficients $AD$ and $BD$ are arbitrary constants, we can let $AD \equiv A$ and $BD \equiv B$ and express the solution as

General Solution

$$y(x, t) = A \sin kx \cos \omega t + B \sin kx \sin \omega t. \tag{8.28}$$

This solution is representative of the *normal mode of vibration* of the string, where each point on the string vibrates at the same frequency

$$\omega_n = 2\pi\nu_n = \frac{2\pi w}{\lambda_n}. \tag{8.29}$$

The initial position and velocity at $t = 0$ is easily obtained from the initial conditions (Equations 8.11a and 8.11b) and Equation 8.28 to be

$$y_0(x) = A \sin k_n x, \tag{8.30}$$
$$v_0(x) = \omega_n B \sin k_n x, \tag{8.31}$$

where the subscript on $\omega$ and $k$ has been introduced to signify their dependence on the *mode of vibration n*. Only with these initial conditions will the string vibrate in one of its normal modes. However, a more general solution may be obtained by invoking the *principle of linear superposition*. That is, by adding solutions of the type given by Equation 8.28, using different constants $A$ and $B$ for each normal mode, we have

$$y(x, t) = \sum_{n=1}^{\infty} (A_n \sin k_n x \cos \omega_n t + B_n \sin k_n x \sin \omega_n t), \quad (8.32)$$

with initial conditions

$$y_0(x) = \sum_{n=1}^{\infty} A_n \sin k_n x, \quad (8.33)$$

$$v_0(x) = \sum_{n=1}^{\infty} \omega_n B_n \sin k_n x, \quad (8.34)$$

as the general solution to Equation 8.10 that satisfies the boundary conditions expressed by Equation 8.12. Actually, the general solution is only completely known when each of the infinite number of coefficients $A_n$ and $B_n$ are known. These may be determined directly from the initial conditions, as given in Equations 8.33 and 8.34, by expressing $k_n$ as $n\pi/L$, multiplying by $\sin(m\pi x/L)\,dx$, and integrating from $x = 0$ to $x = L$ to obtain

$$A_n = \frac{2}{L}\int_0^L y_0(x) \sin k_n x\, dx, \quad (8.35)$$

$$B_n = \frac{2}{\omega_n L}\int_0^L v_0(x) \sin k_n x\, dx. \quad (8.36)$$

## 8.3 Traveling Waves and the Classical Wave Equation

Although the general solution to the classical wave equation has been obtained, it is interesting to note that Equations 8.16 and 8.17 also have complex solutions

$$h(t) = Ae^{\pm j\omega t}, \quad (8.37)$$

$$f(x) = e^{jkx}, \quad (8.38)$$

where $j \equiv \sqrt{-1}$. These solutions are easily verified by direct substitution into the respective partial differential equations and realizing that

$$k \equiv \frac{2\pi}{\lambda} = \frac{2\pi}{w/\nu} = \frac{2\pi\nu}{w} = \frac{\omega}{w}. \tag{8.39}$$

Consequently, the classical wave equation (Equation 8.10) has complex solutions of the form

$$y(x, t) = Ae^{j(kx \pm \omega t)}. \tag{8.40}$$

As only the real part of a complex solution is of physical significance, in the classical case the solutions expressed by Equation 8.40 are of the form

$$y(x, t) = A \cos(kx + \omega t), \tag{8.41a}$$
$$y(x, t) = A \cos(kx - \omega t). \tag{8.41b}$$

Since the sine and cosine are similar mathematical functions, then

$$y(x, t) = A \sin(kx + \omega t), \tag{8.42a}$$
$$y(x, t) = A \sin(kx - \omega t) \tag{8.42b}$$

are also possible solutions. Even though all of these solutions (Equations 8.40 to 8.42) satisfy the equation of motion of a vibrating string, they, unfortunately, do not satisfy all of the *initial* and *boundary* conditions expressed by Equations 8.11 and 8.12. However, they are of considerable interest in that they represent waves traveling down the string, as will be directly illustrated.

According to the solutions given above, any particular point on a vibrating string will move with simple harmonic motion in time with amplitude $A$ and frequency $\omega$. Instead of using the exponential, sine, or cosine functions explicitly, let the solutions be generalized to the form

$$y_+(x, t) = Af(kx + \omega t) \tag{8.43a}$$

and

$$y_-(x, t) = Af(kx - \omega t). \tag{8.43b}$$

Since angular velocity $\omega$ is related to wave number $k$ by the relation (see Equation 8.25 or 8.39)

$$\omega = kw,$$

then the generalized solutions can be expressed as

$$y_+(x, t) = Af(kx + kwt)$$

and

$$y_-(x, t) = Af(kx - kwt).$$

Defining the *phase* of these waves by

$$\alpha \equiv x + wt \tag{8.44a}$$

and

$$\beta \equiv x - wt, \tag{8.44b}$$

our generalized solutions become

$$y_+(x, t) = Af(k\alpha) \tag{8.45a}$$

and

$$y_-(x, t) = Af(k\beta). \tag{8.45b}$$

For a constant value of the phase $\alpha$, $y_+(x, t)$ of Equation 8.45a has a fixed value. That is, for a constant phase $d\alpha = 0$ and Equation 8.44a gives

$$dx + wdt = 0.$$

Solving this result for the *phase velocity*,

$$w = -\frac{dx}{dt}, \tag{8.46a}$$

results in $w$ being *negative* valued. This means that the value of $y_+(x, t)$ at the point $x + dx$ at the time $t + dt$ will be the same as its value at the point $x$ at time $t$. Thus, the wave moves along the string in the *negative* $x$-direction with a constant *phase velocity*. Likewise, from Equation 8.44b

$$dx - wdt = 0$$

for a constant value of the phase $\beta$. Consequently, for the solution given by Equation 8.43b

$$w = \frac{dx}{dt} \tag{8.46b}$$

and the wave is recognized as moving along the string in a *positive* $x$-direction with constant phase $\beta$.

The above results suggest that solutions of the general form given by Equations 8.43a and 8.43b represent *traveling waves* moving in either the *negative* or *positive* $x$-direction, respectively. These solutions can be combined and expressed as

$$y(x, t) = f(kx \pm \omega t), \tag{8.47}$$

where both directions of the phase velocity are included and $f$ represents either a complex exponential, sine, or cosine function. In what follows, a general mathematical procedure is presented for obtaining the *wave equation* for which $y(x, t)$ is a solution. Even though we know the answer, such a procedure will prove useful later in the development of Schrödinger's quantum mechanics. The wave equation for which $y(x, t)$ of Equation 8.47 is a solution is directly obtainable by defining

$$\gamma \equiv kx \pm \omega t \tag{8.48}$$

and taking a few partial derivatives. The second order partial derivative of $y(x, t)$ with respect to $x$ is just

$$\begin{aligned}\frac{\partial^2 y}{dx^2} &= \frac{\partial}{\partial x}\frac{\partial f}{\partial x} \\ &= \frac{\partial}{\partial x}\left(\frac{df}{d\gamma}\frac{\partial \gamma}{\partial x}\right) \\ &= k\,\frac{\partial}{\partial x}\frac{df}{d\gamma} \\ &= k\,\frac{d}{d\gamma}\frac{\partial f}{\partial x} \\ &= k\,\frac{d}{d\gamma}\left(\frac{df}{d\gamma}\frac{\partial \gamma}{\partial x}\right) \\ &= k^2\,\frac{d^2 f}{d\gamma^2},\end{aligned} \tag{8.49}$$

where the explicit form of $\gamma$ has been used in reducing $\partial\gamma/\partial x$. The second order partial derivative of $y(x, t)$ with respect to $t$ is obtained in a similar manner as

$$\frac{\partial^2 y}{\partial t^2} = \omega^2\,\frac{d^2 f}{d\gamma^2}. \tag{8.50}$$

From Equations 8.49 and 8.50 we have

$$\frac{d^2 f}{d\gamma^2} = \frac{1}{k^2}\frac{\partial^2 y}{\partial x^2} = \frac{1}{\omega^2}\frac{\partial^2 y}{\partial t^2},$$

which means

$$\frac{\partial^2 y}{\partial x^2} = \frac{k^2}{\omega^2}\frac{\partial^2 y}{\partial t^2}. \tag{8.51}$$

Since $k = \omega/w$ (Equation 8.25), the result of Equation 8.51 reduces exactly to the previously derived classical wave equation. Certainly, then, any wave traveling with a phase velocity $w$ is a solution to the classical wave equation.

## 8.4 De Broglie's Hypothesis

Electromagnetic radiation has been shown to possess dual properties of wave and particle-like behavior. This dual nature of light is difficult to accept, since a wave and a particle are fundamentally different in classical physics. Typically, a wave is characterized by an amplitude $A$, intensity $I$, phase velocity $w$, wavelength $\lambda$, and frequency $\nu$; whereas, mass $m$, velocity $v$, momentum $p$, and energy $E$ are specified for a particle. The conceptual difficulty arises because at any instant in time a particle is considered to occupy a very definite position in space, but a wave is necessarily extended over a relatively large region of space. The acceptance of this dual property is necessary, however, for the satisfactory explanation of all physical phenomena observed and quantitatively measured for the radiation of the electromagnetic spectrum.

The connection between the wave and particle behavior of *electromagnetic radiation* was postulated by Einstein to be

$$\epsilon = h\nu, \tag{6.53}$$

Einstein—Photon Energy

where $\epsilon$ is the photon's energy and $\nu$ the frequency of the associated wave. We have seen how this fundamental postulate of nature has been of importance in the development of the relativistic Doppler effect (Chapter 6, Section 6.8), the photoelectric equation, the Compton equation, and the Bohr model of the hydrogen atom. In fact, the relationship between a particle-like momentum and wavelength was obtained in our discussion of the Compton effect (Chapter 6, Section 6.7) as

$$p_\epsilon = \frac{h}{\lambda} \tag{6.64}$$

Compton—Photon Momentum

by a utilization of Einstein's photon postulate. Although this *dual* property appears contradictory, these two equations allow the particle-like properties of energy and momentum for a *photon* to be directly obtained from the wave characteristics of frequency and wavelength, respectively.

Believing symmetry and simplicity in physical phenomena to be fundamental in nature, Louis de Broglie considered ordinary particles like electrons and alpha particles as manifesting wave characteristics. In his 1924 dissertation de Broglie considered **matter waves** (originally called *pilot waves*) as being associated with particles and the motion of a particle as being governed by the wave propagation properties of the associated *matter waves*. By analogy with the *modern view* of the dual nature of electromagnetic radiation, de Broglie postulated the frequency and wavelength of a particle's associated *matter wave* to be determined by the particle's energy $E$ and momentum $p$, respectively, by

De Broglie—Frequency

$$\nu = \frac{E}{h}, \tag{8.52}$$

De Broglie—Wavelength

$$\lambda = \frac{h}{p}. \tag{8.53}$$

This hypothesis concerning the wave properties of matter was fundamentally necessary for the later development of the modern *quantum theory*.

It should be emphasized that de Broglie's hypothesis was based *entirely* on his physical insight of nature and constituted a *serious* departure from *conventional* thinking. In fact, at that time there was no direct experimental evidence available to support his hypothesis of particles having wave-like characteristics. It was *three* years (1927) after de Broglie published his dissertation on the postulated *wave nature* of particles that experimental confirmation was reported by C. Davisson and L. Germer of Bell Telephone Laboratories. Davisson and Germer were studying the scattering of a beam of electrons from a single large crystal of nickel. The results of their experiment suggested that electrons were being *diffracted* from crystal planes, much like x-ray diffraction obeying the Bragg condition (Chapter 6, Section 6.3) for *constructive interference*. In one particular experiment a beam of 54 V electrons, obtained from a hot cathode, were scattered at a 65° angle with respect to the crystal planes. The atomic spacing of the planes was measured by x-ray diffraction to be 0.91 Å. Thus, the *wavelength* associated with the *electrons* is directly calculable using the equation for *Bragg diffraction*,

Bragg's Law

$$n\lambda = 2d \sin \theta, \tag{6.31}$$

developed in Section 6.3. With $n = 1$, the *electron wavelength* is given by

$$\lambda = 2(0.91 \text{ Å}) \sin 65° = 1.65 \text{ Å}. \tag{8.54}$$

This wavelength reported by Davisson and Germer is very nearly predicted by de Broglie's wavelength postulate. The energy of the beam of electrons is obtained from the electric field between two electrodes. Consequently, using the same reasoning as presented in Chapter 5 leading to Equation 5.13, the electrons have kinetic energy given by

$$T = \frac{p^2}{2m} = eV,$$

where $e$ is the charge of an electron and $V = 54$ V is the electrical potential between the electrodes. From this relation and de Broglie's wavelength postulate we obtain

$$\begin{aligned}\lambda &= \frac{h}{(2meV)^{1/2}} \qquad (8.55)\\ &= h/[2(9.11 \times 10^{-31} \text{ kg})(1.60 \times 10^{-19} \text{ C})(54 \text{ V})]^{1/2}\\ &= \frac{6.63 \times 10^{-34} \text{ J} \cdot \text{s}}{3.97 \times 10^{-24} \text{ kg} \cdot \text{m/s}}\\ &= 1.67 \text{ Å},\end{aligned}$$

which is very close to the Davisson and Germer observed wavelength given in Equation 8.54. The slight difference is attributable to the refraction of electron waves at the air-crystal boundary, which we have ignored in our theoretical calculation. The kinetic energy of an electron actually increases slightly when it enters the nickel crystal, with the net effect of the associated de Broglie wavelength being slightly less than that predicted by Equation 8.55. Despite this slight discrepancy between the theoretical and experimental results, the Davisson and Germer *diffraction* experiment confirmed de Broglie's wave hypothesis for particles. Since that time, the *diffraction* of atomic particles and atoms has been frequently demonstrated and widely used in studying crystal structure.

## Consistency with Bohr's Quantization Hypothesis

Although de Broglie's *frequency* and *wavelength* postulates were not immediately confirmed by experimental observations, they did provide a fundamental explanation for Bohr's apparently arbitrary quantization condition governing the *discrete* energy levels of an electron in a hydrogen atom. In the previous classical consideration of a string of length $L$ that is fastened at both ends, standing waves occurred when the length of the string was exactly equal to an integral value of half wavelengths, as illustrated in Figure 8.2. Now, however, the consideration is to obtain the condition for de Broglie standing waves associated with an electron traveling in a circular orbit of radius $r$ and length $2\pi r$. If a string was formed into a circular loop, waves could be thought of as propagating around the string in both directions, but now there would be no *reflections*. Consequently, as suggested by Figure 8.3, the condition for standing waves in a circular path is

Standing Waves

$$2\pi r = n\lambda, \qquad n = 1, 2, 3, \cdots. \tag{8.56}$$

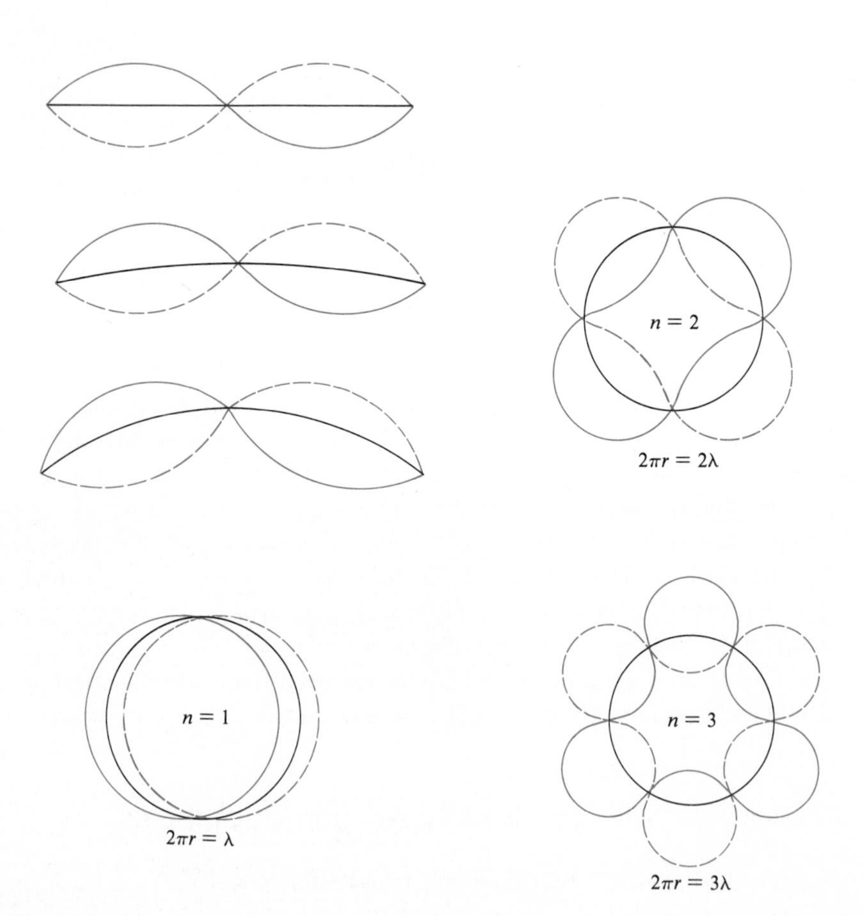


Figure 8.3
The electron orbit (brown line) in a Bohr atom and its associated de Broglie standing waves (black line).

Substitution of the de Broglie *wavelength* postulate (Equation 8.53) into this equation immediately yields

$$2\pi r = \frac{nh}{p}.$$

This result can be rewritten as

$$pr = n\hbar,$$

where $\hbar$ is the normal Planck's constant $h$ divided by $2\pi$. Since the electron momentum is just $mv$, the last equation becomes

$$mvr = n\hbar, \tag{8.57}$$

which is precisely Bohr's quantum hypothesis (Equation 7.24) presented in Chapter 7. In essence we have that Bohr's quantization condition could be replaced by the more fundamental de Broglie *wavelength* postulate.

The de Broglie *frequency* postulate is also completely consistent with the Bohr model of the hydrogen atom. To realize this, we must first recognize that the **phase velocity** $w$, defined by

$$w \equiv \lambda\nu, \tag{8.58}$$

Phase Velocity

can be expressed in terms of the electron's energy $E$ and momentum $p$,

$$w = \frac{E}{p}, \tag{8.59}$$

by direct substitution of de Broglie's wavelength and frequency postulates. Since the *nonrelativistic* energy of a particle is given by

$$E = \tfrac{1}{2}mv^2 = \frac{p^2}{2m}, \tag{8.60}$$

then substitution into Equation 8.59 immediately yields

$$w = \frac{p}{2m}. \tag{8.61}$$

Thus, the *phase velocity* $w$ is directly related to the *particle velocity* $v$ by the relation

$$w = \frac{v}{2}, \tag{8.62}$$

by substituting $p = mv$ in Equation 8.61. Returning to de Broglie's *frequency* postulate, we can express the energy of an electron in a hydrogen atom as

$$E = -h\nu, \tag{8.63}$$

where the negative sign simply indicates the electron is in a stable *bound* circular orbit about the proton. Since $\nu = w/\lambda$ from Equation 8.58, then from Equations 8.62 and 8.63 we have

$$\begin{aligned} E &= -\frac{hw}{\lambda} \\ &= -\frac{hv}{2\lambda}. \end{aligned} \tag{8.64}$$

But, from Equation 8.56 $\lambda = 2\pi r/n$, thus our energy equation becomes

$$\begin{aligned} E &= -\frac{nhv}{2(2\pi r)} \\ &= -\frac{n\hbar v}{2r}, \end{aligned} \tag{8.65a}$$

where the definition for $\hbar$ has been utilized. Clearly, the electron's energy $E$, velocity $v$, and radius $r$ are dependent on the principal quantum number $n$. As such, we rewrite the energy equation using a subscript notation as

$$E_n = -\frac{n\hbar v_n}{2r_n}. \tag{8.65b}$$

From Chapter 7, Section 7.3 $r_n$ and $v_n$ are given by

$$r_n = \frac{n^2}{Z}\frac{\hbar^2}{mke^2} \tag{7.25}$$

and

$$v_n = \frac{Z}{n}\left(\frac{ke^2}{\hbar}\right), \tag{7.28}$$

respectively. Substitution of these quantities into Equation 8.65b immediately yields

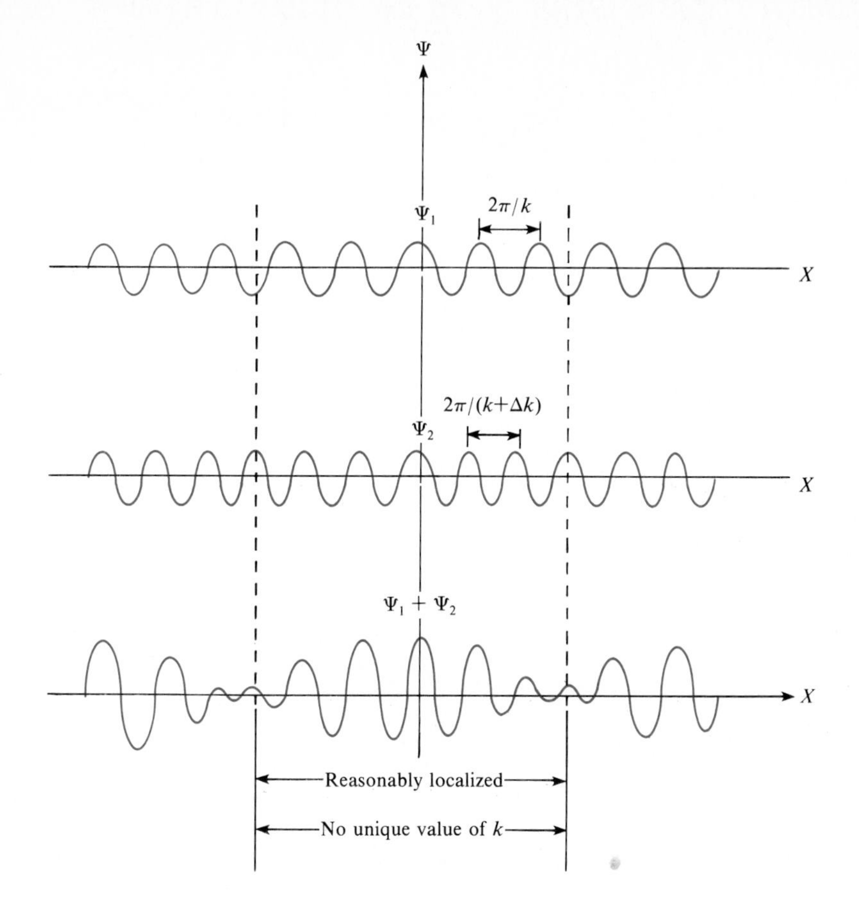


Figure 8.5
The linear superposition of two similar wave functions.

we obtain a rather interesting result of

$$\Psi(x, t) = 2A \cos\left[\left(k + \frac{\Delta k}{2}\right)x - \left(\omega + \frac{\Delta\omega}{2}\right)t\right] \cos\left(-\frac{\Delta k}{2}x + \frac{\Delta\omega}{2}t\right),$$

which immediately reduces to

$$\Psi(x, t) = 2A \cos\left[\left(k + \frac{\Delta k}{2}\right)x - \left(\omega + \frac{\Delta\omega}{2}\right)t\right] \cos\left(\frac{\Delta k}{2}x - \frac{\Delta\omega}{2}t\right). \tag{8.82}$$

Allowing that $\Delta k \ll k$ and $\Delta\omega \ll \omega$, Equation 8.82 becomes

$$\Psi(x, t) = 2A \cos(kx - \omega t) \cos\left(\frac{\Delta k}{2}x - \frac{\Delta\omega}{2}t\right), \tag{8.83}$$

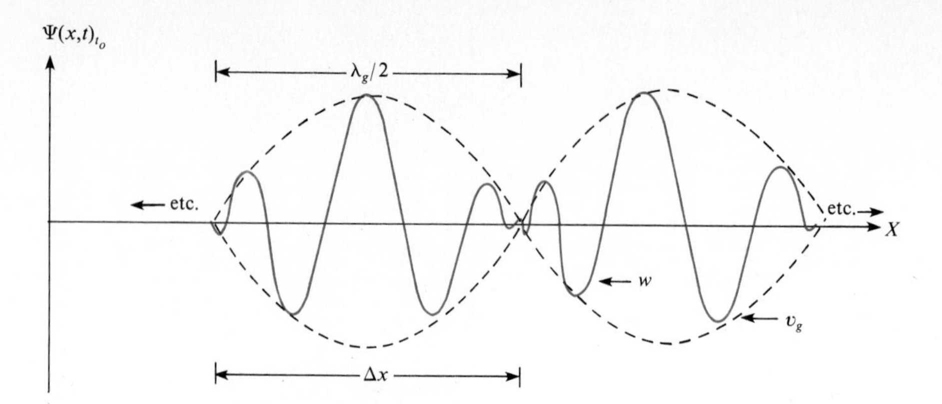


Figure 8.6
The sum of $\Psi_1(x, t)$ and $\Psi_2(x, t)$ at time $t = t_o$.

where $\cos(kx - \omega t)$ represents **phase waves** and $\cos(\frac{1}{2}\Delta kx - \frac{1}{2}\Delta\omega t)$ represents the **group waves** with

Group Wave Number

$$k_g \equiv \frac{\Delta k}{2}, \tag{8.84}$$

Group Angular Frequency

$$\omega_g \equiv \frac{\Delta\omega}{2}. \tag{8.85}$$

The physical meaning of phase and group waves is suggested in Figure 8.6, where $\Psi(x, t)$ of Equation 8.83 is plotted as a function of $x$ for a fixed value of time $t = t_0$.

## 8.6 Group, Phase, and Particle Velocities

From our previous discussion it should be clear that in general we can superpose *any number* of waves to obtain virtually any type of *wave-packet* (see Figure 8.6) desirable. The resulting *matter waves* have associated with them a *phase velocity* $w$ and a *group velocity* $v_g$. To understand the differences between group, phase, and particle velocities, each will be represented by a defining equation and then its relationship to other quantities like energy and momentum will be examined. From these relationships the interrelation between the three velocities can be obtained for a particle traveling at relativistic or nonrelativistic speeds.

**Group velocity** $v_g$ is simply defined by the equation

Group Velocity

$$v_g \equiv \lambda_g \nu_g. \tag{8.86}$$

Its relationship to energy and momentum is easily obtained by considering

$$\begin{aligned} v_g \equiv \lambda_g \nu_g &= \frac{\lambda_g}{2\pi} 2\pi\nu_g \\ &= \frac{\omega_g}{k_g} = \frac{\Delta\omega/2}{\Delta k/2} \qquad (8.87a) \\ &= \frac{\Delta\omega}{\Delta k} = \frac{\hbar\Delta\omega}{\hbar\Delta k} \qquad (8.87b) \\ &= \frac{\Delta E}{\Delta p}. \qquad (8.87c) \end{aligned}$$

From this result the group velocity is also obviously expressable as the differential

$$v_g = \frac{dE}{dp}. \qquad (8.88)$$

**Phase velocity** $w$, defined by Equation 8.58,

$$w \equiv \lambda\nu, \qquad (8.58)$$

Phase Velocity

is also easily expressed in terms of energy and momentum as

$$\begin{aligned} w \equiv \lambda\nu &= \frac{\lambda}{2\pi} 2\pi\nu \\ &= \frac{\omega}{k} = \frac{\hbar\omega}{\hbar k} \qquad (8.25) \\ &= \frac{E}{p}, \qquad (8.59) \end{aligned}$$

which was originally obtained and expressed in Equation 8.59.

**Particle velocity** $v$ has the same defining equations in our present consideration as it does in general physics. With respect to the considerations made herein, we take its defining equation to be

$$v \equiv \frac{\Delta x}{\Delta t}, \qquad (8.89)$$

Average Particle Velocity

because of the one-dimensionality of the de Broglie waves. One interesting relationship between $\Delta x$ and $\Delta k$ can be obtained by referring to Figure 8.6 and Equation 8.84. From Figure 8.6 we have

$$\Delta x = \frac{\lambda_g}{2}$$

$$= \frac{2\pi/k_g}{2}$$

$$= \frac{\pi}{k_g}$$

and substitution from Equation 8.84 immediately gives

Uncertainty in Position

$$\Delta x = \frac{2\pi}{\Delta k}. \tag{8.90}$$

This relationship will prove useful in the problems, as well as, in the development of the *Heisenberg uncertainty principle* in the next section.

We are now in a position to develop the interrelationship between the group, phase, and particle velocities. The first consideration is for a *classical* particle having nonrelativistic energy given by

$$E = \tfrac{1}{2}mv^2 = \frac{p^2}{2m}, \tag{8.60}$$

as given Equation 8.60. We already know its phase velocity is related to its particle velocity from Equation 8.62

Non-Relativistic

$$w = \frac{v}{2}, \tag{8.62}$$

so we need to find how $w$ is related to the group velocity $v_g$. Since Equation 8.88 gives the group velocity $v_g$ as the first order derivative of energy $E$ with respect to momentum $p$, we differentiate Equation 8.60 to obtain

$$dE = \frac{2p\,dp}{2m} = \frac{p\,dp}{m}.$$

Rearranging this last equation and using the definition of linear momentum gives

$$\frac{dE}{dp} = \frac{p}{m} = v,$$

from which we obtain

$$v_g = v. \qquad (8.91)$$ Non-Relativistic

Of course, a comparison of Equations 8.62 and 8.91 gives the relationship between the phase velocity $w$ and group velocity $v_g$,

$$w = \frac{v_g}{2}, \qquad (8.92)$$ Non-Relativistic

for a particle having nonrelativistic energy $E$.

From the classical considerations above, it is reasonable to expect that relationships between $v_g$, $w$, and $v$ for a particle having a *relativistic* energy can be obtained in a similar manner. That is, we start with a relativistic equation involving *energy* and *momentum* and manipulate it to obtain all possible relationships between the three velocities. One equation satisfying this requirement is

$$E^2 = E_0^2 + p^2c^2, \qquad (4.46)$$

which we recall is *invariant* to all *inertial* observers. Proceeding in the same spirit as before, we first need to find an expression for phase velocity $w$ and then one for group velocity $v_g$. Since phase velocity is equal to the ratio of energy $E$ and momentum $p$, then we divide Equation 4.46 by $p^2$ to obtain

$$\frac{E^2}{p^2} = \frac{E_0^2}{p^2} + c^2. \qquad (8.93)$$

This equation can be reduced by using

$$m = \Gamma m_0 \qquad (4.25)$$

to express the relativistic momentum as

$$p = mv = \Gamma m_0 v, \qquad (8.94)$$

and the equations

$$\Gamma \equiv \frac{1}{\sqrt{1 - \frac{v^2}{c^2}}} \qquad (4.23)$$

$$E_0 = m_0c^2. \qquad (4.39)$$

Substitution of these equations into Equation 8.93 gives

$$\begin{aligned}\frac{E^2}{p^2} &= \frac{m_0^2c^4}{\Gamma^2 m_0^2 v^2} + c^2 \\ &= \frac{c^4}{v^2}\left(1 - \frac{v^2}{c^2}\right) + c^2 \\ &= \frac{c^4}{v^2},\end{aligned}$$

and taking the square root of the last equality gives the result

Relativistic

$$w = \frac{c^2}{v}. \tag{8.95}$$

To obtain an expression for the group velocity $v_g$, we note it involves a differential from Equation 8.88 and, accordingly, differentiate Equation 4.46 to obtain

$$2EdE = 2pc^2dp$$

or

$$EdE = pc^2dp.$$

This result is easily rearranged in a more convenient form as

$$\frac{dE}{dp} = \frac{c^2}{E/p},$$

which in terms of group velocity and phase velocity becomes

Relativistic

$$v_g = \frac{c^2}{w}. \tag{8.96}$$

A comparison of Equations 8.95 and 8.96 gives the relationship between $v_g$ and $v$ for a particle having a relativistic energy:

Relativistic

$$v_g = v. \tag{8.97}$$

From Equations 8.91 and 8.97 we see that the group velocity of the de Broglie wave associated with a particle is always equal to the classical particle velocity. The results expressed by Equations 8.95, 8.96, and 8.97 are also obtainable from Einstein's famous energy-mass equivalence relation,

$$E = mc^2 = \Gamma m_0 c^2 \tag{4.40}$$

and the relationships expressed by Equations 4.23, 4.25, and 4.39. This should be conceptually obvious, since Equation 4.46 was *originally* derived from Equation 4.40 in Chapter 4. Although the derivations suggested are straight forward, a bit more mathematics is involved in obtaining an expression for $dE/dp$ and is left as an exercise in the problem section.

## 8.7 Heisenberg's Uncertainty Principle

We have thus far observed that the monochromatic wave (constant angular velocity $\omega$)

$$\Psi(x, t) = A \cos(kx - \omega t),$$

having constant amplitude $A$ (undamped), definite momentum $p = \hbar k$, and energy $E = \hbar\omega$, can be viewed as a *wave packet* of *infinite extent* ($\lambda_g = \infty$). Therefore, the particle's exact location cannot be specified—the wave is not localized along the $X$-axis. We know the particle's momentum exactly, but we know nothing about its position. Another way of saying this is that there is *equal probability* of finding it anywhere in the $x$-domain, $-\infty < x < +\infty$. We have further seen that the *linear superposition* of two waves $\Psi_1(x, t)$ and $\Psi_2(x, t)$ describes a particle and gives rise to *wave groups*. We are no longer totally ignorant of the particle's location; however, we now have some *uncertainty* in the momentum of the particle, since we have a mixture of two momentum states corresponding to $k$ and $k + \Delta k$. In particular, for

$$\Psi_2(x, t) = A \cos[(k + \Delta k)x - (\omega + \Delta\omega)t],$$

the linear superposition of $\Psi_1(x, t)$ and $\Psi_2(x, t)$ gave us the function $\Psi(x, t)$, plotted in Figure 8.6, which represents an *infinite* succession of *groups of waves* traveling in the positive $x$-direction. This particular *matter wave* suggests that the associated particle has equal probability of being located within *any one* of the groups at the time $t = 0$. Even if we consider a single group, our interpretation of $\Psi(x, t)$ suggests that the particle's location within that group is *uncertain* to within a distance comparable to the length $\Delta x$ of the group.

It is possible, by adding a sufficiently large number of the right kind of monochromatic waves, to obtain a resultant *single wave packet*, being quantitatively represented by

$$\Psi(x, t) = \sum_{j=1}^{N} A_j \cos(k_j x - \omega_j t). \tag{8.98}$$

Let us consider three possible wave packets, as illustrated in Figure 8.7. The wave depicted in (a) is highly localized in space; however, its wavelength and thus its momentum is very indefinite. The wave illustrated in (b) is less spatially localized, but its momentum is also less indefinite. Obviously, by comparing the waves of (a) and (b), it appears that we must *sacrifice* our absolute knowledge of the *position* of a particle, if we are to have a reasonably well defined momentum. This situation is suggested in (c) of Figure 8.7.

One should note with particular attention the difference between the *phase wavelength* $\lambda$ depicted in Figure 8.7 and the *group wavelength* $\lambda_g$ illustrated in Figure 8.6. From Equation 8.90,

$$\Delta x = \frac{2\pi}{\Delta k}, \tag{8.90}$$

we have that the *uncertainty* $\Delta x$ in the particle's position is related to the *uncertainty* $\Delta k$ in the particle's wave number. This relationship is also

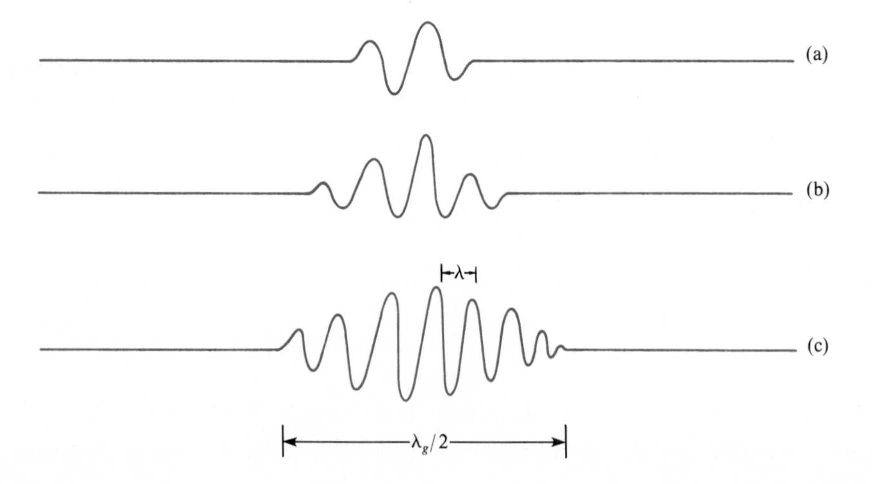


Figure 8.7
Three single-wave packets varying in degrees of localization.

derivable by realizing that $v = v_g$ for a particle having relativistic or nonrelativistic energy. Consequently, substitution from Equations 8.89 and 8.87b and solving for the **spatial spread** $\Delta x$ gives

$$\Delta x = \frac{\Delta\omega}{\Delta k}\Delta t = \frac{2\pi\Delta\nu\Delta t}{\Delta k} = \frac{2\pi}{\Delta k}, \tag{8.90}$$

where the last equality was obtained by equating the **frequency spread** $\Delta\nu$ with the inverse **time spread** $\Delta t$ (i.e., $\Delta\nu = 1/\Delta t$). Now, using de Broglie's postulate in the form

$$\Delta k = \frac{\Delta p}{\hbar}$$

we obtain one form of **Heisenberg's uncertainty principle**

Heisenberg's Uncertainty Principle

$$\Delta x\,\Delta p = h. \tag{8.99a}$$

This equation is interpreted to mean that the *uncertainty in position* times the *uncertainty in momentum* is equal to the universal constant $h$. To obtain an alternative expression for the uncertainty principle, we need only recall that for a relativistic or nonrelativistic particle $v = v_g$. Thus substitution from Equations 8.89 and 8.87c yields

$$\frac{\Delta x}{\Delta t} = \frac{\Delta E}{\Delta p}.$$

Solving this equation for $\Delta x\Delta p$ and using Equation 8.99a results in

$$\Delta x\Delta p = \Delta E\Delta t = h, \tag{8.99b}$$

which illustrates that the *uncertainty in energy* times the *uncertainty in time* is equal to Planck's constant. The latter equality is an alternative form of the *Heisenberg uncertainty principle*.

## Review of Fundamental and Derived Equations

Many fundamental and derived equations of this chapter are listed below, along with new fundamental postulates of modern physics.

## FUNDAMENTAL EQUATIONS—CLASSICAL/MODERN PHYSICS

|  |  |
|---|---|
| $\mu \equiv \frac{M}{L}$ | Linear Mass Density |
| $k \equiv \frac{2\pi}{\lambda}$ | Wave Number |
| $\omega = 2\pi\nu$ | Angular Velocity/Frequency |
| $k_g \equiv \frac{\Delta k}{2}$ | Group Wave Number |
| $\omega_g \equiv \frac{\Delta\omega}{2}$ | Group Angular Frequency |
| $v_g \equiv \lambda_g \nu_g$ | Group Velocity |
| $w \equiv \lambda\nu$ | Phase Velocity |
| $v \equiv \frac{\Delta x}{\Delta t}$ | Average Particle Velocity |
| $\Gamma \equiv \left(1 - \frac{v^2}{c^2}\right)^{-1/2}$ | Special Relativistic Symbol |
| $m = \Gamma m_0$ | Relativistic Mass |
| $E \equiv mc^2 = \Gamma m_0 c^2$ | Relativistic Total Energy |
| $E^2 = E_0^2 + p^2c^2$ | Energy-Momentum Invariant |
| $r_n = \frac{n^2}{Z}\left(\frac{\hbar^2}{mke^2}\right)$ | Bohr Electron Radii |
| $v_n = \frac{Z}{n}\left(\frac{ke^2}{\hbar}\right)$ | Bohr Electron Velocities |
| $\mathbf{F} = m\mathbf{a}, \qquad m \neq m(t)$ | Newton's Second Law |
| $\mathbf{p} \equiv m\mathbf{v}$ | Linear Momentum |
| $T = \frac{1}{2}mv^2$ | Classical Kinetic Energy |
| $\epsilon = h\nu$ | Einstein's Photon Postulate |
| $p_\epsilon = \frac{h}{\lambda}$ | Photon Momentum |

## NEW FUNDAMENTAL POSTULATES

$$\nu = \frac{E}{h}$$ De Broglie's Particle Frequency

$$\lambda = \frac{h}{p}$$ De Broglie's Particle Wavelength

$$\Delta x \Delta p = \Delta E \Delta t = h$$ Heisenberg's Uncertainty Principle

## DERIVED EQUATIONS

### *Vibrating String—Equation of Motion*

$$dF_x = 0$$ Net Longitudinal Force

$$dF_y = T d\theta = \mu dx \frac{\partial^2 y}{\partial t^2}$$ Net Transverse Force

$$\frac{d\theta}{dx} = \frac{\mu}{T} \frac{\partial^2 y}{\partial t^2}$$ Equation of Motion

$$w = \sqrt{\frac{T}{\mu}}$$ Phase Velocity

$$\frac{\partial^2 y}{\partial x^2} = \frac{1}{w^2} \frac{\partial^2 y}{\partial t^2}$$ Classical Wave Equation

$$\left.\begin{aligned} y(x, 0) &= y_0(x) \\ \left(\frac{\partial y}{\partial t}\right)_{t=0} &= v_0(x) \end{aligned}\right\}$$ Initial Conditions

$$y(0, t) = y(L, t) = 0$$ Boundary Conditions

### *Vibrating String—Normal Modes*

$$y(x, t) = f(x)h(t)$$ Separation of Variables

$$-\omega^2 = \frac{w^2}{f} \frac{d^2 f}{dx^2}$$

$$= \frac{1}{h} \frac{d^2 h}{dt^2}$$ Wave Equation

$$L = \frac{n\lambda}{2}$$ Condition for Standing Waves

$$h(t) = A \cos \omega t + B \sin \omega t$$ Time Solution

$$f(x) = C \cos \frac{\omega x}{w} + D \sin \frac{\omega x}{w} = D \sin kx$$ Spatial Solution

$$y(x, t) = AD \sin kx \cos \omega t + BD \sin kx \sin \omega t$$ Normal Mode Solution

### Traveling Waves

$$y_+ (x, t) = A \sin(kx + \omega t) = A \sin(kx + kwt)$$ Negatively Directed Traveling Wave

$$y_- (x, t) = A \sin(kx - \omega t) = A \sin(kx - kwt)$$ Positively Directed Traveling Wave

$$y(x, t) = f(kx \pm \omega t)$$ Generalized Traveling Waves

$$\frac{\partial^2 y}{\partial x^2} = \frac{k^2}{\omega^2} \frac{\partial^2 y}{\partial t^2}$$ General Wave Equation

### De Broglie's Hypothesis and Bohr's Quantization Condition

$$2\pi r = n\lambda$$ Electron Standing Waves

$$mvr = n\hbar$$ Bohr's Angular Momentum Quantization

$$E_n = -h\nu_n = -\frac{hw_n}{\lambda_n} = -\frac{n\hbar v_n}{2r_n} = -\frac{Z^2}{n^2}\left(\frac{mk^2e^4}{2\hbar^2}\right)$$ Bohr's Energy Quantization

### De Broglie's Hypothesis and Einsteinian Relativity

$$\Psi(x, t) = Af(k_x x - \omega t)$$ Wave Function in S

$$\Psi'(x', t') = A'f'(k_x' x' - \omega' t')$$ Wave Function in S′

$$k_x = \gamma\left(k_x' + \frac{\omega' u}{c^2}\right)$$ Wave Number Transformation

$$\omega = \gamma(\omega' + k_x' u)$$ Angular Velocity Transformation

$$p_x = \gamma\left(p'_x + \frac{E'u}{c^2}\right) \quad \text{Momentum Transformation}$$

$$E = \gamma(E' + p'_x u) \quad \text{Energy Transformation}$$

### *Matter Waves*

$$\Psi_1 = A\cos(kx - \omega t)$$

$$\Psi_2 = A\cos[(k + \Delta k)x - (\omega + \Delta\omega)t] \quad \text{Monochromatic Waves}$$

$$\Psi(x, t) = \Psi_1(x, t) + \Psi_2(x, t)$$

$$= 2A\cos(kx - \omega t)\cos\left(\frac{\Delta k}{2}x - \frac{\Delta\omega}{2}t\right) \quad \text{Linear Superposition}$$

### *Group, Phase, and Particle Velocities*

$$v_g = \frac{\Delta\omega}{\Delta k} = \frac{\Delta E}{\Delta p} \quad \text{Group Velocity}$$

$$w = \frac{\omega}{k} = \frac{E}{p} \quad \text{Phase Velocity}$$

$$w = \frac{v}{2} = \frac{v_g}{2} \quad \text{Non-Relativistic}$$

$$w = \frac{c^2}{v} = \frac{c^2}{v_g} \quad \text{Relativistic}$$

### *Heisenberg's Uncertainty Principle*

$$\Delta x = \frac{\lambda_g}{2} = \frac{2\pi}{\Delta k} \quad \text{Uncertainty in Position}$$

$$\Delta x \Delta p = \Delta E \Delta t = h \quad \text{Uncertainty Principle}$$

## Problems

**8.1** Verify that $y(x, t) = A \sin(\omega x/w) \cos \omega t$ is a solution to Equation 8.10. Show whether the boundary conditions of Equation 8.12 and initial conditions of Equations 8.11a and 8.11b are satisfied and find the allowed nodes (values of $n$).

*Solution:*

Taking second order derivatives of $y(x, t) = A \sin(\omega x/w) \cos \omega t$ yields

$$\frac{\partial^2 y}{\partial x^2} = -\left(\frac{\omega}{w}\right)^2 y(x, t),$$

$$\frac{\partial^2 y}{\partial t^2} = -\omega^2 y(x, t),$$

which upon substitution into Equation 8.10,

$$\frac{\partial^2 y}{\partial x^2} = \frac{1}{w^2}\frac{\partial^2 y}{\partial t^2},$$

gives

$$-\frac{\omega^2}{w^2} = -\frac{\omega^2}{w^2}.$$

This result verifies that $y(x, t)$ is a solution to the classical wave equation. The boundary conditions

$$y(0, t) = y(L, t) = 0$$

are satisfied since

$$y(0, t) = A \sin 0° \cos \omega t = 0,$$

$$y(L, t) = A \sin \frac{\omega L}{w} \cos \omega t = 0,$$

where the second condition is valid for

$$\frac{\omega L}{w} = n\pi, \qquad n = 1, 2, 3, \cdots.$$

Further, since $\lambda = 2L/n \rightarrow L = n\lambda/2$, we have

$$\frac{\omega(n\lambda/2)}{w} = n\pi$$

or

$$\frac{\omega}{w} = \frac{2\pi}{\lambda} = k$$

as expected. The initial condition

$$y(x, 0) = y_0(x)$$

is satisfied, since

$$y(x, 0) = A \sin kx \cos 0°$$

gives the initial displacement as

$$y_0(x) = A \sin kx.$$

However, the second initial condition

$$\left(\frac{\partial y}{\partial t}\right)_{t=0} = v_0(x)$$

is not satisfied, since

$$\left(\frac{\partial y}{\partial t}\right)_{t=0} = -\omega A \sin kx \sin 0° = 0.$$

**8.2** Do Problem 8.1 for displacements given by (a) $y(x, t) = A \sin(\omega x/w) \sin \omega t$ and (b) $y(x, t) = A \sin(\omega x/w - \omega t)$.

**8.3** Verify that $y(x, t) = A \cos(\omega x/w) \cos \omega t$ is a solution to Equation 8.10. Show whether the boundary conditions, $y(-L/2, t) = y(L/2, t) = 0$, are satisfied. Find the allowed nodes and demonstrate whether the initial conditions given by Equations 8.11a and 8.11b are satisfied.

*Solution:*

In abbreviated form we have the following:

$$y(x, t) = A \cos\left(\frac{\omega x}{w}\right) \cos \omega t \rightarrow$$

$$\frac{\partial^2 y}{\partial x^2} = \left(-\frac{\omega^2}{w^2}\right)y, \qquad \frac{\partial^2 y}{\partial t^2} = -\omega^2 y.$$

Thus, $y(x, t)$ is a solution to

$$\frac{\partial^2 y}{\partial x^2} = \frac{1}{w^2}\frac{\partial^2 y}{\partial t^2} \rightarrow -\left(\frac{\omega}{w}\right)^2 = -\left(\frac{\omega}{w}\right)^2.$$

*Boundary Conditions:*

$$y\left(-\tfrac{1}{2}L, t\right) = A \cos\left(-\frac{\omega L}{2w}\right) \cos \omega t = 0, \text{ IFF}$$

$$\frac{\omega L}{2w} = \frac{n\pi}{2}, \qquad n = 1, 3, 5, \cdot \cdot \cdot .$$

$$y\left(\tfrac{1}{2}L, t\right) = A \cos\left(\frac{\omega L}{2w}\right) \cos \omega t = 0, \text{ IFF}$$

$$\frac{\omega L}{2w} = \frac{n\pi}{2}, \qquad n = 1, 3, 5, \cdot \cdot \cdot .$$

Since these conditions reduce to

$$\frac{\omega L}{w} = n\pi \rightarrow \frac{\omega(n\lambda/2)}{w} = n\pi \rightarrow \frac{\omega}{w} = k.$$

*Initial Conditions:*

$$y(x, 0) = A \cos\left(\frac{\omega x}{w}\right) \cos 0°$$

$$= A \cos\left(\frac{\omega x}{w}\right) = y_0(x). \qquad \text{(Satisfied)}$$

$$\left(\frac{\partial y}{\partial t}\right)_{t=0} = -\omega A \cos\left(\frac{\omega x}{w}\right) \sin 0°$$

$$= 0 \neq v_0(x). \qquad \text{(Not Satisfied)}$$

**8.4** Do Problem 8.3 for (a) $y(x, t) = A \cos(\omega x/w) \sin \omega t$ and (b) $y(x, t) = A \cos(\omega x/w - \omega t)$.

**8.5** Starting with Equation 8.33 derive Equation 8.35.

*Solution:*

Multiplication of Eq. 8.33,

$$y_0(x) = \sum_{n=1}^{\infty} A_n \sin k_n x,$$

by $\sin(m\pi x/L)\, dx$ and integrating gives

$$\int_0^L y_0(x) \sin\left(\frac{m\pi x}{L}\right) dx = \sum_{n=1}^{\infty} \int_0^L A_n \sin\left(\frac{n\pi x}{L}\right) \sin\left(\frac{m\pi x}{L}\right) dx,$$

where we have used

$$k_n = \frac{n\pi}{L}.$$

With the change of variable

$$y \equiv \frac{\pi x}{L} \rightarrow dy = \frac{\pi}{L} dx,$$

the change in the limits of integration are given by

$$x = 0 \rightarrow y = 0 \quad \text{and} \quad x = L \rightarrow y = \pi.$$

Thus, the integral to be solved is of the form

$$\frac{L}{\pi} \int_0^{\pi} \sin ny \sin my\, dy,$$

since $A_n$ comes outside of the integral. In solving this integral we need to consider the following two cases:

*Case 1:* n = m

$$\frac{L}{\pi}\int_0^{\pi} \sin^2 ny\, dy = \frac{L}{n\pi}\int_0^{\pi} \sin^2 ny\, d\,(ny)$$
$$= \frac{L}{2n\pi}(-\cos ny \sin ny + ny)\Big|_0^{\pi}$$
$$= \frac{L}{2}.$$

Thus, substitution back into our first integral equation gives the result

$$A_n = \frac{2}{L}\int_0^L y_0(x) \sin\left(\frac{n\pi x}{L}\right) dx,$$

where we have let $n \to m$ and $m \to n$ to obtain the exact form of Equation 8.35.

*Case 2:* n ≠ m

$$\frac{L}{\pi}\int_0^{\pi} \sin ny \sin my\, dy$$
$$= \frac{L}{\pi}\left[\frac{\sin(n-m)y}{2(n-m)} - \frac{\sin(n+m)y}{2(n+m)}\right]_0^{\pi}$$

which goes to zero for $y = 0$ or $y = \pi$. We could note that

$$\int_0^L \sin\left(\frac{n\pi x}{L}\right) \sin\left(\frac{m\pi x}{L}\right) dx = \frac{L}{2}\delta_{nm},$$

where the symbol $\delta_{nm}$ is called the **Kronecker delta** and defined by

$$\delta_{nm} = 0 \quad \text{for} \quad n \neq m,$$
$$\delta_{nm} = 1 \quad \text{for} \quad n = m.$$

Accordingly, we could have considered this problem as

$$\int_0^L y_0(x) \sin\left(\frac{m\pi x}{L}\right) dx = \sum_{n=1}^{\infty} A_n \int_0^L \sin\left(\frac{n\pi x}{L}\right) \sin\left(\frac{m\pi x}{L}\right) dx$$
$$= \frac{L}{2}\sum_{n=1}^{\infty} A_n\delta_{nm}$$
$$= \frac{L}{2}A_n.$$

**8.6** Starting with Equation 8.34 derive Equation 8.36 using the Kronecker delta symbol.

**8.7** Find the de Broglie wavelength and frequency associated with a 1 g bullet traveling at 663 m/s.

*Solution:*
With $m = 1 \text{ g} = 10^{-3}$ kg and $v = 663$ m/s, we have the de Broglie wavelength given by (Equation 8.53)

$$\lambda = \frac{h}{p} = \frac{h}{mv}$$

$$= \frac{6.63 \times 10^{-34} \text{ J} \cdot \text{s}}{(10^{-3} \text{ kg})(663 \text{ m/s})}$$

$$= 10^{-33} \text{ m}$$

and the de Broglie frequency given by (Equation 8.52)

$$\nu = \frac{E}{h} = \frac{mv^2}{2h}$$

$$= \frac{(10^{-3} \text{ kg})\,(663 \text{ m/s})^2}{2(6.63 \times 10^{-34} \text{ J} \cdot \text{s})}$$

$$= 3.32 \times 10^{35} \text{ Hz.}$$

**8.8** What is the de Broglie wavelength and frequency of an alpha particle (helium nucleus) of 2 MeV energy?

*Answer:* $1.01 \times 10^{-14}$ m, $4.83 \times 10^{20}$ Hz

**8.9** What is the energy of a proton having a de Broglie wavelength of 1 m?

*Solution:*
Knowing $m_p = 1.673 \times 10^{-27}$ kg and $\lambda = 1$ m, we can find the energy $E$ by using

$$E = \frac{p^2}{2m_p}$$

$$= \frac{h^2}{2m_p\lambda^2}$$

$$= \frac{(6.63 \times 10^{-34} \text{ J} \cdot \text{s})^2}{2(1.673 \times 10^{-27} \text{ kg})(1 \text{ m})^2}$$

$$= 1.31 \times 10^{-40} \text{ J.}$$

**8.10** Electrons are accelerated from rest by an electrical potential $V$. If their wave number is $6.28 \times 10^{10}$/m, what is the accelerating potential $V$?

*Answer:* 149 V

**8.11** Find the wave number associated with sound waves traveling at 314 m/s in a medium, if they have a period of $10^{-2}$ s.

*Solution:*
With $v_s = 314$ m/s and $T = 10^{-2}$ s, the wave number is given by

$$k \equiv \frac{2\pi}{\lambda} = \frac{2\pi}{v_s/\nu} = \frac{2\pi\nu}{v_s}$$
$$= \frac{2\pi}{v_s T}$$
$$= \frac{(2)(3.14)}{(314 \text{ m/s})(10^{-2} \text{ s})}$$
$$= 2 \text{ m}^{-1}.$$

**8.12** Derive the trigonometric identity given by Equation 8.81.

*Answer:* $\cos\alpha + \cos\beta = 2\cos\left(\frac{\alpha+\beta}{2}\right)\cos\left(\frac{\alpha-\beta}{2}\right)$

**8.13** Find the sum of $\Psi_1 = 0.004\cos(5.8x - 280t)$ and $\Psi_2 = .004\cos(6x - 300t)$, where all units are in the SI system. Find the phase velocity, the group velocity, and the uncertainty in position of the associated particle.

*Solution:*
From the form of $\Psi_1 = 4 \times 10^{-3}\cos(5.8x - 280t)$, we obtain (deleting all units)

$$A = 4 \times 10^{-3}, \qquad k = 5.8, \qquad \omega = 280,$$

while $\Psi_2 = 4 \times 10^{-3}\cos(6x - 300t)$ gives additional values

$$k + \Delta k = 6, \qquad \omega + \Delta\omega = 300 \rightarrow \Delta k = 0.2, \qquad \Delta\omega = 20.$$

Thus, we have from Equation 8.82

$$\Psi = 2A\cos\left[\left(k + \frac{\Delta k}{2}\right)x - \left(\omega + \frac{\Delta\omega}{2}\right)t\right]\cos\left(\frac{\Delta k}{2}x - \frac{\Delta\omega}{2}t\right)$$
$$= 8 \times 10^{-3}\cos[(5.8 + 0.1)x - (280 + 10)t]\cos(0.1x - 10t)$$
$$= 8 \times 10^{-3}\cos(5.9x - 290t)\cos(0.1x - 10t).$$

From the linear superposition of $\Psi_1$ and $\Psi_2$ we recognize

$$k = 5.9, \qquad \omega = 290, \qquad k_g = 0.1, \qquad \omega_g = 10,$$

from which the *phase velocity*

$$w = \frac{\omega}{k} = \frac{290}{5.9} = 49.15\ \frac{\text{m}}{\text{s}},$$

the *group velocity*

$$v_g = \frac{\Delta\omega}{\Delta k} = \frac{20}{0.2} = 100\ \frac{\text{m}}{\text{s}},$$

and the *uncertainty in position*

$$\Delta x = \frac{2\pi}{\Delta k} = \frac{2\pi}{0.2} = 31.42\ \text{m}$$

are directly obtained.

**8.14** Repeat Problem 8.13 for $\Psi_1 = 0.005\cos(6x - 300t)$ and $\Psi_2 = 0.005\cos(6.2x - 320t)$.

*Answer:* $w = 50.82\ \frac{\text{m}}{\text{s}}, \quad v_g = 100\ \frac{\text{m}}{\text{s}}, \quad \Delta x = 31.42\ \text{m}$

**8.15** Consider two sound waves of frequencies 510 Hz and 680 Hz traveling at 340 m/s. (a) Find the wave numbers $k_1$ and $k_2$ and the *spatial spread* $\Delta x = 2\pi/(k_2 - k_1)$. (b) Find $\Delta\omega$ and the *time spread* $\Delta t = 2\pi/\Delta\omega$. (c) Compare the spatial spread obtained with that computed using $\Delta x = v_s\Delta t$.

*Solution:*
With $\nu_1 = 510$ Hz, $\nu_2 = 680$ Hz, and $v_s = 340\ \text{m/s} \equiv v_g$, the wave numbers are given by

$$k_1 = \frac{2\pi}{\lambda_1} = \frac{2\pi}{v_s/\nu_1} = \frac{2\pi\nu_1}{v_s}$$
$$= \frac{2\pi(510\ \text{Hz})}{340\ \text{m/s}} = 2\pi(1.5)\ \text{m}^{-1} = 3\pi\ \text{m}^{-1}$$

and

$$k_2 = \frac{2\pi\nu_2}{v_s}$$
$$= \frac{2\pi(640\ \text{Hz})}{340\ \text{m/s}} = 4\pi\ \text{m}^{-1}$$

From these values for $k_1$ and $k_2$, the spatial spread is

$$\Delta x = \frac{2\pi}{k_2 - k_1}$$
$$= \frac{2\pi}{(4\pi - 3\pi)\ \text{m}^{-1}} = 2\ \text{m},$$

while the time spread is just

$$\Delta t = \frac{2\pi}{\Delta\omega} = \frac{2\pi}{\omega_2 - \omega_1} = \frac{2\pi}{2\pi(\nu_2 - \nu_1)}$$

$$= \frac{2\pi}{v_s k_2 - v_s k_1} = \frac{2\pi}{v_s(\pi/\text{m})}$$

$$= (2\ \text{m})/(340\ \text{m/s}) = 5.9 \times 10^{-3}\ \text{s}.$$

Also, the spatial spread is given by

$$\Delta x = v_s \Delta t = (340\ \text{m/s})(5.9 \times 10^{-3}\ \text{s}) = 2\ \text{m}.$$

**8.16** Using the relativistic equations $E = mc^2$ and $p = mv$, where $m = \Gamma m_0$, derive Equations 8.95 and 8.97.

*Answer:* $w = \dfrac{c^2}{v}$, $v_g = v$.

**8.17** Calculate the nonrelativistic and relativistic phase velocity $w$ of the de Broglie waves associated with a neutron of 33.4 eV energy.

*Solution:*

With the nonrelativistic phase velocity given by Equation 8.62 (or Equations 8.91 and 8.92) as

$$w = \frac{v}{2}$$

and the relativistic phase velocity given by Equation 8.95 as

$$w = \frac{c^2}{v},$$

we need to first find the particle velocity $v$ associated with the neutron. Using the energy relation (Equation 8.60)

$$E = \tfrac{1}{2}mv^2,$$

we immediately obtain

$$v = \sqrt{2E/m}$$

$$= \sqrt{2(33.4\ \text{eV})(1.60 \times 10^{-19}\ \text{J/eV})/(1.67 \times 10^{-27}\ \text{kg})}$$

$$= \sqrt{4(16.7)(16.0 \times 10^{-20})(\text{m}^2/\text{s}^2)/(16.7 \times 10^{-28})}$$

$$= 8 \times 10^4\ \text{m/s}.$$

Thus, the nonrelativistic phase velocity is

$$w = \frac{v}{2} = 4 \times 10^4\ \text{m/s},$$

while the relativistic phase velocity is

$$w = \frac{c^2}{v} = \frac{(3 \times 10^8 \text{ m/s})^2}{8 \times 10^4 \text{ m/s}} = 1.125 \times 10^{12} \text{ m/s}$$

for the associated de Broglie waves.

**8.18** Repeat Problem 8.17 for an electron of 182.2 eV energy.

*Answer:* $w = 4 \times 10^6$ m/s, $w = 1.125 \times 10^{10}$ m/s

**8.19** Consider a 2 μg mass traveling with a speed of 10 cm/s. If the particle's speed is uncertain by 1.5%, what is its uncertainty in position?

*Solution:*
With $m = 2 \times 10^{-6}$ g $= 2 \times 10^{-9}$ kg and $v = 10$ cm/s $= 0.1$ m/s, the uncertainty in speed is

$$\Delta v = (0.015)(0.1 \text{ m/s}) = 1.5 \times 10^{-3} \text{ m/s}.$$

Thus, the uncertainty in position is given by

$$\Delta x = \frac{h}{\Delta p} = \frac{h}{m\Delta v}$$

$$= \frac{6.63 \times 10^{-34} \text{ J} \cdot \text{s}}{(2 \times 10^{-9} \text{ kg})(1.5 \times 10^{-3} \text{ m/s})}$$

$$= 2.21 \times 10^{-22} \text{ m}.$$

**8.20** If the energy of a nuclear state is uncertain by 1 eV, what is the lifetime of this state, according to Heisenberg?

*Answer:* $4.14 \times 10^{-15}$ s

**8.21** A matter wave traveling at 500 m/s for 10 s has a 0.1% uncertainty in the position of a particle. Find the uncertainty in momentum for the particle.

*Solution:*
With $v = 500$ m/s, $t = 10$ s, $\Delta x = (0.001)x$, we have the uncertainty in momentum given by

$$\Delta p = \frac{h}{\Delta x} = \frac{h}{\Delta(vt)}$$

$$= \frac{(6.63 \times 10^{-34} \text{ J} \cdot \text{s})}{(0.001)(500 \text{ m/s})(10 \text{ s})}$$

$$= 1.33 \times 10^{-34} \text{ kg} \cdot \text{m/s}.$$

**8.22** What is the de Broglie wave number of a proton of 50 MeV kinetic energy? If the energy is uncertain by 3%, what is the uncertainty in time?

*Answer:* $1.56 \times 10^{15}\ \text{m}^{-1}$, $2.76 \times 10^{-21}$ s

**8.23** If the position of a 1 kg object is measured on a frictionless surface to a precision of 0.100 cm, what velocity has been imparted to the object by the measurement, according to the uncertainty principle?

*Solution:*
With $m = 1$ kg and $\Delta x = 1.00 \times 10^{-3}$ m, the uncertainty in momentum is given by

$$\Delta p = m\Delta v,$$

which upon substitution into the uncertainty relation gives

$$\begin{aligned} \Delta v &= \frac{h}{m\Delta x} \\ &= \frac{(6.63 \times 10^{-34}\ \text{J} \cdot \text{s})}{(1\ \text{kg})(1.00 \times 10^{-3}\ \text{m})} \\ &= 6.63 \times 10^{-31}\ \text{m/s}. \end{aligned}$$

**8.24** If the speeed of an electron in the ground state of the Bohr model is uncertain by 1%, what is the uncertainty in position?

*Answer:* $\Delta x = 3.05 \times 10^{-9}$ m

**8.25** Using the Heisenberg uncertainty relation in the form $pr = \hbar$, derive an expression for the minimum radius $r_1$ and energy $E_1$ of a hydrogen atom.

*Solution:*
The total energy of the hydrogen electron is given by Equation 7.14 as

$$\begin{aligned} E &= \frac{p^2}{2m} - \frac{ke^2}{r} \\ &= \frac{\hbar^2}{2mr^2} - \frac{ke^2}{r}, \end{aligned}$$

where $p = \hbar/r$ has been used in obtaining the second equality. The minimum value of $E$ occurs at $r = r_1$, which can be obtained by taking the first order derivative of $E$ with respect to $r$ and setting the result to zero:

$$\frac{dE}{dr} = 0 = -\frac{\hbar^2}{mr^3} + \frac{ke^2}{r^2}.$$

From this relation we immediately obtain

$$r_1 = \frac{\hbar^2}{mke^2}$$

and substitution into the equation for $E$ gives

$$\begin{aligned} E_1 &= \frac{\hbar^2}{2mr_1^2} - \frac{ke^2}{r_1} \\ &= \frac{\hbar^2 m^2 k^2 e^4}{2m\hbar^4} - \frac{ke^2 mke^2}{\hbar^2} \\ &= -\frac{mk^2 e^4}{2\hbar^2}. \end{aligned}$$

These results are in agreement with those predicted by the Bohr model in Equations 7.27 and 7.40, respectively.

**8.26** Using the Heisenberg uncertainty principle in the form $px = \hbar$, derive an expression for the minimum energy $E_1$ of a linear harmonic oscillator.

*Answer:* $E_1 = h\nu$

# CHAPTER 9

# Schrödinger's Quantum Mechanics I

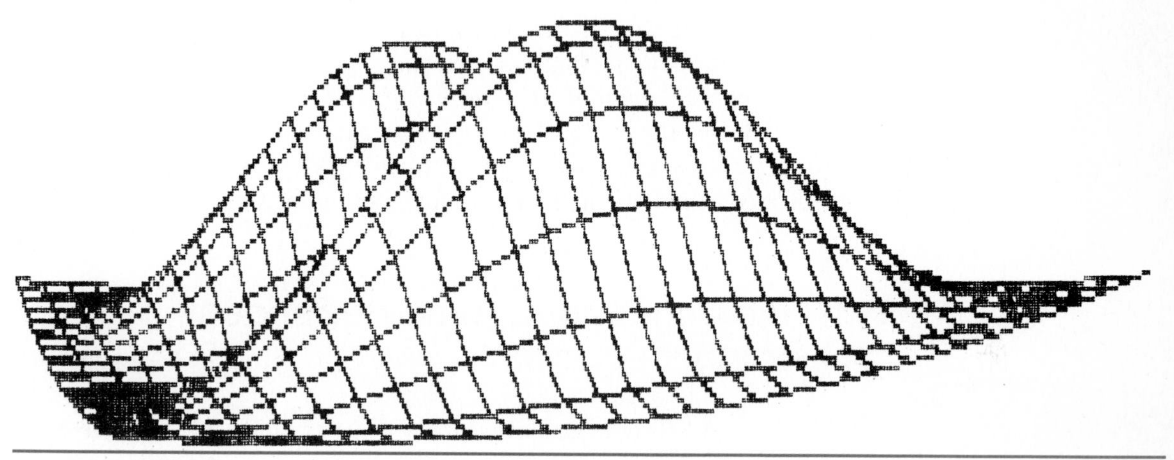


## Introduction

The discussion of matter waves in the last chapter emphasized how the wave-particle duality can be combined in a more general description of nature without any logical contradiction. We considered the problems of a matter wave associated with a *free particle* having an assumed zero potential energy. However, the more general problem involves a particle possessing both kinetic and potential energy that vary in space and time. The de Broglie approach for this more general problem would require the specification of a *matter wave* at every point in space for the propagation of a particle, which would be conceptually and mathematically unattractive. Shortly after de Broglie's hypothesis, Erwin Schrödinger realized this difficulty and circumvented it by *postulating* a nonrelativistic wave equation. Schrödinger's equation combined the total particle energy (kinetic and potential) with de Broglie's energy and momentum postulates, such that a *wave function* solution to the equation would be appropriate for a specified potential energy and consistent with de Broglie's hypothesis.

The Schrödinger equation and its wave function solution are considered in some detail in this chapter. Since Schrödinger postulated the wave equation of quantum mechanics, it represents a fundamental principle that can not be derived. Our approach is a combination of *postulating* and *constructing* the Schrödinger wave equation, by making analogies with the classical wave equation and establishing requirements that the quantum mechanical wave equation and its solutions must satisfy. After constructing the one-dimensional time-dependent Schrödinger equation and generalizing it to three dimensions, we develop the time-independent Schrödinger equation and discuss its general mathematical form in relationship to *eigenvalue* problems. This is followed by a discussion of the *wave function*, its *normalization condition*, and the physical interpretation of its *absolute magnitude squared* as a *probability density*. The chapter is concluded with one-dimensional examples of quantum mechanics, where the Schrödinger wave equation is solved for the *free particle* under the influence of a *constant* potential and the free particle in a box.

## 9.1 One-Dimensional Time-Dependent Schrödinger Equation

In 1925 Erwin Schrödinger developed a wave equation whose wave function solutions were capable of appropriately describing the propagation of de Broglie matter waves. Our study of Schrödinger's fundamental postulate and his theory of quantum mechanics will be primarily concerned with the mathematical details rather than the elusive physical interpretations represented by the mathematics. An alternative formulation of quantum mechanics was developed by W. Heisenberg at about the same time as the Schrödinger theory; however, the latter is more amenable to an introductory treatment than the operational matrix formulation of Heisenberg.

Schrödinger postulated his famous equation as a fundamental principle of nature, based on his remarkable insight into the wave nature of matter and his thorough understanding of wave mechanics. He combined the wave and particle characteristics of matter by adopting de Broglie's postulates,

De Broglie—Momentum

$$p = \frac{h}{\lambda} = \hbar k, \tag{8.53}$$

De Broglie—Energy

$$E = h\nu = \hbar\omega, \tag{8.52}$$

and the classical definition

Schrödinger—Total Energy

$$E = \frac{p^2}{2m} + V \tag{9.1}$$

for the *nonrelativistic* total energy $E$ of a particle having momentum $p$, rest mass $m$, and potential energy $V$. It should be noted that the first term on the right-hand side of Equation 9.1 is just the usual nonrelativistic kinetic energy of a particle. Further, the *group velocity* $v_g$ of a particle's associated *matter wave* is still equal to the velocity $v$ of a particle for the energy expression of Equation 9.1. This is easily verified by using the equation for group velocity,

$$v_g = \frac{dE}{dp}, \tag{8.87}$$

derived in Chapter 8. That is, substitution of Equation 9.1 into this expression for $v_g$ and realizing that the potential energy $V$ is constant for a *free particle* gives

$$\begin{aligned} v_g &= \frac{d}{dp}\left(\frac{p^2}{2m} + V\right) \\ &= \frac{p}{m} = v, \end{aligned}$$

which is the expected result. That Schrödinger chose the energy expression of Equation 9.1 is not surprising, since it represents *conservation of energy*, a fundamental principle of classical physics that has been demonstrated to be applicable in nonclassical considerations such as Einsteinian relativity, the photoelectric effect, the Compton effect, and the Bohr model of the hydrogen atom. Schrödinger felt that the desired wave equation must be consistent with the de Broglie postulates and the conservation of energy principle for a particle or system of particles. Thus, the wave equation must be consistent with

$$\frac{\hbar^2 k^2}{2m} + V(x, t) = \hbar\omega, \tag{9.2}$$

which is directly obtained by combining Equations 8.52, 8.53, and 9.1. One should observe that the potential energy in Equation 9.2 is indicated as having a *general* dependence on space and time coordinates. Schrödinger felt this to be the correct dependence in general, with *conservative force fields* being a special class. In the case of a *conservative force field* the potential energy is *time independent*,

$$V = V(x), \tag{9.3}$$

and the force is given by the classical expression

Conservative Force

$$F(x) = -\frac{dV}{dx}. \tag{9.4}$$

Since the force is independent of time, then so is the momentum, according to Newton's second law of motion

Newton's Second Law

$$\mathbf{F} \equiv \frac{d\mathbf{p}}{dt}, \tag{1.16}$$

and, of course, the total energy given by Equation 9.1.

In addition to the wave equation being consistent with Equation 9.2, Schrödinger felt that it must be *linear* in the *wave function* $\Psi(x, t)$, such that the *wave function solution* to the equation will have the *superposition* property. This suggests that the equation could be of the form

$$\frac{\hbar^2 k^2}{2m}\Psi(x, t) + V(x, t)\Psi(x, t) = \hbar\omega\Psi(x, t), \tag{9.5}$$

where Equation 9.2 has been simply multiplied by $\Psi(x, t)$ from the right-hand side. This symbolic equation is not a wave equation per se, as we should expect it to contain spatial and temporal derivatives of the wave function by analogy with the *classical wave equation*.

In an attempt to identify the partial differential equation that is consistent with Equation 9.5, we will work backward by considering a possible solution to the equation. Since a *conservative force field* constitutes a special case of the general equation, we can consider a *free particle* and its associated wave function as a possible solution. We could select the simple wave function

$$\Psi(x, t) = A\cos(kx - \omega t) \tag{8.41b}$$

discussed in Chapter 8, but its tendency to change in functional form upon differentiation makes it impossible to obtain a general equation involving $\Psi(x, t)$ and its derivatives that is in agreement for all values of $x$ and $t$ with Equation 9.5. In an effort to circumvent this difficulty, we select the *plane monochromatic wave* for a free particle given by

Free Particle
Wave Function

$$\Psi(x, t) = Ae^{\frac{j}{\hbar}(px - Et)}, \tag{9.6}$$

which is similar to the displacement given by Equation 8.40 with de Broglie's postulates being incorporated. This wave function represents a *free particle* traveling in the positive $x$-direction with a momentum and energy known to be exactly $p = \hbar k$ and $E = \hbar\omega$, respectively. Assuming a constant potential energy $V(x) = V_0$ for the free particle, then the force (Equation 9.4) is equivalent to zero, the momentum $p$ is constant from Equation 1.16 above, and consequently, from Equation 9.1 the energy $E$ is constant. Thus $k$ and $\omega$ are constants in accordance with de Broglie's relations (Equations 8.52 and 8.53) and the free particle wave function (Equation 9.6) can be expressed as

$$\Psi(x, t) = Ae^{j(kx - \omega t)}, \tag{9.7}$$

Free Particle Wave Function

in terms of the particle's constant wave number and angular velocity. The first and second order spatial derivatives of this wave function are

$$\frac{\partial}{\partial x}\Psi(x, t) = jk\Psi(x, t), \tag{9.8a}$$

$$\frac{\partial^2}{\partial x^2}\Psi(x, t) = -k^2\Psi(x, t), \tag{9.8b}$$

whereas the first order time derivative yields

$$\frac{\partial}{\partial t}\Psi(x, t) = -j\omega\Psi(x, t). \tag{9.9}$$

From the last two equations we obtain

$$k^2\Psi(x, t) = -\frac{\partial^2}{\partial x^2}\Psi(x, t)$$

and

$$\omega\Psi(x, t) = j\frac{\partial}{\partial t}\Psi(x, t),$$

respectively, which upon substitution into Equation 9.5 yields

$$-\frac{\hbar^2}{2m}\frac{\partial^2\Psi}{\partial x^2} + V(x, t)\Psi(x, t) = j\hbar\frac{\partial\Psi}{\partial t}. \tag{9.10}$$

One-Dimensional Time-Dependent Schrödinger Equation

This equation represents the famous **one-dimensional time-dependent Schrödinger equation.** Even though our consideration was for a *free par-*

*ticle* having a time-independent or constant potential energy $V_0$, we postulate Equation 9.10 to be the correct wave equation in general for $V(x, t) \neq V_0$. It needs to be emphasized that this equation has *not been derived* but merely *constructed* through postulates, as the Schrödinger equation is a *fundamental first principle* of quantum mechanics, much like the first law of thermodynamics or Newton's second law of motion. Our construction of this result utilized a free particle wave function as an assumed solution to a *linear* energy equation, with Schrödinger's nonrelativistic *energy conservation* and *de Broglie's postulates* imposed. Under these constraints, other differential equations could be constructed; however, only Equation 9.10 is compatible with nature and capable of predicting experimental observations for different physical phenomena. A comparison of Schrödinger's equation with the *classical wave equation* (Equation 8.10) shows that they are similar only in that both contain second order derivatives with respect to the spatial coordinate. A major difference between these equations is the presence of only a first order time derivative and an *interaction potential* in Schrödinger's equation. Further, and of great significance is the fact that the Schrödinger equation is *imaginary* on the right-hand-side, whereas the classical wave equation is *real*. Although Equation 9.10 is *one-dimensional*, it is easily generalized to three dimensions, as will be seen in the next section. Incidentally, it is now straight forward and left as an exercise to show that the plane monochromatic wave function of a free particle in one dimension (Equation 9.6) is a solution to the Schrödinger wave equation.

## 9.2 Three-Dimensional Time-Dependent Schrödinger Equation

The generalization of the previous section to include all three spatial dimensions is rather straight forward from the fundamentals of classical physics. A free particle moving in a region of constant potential energy

$$V(\mathbf{r}, t) \equiv V(\mathbf{r}) = V_0 \tag{9.11}$$

has a position vector given by

$$\mathbf{r} = x\mathbf{i} + y\mathbf{j} + z\mathbf{k} \tag{1.1}$$

and a momentum vector

$$\mathbf{p} = p_x\mathbf{i} + p_y\mathbf{j} + p_z\mathbf{k}, \tag{4.54a}$$

where **i, j,** and **k** are the usual set of *orthonormal unit vectors* for a Cartesian coordinate system. The three-dimensional wave function associated with the free particle is given by

$$\Psi(x, y, z, t) = Ae^{\frac{j}{\hbar}(p_x x + p_y y + p_z z - Et)}, \tag{9.12}$$

which can be more compactly written as

$$\Psi(\mathbf{r}, t) = Ae^{\frac{j}{\hbar}(\mathbf{p}\cdot\mathbf{r} - Et)} \tag{9.13}$$

Free Particle Wave Function

by making use of the definition for the *inner* or *dot product* of vectors **p** and **r.** To simplify the mathematical form of the three dimensional Schrödinger equation, we introduce the **del operator** (see Appendix A, Section A.9)

$$\nabla \equiv \frac{\partial}{\partial x}\mathbf{i} + \frac{\partial}{\partial y}\mathbf{j} + \frac{\partial}{\partial z}\mathbf{k}, \tag{9.14}$$

Del Operator

which is a *vector differential operator* that must satisfy the mathematical rules for both *vectors* and *partial differentials*. The *scalar product* of this vector operator with itself is called the **Laplacian operator** and is given by

$$\nabla^2 = \frac{\partial^2}{\partial x^2} + \frac{\partial^2}{\partial y^2} + \frac{\partial^2}{dz^2}. \tag{9.15}$$

Laplacian Operator

The construction of the three-dimensional Schrödinger equation can now proceed in a similar manner to that presented in the previous section, with the new operators given above being utilized. Multiplying Schrödinger's *energy conservation equation* by $\Psi(\mathbf{r}, t)$,

$$\frac{p^2}{2m}\Psi(\mathbf{r}, t) + V(\mathbf{r}, t)\Psi(\mathbf{r}, t) = E\Psi(\mathbf{r}, t), \tag{9.16}$$

we realize that $p^2\Psi$ and $E\Psi$ can be expressed in terms of partial differentials in view of Equation 9.12. That is, operating on Equation 9.12 with the *Laplacian* gives

$$\nabla^2\Psi(\mathbf{r}, t) = -\frac{p^2}{\hbar^2}\Psi(\mathbf{r}, t), \tag{9.17}$$

which is directly solved for

$$p^2\Psi(\mathbf{r}, t) = -\hbar^2\nabla^2\Psi(\mathbf{r}, t). \tag{9.18}$$

Further, the time derivative of the free particle wave function (Equation 9.12 or 9.13) immediately yields

$$E\Psi(\mathbf{r}, t) = j\hbar \frac{\partial}{\partial t} \Psi(\mathbf{r}, t). \tag{9.19}$$

It must be emphasized that these two results are contingent upon the momentum and energy of the free particle being time-independent, as was previously discussed for the free particle. Now, we postulate that the results given by Equations 9.18 and 9.19 are valid in general, even for a potential energy $V(\mathbf{r}, t) \neq V_0$, and substitute them directly into Equation 9.16 to obtain

Three-Dimensional Time-Dependent Schrödinger Equation

$$-\frac{\hbar^2}{2m} \nabla^2\Psi + V(\mathbf{r}, t)\Psi(\mathbf{r}, t) = j\hbar\frac{\partial\Psi}{\partial t}. \tag{9.20}$$

This equation represents the generalized **three-dimensional time-dependent Schrödinger equation** of quantum mechanics. Even though it has been constructed and postulated herein, its form should be obvious from the definition of the Laplacian operator and the one-dimensional Schrödinger equation. One advantage of using the Laplacian operator in the formulation of Equation 9.20, instead of the second order spatial derivatives in Cartesian coordinates, is that the equation is *valid in all coordinate systems*, provided that $\nabla^2$ is appropriately defined in each system. It should be emphasized that the Schrödinger equation can only be used for nonrelativistic problems, where it has been found to be completely accurate in predicting observed phenomena. Further, it is important to recognize that the Schrödinger equation does not represent an additional postulate of nature, because with it Newton's second law of motion can be derived. The details of this derivation are presented in Chapter 10, Section 10.6, where we recognize Newtonian mechanics to be nothing more than a limited approximation to quantum mechanics.

## 9.3 Time-Independent Schrödinger Equation

The time-independent or *steady state* Schrödinger equation is directly obtainable from the time-dependent equation. As before, we consider one-dimensional motion first and then generalize to three dimensions. For a particle restricted to one-dimensional motion having a potential energy which is *time-independent*,

$$V = V(x), \tag{9.3}$$

Conservative Field

solutions to the Schrödinger equation are assumed to be of the form

$$\Psi(x, t) = \psi(x)\phi(t). \tag{9.21}$$

Separation of Variables

In this equation the lower case Greek letters $\psi$ (psi) and $\phi$ (phi) represent the spatial and temporal functions, respectively. By analogy with the *separation of variables* method presented for the classical vibrating string, and recognizing the type of differentials that occur in Schrödinger's one-dimensional equation we have

$$\frac{\partial^2}{\partial x^2}\Psi(x, t) = \phi(t)\frac{d^2}{dx^2}\psi(x) \tag{9.22}$$

and

$$\frac{\partial}{\partial t}\Psi(x, t) = \psi(x)\frac{d}{dt}\phi(t). \tag{9.23}$$

Substitution of these equations into Equation 9.10 yields

$$-\frac{\hbar^2}{2m}\phi\frac{d^2\psi}{dx^2} + V(x)\psi(x)\phi(t) = j\hbar\psi\frac{d\phi}{dt}, \tag{9.24}$$

which can be rewritten in the form

$$-\frac{\hbar^2}{2m}\frac{1}{\psi(x)}\frac{d^2\psi}{dx^2} + V(x) = j\hbar\frac{1}{\phi(t)}\frac{d\phi}{dt} \tag{9.25}$$

by dividing both sides of the equation by $\psi(x)\phi(t)$. Since the spatial and temporal coordinates are independent variables, Equation 9.25 is only valid if each side of the equation is equal to the same constant. From Schrödinger's energy conservation postulate (Equation 9.1) it is obvious that the *separation constant* for Equation 9.25 must be the total energy $E$. Thus, from Equation 9.25

$$j\hbar\frac{1}{\phi(\mathrm{t})}\frac{d\phi}{dt} = E, \tag{9.26}$$

which suggests that

$$j\hbar \int_0^{\phi} \frac{d\phi}{\phi} = E \int_0^{t} dt,$$

since $E$ is a constant. This integral equation is easily solved for

$$1\mathrm{n}|\phi(t)| = \frac{Et}{j\hbar}$$

and exponentiated to give

$$\phi(t) = e^{-\frac{j}{\hbar} ET}. \tag{9.27}$$

The result expressed by Equation 9.27 is perfectly general for a time-independent potential energy and, consequently, the *wave function* solution to Schrödinger's one-dimensional equation is of the form

Wave Function

$$\Psi(x, t) = \psi(x) e^{-\frac{j}{\hbar} ET}, \tag{9.28}$$

where Equation 9.27 has been substituted into Equation 9.21. Now, substitution of this wave function into Schrödinger's one-dimensional time-dependent equation (Equation 9.10) and performing the indicated differentials yields

$$-\frac{\hbar^2}{2m} \phi \frac{d^2\psi}{dx^2} + V\psi\phi = j\hbar \left( -\frac{j}{\hbar} E \right) \phi\psi,$$

which should be obvious from Equations 9.24 and 9.27. Cancelling the common factor $\phi$ and simplifying the right-hand side of this equation gives the sought after **one-dimensional time-independent Schrödinger equation,**

One-Dimensional Time-Independent Schrödinger Equation

$$-\frac{\hbar^2}{2m} \frac{d^2\psi}{dx^2} + V(x)\psi(x) = E\psi(x). \tag{9.29}$$

This result is also obvious from Equation 9.25, since the right-hand side of that equation is equal to the *separation constant E*. As this is an ordinary second order differential equation in the position variable with no *imaginary* factors, its solutions $\psi(x)$ need not be, necessarily, *complex functions*.

The generalization of Equation 9.29 to three dimensions is

$$\left[-\frac{\hbar^2}{2m}\nabla^2 + V(\mathbf{r})\right]\psi(\mathbf{r}) = E\psi(\mathbf{r}), \tag{9.30}$$

Three-Dimensional Time-Independent Schrödinger Equation

which should be obvious from the considerations of the previous section and the definition of the Laplacian. Schrödinger's *time-independent* equation (Equation 9.29 or 9.30) is of the general form of what is commonly called an **eigenvalue equation.** Its solutions $\psi(\mathbf{r})$, referred to as **eigenfunctions,** are normally obtained for only certain values of $E$, which are commonly referred to as the **energy eigenvalues.** Many of the mathematical techniques for solving an *eigenvalue equation* have already been presented in the discussion of the classical vibrating string, where the *eigenvalues* of Equation 8.17 corresponded to $k_n^2$ with *eigenfunctions* given by Equation 8.26. Eigenvalue problems represent one of the most important types of problems in mathematical physics and the solution techniques will be further illustrated by way of examples in Sections 9.6 and 9.7.

## 9.4 Probability Interpretation of the Wave Function

In writing the plane monochromatic wave of Equation 9.6 we realized it to be the associated matter wave of a particle of definite momentum and energy. If it is to be viewed as a *wave packet*, then clearly, because it is of infinite extent, we know nothing of the exact location of the particle. This corresponds to saying that there is *equal probability* of finding the particle anywhere in the domain $-\infty < x < +\infty$. As suggested in Chapter 8, if we add two such waves,

$$\Psi_1(x, t) = A_1 e^{\frac{j}{\hbar}(p_1 x - E_1 t)} \tag{9.31}$$

and

$$\Psi_2(x, t) = A_2 e^{\frac{j}{\hbar}(p_2 x - E_2 t)}, \tag{9.32}$$

together to form the linear superposition

$$\Psi(x, t) = \Psi_1(x, t) + \Psi_2(x, t), \tag{9.33}$$

the resulting wave function describes a particle about whose location we are no longer *totally* ignorant. Of course we no longer know the particle's momentum exactly, since it is now a *mixture* of the two momentum states $p_1$ and $p_2$. In an attempt to obtain a *wave packet* that is rather well *local-*

*ized* at any time $t$, we can generalize the linear superposition to include an infinite number of monochromatic waves. This is easily accomplished by defining

General Wave Function

$$\Psi(x, t) = \sum_{i=1}^{\infty} A_i \Psi_i(x, t), \tag{9.34}$$

where the $A_i$ are arbitrary constants. Actually, this wave function is the *most general form* of the solution to the Schrödinger equation for the potential $V(x)$.

For a free particle the generalized wave function of Equation 9.34 takes the form

$$\Psi(x, t) = \sum_{i=1}^{\infty} A_i e^{\frac{j}{\hbar}(p_i x - E_i t)}, \tag{9.35}$$

where the position $x$ of the associated particle is known to within a fairly narrow range along the $X$-axis. But the momentum of the particle is now virtually unknown. It is neither $p_1$, $p_2$, nor $p_3$ but rather is a *mixture* of the *infinite* number of momentum states. The *state function* defined by Equation 9.35 is *not* the *most general* wave packet for a *free particle*, as a free particle could have any one value of a *continuous range* of possible momentum values from $-\infty$ to $+\infty$. Thus, the *associated wave* could have any value of *wave number* in the domain $-\infty < k < +\infty$. Further, the amplitude symbolized by the letter $A_i$ in Equation 9.35 is generally a function (commonly an exponential) of the wave number. With this in mind we define

$$\Phi(k, t) \equiv A \tag{9.36}$$

and replace the *summation* with an *integral* to obtain

Free Particle

$$\Psi(x, t) = \int_{-\infty}^{+\infty} \Phi(k, t) e^{j(kx - \omega(k)t)} dk, \tag{9.37}$$

as the *ultimate* generalization in one dimension to Equation 9.35. Clearly, de Broglie's postulates have been utilized in obtaining this last equation and the angular frequency $\omega$ has been indicated as a function of the wave number $k$.

One can easily prove that $\Psi(x, t)$ of Equation 9.37 is indeed a concentrated wave packet. That is, $\Psi(x, t)$ at $t \equiv 0$ is large near $x = 0$, since for $x = 0$, $e^{jkx} = 1$ for all $k$, and the contribution to the integral coming from different $k$ add up in phase, making the sum of Equation 9.37 large near $x = 0$. On the other hand, for large $x$, $e^{jkx}$ is a rapidly oscillating function of $k$ and its integral tends to cancel itself out. In other words, it appears that the probability of finding the particle in a given region of space is large where the *wave function* is large and small where it is small. The wave function is commonly called the **probability amplitude** and many *mistakenly* use it as a probability. However, the wave function should not be considered as a probability, since $\Psi(x, t)$ can be both positive and negative valued; whereas, *negative* probabilities are *meaningless*. Further, the probability per unit length or *probability density* $P(x, t)$ of finding a particle in a region of space at a particular time is a *real quantity*; whereas, the wave function $\Psi(x, t)$ is an inherently *complex* function.

In 1926 Max Born postulated a probability interpretation. He proposed that the **relative probability density** or *relative probability per unit length* of finding the particle between $x$ and $x + dx$ at an instant in time $t$ be defined by

$$\begin{aligned} P(x, t) &\equiv |\Psi(x, t)|^2 \\ &\equiv \Psi^*(x, t)\Psi(x, t), \end{aligned} \tag{9.38}$$

Relative Probability Density

where the *asterisk* denotes the **complex conjugate** of $\Psi$. The **absolute probability density** is defined by

$$\rho(x, t) \equiv \frac{P(x, t)}{\int_{-\infty}^{+\infty} P(x, t)dx}, \tag{9.39}$$

Probability Density

which presupposes that the wave function be **square integrable,** that is, the integral of *psi-star-psi*,

$$\int_{-\infty}^{+\infty} \Psi^*(x, t)\Psi(x, t)dx,$$

*must be finite* at the instant $t$. The point to understand is that the wave function $\Psi(x, t)$ has only indirect physical significance, while the quantity $|\Psi(x, t)|^2$ has direct physical meaning.

It has been pointed out that the Schrödinger equation must be linear. Consequently, for any solution of the Schrödinger equation, another solution can be obtained by multiplying it by a *phase factor* $e^{j\theta}$, or a constant, without affecting the physics. So long as the *square-integrability require-*

*ment* is satisfied, we can select the constant such as to *normalize* the wave function. For properly normalized wave functions the **normalization condition** at an instant $t$ is

Normalization Condition

$$\int_{-\infty}^{+\infty} \Psi^*(x, t)\Psi(x, t)dx = 1 \tag{9.40a}$$

for one dimension or

$$\iiint_{-\infty}^{+\infty} \Psi^*(\mathbf{r}, t)\Psi(\mathbf{r}, t)dxdydz = 1 \tag{9.40b}$$

for three dimensions. If we consider the most general form of the wave function as given by Equation 9.34, then it appears reasonable that the normalization condition can be satisfied at any given instant in time by the proper choice of the arbitrary constants $A_i$. Of course, we will want to select the constants carefully so that the wave function is normalized in accordance with Equation 9.40 at all values of time. Equation 9.40a simply means that the total probability of finding the particle associated with the wave function somewhere along the spatial domain is one at time $t$, and all values of $t$ must give this probability. With the *normalization condition* always required, it is apparent from Equation 9.39 that the probability density $\rho(x, t)$ is identical to the *relative* probability density, $P(x, t)$. That is, the quantity $|\Psi(x, t)|^2$ can be interpreted as a *probability density* rather than a *relative* probability density.

In addition to $\Psi$ being normalizable, nature tends to support a *wave function* and *its partial derivatives* (both spatial and temporal) that are *linear*, *single valued*, *finite*, and *continuous* (i.e., well behaved). Further, in the limit as the spatial coordinates approach $+\infty$ or $-\infty$, the *wave function must approach zero* for a *bound state*. Occasionally, we relax one or more of these conditions for mathematical simplification. For example, the free particle wave function is not square integrable (see Section 9.6), but we have found it to be most useful as an example to illustrate the concepts of Schrödinger's quantum mechanics. Clearly, it is an idealization for exact knowledge of the particle's momentum is not obtainable in *wave mechanics*.

## 9.5 Conservation of Probability

With the *normalization condition* (Equation 9.40a) always required, we always have

$$P(x, t) = \rho(x, t) = \Psi^*(x, t)\Psi(x, t) \tag{9.41}$$

as the *probability per unit length* of finding the particle between $x$ and $x + dx$. Consequently,

$$P(x, t)dx = \rho(x, t)dx = \Psi^*(x, t)\Psi(x, t)dx \tag{9.42}$$

is the *probability*, a dimensionless quantity, of finding the particle at a particular point along the $X$-axis. Integrating Equation 9.42 over the range $x = -\infty$ to $x = +\infty$ gives the **total probability** of finding the particle anywhere along the $X$-axis. We can define this probability as

$$P(t) \equiv \int_{-\infty}^{+\infty} \rho(x, t)dx, \tag{9.43a}$$

Total Probability

which reduces to

$$P(t) = \int_{-\infty}^{+\infty} \Psi^*(x, t)\Psi(x, t)dx \tag{9.43b}$$

for properly *normalized* wave functions. The question now is whether or not the Schrödinger equation is compatible with our interpretation of the wave function. More specifically, is $P(t)$ really dependent on time? To obtain the answer, we assume the *normalization condition* to be required and differentiate Equation 9.43b with respect to time, which results in

$$\begin{aligned} \frac{dP}{dt} &= \frac{d}{dt}\int_{-\infty}^{+\infty} \Psi^*(x, t)\Psi(x, t)dx \\ &= \int_{-\infty}^{+\infty} \left(\Psi\frac{\partial\Psi^*}{\partial t} + \Psi^*\frac{\partial\Psi}{\partial t}\right)dx. \end{aligned} \tag{9.44}$$

From the one-dimensional time-dependent Schrödinger equation (Equation 9.10) we obtain

$$\frac{\partial\Psi}{\partial t} = \frac{j\hbar}{2m}\frac{\partial^2\Psi}{\partial x^2} - \frac{j}{\hbar}V\Psi, \tag{9.45a}$$

and taking the *complex conjugate* of this equation gives

$$\frac{\partial\Psi^*}{\partial t} = -\frac{j\hbar}{2m}\frac{\partial^2\Psi^*}{\partial x^2} + \frac{j}{\hbar}V\Psi^*. \tag{9.45b}$$

Now, substitution of the Schrödinger equation in these two forms into Equation 9.44 yields

$$\begin{aligned}\frac{dP}{dt} &= \frac{j\hbar}{2m}\int_{-\infty}^{+\infty}\left(\Psi^*\frac{\partial^2\Psi}{\partial x^2} - \Psi\frac{\partial^2\Psi^*}{\partial x^2}\right)dx \\ &= \frac{j\hbar}{2m}\int_{-\infty}^{+\infty}\frac{\partial}{\partial x}\left(\Psi^*\frac{\partial\Psi}{\partial x} - \Psi\frac{\partial\Psi^*}{\partial x}\right)dx \\ &= \frac{j\hbar}{2m}\left(\Psi^*\frac{\partial\Psi}{\partial x} - \Psi\frac{\partial\psi^*}{\partial x}\right)\Bigg|_{-\infty}^{+\infty}. \end{aligned} \tag{9.46}$$

Recalling that if the wave function is bounded and sufficiently well behaved to be normalized, then $\Psi(x, t)$ and $\Psi^*(x, t)$ must vanish at $x = \pm\infty$. Thus, Equation 9.46 reduces to

Conservation of Probability

$$\frac{dP}{dt} = 0, \tag{9.47}$$

which means that the *total probabililty P* is *constant*. That is, if $P = 1$ at the zero of time ($t \equiv 0$), then $P = 1$ for all of time ($t > 0$). This is called the **conservation of probability** or *norm-preservation*. This result originates in the fortunate circumstance that the temporal derivative in the Schrödinger equation is only a first order partial derivative. If the Schrödinger equation contained a second order partial derivative with respect to time, like the *classical wave equation*, the *normalization preservation condition* would simply not be obtained.

The *conservation of probability* is also required for a particle in motion along the $X$-axis between any two coordinate positions, say $x_1$ and $x_2$ in Figure 9.1. At positive $x_1$ the *probability density flux*, or *probability current*, has a value $S_1$, while at coordinate position $x_2$ it has the value $S_2$. In order to conserve *probability density* $\rho(x, t)$, any difference between the *probability currents* $S_1$ and $S_2$ must correspond to the time rate at which the *total probability* $P(t)$ changes between $x_1$ and $x_2$. That is,

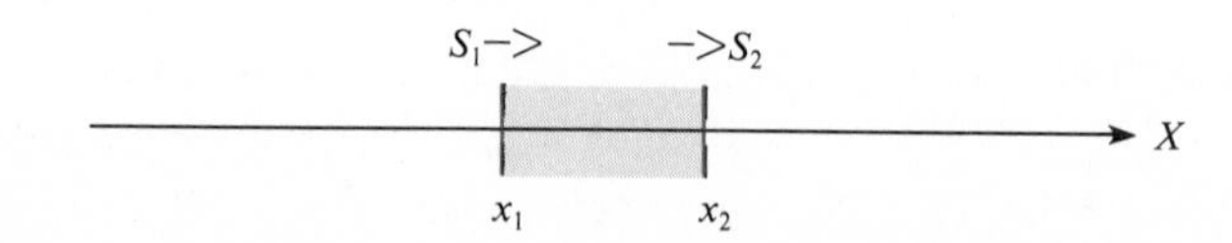


Figure 9.1
The time rate of change of the total probability $P(t)$ in a segment $\Delta x = x_2 - x_1$ equals $S_1 - S_2$, where $S_1$ is the *inward flow* and $S_2$ is the *outward flow* of probability density.

$$\frac{dP}{dt} = \frac{d}{dt}\int_{x_1}^{x_2} \rho(x, t)\, dx = S_1 - S_2, \tag{9.48a}$$

which for properly *normalized* wave functions reduces to

$$\frac{dP}{dt} = \frac{d}{dt}\int_{x_1}^{x_2} \Psi^*\Psi dx = S_1 - S_2. \tag{9.48b}$$

With the time rate of change of the *total probability* given by Equation 9.46, a comparison with Equation 9.48b immediately yields

$$\frac{dP}{dt} = \frac{j\hbar}{2m}\left(\Psi^*\frac{\partial\Psi}{\partial x} - \Psi\frac{\partial\Psi^*}{\partial x}\right)\Bigg|_{x_1}^{x_2}. \tag{9.49}$$

For this result (note the limits are from $x_1$ to $x_2$) to equal $S_1 - S_2$ of Equation 9.48b, we define the **probability density flux,** or **probability current,** with a *negative sign* as

$$S(x, t) \equiv -\frac{j\hbar}{2m}\left(\Psi^*\frac{\partial\Psi}{\partial x} - \Psi\frac{\partial\Psi^*}{\partial x}\right). \tag{9.50}$$

Problem Current

This is the probability per unit time that a particle associated with the wave function $\Psi(x, t)$ will cross the point $x$, in the direction of increasing values of the $x$-coordinate. The generalization to three dimensions is easily accomplished by replacing $x$ with $\mathbf{r}$. The interpretation of $\mathbf{S}(\mathbf{r}, t)$ pertains to the probability flux through a *region of space* that is bounded by a *surface area A*. The definition of the *probability current*, along with the other concepts of Schrödinger's quantum mechanics introduced in this chapter, will be made plausible in the next section, where we consider the case of a *free particle* of energy $E$ and momentum $p$.

## 9.6 Free Particle and a Constant Potential

Having constructed the Schrödinger equation by using the free particle wave function, it seems only appropriate that we use this example to illustrate some other features of the theory of quantum mechanics, such as eigenfunctions, probability density, and probability currrent. For the case of a *free particle* moving under the influence of a *constant potential* ($V(x, t) \equiv$ CONSTANT), we know the force, given by Equation 9.4, will vanish irrespective of the *constant value* assumed by the potential energy

$V(x, t)$. Thus, no generality is sacrificed by defining the potential energy to be zero

Constant Potential

$$V(x, t) \equiv 0. \tag{9.51}$$

This corresponds to the classical situation where the particle may be at rest or in a state of uniform motion with constant momentum $p$, such that in either case its total energy $E$ is a constant.

Quantum mechanically the behavior of the free particle is predicted by the eigenfunction solutions to Schrödinger's time-independent equation (Equation 9.29) for $V(x) = 0$, that is

$$-\frac{\hbar^2}{2m}\frac{d^2\psi}{dx^2} = E\psi(x). \tag{9.52}$$

We know that a solution to this ordinary second order differential equation is of the form

$$\psi(x) = Ae^{\frac{j}{\hbar}px}, \tag{9.53}$$

which is easily verified by direct substitution into Equation 9.52 and realizing that

$$p = (2mE)^{1/2}. \tag{9.54}$$

The *wave function* associated with the eigenfunction solution (Equation 9.53) is easily obtained from Equation 9.28,

$$\Psi(x, t) = \psi(x)e^{-\frac{j}{\hbar}Et}, \tag{9.28}$$

in the form

$$\Psi(x, t) = Ae^{\frac{j}{\hbar}(px - Et)}, \tag{9.6}$$

as given previously by Equation 9.6. As discussed in Chapter 8, the free particle wave function is recognized as a *traveling wave* that oscillates with angular frequency $\omega$ and travels in the positive $x$-direction with a *phase velocity*

$$w = \frac{\omega}{k} = \frac{E}{p}. \tag{9.55}$$

That the free particle characterized by Equation 9.6 is travelling in the positive $x$-direction can now be realized by evaluating the *probability current* $S(x, t)$, since the sign of $S(x, t)$ indicates the direction of motion of the particle. With

$$\frac{\partial \Psi}{\partial x} = \frac{jp}{\hbar}\Psi$$

and

$$\frac{\partial \Psi^*}{\partial x} = -\frac{jp}{\hbar}\Psi^*$$

substituted into Equation (9.50), we obtain

$$\begin{aligned} S(x, t) &\equiv -\frac{j\hbar}{2m}\left(\Psi^*\frac{\partial \Psi}{\partial x} - \Psi\frac{\partial \Psi^*}{\partial x}\right) \\ &= \frac{p}{m}\Psi^*\Psi \\ &= v|\Psi|^2, \end{aligned} \tag{9.56}$$

where the classical linear momentum $p = mv$ has been used in obtaining the last equality. Thus, the probability current for the free particle is just the product of its constant speed $v$ and probability density $P(x, t)$. The *conservation of probability* for the free particle is given by Equation 9.48b as

$$\begin{aligned} \frac{dP}{dt} &= \frac{d}{dt}\int_{x_1}^{x_2} \Psi^*\Psi dx \\ &= S_1 - S_2 \\ &= v_1|\Psi|^2_{x_1} - v_2|\Psi|^2_{x_2}. \end{aligned} \tag{9.57}$$

For the free particle $v_1 = v_2 \equiv v$ and $P(x, t)$ is given by

$$P(x, t) \equiv \Psi^*(x, t)\Psi(x, t) = A^*A. \tag{9.58}$$

Under these substitutions, Equation 9.57 obviously reduces to zero,

$$\frac{dP}{dt} = vA^*A - vA^*A = 0,$$

irrespective of the values chosen for $x_1$ and $x_2$. This result is consistent with the fact that for our free particle $v$, $A$, and $A^*$ are constant in the temporal and position coordinates.

What if the particle is traveling in the negative $x$-direction? Clearly, the associated wave function could be represented by

$$\Psi(x,\,t) = Be^{-\frac{j}{\hbar}(px-Et)}, \tag{9.59}$$

since for this wave function the probability current has a *negative valued* result given by

$$S(x,\,t) = -v|\Psi|^2 = -vB^*B. \tag{9.60}$$

As before, the probability current $S(x, t)$ and probability density $P(x, t)$,

$$P(x,\,t) = |\Psi|^2 = B^*B, \tag{9.61}$$

are constants in time and independent of the particle's position, so that conservation of probability,

$$\begin{aligned}\frac{dP}{dt} &= S_1 - S_2\\ &= (-vB^*B) - (-vB^*B) = 0,\end{aligned} \tag{9.62}$$

is obtained. For the particle described by the wave function of Equation 9.59 (or Equation 9.6), the momentum is known to be a constant $p$, but we have no knowledge whatsoever of its position along the $X$-axis. That is, the particle can be *anywhere* in the domain from $x = -\infty$ to $x = +\infty$.

Since the Schrödinger equation is a linear differential equation, and since the wave functions given by Equations 9.6 and 9.59 are solutions, then so is their sum

$$\Psi(x,\,t) = e^{-\frac{j}{\hbar}Et}\,(Ae^{\frac{j}{\hbar}px} + Be^{-\frac{j}{\hbar}px}). \tag{9.63}$$

A comparison of this equation with Equation 9.28 suggests that the eigenfunction

$$\psi(x) = Ae^{\frac{j}{\hbar}px} + Be^{-\frac{j}{\hbar}px}. \tag{9.64}$$

is also a solution to Equation 9.52. In fact, since the eigenfunction solution of Equation 9.64 involves *two arbitrary constants*, it is the *general form* of the solution to the ordinary second order differential equation (Equation 9.52).

There is one point of difficulty with the solutions represented by Equations 9.6, 9.59, and 9.63 and it is that the normalization condition can not be satisfied for finite values of the constants $A$ and $B$. As a specific example, the normalization condition for the wave function of Equation 9.6 diverges unless $A = 0$, that is

$$\int_{-\infty}^{+\infty} \Psi^*(x, t)\Psi(x, t)dx = \int_{-\infty}^{+\infty} A^*A dx = A^*A \int_{-\infty}^{+\infty} dx.$$

It must be emphasized that the free particle wave function of Equation 9.6 represents the highly idealized situation of a particle traveling in a beam of infinite length, whose momentum is known exactly and whose position is completely unknown. A physically meaningful wave function would be of essentially uniform amplitude over the length of the beam but would vanish for values of $x$ that are very large or very small. It would constitute a *group* of length $\Delta x$ which is finite. Further, Heisenberg's uncertainty principle requires the realistic wave function to have not only the single momentum $p = \hbar k$ but also a momenta distribution $\Delta p = \hbar \Delta k = h/\Delta x$ that is centered about the single momentum $p$. The idealized wave function of Equation 9.6 can be considered as representing the realistic wave function in the *limit* $\Delta x \rightarrow \infty$. In this limiting case the normalization condition is not obtained; however, the idealized wave function can be employed to calculate physically meaningful quantities that *do not depend* on the value of the multiplicative constant.

One last observation is in order before concluding our discussion of the free particle wave functions. The problem of the *normalization condition* not vanishing unless the constants $A$ and $B$ in Equation 9.63 are equal to zero can be *removed* by completely suppressing the wave function outside of a large but finite region. That is we construct a more realistic wave function by allowing the eigenfunction of Equation 9.64 to be

$$\psi(x) = Ae^{\frac{j}{\hbar}px} + Be^{-\frac{j}{\hbar}px}, \qquad 0 < x < L \tag{9.65}$$

and

$$\psi(x) = 0, \qquad L \leq x \leq 0. \tag{9.66}$$

This particular problem will be the topic of discussion in the next section.

## 9.7 Free Particle in a Box (Infinite Potential Well)

Consider a particle of mass $m$ confined to one dimensional motion in a *box* of size $L$. The motion of the particle takes place along the $X$-axis in the domain $0 < x < L$, which constitutes the *box* illustrated in Figure 9.2. The particle is *free* inside this domain, but not *really free* since a truly free particle must have the domain $-\infty < x < +\infty$. We further consider the box to have *impenetrable* walls at $x = 0$ and $x = L$, by assuming an infinite potential energy to exist at these coordinate positions. In the domain $0 < x < L$ the potential energy is of course constant and, as before, we will choose the constant to be *zero*. Thus, the potential energy is described as

$$V(x) = \begin{cases} 0, & 0 < x < L\,, \\ \infty, & L \leq x \leq 0\,. \end{cases} \tag{9.67}$$

Under the constraints of this potential, the particle moves back and forth in the one dimensional box having perfectly elastic collisions with the infinitely hard potential *walls*. Clearly, the particle cannot be found outside the box, since it cannot have an infinite energy. Consequently, we have the *boundary conditions* of Equation 9.66,

Boundary Conditions

$$\Psi(x, t) = 0, \qquad L \leq x \leq 0 \tag{9.66}$$

imposed on the particle's associated wave function.

Since within the box the potential energy is independent of time, Schrödinger's time-independent equation (Equation 9.29) is applicable and becomes

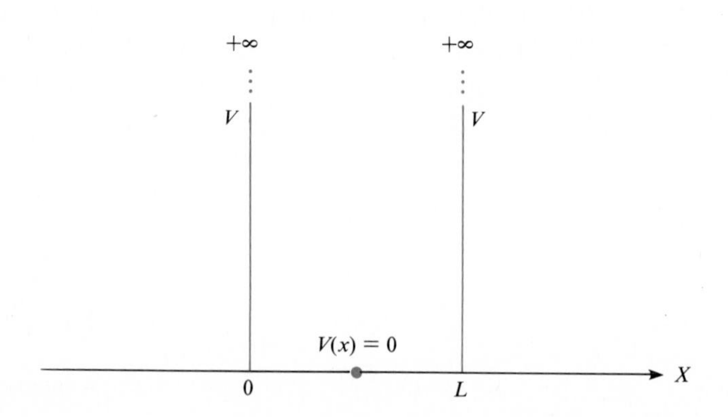


Figure 9.2
A "free" particle confined to a box of length $L$.

$$\frac{d^2\psi}{dx^2} + \frac{2m}{\hbar^2} E\psi = 0. \tag{9.68}$$

We know from the previous section that this ordinary second order differential equation has a *general solution* of the form

$$\psi(x) = Ae^{jkx} + Be^{-jkx}, \tag{9.69}$$

where de Broglie's momentum postulate $p = \hbar k$ has been substituted into Equation 9.64. Expanding this eigenfunction solution by using Euler's relation (see Appendix A, Section A.7) results in

$$\psi(x) = A \cos kx + jA \sin kx + B \cos kx - jB \sin kx, \tag{9.70}$$

which under the *boundary condition* $\psi(x = 0) = 0$ yields

$$0 = A \cos 0° + B \cos 0°.$$

Thus, for the eigenfunction to vanish at $x = 0$ we must require

$$B = -A \tag{9.71}$$

and the eigenfunction of Equation 9.70 reduces to the form

$$\psi(x) = 2jA \sin kx. \tag{9.72}$$

Now, imposing the second boundary condition, $\psi(x = L) = 0$, on this eigenfunction gives

$$0 = 2jA \sin kL. \tag{9.73}$$

Since $A \neq 0$, Equation 9.73 can be satisfied by requiring

$$kL = n\pi, \qquad n = 1, 2, 3, \cdots,$$

where $n$ is interpreted as the *principal quantum number*. Recognizing that the value of the wave number $k$ will depend upon the value assumed by $n$, we rewrite the above *wave number quantization* relation with a subscript notation as

$$k_n = \frac{n\pi}{L}, \qquad n = 1, 2, 3, \cdots. \tag{9.74}$$

Wave Number Quantization

Now, the eigenfunction of Equation 9.72 becomes

$$\psi_n(x) = 2jA \sin k_n x = 2jA \sin\frac{n\pi x}{L}, \tag{9.75}$$

where the $n$-subscript on $\psi$ denotes the dependence of the *eigenfunctions* on the principal quantum number.

The free particle in a box eigenfunction of Equation 9.75 could be substituted into the eigenvalue equation (Equation 9.68) to obtain a relation for the energy eigenvalues $E_n$. But, as a general rule, it is always prudent to first *normalize* the eigenfunction. This is easily accomplished by realizing that the *normalization condition*,

$$\int_{-\infty}^{+\infty} \Psi^*(x, t)\Psi(x, t)dx = 1, \tag{9.40a}$$

reduces to

Eigenfunction Normalization

$$\int_{-\infty}^{+\infty} \psi^*(x)\psi(x)dx = 1 \tag{9.76}$$

as a result of Equation 9.28,

$$\Psi(x, t) = \psi(x)e^{-\frac{j}{\hbar}Et}. \tag{9.28}$$

For our particular case of a particle confined to an infinitely deep potential well, the limits of integration on Equation 9.76 must be from $x = 0$ to $x = L$. Thus, the eigenfunctions of Equation 9.75 are normalized by considering

$$\begin{aligned} 1 &= \int_0^L |\psi_n|^2 dx \\ &= 4A^2 \int_0^L \sin^2\left(\frac{n\pi x}{L}\right)dx \\ &= 4A^2 \frac{L}{n\pi}\int_0^{n\pi} \sin^2 \theta d\theta \\ &= 4A^2 \frac{L}{n\pi}\int_0^{n\pi} \tfrac{1}{2}(1 - \cos 2\theta)d\theta \end{aligned}$$

$$= 4A^2 \frac{L}{n\pi} \frac{n\pi}{2}$$

$$= 2A^2 L, \qquad (9.77)$$

where the variable of integration has been changed from $x$ to $\theta \equiv n\pi x/L$ and the trigonometric identity

$$\sin^2 \theta = \tfrac{1}{2}(1 - \cos 2\theta)$$

from Appendix A, Section A.6 has been employed. Now, solving Equation 9.77 for $A$,

$$A = \sqrt{\frac{1}{2L}}, \qquad (9.78)$$

and substituting into Equation 9.75 gives

$$\psi_n(x) = j\sqrt{\frac{2}{L}} \sin \frac{n\pi x}{L} \qquad (9.79)$$

Normalized Energy Eigenfunctions

as the *normalized eigenfunctions* for the free particle in an infinite potential well.

With the *normalized eigenfunctions* given by Equation 9.79, it is now an easy task to substitute into the *eigenvalue equation* (Equation 9.68) and obtain the relation

$$E_n = n^2 \frac{\pi^2 \hbar^2}{2mL^2}$$

$$\equiv n^2 E_1 \qquad (9.80)$$

Energy Eigenvalues

for the *allowed energy eigenvalues* of the particle in a box. This result is also easily obtained by noting that the allowed energy of the free particle in a box is given by

$$E_n = \frac{p_n^2}{2m} = \frac{\hbar^2 k_n^2}{2m} \qquad (9.81)$$

and using Equation 9.74 for the *quantized wave numbers* $k_n$. From the form of Equation 9.80, it is immediately obvious that the particle's energy is quantized into discrete values or levels and that the particle can be in any one of a number of discrete energy states available. Also, we note that the allowed energy of the particle is a quadratic function of the principal quantum number $n$, as illustrated in Figure 9.3. The particle can lose or absorb energy; however, the amount lost or absorbed must be exactly equal to the *energy difference* between two allowable energy states. Further, we note that the particle can not have an energy of zero, since its *ground state energy* or *zero point energy* is given by Equation 9.80 for $n = 1$ as

Ground State Energy

$$E_1 = \frac{\pi^2 \hbar^2}{2mL^2}. \tag{9.82}$$

Clearly, this result is in contradiction to the predictions of classical physics, since according to Newtonian mechanics the particle could have any value, including zero for its energy. This contradiction is, however, consistent with Heisenberg's *uncertainty principle*. That is, the *uncertainty in position* is known to be

$$\Delta x \approx L,$$

which implies that

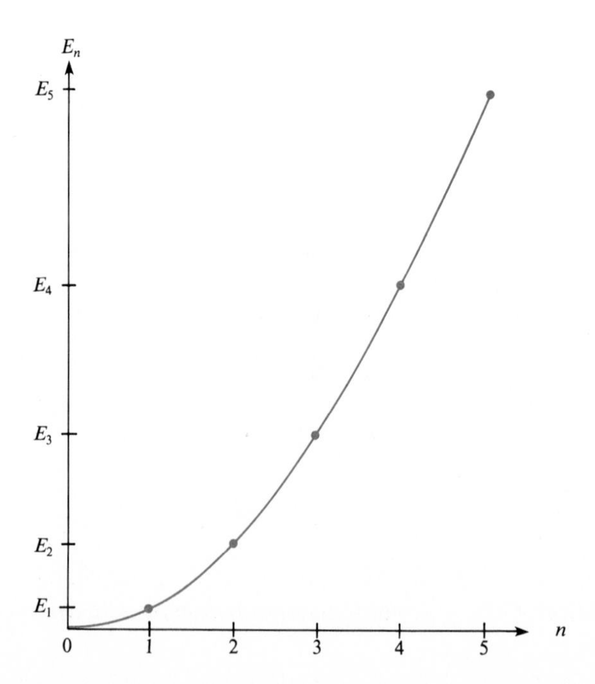


Figure 9.3
The quadratic dependence of the energy eigenvalues $E_n$ on the principal quantum number $n$ for a particle in a box.

$$\Delta p \approx \frac{h}{L}$$

from Heisenberg's principle of Equation 8.99a. Thus, the energy can *never* be equal to zero, as this would require the uncertainty in momentum to be zero ($\Delta p = 0$) and complete ignorance in the position of the particle, i.e.,

$$\Delta x = \frac{h}{\Delta p} = \frac{h}{0} = \infty.$$

Consider the quantum-mechanical implications for a 10 g macro-size particle confined to a box of length $L = 10$ cm. In this case Equation 9.82 predicts that $E_1 \approx 5.5 \times 10^{-64}$ J, which suggests the particle has an approximate speed of $3.3 \times 10^{-31}$ m/s. If you were observing such a *macro-particle*, you would be convinced that it was stationary. On the other hand, if the *macro-particle* were moving with a distinguishable speed, say about 0.2 m/s, the corresponding principal quantum number would be approximately $6 \times 10^{29}$. It is not difficult to understand why we will never directly *observe* a quantum-mechanical particle in a box. However, an electron of mass $m_e = 9.11 \times 10^{-31}$ kg in a box of size $L = 10^{-10}$ m has energy levels given by roughly $E_n \approx 38n^2$ eV. Thus, we realize that in the micro-world the energy-level spacing is easily perceptible, but in the macro-world the quantum-mechanical aspects are physically infinitesimal.

It needs to be emphasized that the normalized eigenfunctions of Equation 9.79 represent *only one* possible solution to the eigenvalue equation (Equation 9.68). We could have assumed a general solution of the form

$$\psi(x) = A \sin kx + B \cos kx, \tag{9.83}$$

which reduces to

$$\psi(x) = A \sin kx \tag{9.84}$$

under the boundary condition (see Equation 9.66) $\psi(x) = 0$ at $x = 0$. From the second boundary condition, $\psi(x) = 0$ at $x = L$, the *quantization* of the wave number is obtained,

$$k_n = \frac{n\pi}{L}, \qquad n = 1, 2, 3, \cdot \cdot \cdot, \tag{9.74}$$

which is identical to that determined previously. The normalized eigenfunctions in this case are given by

Normalized Energy Eigenfunctions

$$\psi_n(x) = \sqrt{\frac{2}{L}} \sin \frac{n\pi x}{L}, \tag{9.85}$$

which is easily verified by arguments similar to those presented in the derivation leading to Equations 9.77 and 9.78. Also, the energy eigenvalues given in Equation 9.80 are appropriate for these eigenfunctions, which should be obvious since Equation 9.81 ($E_n = \hbar^2 k_n^2/2m$) is still applicable.

As a few other observations about the quantum-mechanical particle in a box, we note that the *normalized wave functions* (see Equations 9.28 and 9.85)

$$\begin{aligned} \Psi_n(x, t) &= \psi_n(x) e^{-\frac{j}{\hbar} E_n t} \\ &= \sqrt{\frac{2}{L}} \sin \frac{n\pi x}{L}\, e^{-\frac{j}{\hbar} E_n t} \end{aligned} \tag{9.86}$$

satisfy the requirements of Schrödinger's quantum mechanics. That is, for every value of the principal quantum number $n$, the eigenfunction $\psi_n(x)$ and its spatial derivatives are continuous and single valued in $x$. Certainly, the wave function $\Psi_n(x, t)$ is *square integrable*, since the normalization condition (see Equation 9.76) is satisfied. Thus, the *absolute* probability density $\rho(x, t)$ is equal to the *relative* probability density $P(x, t)$ for each quantum number $n$,

$$\begin{aligned} \rho_n(x, t) &= \frac{\Psi_n^*(x, t)\Psi_n(x, t)}{\int_{-\infty}^{+\infty} |\Psi_n(x, t)|^2 dx} \\ &= \Psi_n^*(x, t)\Psi_n(x, t) = P_n(x, t). \end{aligned} \tag{9.87}$$

It is instructive to plot a few of the *probability densities* $P_n$,

Probability Densities

$$\begin{aligned} P_n(x, t) &\equiv \Psi_n^*(x, t)\Psi(x, t) = \psi_n^*(x)\psi_n(x) \\ &= \frac{2}{L} \sin^2 \frac{n\pi x}{L}, \end{aligned} \tag{9.88}$$

from $x = 0$ to $x = L$, as illustrated in Figure 9.4. Clearly, for $n = 1$ (Figure 9.4), the particle has the *greatest probability* of being located at

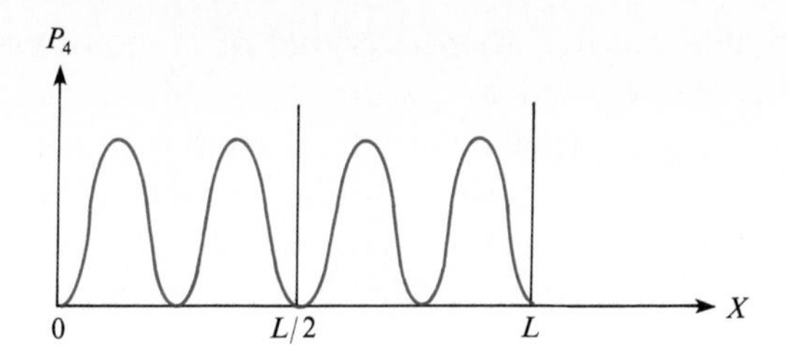


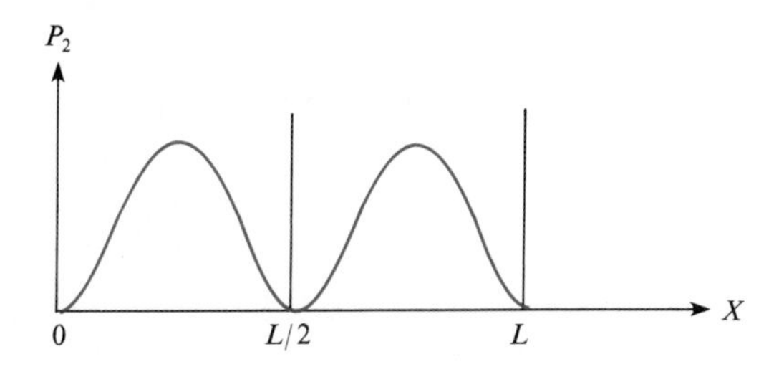


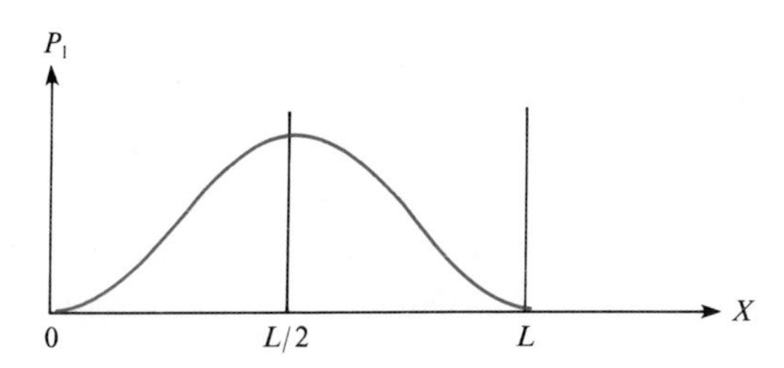


Figure 9.4
Probability densities $P_n(x,t)$ for $n = 1$, 2, and 4 for a particle confined to a box of infinite potential walls.

$x = L/2$. However, for $n = 2$ and $n = 4$, the particle has a *zero probability* of being located at $x = L/2$. Also, note that this same result (Equation 9.88) is obtained for either *set* of eigenfunctions given by Equation 9.79 or Equation 9.85.

## Conduction Electrons in One Dimension

As a particular application of the quantum mechanics for a particle in a box, suppose we have a large number $N$ of free electrons to be accommodated in the one-dimensional box. This situation corresponds to the **valence electrons** of the atoms in a metal, which become *conductors of electricity* and, hence, are called **conduction electrons.** In this *free electron model*, we consider the *conduction electrons* as being *noninteracting* and ignore the electrostatic potential of the *ion cores*. This approximating model is particularly useful in describing and understanding the electrical properties of simple metals, where a crystal of $N$ atoms contains $N$ *conduction electrons* and $N$ positive *ion cores* (e.g., alkali metals).

Certainly, the $N$-electrons can not *all* have the same quantum mechanical energy, for then they would all have identical principal quantum numbers $n$, according to Equation 9.80. Recall that the *Pauli exclusion principle* introduced in Chapter 7, Section 7.7, which applies to electrons in any *quantum state* in atoms, molecules, and solids, does not allow any two electrons to have *identical sets of quantum numbers*. Accordingly, each quantum state can accommodate at most only one electron, although from the Bohr-Sommerfeld model we know that a *degeneracy* of the energy level can occur when more than one *quantum state* has the same *energy*. Since all electrons have an *intrinsic spin*, given by the *spin magnetic quantum number* as $m_s = \pm\frac{1}{2}$, the *quantum state* of a *free electron* is completely specified by a numeration of its principal and spin quantum numbers $n$ and $m_s$. Thus, each *energy level* specified by the principal quantum number $n$ can accommodate *two* electrons, one with $m_s = +\frac{1}{2}$ and the other with $m_s = -\frac{1}{2}$. If $N = 10$ in our system of $N$-electrons, then in the *ground state*, the *energy levels* corresponding to $n = 1, 2, 3, 4$, and 5 are completely filled with electrons, while the levels for $n > 5$ are completely empty.

With this understanding of the *energy levels* and the *quantum states* associated with *each energy level*, we can easily calculate the density of electronic states in our free electron system. The **density of states** $D(E)$ can be simply thought of as *the number of electronic states per unit energy range* and defined by the relation.

Density of States

$$D(E) \equiv \frac{d}{dE}N(E), \tag{9.89}$$

where $N(E)$ is the *total allowed number of quantum states*. With two spin states associated with each energy level, then

$$N(E) = 2n \tag{9.90}$$

for a ground state configuration, where $n$ is (from Equation 9.80)

$$\begin{aligned} n &= \left(\frac{2mL^2E}{\pi^2\hbar^2}\right)^{1/2} = \left(\frac{8mL^2E}{h^2}\right)^{1/2} \\ &= \frac{2L}{h}(2m)^{1/2}E^{1/2}. \end{aligned} \tag{9.91}$$

Thus, substitution into Equation 9.50 gives

$$N(E) = \frac{4L}{h}(2m)^{1/2}E^{1/2} \tag{9.92}$$

for the total allowed number of electronic quantum states in our system of $N$-electrons. Now, substitution into the defining equation for the density of states (Equation 9.89) gives

$$\begin{aligned} D(E) &= \frac{4L}{h}(2m)^{1/2}\,\tfrac{1}{2}E^{-1/2} \\ &= \frac{4L}{h}\left(\frac{m}{2E}\right)^{1/2}. \end{aligned} \tag{9.93}$$

This result is particularly important in the study of the theory of metals for the electrical properties resulting from essentially free *conduction electrons*. Actually, the more useful result is the *density of electronic states* in three dimensions, which will be considered in detail in Section 10.7.

## Review of Fundamental and Derived Equations

The fundamental and derived equations of this chapter are listed below, along with the fundamental postulates of quantum mechanics.

### FUNDAMENTAL EQUATIONS—CLASSICAL PHYSICS

| | |
|---|---|
| $\mathbf{r} = x\mathbf{i} + y\mathbf{j} + z\mathbf{k}$ | Displacement Vector |
| $k \equiv \frac{2\pi}{\lambda}$ | Wave Number |
| $\omega = 2\pi\nu$ | Angular Velocity/Frequency |
| $v_g = \frac{dE}{dp}$ | Group Velocity |
| $\mathbf{p} = p_x\mathbf{i} + p_y\mathbf{j} + p_z\mathbf{k}$ | Components of Linear Momentum |
| $\mathbf{p} \equiv m\mathbf{v}$ | Linear Momentum |
| $\mathrm{F} \equiv \frac{d\mathbf{p}}{dt}$ | Newton's Second Law |
| $F(x) = -\frac{dV}{dx}$ | Conservative Force |
| $V(\mathbf{r}, t) \equiv V(\mathbf{r})$ | Constant Potential Energy |
| $E = \frac{1}{2}mv^2 = \frac{p^2}{2m}$ | Free Particle Energy |

## MATHEMATICAL OPERATORS AND RELATIONS

$$\nabla \equiv \frac{\partial}{\partial x}\mathbf{i} + \frac{\partial}{\partial y}\mathbf{j} + \frac{\partial}{\partial z}\mathbf{k}$$ Del Operator

$$\nabla^2 = \frac{\partial^2}{\partial x^2} + \frac{\partial^2}{\partial y^2} + \frac{\partial^2}{\partial z^2}$$ Laplacian Operator

$$e^{j\theta} = \cos\theta + j\sin\theta$$ Euler's Relation

## FUNDAMENTAL POSTULATES—QUANTUM MECHANICS

$$p = \frac{h}{\lambda} = \hbar k$$ De Broglie's Momentum Postulate

$$E = h\nu = \hbar\omega$$ De Broglie's Energy Postulate

$$E = \frac{p^2}{2m} + V$$ Schrödinger's Energy Postulate

$$\Delta x \Delta p = h$$ Heisenberg's Uncertainty Principle

$$\left.\begin{aligned} -\frac{\hbar^2}{2m}\frac{\partial^2\Psi}{\partial x^2} + V(x, t)\Psi(x, t) &= j\hbar\frac{\partial\Psi}{\partial t} \\ \left[-\frac{\hbar^2}{2m}\nabla^2 + V(\mathbf{r}, t)\right]\Psi(\mathbf{r}, t) &= j\hbar\frac{\partial\Psi}{\partial t}\end{aligned}\right\}$$ Schrödinger's Time-Dependent Equations

$$\left.\begin{aligned} -\frac{\hbar^2}{2m}\frac{d^2\psi}{dx^2} + V(x)\psi(x) &= E\psi(x) \\ \left[-\frac{\hbar^2}{2m}\nabla^2 + V(\mathbf{r})\right]\psi(\mathbf{r}) &= E\psi(\mathbf{r})\end{aligned}\right\}$$ Schrödinger's Time-Independent Equations

## FUNDAMENTAL EQUATIONS—QUANTUM MECHANICS

$$\left.\begin{aligned} \Psi(x, t) &= Ae^{\frac{j}{\hbar}(px - Et)} = Ae^{j(kx - \omega t)} \\ \Psi(\mathbf{r}, t) &= Ae^{\frac{j}{\hbar}(\mathbf{p}\cdot\mathbf{r} - Et)} = Ae^{j(\mathbf{k}\cdot\mathbf{r} - \omega t)}\end{aligned}\right\}$$ Free Particle Wave Function

$$\Psi(x, t) = \int_{-\infty}^{+\infty} \Phi(k, t)e^{j(kx - \omega t)}dk$$ General Free Particle Wave Function

$$\left.\begin{aligned} \int_{-\infty}^{+\infty} \Psi^*(x, t)\Psi(x, t)dx &= 1 \\ \iiint_{-\infty}^{+\infty} \Psi^*(\mathbf{r}, t)\Psi(\mathbf{r}, t)dxdydz &= 1\end{aligned}\right\}$$ Normalization Condition

$$P(x,t) \equiv |\Psi(x,t)|^2 = \Psi^*(x,t)\Psi(x,t)$$ Relative Probability Density

$$\rho(x,t) \equiv \frac{P(x,t)}{\int_{-\infty}^{+\infty} P(x,t)dx}$$ Absolute Probability Density

$$P(t) \equiv \int_{-\infty}^{+\infty} \rho(x,t)dx$$ Total Probability

$$\frac{d}{dt}P(t) = 0$$ Conservation of Probability

$$S(x,t) \equiv -\frac{j\hbar}{2m}\left(\Psi^*\frac{\partial\Psi}{\partial x} - \Psi\frac{\partial\Psi^*}{\partial x}\right)$$ Probability Density Flux

$$D(E) \equiv \frac{d}{dE}N(E)$$ Density of States

## DERIVED EQUATIONS

### *Time-Independent Schrödinger Equation*

$$\Psi(x,t) = \psi(x)\phi(t) = \psi(x)e^{-\frac{j}{\hbar}Et}$$ Separation of Variables

$$\left(-\frac{\hbar^2}{2m}\frac{d^2}{dx^2} + V\right)\psi = E\psi$$ Eigenvalue Equation

### *Conservation of Probability*

$$\frac{d}{dt}P(t) = \frac{j\hbar}{2m}\left(\Psi^*\frac{\partial\Psi}{\partial x} - \Psi\frac{\partial\Psi^*}{\partial x}\right) = 0$$ Conservation of Probability

### *Free Particle and a Constant Potential*

$$\Psi(x,t) = Ae^{\pm\frac{j}{\hbar}(px - Et)}$$ Free Particle Wave Function

$$P(x,t) = A^*A$$ Relative Probability Density

$$S(x,t) = \pm vA^*A$$ Probability Current

$$\frac{d}{dt}P(t) = S_1 - S_2 = 0$$ Conservation of Probability

$$\psi(x) = Ae^{\frac{j}{\hbar}px} + Be^{-\frac{j}{\hbar}px}$$ General Eigenfunction

### Free Particle and an Infinite Potential Well

| | |
|---|---|
| $V(x) = \begin{cases} 0, & 0 < x < L \\ \infty, & L \leq x \leq 0 \end{cases}$ | Potential Energy |
| $\Psi(x, t) = 0, \quad L \leq x \leq 0$ | Boundary Conditions |
| $\dfrac{d^2\psi}{dx^2} + \dfrac{2m}{\hbar^2} E\psi = 0$ | Eigenvalue Equation |
| $\psi(x) = Ae^{jkx} + Be^{-jkx}$ | Assumed Eigensolution |
| $\psi_n(x) = j\sqrt{\dfrac{2}{L}} \sin\dfrac{n\pi x}{L}$ | Normalized Energy Eigenfunctions |
| $\psi(x) = A \sin kx + B \cos kx$ | Assumed Eigensolution |
| $\psi_n(x) = \sqrt{\dfrac{2}{L}} \sin\dfrac{n\pi x}{L}$ | Normalized Energy Eigenfunctions |
| $k_n = \dfrac{n\pi}{L}, \quad n = 1, 2, 3, \cdots$ | Wave Number Quantization |
| $E_n = n^2 \dfrac{\pi^2\hbar^2}{2mL^2} \equiv n^2 E_1$ | Energy Eigenvalues |
| $\rho_n(x, t) = P_n(x, t) = \dfrac{2}{L} \sin^2 k_n x$ | Probability Densities |
| $N(E) = \dfrac{4L}{h} (2m)^{1/2} E^{1/2}$ | Total Quantum States |
| $D(E) = \dfrac{4L}{h} \left(\dfrac{m}{2E}\right)^{1/2}$ | Electronic Density of States |

## Problems

**9.1** By combining de Broglie's and Schrödinger's postulates and using the free particle wave function $\Psi(x, t) = \mathrm{A}e^{j(kx - \omega t)}$, construct Schrödinger's one-dimensional time-dependent wave equation.

*Solution:*

Combining Schrödinger's postulate ($E = p^2/2m + V$) with de Broglie's postulates ($p = \hbar k$, $E = \hbar\omega$) yields

$$\hbar\omega = \frac{\hbar^2 k^2}{2m} + V(x, t).$$

The quantities $\omega$ and $k^2$ in this equation can be replaced by partial derivatives of the free particle wave function. That is, with

$$\frac{\partial \Psi}{\partial t} = -j\omega\Psi \rightarrow \omega = \frac{j}{\Psi}\frac{\partial \Psi}{\partial t},$$

$$\frac{\partial^2 \Psi}{\partial x^2} = -k^2\Psi \rightarrow k^2 = -\frac{1}{\Psi}\frac{\partial^2 \Psi}{\partial x^2}$$

substituted into the above equation, we immediately obtain Schrödinger's one-dimensional time-dependent wave equation:

$$j\hbar\frac{\partial \Psi}{\partial t} = -\frac{\hbar^2}{2m}\frac{\partial^2 \Psi}{\partial x^2} + V(x, t)\Psi(x, t).$$

**9.2** Show that Schrödinger's energy conservation postulate (Equation 9.1) is directly obtained, by using the free particle wave function of Equation 9.6 as a solution to the one-dimensional time-dependent Schrödinger equation.

*Answer:*

$$\frac{p^2}{2m} + V = E$$

**9.3** Show conclusively that the wave functions $\Psi(x, t) = A\cos(kx - \omega t)$ and $\Psi(x, t) = A\sin(kx - \omega t)$ are not solutions to the one-dimensional time-dependent Schrödinger equation.

*Solution:*
Assuming $V(x, t) \equiv 0$, direct substitution of $\Psi(x, t) = A\cos(kx - \omega t)$ into Equation 9.10 yields

$$-\frac{\hbar^2}{2m}[-k^2 A\cos(kx - \omega t)] = j\hbar[\omega A\sin(kx - \omega t)].$$

With $\hbar^2 k^2/2m = p^2/2m = E$ and $\hbar\omega = E$ substituted into this result, we obtain

$$\cos(kx - \omega t) = j\sin(kx - \omega t),$$

which is clearly not valid. In a similar manner, substitution of $\Psi(x, t) = A\sin(kx - \omega t)$ into the Schrödinger equation gives

$$\sin(kx - \omega t) = -j\cos(kx - \omega t),$$

after de Broglie's postulates are employed.

**9.4** Assuming $\Psi(x, t) = \psi(x)\phi(t)$ is a solution to Schrödinger's time-dependent equation, obtain Schrödinger's one-dimensional time-independent equation. Further, show that in general the time-dependent eigenfunction is given by $\phi(t) = e^{-(j/\hbar)Et}$, where the total energy E is the separation constant.

*Answer:* $$-\frac{\hbar^2}{2m}\frac{d^2\psi}{dx^2} + V(x)\psi(x) = E\psi(x)$$

**9.5** Show that if $\Psi(x, t)$ is in general a *complex function*, then the product of $\Psi^*(x, t)$ and $\Psi(x, t)$ is *always* a real function.

*Solution:*
Consider a general complex wave function to be of the form

$$\Psi(x, t) = A + jB,$$

where $A$ and $B$ are real functions. With the complex conjugate of $\Psi(x, t)$ taken as

$$\Psi^*(x, t) = A - jB,$$

then the product

$$\Psi^*(x, t)\Psi(x, t) = A^2 + B^2$$

is *always real.*

**9.6** Show that the probability density flux $S(x, t)$ is always real for the general case where $\Psi(x, t) = A(x, t) + jB(x, t)$ is a complex function.

*Answer:* $$S(x, t) = \frac{\hbar}{m}\left(A\frac{\partial B}{\partial x} - B\frac{\partial A}{\partial x}\right)$$

**9.7** Show that $S(x, t)$ is real for the free particle wave function of Equation 9.6.

*Solution:*
With $\Psi(x, t) = Ae^{\frac{j}{\hbar}(px - Et)}$ substituted into Equation 9.50,

$$S(x, t) = -\frac{j\hbar}{2m}\left(\Psi^*\frac{\partial\Psi}{\partial x} - \Psi\frac{\partial\Psi^*}{\partial x}\right),$$

we obtain

$$S(x, t) = -\frac{j\hbar}{2m}\left[\frac{jp}{\hbar}\Psi^*\Psi - \left(-\frac{jp}{\hbar}\right)\Psi\Psi^*\right]$$

$$= -\frac{j\hbar}{2m}\frac{2jp}{\hbar}A^*A$$

$$= \frac{p}{m}A^2 = vA^2,$$

where the second equality is obvious since $\Psi^*\Psi = \Psi\Psi^* = A^2$. Since $m$, $v$, and $A$ are real, then $S(x, t)$ is real for the free particle wave function.

**9.8** Verify the result given in Equation 9.60 by employing the wave function given in Equation 9.59 with the defining equation for the probability flux density.

*Answer:* $S(x, t) = -vB^*B$

**9.9** Consider the three dimensional plane wave

$$\Psi(\mathbf{r}, 0) = e^{\frac{j}{\hbar}\mathbf{p}\cdot\mathbf{r}}$$

at time defined to be zero, and show that $\mathbf{S}(\mathbf{r}, 0) = \mathbf{v}$.

*Solution:*

With $\Psi(\mathbf{r}, 0) = e^{\frac{j}{\hbar}\mathbf{p}\cdot\mathbf{r}}$, $\mathbf{S}(\mathbf{r}, 0)$ becomes

$$\begin{aligned}\mathbf{S}(\mathbf{r}, 0) &= -\frac{j\hbar}{2m}(\Psi^*\nabla\Psi - \Psi\nabla\Psi^*) \\ &= -\frac{j\hbar}{2m}\left[\Psi^*\left(\frac{j\mathbf{p}}{\hbar}\right)\Psi - \Psi\left(-\frac{j\mathbf{p}}{\hbar}\right)\Psi^*\right] \\ &= -\frac{j\hbar}{2m}\frac{2j\mathbf{p}}{\hbar} \\ &= \frac{\mathbf{p}}{m} = \mathbf{v}.\end{aligned}$$

**9.10** By considering the three dimensional Schrödinger equation and probability current $S(\mathbf{r}, t)$, show that the *divergence* of $\mathbf{S}(\mathbf{r}, t)$ results in a relation that is similar in form to the classical *equation of continuity*.

*Answer:* $\nabla\cdot\mathbf{S} = -\frac{\partial}{\partial t}\Psi^*\Psi.$

**9.11** Verify that the eigenfunction given in Equation 9.69 is a solution to the time-independent Schrödinger equation.

*Solution:*
With the eigenfunction $\psi(x) = Ae^{jkx} + Be^{-jkx}$ substituted into Equation 9.29,

$$-\frac{\hbar^2}{2m}\frac{d^2\psi}{dx^2} + V(x)\psi(x) = E\psi(x),$$

we obtain

$$
\begin{aligned}
E\psi &= -\frac{\hbar^2}{2m}\,[(jk)^2 A e^{jkx} + (-jk)^2 B e^{-jkx}] + V\psi \\
&= -\frac{\hbar^2}{2m}\,(-k^2)\,\psi + V\psi \\
&= \frac{\hbar^2 k^2}{2m}\,\psi + V\psi.
\end{aligned}
$$

This result is identical to Schrödinger's conservation of energy postulate (Equation 9.1), after cancelling out the common eigenfunction $\psi(x)$ and using de Broglie's momentum postulate $p = \hbar k$.

**9.12** A free particle in the ground state is confined to a one dimensional box of size $L$. What is the *probability* of finding the particle in an interval $\Delta x = 10^{-3}L$ at $x = (1/4)L$, $x = (1/2)L$, $x = (3/4)L$, and $x = L$?

*Answer:*

$$1 \times 10^{-3},\ 2 \times 10^{-3},\ 1 \times 10^{-3},\ 0$$

**9.13** Repeat Problem 9.12 for a free particle in a box in the first excited state.

*Solution:*

In the first excited state the *normalized eigenfunction* for a free particle in a box (see Equation 9.85) is

$$\psi_2(x) = \sqrt{\frac{2}{L}}\,\sin\frac{2\pi x}{L},$$

so the probability is given by

$$
\begin{aligned}
P(x)\Delta x &= \psi^*(x)\psi(x)\Delta x \\
&= \frac{2\Delta x}{L}\,\sin^2\frac{2\pi x}{L} \\
&= (2 \times 10^{-3})\,\sin^2\frac{2\pi x}{L},
\end{aligned}
$$

where the third equality was obtained by substituting $\Delta x = 10^{-3}L$. Thus, for the different values of $x$ from Problem 9.13 we obtain the following:

$$
\begin{aligned}
x &= \frac{1}{4}L \rightarrow P\Delta x = (2 \times 10^{-3})\,\sin^2\frac{\pi}{2} = 2 \times 10^{-3}, \\
x &= \frac{1}{2}L \rightarrow P\Delta x = (2 \times 10^{-3})\,\sin^2\pi = 0,
\end{aligned}
$$

$$x = \frac{3}{4}L \rightarrow P\Delta x = (2 \times 10^{-3}) \sin^2 \frac{3\pi}{2} = 2 \times 10^{-3},$$

$$x = L \quad \rightarrow P\Delta x = (2 \times 10^{-3}) \sin^2 2\pi = 0.$$

**9.14** Verify that the eigenfunctions $\psi_n(x) = \sqrt{2/L} \sin k_n x$ and $\psi_n(x) = \sqrt{2/L} \cos k_n x$, where $k_n = n\pi/L$, are both solutions to the eigenvalue equation for the free particle in a box (Equation 9.68), by showing that $E_n = n^2h^2/8mL^2$ are the energy eigenvalues obtained for both.

**9.15** Consider a proton to be a free particle in a box of the size of a nucleus. Find the energy released when a proton makes a transition from the quantum state $E_2$ to the ground state $E_1$.

*Solution:*

With the nucleus of an atom being thought of as a one-dimensional box of length $L = 1 \times 10^{-14}$ m (see the discussion preceding Equation 5.80) and $m = m_p = 1.67 \times 10^{-27}$ kg, the energy released is given by

$$E_2 - E_1 = 2^2\text{E}_1 - \text{E}_1 = 3\text{E}_1,$$

where the ground state energy

$$E_1 = 1^2 \frac{\pi^2\hbar^2}{2m_p L^2}$$

is obtained from Equation 9.80. Combining the above two equations and substituting the physical data gives

$$\begin{aligned} E_2 - E_1 &= \frac{3\pi^2(1.05 \times 10^{-34} \text{ J} \cdot \text{s})^2}{2(1.67 \times 10^{-27} \text{ kg})(10^{-14} \text{ m})^2} \\ &= 9.76 \times 10^{-13} \text{ J} \\ &= 6.10 \text{ MeV}. \end{aligned}$$

This result is in approximate agreement with the observed energy differences between stationary states of protons (and neutrons) in a nucleus.

**9.16** Consider a particle of mass $10^{-3}$ g confined to a box of length 1 cm. Treating this problem quantum mechanically, find the ground state energy and speed of the particle.

*Answer:*

$$E_1 = 5.44 \times 10^{-60} \text{ J}, \ v_1 = 3.30 \times 10^{-27} \text{ m/s}$$

**9.17** If the particle of Problem 9.16 were moving with a distinguishable speed of $10^{-1}$ cm/s, what is the approximate value of the principal quantum number $n$?

*Solution:*

Equating the quantized energy of Equation 9.80 with kinetic energy $\frac{1}{2}mv^2$,

$$n^2\frac{\pi^2\hbar^2}{2mL^2} = \tfrac{1}{2}mv^2,$$

and solving for $n$ gives

$$n = \frac{mvL}{\pi\hbar},$$

which with the physical data yields

$$n = \frac{(10^{-6}\text{ kg})(10^{-3}\text{ m/s})(10^{-2}\text{ m})}{\pi(1.05 \times 10^{-34}\text{ J}\cdot\text{s})}$$
$$\approx 3 \times 10^{22}.$$

**9.18** Evaluate the probability density flux for the wave function given by Equation 9.86.

*Answer:* $S(x, t) = 0$

**9.19** Consider a particle confined to a two dimensional box of edge $L$, where the potential energy $V(x, y)$ is restricted to the values $V = 0$ inside and $V = \infty$ outside of the box. If the eigenfunction solution $\psi(x, y)$ to Schrödinger's two dimensional time-independent equation has the boundary conditions $\psi(x, y) = 0$ at $x = 0$ and $x = L$ (for all $y$) and $\psi(x, y) = 0$ at $y = 0$ and $y = L$ (for all $x$), find the *normalized eigenfunction.*

*Solution:*

From Equations 9.29 and 9.30 it is apparent that Schrödinger's two-dimensional time-independent equation is of the form (with $V(x, y) = 0$)

$$-\frac{\hbar^2}{2m}\left[\frac{\partial^2}{\partial x^2}\psi(x, y) + \frac{\partial^2}{\partial y^2}\psi(x, y)\right] = E\psi(x, y)$$

and assuming separation of variables for $\psi(x, y)$,

$$\psi(x, y) = \psi_x(x)\psi_y(y),$$

we immediately obtain

$$-\frac{\hbar^2}{2m}\left(\frac{1}{\psi_x}\frac{d^2\psi_x}{dx^2} + \frac{1}{\psi_y}\frac{d^2\psi_y}{dy^2}\right) = E.$$

Since the energy for a free particle in a two-dimensional box can be expressed as

$$E = \frac{p^2}{2m} = \frac{p_x^2 + p_y^2}{2m}$$
$$\equiv E_x + E_y,$$

then substitution into the above wave equation yields

$$\frac{d^2\psi_x}{dx^2} + \frac{2mE_x}{\hbar^2}\psi_x = 0,$$

$$\frac{d^2\psi_y}{dy^2} + \frac{2mE_y}{\hbar^2}\psi_y = 0.$$

These two ordinary differential equations are of identical form to that given by Equation 9.68 for the one-dimensional free particle in a box. Thus, the normalized eigenfunction solutions can be chosen as (see Equation 9.85)

$$\psi_{n_x} = \sqrt{\frac{2}{L}}\sin\frac{n_x\pi x}{L},$$

$$\psi_{n_y} = \sqrt{\frac{2}{L}}\sin\frac{n_y\pi y}{L},$$

and the normalized eigenfunction associated with a particle in a square box becomes

$$\psi_{n_x n_y} = \frac{2}{L}\sin\frac{n_x\pi x}{L}\sin\frac{n_y\pi y}{L},$$

where the symbolic notation

$$\psi_{n_x n_y} \equiv \psi(x, y) = \psi_x(x)\psi_y(y)$$

has been introduced.

**9.20** Find the energy eigenvalues for the free particle in a square box of Problem 9.19, and show that the wave number quantization conditions are given by $k_x = n_x\pi/L$ and $k_y = n_y\pi/L$.

*Answer:* $E_{n_x n_y} = (n_x^2 + n_y^2)\dfrac{\pi^2\hbar^2}{2mL^2}$

**9.21** Consider a proton to be a free particle in a square box of edge $L = 1 \times 10^{-14}$ m. Find the energy released when a proton makes a transition from the first excited state to the ground state.

*Solution:*
From the energy eigenvalue equation obtained in Problem 9.20, the ground state energy for a particle in a square box corresponds to $n_x = 1$ and $n_y =$

1, while the first excited state is specified by the set of quantum numbers $(n_x, n_y) = (1, 2)$ or $(n_x, n_y) = (2, 1)$. Consequently, using the energy eigenvalue equation

$$E_{n_x n_y} = (n_x^2 + n_y^2)\frac{\pi^2\hbar^2}{2mL^2},$$

we immediately obtain

$$E_{11} = 2\frac{\pi^2\hbar^2}{2mL^2},$$

$$E_{12} = 5\frac{\pi^2\hbar^2}{2mL^2}.$$

Thus, the energy released is given by

$$E_{12} - E_{11} = 3\frac{\pi^2\hbar^2}{2mL^2},$$

which is identical to the equation obtained in Problem 9.15 for the proton confined to a one dimensional box of length $L = 1 \times 10^{-14}$ m.

**9.22** Verify that $\Psi(x, 0) = A\int_{k_0-a}^{k_0+a} e^{jkx}dk = \dfrac{2A}{x}\sin ax\, e^{jk_0x}$. For $a \ll k_0$, what does the graph of this wave function resemble?

*Answer:* $\Psi(0, 0) = 2aA,\ \Psi\left(\pm\dfrac{\pi}{a}, 0\right) = 0$

**9.23** Consider the wave function $\Psi(x, 0) = \int_{-\infty}^{+\infty} A(k)e^{jkx}dk$ defined at $t = 0$. If the amplitude is given by $A(k) = e^{-k^2/2s^2}$, show that $\Psi(x, 0) = s\sqrt{2\pi}\, e^{-s^2x^2/2}$, by evaluating the, so called, **Gaussian integral.**

*Solution:*

The integral can be expressed as

$$\Psi(x, 0) = \int_{-\infty}^{+\infty} e^{-k^2/2s^2}e^{jkx}dk$$

$$= \int_{-\infty}^{+\infty} e^{-ak^2}e^{ck}dk,$$

where $a \equiv 1/2s^2$ and $c \equiv jx$. Since this integral is of the general form

$$\int_{-\infty}^{+\infty} e^{-au^2}e^{cu}du = e^{c^2/4a}\sqrt{\frac{\pi}{a}}$$

given in Section A.10 (see also the discussion presented in Section 10.4), we have

$$\Psi(x, 0) = e^{(jx)^2/4(1/2s^2)} \sqrt{\frac{\pi}{1/2s^2}}$$

$$= s\sqrt{2\pi}\, e^{-s^2x^2/2}.$$

**9.24** Normalize the wave function obtained in Problem 9.23.

*Answer:* $\Psi(x, 0) = (s^2/\pi)^{1/4} e^{-s^2x^2/2}$.

CHAPTER 10

# Schrödinger's Quantum Mechanics II

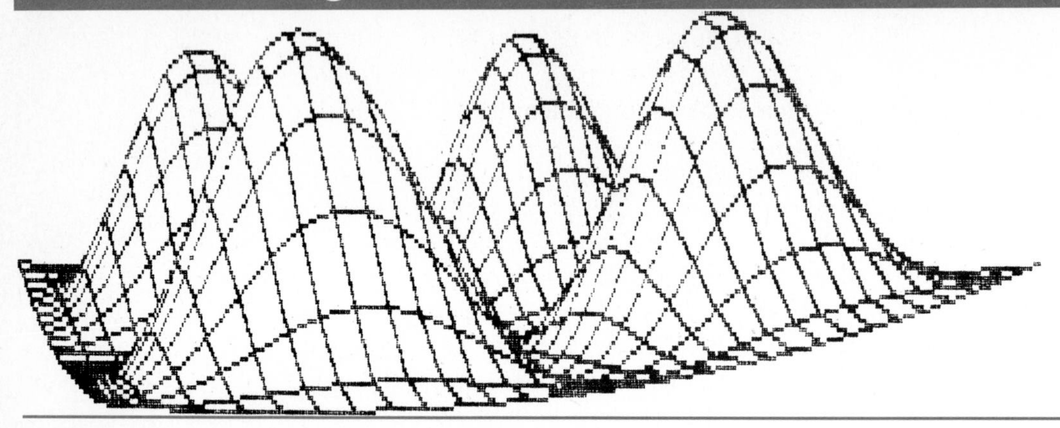

## Introduction

Many of the fundamentals of Schrödinger's theory of quantum mechanics were introduced in the last chapter, including the time-dependent Schrödinger equation and the interpretation of its wave function solution as a probability amplitude, the steady-state Schrödinger equation and its eigenfunction solution, the definition of probability requirements for conservation of probability and wave function (or eigenfunction) normalization, and the concept of a probability density flux. These fundamentals of quantum mechanics were applied to two examples involving *free* particles, with the *free particle* in a *box* example being most noteworthy as a one-dimensional description of conduction band electrons in a solid. In all of these discussions the wave function and eigenfunction solutions to Schrödinger's equations were represented as being intrinsically dependent on the position variable $x$. There exist, however, physical properties of a quantum mechanical particle or system that are more directly dependent on the momentum variable $p$ instead of the position variable. In this concluding chapter of Schrödinger's quantum mechanics, we first investigate the representation of wave functions and eigenfunctions in both position-space and momentum-space, and then consider additional fundamentals like *expectation values* and quantum mechanical *differential operators*.

We begin this chapter by considering the problem of finding the *Fourier transform* of a position-space wave function, $\Psi(x, t)$. To facilitate this problem and find the wave function $\Phi(p, t)$ in momentum-space, we introduce the *Dirac delta function* and capitalize on its mathematical properties. The problem is then particularized and the *Fourier transform* obtained for the generalized free particle wave function. We then consider how the *average value* of a physical observable is mathematically described in quantum mechanics. This average value of an observable for a quantum mechanical particle, corresponding to the value we would *expect* to obtain by averaging actual experimental measurements, is called the *expectation value*, which is carefully introduced and defined in terms of the particle's associated wave function. The discussion of a quantum mechanical *expectation value* is followed by a consideration of how the momentum variable $p$ behaves like a *differential operator* in position-space, and likewise for the position variable in momentum-space. Our discussion of quantum mechanical operators continues with an interpretation of energy as an operator in both position and momentum space and the definition of the *Hamiltonian* operator. Equipped with the fundamental relations for expectation values and quantum mechanical operators, we next consider the formal mathematical *correspondence* between quantum mechanics and classical mechanics. In this consideration, *Bohr's correspondence principle* is illustrated by way of two examples, using rather straight forward, although somewhat involved, mathematical arguments. In addition, we introduce a more elegant method for demonstrating the correspondence principle, which employs *operational algebra* and the definition of the *commutator*. We conclude our introductory treatment of quantum mechanics by considering the problem of a *free particle* in a *three-dimensional box*. Further, the generalization of this problem to describe conduction electrons in a solid is considered, by deriving the very important electronic *density of states* formula.

## 10.1 Wave Functions in Position and Momentum Representations

In the last chapter the wave function of Schrödinger's equation was discussed in terms of its functional dependence on the spatial and temporal coordinates. Because of the *spatial dependence*, the wave function, often referred to as a **state function,** is said to be expressed in the *position representation*. For example, the ultimate generalization of the one-dimensional free particle *state function* is given in the *position representation* as

$$\Psi(x, t) = \sqrt{\frac{1}{2\pi\hbar}} \int_{-\infty}^{+\infty} \Phi(p, t) e^{\frac{j}{\hbar}(px - Et)} dp, \qquad (10.1)$$

Position-Space Free Particle Wave Function

which is, essentially, the same as that given by Equation 9.37. This state function in the position representation has been directly obtained from Equation 9.37 by employing de Broglie's energy and momentum postulates (Equations 8.52 and 8.53) and introducing a *convenient* multiplicative constant of $(1/2\pi\hbar)^{1/2}$. Remember, the *physics* has not been affected by introducing this constant, but its usefulness will shortly become apparent in the development of this section. Since $\Psi(x, t)$ is a state function in the *position representation*, $\Phi(p, t)$ could be interpreted as the free particle state function in the *momentum representation*. Our immediate objective is to mathematically operate on Equation 10.1 and solve it for $\Phi(p, t)$, which in mathematical language is equivalent to finding the **Fourier transform** $\Phi(p, t)$ of the *position state function* $\Psi(x, t)$. This task is reminiscent of the problem encountered with the vibrating string in Chapter 8, Section 8.2, where the relations for the coefficients $A_n$ and $B_n$ (Equations 8.35 and 8.36) were determined (recall Problems 8.5 and 8.6) from the two equations representing the *initial conditions*. Although we could proceed with the problem of determining $\Phi(p, t)$ in a similar manner using ordinary calculus, it is both convenient and advantageous to introduce a new mathematical function that provides an easy *operational* approach for finding *Fourier transforms* in quantum mechanics.

## Dirac Delta Function

The mathematical properties of the *Fourier transform* of a *function* suggested the definition of a new *singular function* to P. A. M. Dirac, which has become known as the *Dirac delta function*. In terms of the one-dimensional wave number $k_x$, which we will simply call $k$ for convenience, the **Dirac delta function** is defined as

Dirac Delta Function

$$\delta(k - k') \equiv \frac{1}{2\pi} \int_{-\infty}^{+\infty} e^{j(k-k')x} dx, \tag{10.2}$$

where $k'$ represents a wave number whose value is only slightly different from that of $k$. This function could also be defined in terms of the one-dimensional position coordinate as

Dirac Delta Function

$$\delta(x - x') \equiv \frac{1}{2\pi} \int_{-\infty}^{+\infty} e^{j(x-x')k} dk, \tag{10.3}$$

which in terms of momentum (de Broglie's postulate) becomes

$$\delta(x - x') = \frac{1}{2\pi\hbar} \int_{-\infty}^{+\infty} e^{\frac{j}{\hbar}(x-x')p} dp. \quad (10.4)$$

In a similar manner Equation 10.2 can be expressed in terms of momenta $p$ and $p'$ by

$$\delta(p - p') = \frac{1}{2\pi\hbar} \int_{-\infty}^{+\infty} e^{\frac{j}{\hbar}(p-p')x} dx, \quad (10.5)$$

where a property of the Dirac delta function given by Equation 10.11 has been employed in addition to de Broglie's momentum postulate. Considering $k$ to be a *dummy variable* for the moment, the definition of the function $\delta(k - k')$ by Equation 10.2 means that the function has the properties

$$\delta(k - k') = \begin{cases} 0, & \text{if } k \neq k', \\ \infty, & \text{if } k = k'. \end{cases} \quad (10.6)$$

The Dirac delta function is meaningful only when it appears under integral signs, where it has the property

$$\int_{-\infty}^{+\infty} \delta(k - k')dk = 1. \quad (10.7)$$

Two other properties which are of particular usefulness are

$$\delta(k - k') = \delta(k' - k), \quad (10.8)$$

and

$$f(k)\delta(k - k') = f(k')\delta(k - k'), \quad (10.9)$$

where $f(k)$ is just a *general function* of the *dummy variable* $k$. This last equality allows us to obtain the most commonly used property of the Dirac delta function, namely

$$\begin{aligned} \int_{-\infty}^{+\infty} f(k)\delta(k - k')dk &= \int_{-\infty}^{+\infty} f(k')\delta(k - k')dk \\ &= f(k') \int_{-\infty}^{+\infty} \delta(k - k')dk \\ &= f(k'), \end{aligned} \quad (10.10)$$

where the property given by Equation 10.7 has been used in obtaining the last equality. It should be emphasized that $f(k)$ can be *any* function of the variable $k$, real or complex, and that as a result of the property given by Equation 10.9, $f(k')$ could be taken out of the integral in the above derivation. Other useful properties of the Dirac delta function include

$$\delta[a(k - k')] = \frac{1}{a}\,\delta(k - k'), \tag{10.11}$$

where $a$ is a constant, and

$$\int_{-\infty}^{+\infty} \delta(k - k_1)\delta(k_1 - k_2)\, dk_1 = \delta(k - k_2). \tag{10.12}$$

It should be observed that the property given by Equation 10.11 has already been used in obtaining Equation 10.5 from Equation 10.2, that is,

$$\delta(k - k') = \delta\left[\frac{1}{\hbar}(p - p')\right] = \hbar\delta(p - p').$$

Although other properties of the Dirac delta function involving derivatives could be listed, those listed above are the most useful and are completely adequate for our purposes. We need to remember, however, that in general problems of quantum mechanics are expressed in a three-dimensional formulation. The definition of the Dirac delta function and its properties can be easily generalized to three dimensions, for example,

$$\delta(\mathbf{k} - \mathbf{k}') = \delta(k_x - k_x')\delta(k_y - k_y')\delta(k_z - k_z') \tag{10.13}$$

$$\delta(\mathbf{k} - \mathbf{k}') = \left(\frac{1}{2\pi}\right)^3 \iiint_{-\infty}^{+\infty} e^{j[(k_x - k_x')x + (k_y - k_y')y + (k_z - k_z')z]}\, dxdydz. \tag{10.14}$$

Under the initial conditions of

$$k_x' = k_y' = k_z' = 0,$$

these equations reduce to

$$\begin{aligned}\delta(\mathbf{k}) &= \delta(k_x)\delta(k_y)\delta(k_z)\\ &= \left(\frac{1}{2\pi}\right)^3 \iiint_{-\infty}^{+\infty} e^{j\mathbf{k}\cdot\mathbf{r}}\, dxdydz.\end{aligned} \tag{10.15}$$

As a last point of interest, it needs to be emphasized that the definitions and properties of the Dirac delta function can be generalized and particularized for any useful and important *physical variable*. We will continue to emphasize one-dimensional considerations and use $p_x \equiv p$ and $k_x \equiv k$; however, the generalizations to three dimensions should be rather apparent.

## Free Particle Position and Momentum Wave Functions

As an example of the operational utilization of the Dirac delta function, let us return to the problem of obtaining an expression in the *momentum representation* for the *state function* of a one-dimensional free particle. To simplify matters consider an instant in time, say at $t \equiv 0$, such that Equation 10.1 can be expressed in terms of the *eigenfunctions* $\psi(x) = \Psi(x, 0)$ and $\phi(p) = \Phi(p, 0)$ as

$$\psi(x) = \sqrt{\frac{1}{2\pi\hbar}} \int_{-\infty}^{+\infty} \phi(p) e^{\frac{j}{\hbar}px}\, dp. \tag{10.16}$$

Position-Space Free Particle Eigenfunction

Now, multiplying both sides of this equation by $\sqrt{1/2\pi\hbar}\, e^{-(j/\hbar)p'x}$ and integrating over the position variable gives

$$\sqrt{\frac{1}{2\pi\hbar}} \int_{-\infty}^{+\infty} \psi(x) e^{-\frac{j}{\hbar}p'x}\, dx = \frac{1}{2\pi\hbar} \iint_{-\infty}^{+\infty} \phi(p) e^{\frac{j}{\hbar}(p-p')x}\, dpdx$$
$$= \int_{-\infty}^{+\infty} \phi(p) \left( \frac{1}{2\pi\hbar} \int_{-\infty}^{+\infty} e^{\frac{j}{\hbar}(p-p')x}\, dx \right) dp,$$

where the second equality has been obtained by *interchanging* the order of integration. Invoking the properties of the Dirac delta function, we obtain

$$\sqrt{\frac{1}{2\pi\hbar}} \int_{-\infty}^{+\infty} \psi(x) e^{-\frac{j}{\hbar}p'x}\, dx = \int_{-\infty}^{+\infty} \phi(p)\delta(p - p')\, dp$$
$$= \int_{-\infty}^{+\infty} \phi(p')\delta(p - p')\, dp$$
$$= \phi(p') \int_{-\infty}^{+\infty} \delta(p - p')\, dp$$
$$= \phi(p').$$

Dropping the *prime* in our result, as it no longer serves any useful purpose, we have

Momentum-Space
Free Particle
Eigenfunction

$$\phi(p) = \sqrt{\frac{1}{2\pi\hbar}} \int_{-\infty}^{+\infty} \psi(x) e^{-\frac{j}{\hbar}px} dx \tag{10.17}$$

for the *Fourier transform* of Equation 10.16. This result represents the eigenfunction in the *momentum representation* for the one-dimensional free particle. Since for the *free particle* the wave functions in the *position* and *momentum* representations are given by (see Equation 9.28)

Free Particle

$$\Psi(x, t) = \psi(x) e^{-\frac{j}{\hbar}Et} \tag{10.18}$$

and (note the exponential in Equation 10.1)

Free Particle

$$\Phi(p, t) = \phi(p) e^{\frac{j}{\hbar}Et}, \tag{10.19}$$

then the *Fourier transform* to Equation 10.1 is just

Momentum-Space
Free Particle
Wave Function

$$\Phi(p, t) = \sqrt{\frac{1}{2\pi\hbar}} \int_{-\infty}^{+\infty} \Psi(x, t) e^{-\frac{j}{\hbar}(px - Et)} dx. \tag{10.20}$$

This equation gives the *state function* for a one-dimensional free particle in the *momentum representation*. Usually little or no emphasis is placed on the momentum representation in quantum mechanics, since, as we shall see in the following sections, solutions to quantum mechanical problems are *invariant* to the representation formulation.

## 10.2 Expectation Values

Consider a particle being in a quantum mechanical state described by its associated wave function, which contains all of the physical information of the particle allowed by the Heisenberg uncertainty principle. In an attempt to find a suitable method for *averaging the physical properties* of this system, we could make a number of measurements on a large number $N$ of identical particles in similarly prepared systems. Alternatively, we could make a large number of repeated measurements on the same particle over a considerable period of time. After J. Willard Gibbs (see Chapter 11, Section 11.1), we choose to average over a large number $N$ of copies of the system, with each particle being in the quantum state associated with the wave function, to obtain what is called an *ensemble average*. Now, allow any *physical observable*, like position, momentum, energy, and so on, to be denoted by the letter $Q$. If a measurement is made on one of the systems in the *ensemble* to determine whether the particle has a particular value for the *physical observable*, the result will be definite—either it has that particular value for $Q$ or it has some other value. If this measurement is repeated on all of the systems in the ensemble, the *relative* number of times that the particle is found to have the same value for $Q$ is taken as a measure of the *probability* that the particle will have that *particular value* for the *physical observable* $Q$. As a more specific example, consider an ensemble of systems containing a classical particle and the measurements for the particle's position $x$ at an instant in time. If our measurements revealed the particle to have the position $x_1$ in $n_1$ systems, the position $x_2$ in $n_2$ systems, and so on, then a common sense definition for the *average* or *expected* value of the particle's position in our ensemble is given by

$$\langle x \rangle = \frac{n_1 x_1 + n_2 x_2 + n_3 x_3 + \cdots}{N}, \tag{10.21}$$

where the total number of systems $N$ is simply

$$N = n_1 + n_2 + n_3 + \cdots. \tag{10.22}$$

In this case, $n_1/N$ is interpreted as the *probability* of the particle having the particular value $x_1$. In quantum theory this *probability* of finding the particle at the particular point $x_1$ along the $X$-axis is given by (see Equations 9.41 and 9.42)

$$\rho_1(x, t)dx = \frac{|\Psi_1|^2\, dx}{\int_{-\infty}^{+\infty} |\Psi|^2\, dx}, \tag{10.23}$$

where $\Psi_1$ is the particle's associated wave function evaluated at $x = x_1$. As our quantum mechanical probability is analogous to the first term in Equation 10.21 for the classical particle, it is apparent that in quantum mechanics the summation ($N = n_1 + n_2 + n_3 + \cdots$) is replaced by an integral and the, so called, *occupation numbers* $n_i$ are replaced by the *relative probabilities*

$$\mathrm{P}_i(x, t)dx = |\Psi_i|^2 dx. \tag{10.24}$$

Thus, by these analogies and the form of Equation 10.21, we can define the quantum mechanical *expectation value* for a particle's position as

$$\langle x \rangle \equiv \frac{\int_{-\infty}^{+\infty} x|\Psi|^2 \, dx}{\int_{-\infty}^{+\infty} |\Psi|^2 \, dx}, \tag{10.25}$$

with an integral in the numerator instead of the summation in Equation 10.21. The form of Equation 10.25 and the definition of the absolute probability density,

Probability Density

$$\rho(x, t) \equiv \frac{P(x, t)}{\int_{-\infty}^{+\infty} P(x, t) \, dx} = \frac{|\Psi|^2}{\int_{-\infty}^{+\infty} |\Psi|^2 \, dx}, \tag{9.39}$$

suggest that the position expectation value can be expressed as

$$\langle x \rangle = \int_{-\infty}^{+\infty} x\rho(x, t) \, dx, \tag{10.26}$$

which reduces to

$$\langle x \rangle = \int_{-\infty}^{+\infty} xP(x, t) \, dx \tag{10.27}$$

for properly *normalized* wave functions.

The discussion above can be generalized and used to define the expectation value of *any physical variable* that is a function of the $x$-coordinate of a particle described by its associated wave function. As such, the quantum mechanical **expectation value** of a variable $Q(x)$ is defined by

Expectation Value

$$\langle Q(x) \rangle \equiv \int_{-\infty}^{+\infty} Q(x)\rho(x, t) \, dx, \tag{10.28}$$

which reduces to

$$\langle Q(x)\rangle = \int_{-\infty}^{+\infty} Q(x)P(x, \mathrm{t})\, dx$$

$$\equiv \int_{-\infty}^{+\infty} \Psi^*(x, t)Q(x)\Psi(x, t)\, dx \qquad (10.29)$$

Expectation Value for Normalized Wave Functions

for properly *normalized* wave functions. These formulas are valid even for variables that are dependent on the time coordinate, since the *expectation value* of $Q$, $\langle Q\rangle$, must be evaluated at an instant, usually $t \equiv 0$, in time. The reasons for the form of the integrand in Equation 10.29 as $\Psi^*Q\Psi$ instead of $Q\Psi^*\Psi$ or $\Psi^*\Psi Q$ will be fully discussed in the next section. It should be emphasized that $\langle Q\rangle$ is the value of the physical observable that we *expect* to obtain, if we average the experimental values of $Q$ that are measured at an instant in time for a large number of particles described by the same wave function.

The form of Equations 10.28 and 10.29 are clearly appropriate for a *position-dependent variable* of a particle described by a state function in the *position representation*. But, suppose the physical observable in question is directly dependent on momentum instead of position, that is $Q = Q(p)$. In this case it would seem more appropriate to specify the *state* of the particle in the *momentum representation* and define the *expectation value* of $Q$ by

$$\langle Q(p)\rangle \equiv \int_{-\infty}^{+\infty} Q(p)\rho(p, t)\, dp, \qquad (10.30)$$

Expectation Value

where the *absolute probability density* $\rho(p, t)$ in the *momentum representation* is given by

$$\rho(p, t) = \frac{|\Phi(p, t)|^2}{\int_{-\infty}^{+\infty} |\Phi(p, t)|^2\, dp}. \qquad (10.31)$$

Probability Density

For properly *normalized* state functions $\Phi(p, t)$, our definition for $\langle Q(p)\rangle$ reduces to

$$\langle Q(p)\rangle = \int_{-\infty}^{+\infty} Q(p)P(p, t)\, dp$$

$$\equiv \int_{-\infty}^{+\infty} \Phi^*(p, t)Q(p)\Phi(p, t)\, dp. \qquad (10.32)$$

Expectation Value for Normalized Wave Functions

At this point an important question must be raised. If the *physical observable* in Equations 10.29 and 10.32 is the same, will the value obtained by using Equation 10.29 be identical to the value obtained for $\langle Q\rangle$ by using Equation 10.32? Clearly, we must have an affirmative answer to this question, if the formalism introduced is to be consistent. We *must* be able to calculate the expectation value of any physical observable with *either* the position or momentum representations of the state function and obtain the exact same answers. That is, at any particular value of time $t$ (say at the value $t \equiv 0$), we must have

Expectation Value in Position or Momentum Space

$$\langle Q\rangle = \int_{-\infty}^{+\infty} \Psi^*(x, 0)Q\Psi(x, 0)dx = \int_{-\infty}^{+\infty} \Phi^*(p, 0)Q\Phi(p, 0)dp \quad (10.33)$$

for *normalized* wave functions.

As a verification of Equation 10.33 and an example of the operational properties of the Dirac delta function, let us employ the one-dimensional wave function of a free particle given by Equation 10.1. At an instant in time, say $t \equiv 0$, the state function in the position representation given by Equation 10.1 reduces to the eigenfunction given in Equation 10.16. Thus, direct substitution into Equation 10.33 gives

$$\begin{aligned}\langle Q\rangle &= \int_{-\infty}^{+\infty} \Psi^*(x, 0)Q\Psi(x, 0)\,dx \\ &= \frac{1}{2\pi\hbar}\int_{-\infty}^{+\infty}\left[\int_{-\infty}^{+\infty} \phi^*(p')e^{-\frac{j}{\hbar}p'x}dp'\right] Q\left[\int_{-\infty}^{+\infty} \phi(p)e^{\frac{j}{\hbar}px}dp\right] dx \\ &= \int_{-\infty}^{+\infty} \phi^*(p')\left(\frac{1}{2\pi\hbar}\int_{-\infty}^{+\infty} e^{\frac{j}{\hbar}(p-p')x}dx\right) Q\phi(p)\,dp'dp, \quad (10.34)\end{aligned}$$

where we have rearranged factors and interchanged the order of integration in obtaining the last equality. Now, employing the definition of the Dirac delta function (in the form of Equation 10.5) and its properties in Equation 10.34 yields

$$\begin{aligned}\langle Q\rangle &= \iint_{-\infty}^{+\infty} \phi^*(p')\delta(p - p')Q\phi(p)\,dp'dp \\ &= \int_{-\infty}^{+\infty}\left[\int_{-\infty}^{+\infty} \phi^*(p')\delta(p - p')dp'\right] Q\phi(p)\,dp \\ &= \int_{-\infty}^{+\infty} \phi^*(p)\left[\int_{-\infty}^{+\infty} \delta\,(p - p')dp'\right] Q\phi(p)\,dp\end{aligned}$$

$$= \int_{-\infty}^{+\infty} \phi^*(p) Q \phi(p)\, dp$$

$$= \int_{-\infty}^{+\infty} \Phi^*(p, 0) Q \Phi(p, 0)\, dp, \tag{10.35}$$

where the last equality should be obvious for $t = 0$ from Equation 10.19. A comparison of Equations 10.34 and 10.35 verifies the validity of Equation 10.33 for the free particle wave functions. Particular attention should be given to the properties of the Dirac delta function as used in this verification, as they will be further employed in the next section. Also, we should note that since the free particle wave functions of Equations 10.1 and 10.2 are not normalized, we should have verified the equality of Equations 10.28 and 10.30, i.e.,

$$\langle Q \rangle = \frac{\displaystyle\int_{-\infty}^{+\infty} \Psi^* Q \Psi\, dx}{\displaystyle\int_{-\infty}^{+\infty} \Psi^* \Psi\, dx} = \frac{\displaystyle\int_{-\infty}^{+\infty} \Phi^* Q \Phi\, dp}{\displaystyle\int_{-\infty}^{+\infty} \Phi^* \Phi\, dp}. \tag{10.36}$$

The numerators in this equation are clearly equivalent from the above verification, and the equivalence of the denominators,

$$\int_{-\infty}^{+\infty} \Psi^*(x, t) \Psi(x, t)\, dx = \int_{-\infty}^{+\infty} \Phi^*(p, t) \Phi(p, t)\, dp, \tag{10.37}$$

can be easily demonstrated by similar arguments.

## 10.3 Momentum and Position Operators

In most quantum mechanical problems Schrödinger's equation is solved for the wave function and then the expectation value of physical variables like position, momentum, and energy are determined. This poses a problem, however, since the wave function solution to Schrödinger's equation (Equation 9.10) is necessarily in the *position representation*. How, then, is the expectation value of momentum to be determined? It would seem most appropriate to use the wave function in the *momentum representation* for this problem and begin with the equation

$$\langle p \rangle = \int_{-\infty}^{+\infty} \Phi^*(p, t) p \Phi(p, t)\, dp, \tag{10.38}$$

which would necessitate finding the Fourier transform of the solution $\Psi(x, t)$ to Schrödinger's equation. This should not be necessary, however, since the result expressed by Equation 10.33 indicates that the problem can be formulated in either the *position* or the *momentum representation*. But in the position representation,

$$\langle p \rangle = \int_{-\infty}^{+\infty} \Psi^*(x, t) p \Psi(x, t) dx, \tag{10.39}$$

an immediate difficulty is encountered, as the momentum $p$ must be expressed as a function of $x$ and $t$ before the integration can be performed. If, however, the position $x$ is specified, the Heisenberg uncertainty principle,

Heisenberg Uncertainty Principle

$$\Delta x \Delta p = h, \tag{8.99a}$$

indicates that no such function as $p = p(x, t)$ exists, since an exact determination of $p$ is not possible.

To obtain some insight as to the proper way of evaluating $\langle p \rangle$ in the position representation, we will consider Equation 10.38 and use the one-dimensional free particle wave function given by Equation 10.20. At a particular instant in time, say $t = 0$, the wave function of Equation 10.20 becomes identical to the eigenfunction of Equation 10.17, $\Phi(p, 0) = \phi(p)$, so direct substitution into Equation 10.38 yields

$$\begin{aligned} \langle p \rangle &= \int_{-\infty}^{+\infty} \Phi^*(p, 0) p \Phi(p, 0) dp = \int_{-\infty}^{+\infty} \phi^*(p) p \phi(p) dp \\ &= \frac{1}{2\pi\hbar} \iiint_{-\infty}^{+\infty} \psi^*(x') e^{\frac{j}{\hbar}px'} p\psi(x) e^{-\frac{j}{\hbar}px} dx'\, dx\, dp. \end{aligned} \tag{10.40}$$

The reduction of this equation is facilitated by observing that

$$\frac{d}{dx}\left[\psi(x) e^{-\frac{j}{\hbar}px}\right] = e^{-\frac{j}{\hbar}px} \frac{d\psi}{dx} - \frac{j}{\hbar} p\, \psi e^{-\frac{j}{\hbar}px},$$

which can be solved for

$$p\psi(x) e^{-\frac{j}{\hbar}px} = j\hbar \frac{d}{dx}(\psi e^{-\frac{j}{\hbar}px}) + e^{-\frac{j}{\hbar}px}\left(-j\hbar \frac{d}{dx}\right)\psi. \tag{10.41}$$

Substitution of this result into Equation 10.40 results in

$$\langle p\rangle = \frac{1}{2\pi\hbar}\iiint_{-\infty}^{+\infty} \psi^*(x')e^{\frac{j}{\hbar}px'}e^{-\frac{j}{\hbar}px}\left(-j\hbar\frac{d}{dx}\right)\psi(x)\,dx'\,dx\,dp$$

$$+\frac{j}{2\pi}\iint_{-\infty}^{+\infty} \psi^*(x')e^{\frac{j}{\hbar}px'}\left[\int_{-\infty}^{+\infty} d(\psi e^{-\frac{j}{\hbar}px})\right]dx'dp, \qquad (10.42)$$

where the integral in the bracket of the second term must vanish,

$$\int_{-\infty}^{+\infty} d\left[\psi(x)e^{-\frac{j}{\hbar}px}\right] = \left[\psi(x)e^{-\frac{j}{\hbar}px}\right]_{-\infty}^{+\infty} = 0, \qquad (10.43)$$

for a well behaved eigenfunction. Now, substituting Equation 10.43 into Equation 10.42 and using the properties of the Dirac delta function, we obtain

$$\begin{aligned}
\langle p\rangle &= \iint_{-\infty}^{+\infty} \psi^*(x')\left(\frac{1}{2\pi\hbar}\int_{-\infty}^{+\infty} e^{\frac{j}{\hbar}(x'-x)p}dp\right)\left(-j\hbar\frac{d}{dx}\right)\psi(x)\,dx'\,dx \\
&= \int_{-\infty}^{+\infty}\left[\int_{-\infty}^{+\infty} \psi^*(x')\delta(x'-x)dx'\right]\left(-j\hbar\frac{d}{dx}\right)\psi(x)\,dx \\
&= \int_{-\infty}^{+\infty} \psi^*(x)\left[\int_{-\infty}^{+\infty} \delta(x'-x)dx'\right]\left(-j\hbar\frac{d}{dx}\right)\psi(x)\,dx \\
&= \int_{-\infty}^{+\infty} \psi^*(x)\left(-j\hbar\frac{d}{dx}\right)\psi(x)\,dx \\
&= \int_{-\infty}^{+\infty} \Psi^*(x,0)\left(-j\hbar\frac{\partial}{\partial x}\right)\Psi(x,0)\,dx, \qquad (10.44)
\end{aligned}$$

where the last equality should be obvious from Equation 10.18 for $t = 0$. A comparison of Equations 10.40 and 10.44 gives

$$\begin{aligned}
\langle p\rangle &= \int_{-\infty}^{+\infty} \Phi^*(p,0)p\Phi(p,0)\,dp \\
&= \int_{-\infty}^{+\infty} \Psi^*(x,0)\left(-j\hbar\frac{\partial}{\partial x}\right)\Psi(x,0)\,dx, \qquad (10.45)
\end{aligned}$$

which suggests that momentum $p$ behaves like a *differential operator* in the *position representation*. That is, we should associate the **differential momentum operator**

Momentum Operator

$$\hat{p}_x \equiv -j\hbar \frac{\partial}{\partial x} \tag{10.46}$$

with the $x$-component of momentum, when we attempt a calculation for the expectation value of momentum in the *position representation*. Our discussion also indicates that in *momentum-space* the *momentum operator* behaves like

$$\hat{p}\phi(p) = p\phi(p) \tag{10.47}$$

while in *real-space* it behaves like

$$\hat{p}_x\psi(x) = -j\hbar \frac{d\psi}{dx}. \tag{10.48}$$

Generalizing to three dimensions, the **vector momentum operator** can be defined by

Vector Momentum Operator

$$\hat{\mathbf{p}} \equiv -j\hbar \boldsymbol{\nabla}, \tag{10.49}$$

which suggests that the momentum *eigenvalue equation* is of the form

Momentum Eigenvalue Equation

$$-j\hbar \boldsymbol{\nabla}\psi(\mathbf{r}) = \mathbf{p}\psi(\mathbf{r}) \tag{10.50}$$

with a solution

$$\psi(\mathbf{r}) = e^{\frac{j}{\hbar}\mathbf{p}\cdot\mathbf{r}}. \tag{10.51}$$

Of course the eigenfunction of Equation 10.51 could be multiplied by a constant or a function of time and the *eigenvalue equation* (Equation 10.50) would still be satisfied. Consequently, the plane monochromatic waves introduced in Chapter 8 are the *eigenfunctions* of the *momentum operator*.

It is rather straight forward to reproduce arguments similar to the ones used above to show that the position variable $x$ should be treated as an operator $\hat{x}$ in the *momentum representation*. The result obtained using the free particle wave function given by Equation 10.1 and the properties of the Dirac delta function is

$$\langle x \rangle = \int_{-\infty}^{+\infty} \Psi^*(x, 0) x \Psi(x, 0)\, dx$$

$$= \int_{-\infty}^{+\infty} \psi^*(x) x \psi(x) dx$$

$$\vdots \qquad \vdots$$

$$= \int_{-\infty}^{+\infty} \phi^*(p) \left( j\hbar \frac{d}{dp} \right) \phi(p) dp$$

$$= \int_{-\infty}^{+\infty} \Phi^*(p, 0) \left( j\hbar \frac{\partial}{\partial p} \right) \Phi(p, 0)\, dp, \qquad (10.52)$$

which suggests that position $x$ behaves like a *differential operator* in the *momentum representation*. Thus, we define the differential **position operator** by

$$\hat{x} \equiv j\hbar \frac{\partial}{\partial p}, \qquad (10.53)$$

Position Operator

where it is understood that $p$ is the $x$-component of the vector momentum **p**. Of course $\hat{x} \rightarrow x$ in position space just like $\hat{p}_x \rightarrow p_x$ in momentum space. Although the derivation leading to Equation 10.52 is rigorous using the properties of the Dirac delta function, the relation for the *position operator* is easily inferred by considering a plane traveling wave of the form (see Equations 10.17 and 10.19)

$$\Phi(p, t) = A e^{-\frac{j}{\hbar}(px - Et)}.$$

Taking the derivative of this wave function with respect to $p$ gives

$$\frac{\partial \Phi}{\partial p} = -\frac{j}{\hbar} x\, \Phi \rightarrow \hat{x}\Phi(p, t) = \left( j\hbar \frac{\partial}{\partial p} \right) \Phi(p, t). \qquad (10.54)$$

In a similar manner, a first order derivative of

$$\Psi(x, t) = A e^{\frac{j}{\hbar}(px - Et)} \qquad (9.6)$$

with respect to $x$ gives

$$\frac{\partial \Psi}{\partial x} = \frac{j}{\hbar} p\Psi \rightarrow \hat{p}\Psi(x, t) = \left(-j\hbar \frac{\partial}{\partial x}\right) \Psi(x, t), \tag{10.55}$$

which suggests the *momentum operator* defined by Equation 10.46. As before, we can write down an eigenvalue equation for $x$ in the *position representation* as

$$\hat{x}\psi(x) = x\psi(x), \tag{10.56}$$

while in the *momentum representation* we have

$$\hat{x}\phi(p) = j\hbar \frac{d}{dp} \phi(p). \tag{10.57}$$

Although the results of this section have been obtained for free particles, they are perfectly general and as valid as Schrödinger's equation, which will be convincingly argued in the next section. To reiterate in a slightly different form, the *eigenvalue* equations in the *position representation* are

Position-Space Eigenvalue Equations

$$\hat{x}\psi(x) \rightarrow x\psi(x) = x_0\psi(x), \tag{10.56}$$

$$\hat{p}\psi(x) \rightarrow -j\hbar \frac{d}{dx} \psi(x) = p\psi(x), \tag{10.48}$$

while in the *momentum representation* they are

Momentum-Space Eigenvalue Equations

$$\hat{x}\phi(p) \rightarrow j\hbar \frac{d}{dp} \phi(p) = x\phi(p) \tag{10.57}$$

$$\hat{p}\phi(p) \rightarrow p\phi(p) = p_0\phi(p). \tag{10.47}$$

Incidentally, the *eigensolutions* to Equations 10.48 and 10.57 are Fourier transforms of one another,

$$\psi(x) = e^{\frac{j}{\hbar}px} \quad \text{and} \quad \phi(p) = e^{-\frac{j}{\hbar}px}, \tag{10.58}$$

while the solutions to Equations 10.56 and 10.47 are *improper functions*,

$$\psi(x) = \delta(x - x_0) \quad \text{and} \quad \phi(p) = \delta(p - p_0), \tag{10.59}$$

which vanish everywhere except for $x = x_0$ and $p = p_0$. The eigenfunctions expressed by Equation 10.59 are exactly what we need for a wave function where the square of the *probability amplitude* is to represent the relative probability that the associated particle will be found at a particular point $x_0$ or have a particular momentum $p_0$. If there is no *uncertainty* in our knowledge that the particle is at the position $x_0$, then the *state function* must vanish at all other points for which $x \neq x_0$.

It is now possible to show why the expectation value of a physical variable is defined by

$$\langle Q \rangle \equiv \int_{-\infty}^{+\infty} \Psi^* Q \Psi \, dx \tag{10.29}$$

instead of the alternatives

$$\langle Q \rangle = \int_{-\infty}^{+\infty} Q \Psi^* \Psi \, dx \tag{10.60a}$$

or

$$\langle Q \rangle = \int_{-\infty}^{+\infty} \Psi^* \Psi Q \, dx. \tag{10.60b}$$

Clearly, if $Q = p$, Equation 10.60a gives

$$\begin{aligned}
\langle p \rangle &= -j\hbar \int_{-\infty}^{+\infty} \frac{\partial}{\partial x} (\Psi^* \Psi) dx \\
&= -j\hbar \int_{-\infty}^{+\infty} d(\Psi^* \Psi) \\
&= -j\hbar \left( \Psi^* \Psi \right) \Big|_{-\infty}^{+\infty} \\
&= 0,
\end{aligned}$$

since well behaved wave functions and their complex conjugates must vanish at $x = \pm\infty$. In the case of the second alternative (Equation 10.60b) we have

$$\langle p \rangle = -j\hbar \int_{-\infty}^{+\infty} \Psi^* \Psi \frac{\partial}{\partial x} dx,$$

which is mathematically absurd. Thus, the expectation value given by Equations 10.29 and 10.32 are the only appropriate definitions for variables that behave as operators in the position or momentum representations. Of course, for quantities like $x$ and $V(x)$ in the position representation or $p$ in the momentum representation, the order of factors in the integrand is immaterial, but for differential operators, the defining order for the wave functions and the operator must be preserved.

## Momentum Eigenvalues of a Free Particle in a One-Dimensional Box

Before leaving this section to consider *energy operators* in quantum mechanics, it might prove beneficial to apply our results for the one-dimensional *momentum operator* (Equation 10.46) and the associated *momentum eigenvalue* (Equation 10.48) to the problem of a *free particle in a box* discussed in Section 9.7. For this problem we found the *normalized eigenfunctions* to Schrödinger's eigenvalue equation to be of the form

$$\psi_n(x) = j\sqrt{\frac{2}{L}}\sin\frac{n\pi x}{L}, \tag{9.79}$$

with associated quantized energy eigenvalues given by

$$E_n = n^2\frac{\pi^2\hbar^2}{2mL^2}. \tag{9.80}$$

It should be apparent that the momentum eigenvalue equation (Equation 10.48) can be expressed as

$$-j\hbar\frac{d\psi_n}{dx} = p_n\psi_n, \tag{10.61}$$

where $p_n$ denotes the *quantized momentum eigenvalues*. Clearly, the *energy eigenfunctions* of Equation 9.79 are not solutions to Equation 10.61, as

$$-j\hbar\frac{d\psi_n}{dx} = -j\hbar j\sqrt{\frac{2}{L}}\frac{n\pi}{L}\cos\frac{n\pi x}{L} \neq p_n\psi_n.$$

The correct solution to Equation 10.61 is suggested by Equation 9.80 and the relation $E_n = p_n^2/2m$. Since $E_n$ in Equation 9.80 is constant for any given value of $n$, then the *quantized momentum eigenvalues* given by

$$p_n = \pm\sqrt{2mE_n} = \pm\frac{n\pi\hbar}{L} \tag{10.62}$$

Momentum Eigenvalues

are constant for a given value of $n$. Thus, Equation 10.61 can be expressed as

$$\int_0^{\psi_n} \frac{1}{\psi_n}\, d\psi_n = \frac{j}{\hbar} p_n \int_0^x dx$$

$$= \pm\frac{jn\pi}{L}\int_0^x dx,$$

which upon integration and exponentiation yields

$$\psi_n = e^{\pm jn\pi x/L}.$$

The *normalized* momentum eigenfunctions are of the form

$$\psi_n(x) = \sqrt{\frac{1}{L}}\, e^{\pm jn\pi x/L}, \tag{10.63}$$

Normalized Momentum Eigenfunctions

since the normalization condition

$$\int_0^L \psi_n^*(x)\psi_n(x)\,dx = 1$$

is obviously satisfied. The $\pm$ sign in the exponential of Equation 10.63 means that we have the two eigenfunctions

$$\psi_n^+ = \sqrt{\frac{1}{L}}\, e^{+jn\pi x/L} \tag{10.64a}$$

and

$$\psi_n^- = \sqrt{\frac{1}{L}}\, e^{-jn\pi x/L} \tag{10.64b}$$

Normalized Momentum Eigenfunctions

as solutions to the momentum eigenvalue equation (Equation 10.61). These eigenfunctions represent plane waves traveling in the positive and negative $x$-directions, respectively, and they are obvious solutions to Equation 10.61. That is, substitution of Equations 10.64a and 10.64b separately into Equation 10.61 yields

$$p_n^+ = +\frac{n\pi\hbar}{L},$$
$$p_n^- = -\frac{n\pi\hbar}{L},$$

which are in agreement with the *momentum eigenvalues* for the *free particle* in a *box* expressed by Equation 10.62. These momentum eigenvalues reflect the fact that the particle is moving back and forth between $x = 0$ and $x = L$. This means that the particle's momentum expectation value $\langle p \rangle$ should be zero, since its *average momentum* for a given value of $n$ is clearly

$$p_{\text{AVG}} = \frac{1}{2}\left(\frac{+n\pi\hbar}{L} + \frac{-n\pi\hbar}{L}\right) = 0.$$

As a last point of interest, it should be noted that the *energy eigenfunction* (Equation 9.79) can be represented by a linear combination of the two *momentum eigenfunctions* by the relation

$$\frac{\sqrt{2}}{2j}(\psi_n^+ - \psi_n^-) = \sqrt{\frac{2}{L}}\sin\frac{n\pi x}{L},$$

which along with Equations 10.64a and 10.64b suggests an alternative form for the *momentum eigenfunctions*

Normalized Momentum Eigenfunctions

$$\psi_n^+ = \frac{1}{2j}\sqrt{\frac{2}{L}}\,e^{+jn\pi x/L}, \quad (10.65a)$$

$$\psi_n^- = \frac{1}{2j}\sqrt{\frac{2}{L}}\,e^{-jn\pi x/L}. \quad (10.65b)$$

## 10.4 Example: Expectation Values in Position and Momentum Space

Before going any further in the development of the theory of quantum mechanics, an example of position and momentum expectation values in both representations will be considered, which should clarify the meaning of the previous sections. A wave function in the momentum representation is considered to be given by

$$\Phi_u(p, 0) = \phi_u(p) = e^{-p^2/2d^2\hbar^2}, \tag{10.66}$$

where time has been suppressed ($t \equiv 0$) and the $u$-subscript denotes an unnormalized state function. Preferring to work with normalized wave functions, we write down the *normalization integral* and perform the obvious substitution from Equation 10.66:

$$\begin{aligned} 1 &= \int_{-\infty}^{+\infty} \phi_u^*(p)\, \phi_u(p)\, dp \\ &= \int_{-\infty}^{+\infty} e^{-p^2/d^2\hbar^2} dp. \end{aligned}$$

This integral is of the form

$$\int_{-\infty}^{+\infty} e^{-au^2} du = \sqrt{\frac{\pi}{a}} \tag{10.67}$$

given in Appendix A, Section A.10, whence the integral above yields

$$\int_{-\infty}^{+\infty} e^{-p^2/d^2\hbar^2} dp = d\hbar\pi^{1/2}.$$

A comparison of this result with the *normalization integral* gives

Normalized Momentum-Space Eigenfunction

$$\phi(p) = \left(\frac{1}{d\hbar\pi^{1/2}}\right)^{1/2} e^{-p^2/2d^2\hbar^2} \tag{10.68}$$

for the properly *normalized eigenfunction*. Since this eigenfunction is appropriate for the *momentum representation*, the expectation values of momentum and the square of momentum can now be easily determined. We have for the former from Equations 10.32 and 10.68

$$\begin{aligned} \langle p \rangle &= \int_{-\infty}^{+\infty} \Phi^*(p, 0)\, p\, \Phi(p, 0)\, dp \\ &= \int_{-\infty}^{+\infty} \phi^*(p)\, p\, \phi(p)\, dp \end{aligned}$$

$$= \frac{1}{d\hbar\pi^{1/2}} \int_{-\infty}^{+\infty} p e^{-p^2/d^2\hbar^2} dp$$

$$= 0. \tag{10.69}$$

The integral above was evaluated by recalling that *every odd function integrated between symmetric limits vanishes*, where a function $f(x)$ is *odd* if and only if

Odd Function

$$f(x) = -f(-x) \tag{10.70a}$$

and *even* if and only if

Even Function

$$f(x) = +f(-x). \tag{10.70b}$$

The determination of the expectation value of $p^2$ is also easily accomplished by using Equations 10.32 and 10.68. That is,

$$\langle p^2 \rangle = \int_{-\infty}^{+\infty} \Phi^*(p, 0) p^2 \Phi(p, 0)\, dp$$

$$= \int_{-\infty}^{+\infty} \phi^*(p) p^2 \phi(p)\, dp$$

$$= \frac{1}{d\hbar\pi^{1/2}} \int_{-\infty}^{+\infty} p^2 e^{-p^2/d^2\hbar^2} dp, \tag{10.71}$$

where the integral is of the general form (see Appendix A, Section A.10)

$$\int_{-\infty}^{+\infty} u^{2n} e^{-au^2} du = \sqrt{\frac{\pi}{a}} \left( \frac{1}{2} \cdot \frac{3}{2} \cdots \frac{2n-1}{2} \right) a^{-n}. \tag{10.72}$$

With $n = 1$ and $a = (1/d\hbar)^2$ in this equation, Equation 10.71 becomes

$$\langle p^2 \rangle = \frac{1}{d\hbar\pi^{1/2}} (\pi d^2\hbar^2)^{1/2} \left( \frac{1}{2} \right) (\mathrm{d}^2\hbar^2)$$

$$= \tfrac{1}{2} d^2\hbar^2. \tag{10.73}$$

We could go further and determine the expectation values of $x$ and $x^2$ by using Equations 10.29, 10.53, and 10.68, but we will attempt to edify by first obtaining the appropriate eigenfunction in the *position representation*. The problem is simply to obtain $\psi(x)$ by finding the Fourier transform of $\phi(p)$. We begin by substituting Equation 10.68 into Equation 10.16 to obtain

$$\psi(x) = \left(\frac{1}{2d\pi^{3/2}\hbar^2}\right)^{1/2} \int_{-\infty}^{+\infty} e^{-p^2/2d^2\hbar^2} e^{\frac{j}{\hbar}px}\, dp.$$

This integral is of the form

$$\int_{-\infty}^{+\infty} e^{-au^2} e^{cu} du = \sqrt{\frac{\pi}{a}}\, e^{c^2/4a} \tag{10.74}$$

given in Appendix A, Section A.10, so with $a = 1/2d^2\hbar^2$, $c = jx/\hbar$, and a bit of algebraic simplification we obtain

Normalized Position-Space Eigenfunction

$$\psi(x) = \left(\frac{d}{\pi^{1/2}}\right)^{1/2} e^{-d^2x^2/2}. \tag{10.75}$$

This eigenfunction in the position representation is the Fourier transform of the eigenfunction $\phi(p)$ given in Equation 10.68. Since $\phi(p)$ is normalized, $\psi(x)$ should also be normalized. As a check we consider

$$\begin{aligned} 1 &= \int_{-\infty}^{+\infty} \Psi^*(x, 0)\Psi(x, 0)\, dx \\ &= \int_{-\infty}^{+\infty} \psi^*(x)\psi(x)\, dx \\ &= \frac{d}{\pi^{1/2}} \int_{-\infty}^{+\infty} e^{-d^2x^2} dx \\ &= \frac{d}{\pi^{1/2}} \sqrt{\frac{\pi}{d^2}} \\ &= 1, \end{aligned}$$

where Equation 10.67 has been used in evaluating the integral.

With the normalized eigenfunction in position-space given by Equation 10.75, it becomes an easy task to determine expectation values of $x$ and $x^2$. Using Equations 10.29 and 10.75, we obtain

$$\begin{aligned}\langle x\rangle &= \int_{-\infty}^{+\infty} \Psi^*(x, 0)\, x\, \Psi(x, 0)dx \\ &= \int_{-\infty}^{+\infty} \psi^*(x)\, x\, \psi(x)dx \\ &= \frac{d}{\pi^{1/2}} \int_{-\infty}^{+\infty} x e^{-d^2x^2} dx \\ &= 0, \end{aligned} \tag{10.76}$$

where the integrand has been observed to be an *odd* function of $x$. In a similar manner, $\langle x^2\rangle$ is given by

$$\begin{aligned}\langle x^2\rangle &= \int_{-\infty}^{+\infty} \psi^*(x)\, x^2\, \psi(x)dx \\ &= \frac{d}{\pi^{1/2}} \int_{-\infty}^{+\infty} x^2 e^{-d^2x^2} dx \\ &= \frac{d}{\pi^{1/2}} \left(\frac{\pi}{d^2}\right)^{1/2} \left(\frac{1}{2}\right) \left(\frac{1}{d^2}\right) \\ &= \frac{1}{2d^2}, \end{aligned} \tag{10.77}$$

where the integral was evaluated using Equation 10.72. As a point of interest, for $\langle x\rangle = 0$ the *root-mean-square*, $\langle x^2\rangle^{1/2}$, of the position variable indicates the *deviation about the average*, which would be observed in measuring the position of the particle. In a case where $\langle x\rangle \neq 0$, the **standard deviation,** $(\langle x^2\rangle - \langle x\rangle^2)^{1/2}$, is the measure of such *deviations* about the average. Such comments are equally applicable for the expectation value of momentum.

From the above considerations it should be obvious that the expectation values of $x$ and $x^2$ are most easily determined by using a normalized eigenfunction in position space, while the momentum representation is more appropriate in determining $\langle p\rangle$ and $\langle p^2\rangle$. From the last section, however, we know that the expectation value of *any physical variable* can be found with either position-space or momentum-space wave functions. As a last example, illustrating the manner in which expectation values are calculated, we will consider the determination of $\langle p^2\rangle$ in the *position representation*. Clearly, from the previous sections we have

$$\langle p^2 \rangle = \int_{-\infty}^{+\infty} \psi^*(x) \hat{p}^2 \psi(x) dx, \tag{10.78}$$

where the *momentum operator* is defined by Equation 10.46 and $\psi(x)$ is given by Equation 10.75. With these substitutions, Equation 10.78 becomes

$$\begin{aligned}
\langle p^2 \rangle &= -\hbar^2 \int_{-\infty}^{+\infty} \psi^*(x) \frac{d^2}{dx^2} \psi(x) dx \\
&= -\frac{d\hbar^2}{\pi^{1/2}} \int_{-\infty}^{+\infty} e^{-d^2x^2/2} \frac{d^2}{dx^2} e^{-d^2x^2/2} dx \\
&= \frac{d^3\hbar^2}{\pi^{1/2}} \int_{-\infty}^{+\infty} e^{-d^2x^2/2} \frac{d}{dx} (xe^{-d^2x^2/2}) dx \\
&= \frac{d^3\hbar^2}{\pi^{1/2}} \left( \int_{-\infty}^{+\infty} e^{-d^2x^2} dx - d^2 \int_{-\infty}^{+\infty} x^2 e^{-d^2x^2} dx \right) \\
&= \frac{d^3\hbar^2}{\pi^{1/2}} \left( \frac{\pi^{1/2}}{d} - d^2 \frac{\pi^{1/2}}{d} \frac{1}{2} \frac{1}{d^2} \right) \\
&= \tfrac{1}{2} d^2 \hbar^2,
\end{aligned} \tag{10.79}$$

where the integrals were evaluated using Equations 10.67 and 10.72. A quick comparison of this result with Equation 10.73 shows that the expectation value of $p^2$ when determined in the *position representation* is the same as when it is determined in the *momentum representation*. However, it should be observed that the *former* evaluation was considerably easier, mathematically, than the *latter*.

## Linear Harmonic Oscillator

Although the eigenfunction of this section has been most useful in illustrative examples for finding expectation values and a Fourier transform, it was not chosen arbitrarily. It is in fact the *ground state* eigenfunction for the *linear harmonic oscillator*. To verify this fact, recall from Chapter 7, Section 7.6 that

$$V = \tfrac{1}{2} kx^2, \tag{7.91}$$

where the wave number $k$ is related to the angular frequency $\omega$ of the oscillator by

$$k = m\omega^2. \qquad (7.95)$$

This equation is easily determined, since for the conservative force $F = -kx$ (Hooke's law) of the harmonic oscillator, $T_{max} = V_{max} \rightarrow \frac{1}{2}mv^2 = \frac{1}{2}kx^2 \rightarrow mx^2\omega^2 = kx^2 \rightarrow k = m\omega^2$. Using the two relations given above, the steady-state Schrödinger equation takes the form

$$-\frac{\hbar^2}{2m}\frac{d^2\psi}{dx^2} + \frac{m\omega^2 x^2}{2}\psi(x) = E\psi(x),$$

which can be rearranged in the form

$$\frac{d^2\psi}{dx^2} - \left(\frac{m\omega}{\hbar}\right)^2 x^2\psi(x) = -\frac{2mE}{\hbar^2}\psi(x). \qquad (10.80)$$

Now, direct substitution of $\psi(x)$ given in Equation 10.75 should yield a relation for the *ground state* energy eigenvalue. That is, with

$$\begin{aligned}
\frac{d^2\psi}{dx^2} &= \frac{d}{dx}\frac{d}{dx}\left[\left(\frac{d}{\pi^{1/2}}\right)^{1/2} e^{-d^2x^2/2}\right] \\
&= \frac{d}{dx}[-d^2x\psi(x)] \\
&= -d^2\psi(x) + d^4x^2\psi(x)
\end{aligned}$$

substituted into Equation 10.80, we obtain

$$-d^2\psi + d^4x^2\psi - \left(\frac{m\omega}{\hbar}\right)^2 x^2\psi = -\frac{2mE}{\hbar^2}\psi\cdot \qquad (10.81)$$

For this equation to be satisfied for all values of $x$, the terms involving $x^2$ must sum to zero. This requires the parameter $d$ to be equated with $(m\omega/\hbar)^{1/2}$, that is

$$d^2 = \frac{m\omega}{\hbar}\cdot \qquad (10.82)$$

Thus, cancelling the common factor $\psi(x)$ in Equation 10.81 and substituting Equation 10.82 yields

$$-d^2 = -\frac{2mE}{\hbar^2},$$

which when solved for $E$ gives

$$\begin{aligned} E &= \frac{d^2\hbar^2}{2m} \\ &= \frac{m\omega}{\hbar}\frac{\hbar^2}{2m} \\ &= \frac{\hbar\omega}{2} \\ &= \tfrac{1}{2}h\nu. \end{aligned} \tag{10.83}$$

This result is identical to that obtained in Chapter 7, Section 7.3 (Equation 7.42) for an orbiting electron about an atomic nucleus. It is different than Equation 7.101, obtained by the *Wilson-Sommerfeld quantization rule* in Section 7.6, since in that derivation we were considering a purely *classical* harmonic oscillator.

The quantum mechanical harmonic oscillator has not been considered in rigorous detail because the general methods of finding the eigenfunction solutions to Schrödinger's equation (Equation 10.80) are beyond the mathematical level of this text. Although not terribly difficult, the general solution to Equation 10.80 involves *Hermite polynomials*, which are normally introduced to students in advanced mathematics. We can, however, guess at the general form of the *energy eigenvalues*, by realizing the energy levels are equally spaced (see Equation 7.101) for the *classical* oscillator. Assuming this to be equally valid for the *quantum mechanical oscillator*, the relation

$$E_n = \left(n + \tfrac{1}{2}\right)h\nu \tag{10.84}$$

Harmonic Oscillator Energy Eigenvalues

satisfies this requirement and Equation 10.83 for $n = 0, 1, 2, 3, \cdots$. That is, at $n = 0$, the *ground state* or *zero point* energy is correctly predicted by Equation 10.84 to be that given by Equation 10.83.

## 10.5 Energy Operators

The total energy $E$ for a quantum mechanical particle is expected to behave like a differential operator, since for a free particle of constant energy, $E$ is directly proportional to the square of momentum, which is a quantum mechanical operator. We could find the *energy operator* of quantum mechanics by determining its expectation value $\langle E \rangle$ using a method similar to that detailed for $\langle p \rangle$, where the free particle wave function in momentum space (Equation 10.20) was employed along with the properties of the Dirac delta function. Now, however, it is considerably easier to capitalize on the wave functions of position and momentum space given by Equations 10.18 and 10.19. That is, for the one-dimensional free particle wave function in the position representation given by Equation 10.18, $\langle E \rangle$ is just

$$\begin{aligned}\langle E \rangle &= \int_{-\infty}^{+\infty} \Psi^*(x, t) E \Psi(x, t) dx \\ &= \int_{-\infty}^{+\infty} \psi^*(x) e^{+\frac{j}{\hbar}Et} E\psi(x) e^{-\frac{j}{\hbar}Et} dx \\ &= \int_{-\infty}^{+\infty} \psi^*(x) e^{+\frac{j}{\hbar}Et} \left( j\hbar \frac{\partial}{\partial t} \right) \psi(x) e^{-\frac{j}{\hbar}Et} dx \\ &= \int_{-\infty}^{+\infty} \Psi^*(x, t) \left( j\hbar \frac{\partial}{\partial t} \right) \Psi(x, t) dx, \end{aligned} \tag{10.85}$$

where the third equality has been obtained from

$$\begin{aligned}\frac{\partial}{\partial t} \Psi(x, t) &= \frac{\partial}{\partial t} \left( \psi(x) e^{-\frac{j}{\hbar}Et} \right) \\ &= -\frac{j}{\hbar} E\psi(x) e^{-\frac{j}{\hbar}Et}\end{aligned}$$

by solving for

$$E\psi(x) e^{-\frac{j}{\hbar}Et} = \left( j\hbar \frac{\partial}{\partial t} \right) \Psi(x, t). \tag{10.86}$$

The result of Equation 10.85 suggests that the total energy $E$ behaves like a differential operator in quantum mechanics. Thus, we define the **energy operator** in *position space* by

Energy Operator in Position-Space

$$\hat{E} \equiv j\hbar \frac{\partial}{\partial t}. \tag{10.87}$$

Clearly, the *energy eigenvalue equation* in the *position representation* is given by

$$\hat{E}\psi'(t) \rightarrow j\hbar \frac{d}{dt} \psi'(t) = E\psi'(t), \tag{10.88}$$

where the eigenfunction solution is of the form

$$\psi'(t) = e^{-\frac{j}{\hbar}Et}. \tag{10.89}$$

The energy operator in the *momentum representation* can be found by using similar arguments to those presented above. That is, using Equation 10.19 we have

$$\begin{aligned}
\langle E \rangle &= \int_{-\infty}^{+\infty} \Phi^*(p, t) E \Phi(p, t) dp \\
&= \int_{-\infty}^{+\infty} \phi^*(p) e^{-\frac{j}{\hbar}Et} E \phi(p) e^{\frac{j}{\hbar}Et} dp \\
&= \int_{-\infty}^{+\infty} \Phi^*(p, t) \left( - j\hbar \frac{\partial}{\partial t} \right) \Phi(p, t) dp,
\end{aligned} \tag{10.90}$$

so the *momentum-space* **energy operator** is

$$\hat{E} \equiv -j\hbar \frac{\partial}{\partial t}. \tag{10.91}$$

Energy Operator in Momentum-Space

In this case the corresponding eigenvalue equation is given by

$$\hat{E}\phi'(t) \rightarrow -j\hbar \frac{d}{dt} \phi'(t) = E\phi'(t), \tag{10.92}$$

having the solution

$$\phi'(t) = e^{+\frac{j}{\hbar}Et}. \tag{10.93}$$

## Hamiltonian Operator

In advanced classical mechanics the total energy for a conservative system is called the **Hamiltonian** and given by

Hamiltonian

$$H = T + V, \tag{10.94}$$

where $T$ is the kinetic energy and $V$ is the potential energy. Since kinetic energy can be expressed in terms of the momentum ($T = p^2/2m$), we can obtain the **kinetic energy operator** in *position space* from the *momentum operator* given in Equation 10.46 as

Kinetic Energy Operator

$$\begin{aligned} \hat{T} &\equiv \frac{\hat{p}^2}{2m} \\ &= \frac{1}{2m}\left(-j\hbar\frac{\partial}{\partial x}\right)^2 \\ &= -\frac{\hbar^2}{2m}\frac{\partial^2}{\partial x^2}. \end{aligned} \tag{10.95}$$

Thus, the Hamiltonian of Equation 10.94 can be regarded as an operator defined by

Hamiltonian Operator

$$\hat{H} \equiv \frac{\hat{p}^2}{2m} + \hat{V}(\hat{x}, t) \tag{10.96}$$

in terms of the *momentum* and *position* operators. It should be emphasized that the *potential energy* is an *operator* because of its dependence on the *position operator*. Further, since the Hamiltonian of Equation 10.94 is in general equivalent to the total *mechanical energy*,

$$H = E, \tag{10.97}$$

it is evident that the time-dependent Schrödinger equation can be expressed as

$$\hat{H}\Psi(x, t) = \hat{E}\Psi(x, t). \tag{10.98}$$

That is, from the operator properties for $H$ and $E$ above we have the one-dimensional equation

$$\left[\frac{\hat{p}^2}{2m} + \hat{V}(x, t)\right]\Psi(x, t) = \hat{E}\Psi(x, t),$$

which from Equations 10.95 and 10.87 becomes

$$\left[-\frac{\hbar^2}{2m}\frac{\partial^2}{\partial x^2} + V(x, t)\right]\Psi(x, t) = j\hbar\frac{\partial\Psi}{\partial t}. \qquad (9.10)$$

Further, if the potential energy is time-independent, $V = V(x)$, then Equation 10.97 suggests that

$$\hat{H}\psi(x) = E\psi(x). \qquad (10.99)$$

Thus, the time-independent or *steady-state* Schrödinger equation is directly obtained by substituting from Equation 10.96 for $V(x, t) = V(x)$,

$$\left[\frac{\hat{p}^2}{2m} + \hat{V}(x)\right]\psi(x) = E\psi(x),$$

and realizing the operator equivalence of $p$:

$$\left[-\frac{\hbar^2}{2m}\frac{d^2}{dx^2} + V(x)\right]\psi(x) = E\psi(x). \qquad (9.29)$$

Since Schrödinger's equations are regarded as fundamental postulates of quantum mechanics, the above arguments demonstrate that the energy and momentum operator definitions are fundamental postulates. It is important in this and the last section to realize that energy $E$, momentum $p$, and position $x$ can be replaced by their corresponding differential operators in an equation. These *substitutional properties* will be fully utilized in the next section to demonstrate *Bohr's correspondence principle* in quantum mechanics.

## 10.6 Correspondence between Quantum and Classical Mechanics

In 1924 Niels Bohr proposed that all new theories of physics must reduce to the well-known corresponding classical theory in the limit to which the classical theory is known to be valid. This requirement is known as *Bohr's correspondence principle*, which was originally introduced in Chapter 1, Section 1.6. As we have already seen in Chapters 2 to 4, Einstein's *special theory of relativity* constitutes a *new theory* of physics which obeys Bohr's principle. Originally, Bohr proposed that this principle must be obeyed by *quantum mechanics* in the limit that the objects of consideration are *macro* instead of *micro* in size. That is, in the limit of *large* objects, the theory of quantum mechanics must reduce to classical physics. Indeed, as will be demonstrated by examples below, Newtonian mechanics is only an approximation of quantum mechanics, when the *average* motion of a wave packet described by a wave function solution to Schrödinger's equation is considered.

As a first example of the *correspondence principle*, we will utilize the Schrödinger equation and the concept of *expectation* (average) values to derive the classical equation defining momentum in the form

$$\langle p \rangle = m \frac{d\langle x \rangle}{dt}. \tag{10.100}$$

Clearly, from the definition of momentum as an operator in position space (Equation 10.46) and the defining equation for an expectation value (Equation 10.29 or Equation 10.33) we have

$$\begin{aligned} \langle p \rangle &= \int_{-\infty}^{+\infty} \Psi^*(x, t) \hat{p} \Psi(x, t)\, dx \\ &= -j\hbar \int_{-\infty}^{+\infty} \Psi^*(x, t) \frac{\partial}{\partial x} \Psi(x, t)\, dx, \end{aligned} \tag{10.101}$$

where we have assumed *normalized* wave functions for simplicity in writing $\langle p \rangle$. This is as far as we need to go with the left-hand side of Equation 10.100, so we turn our attention to the right-hand side. Considering the first order time derivative of the expectation value of position, we have

$$\begin{aligned} \frac{d\langle x \rangle}{dt} &= \frac{d}{dt} \int_{-\infty}^{+\infty} \Psi^*(x, t)\, x\, \Psi(x, t)\, dx \\ &= \int_{-\infty}^{+\infty} \left( \frac{\partial \Psi^*}{\partial t} x \Psi + \Psi^* x \frac{\partial \Psi}{\partial t} \right) dx, \end{aligned} \tag{10.102}$$

where the *partial derivative* with respect to time has been used in the second equality, since the wave function depends on both position and time variables. This equation can be transformed by using Schrödinger's equation and its complex conjugate,

$$\frac{\partial \Psi}{\partial t} = \frac{j\hbar}{2m} \frac{\partial^2 \Psi}{\partial x^2} - \frac{j}{\hbar} V\Psi, \tag{9.45a}$$

$$\frac{\partial \Psi^*}{\partial t} = -\frac{j\hbar}{2m} \frac{\partial^2 \Psi^*}{\partial x^2} + \frac{j}{\hbar} V\Psi^*, \tag{9.45b}$$

into the form

$$\frac{d\langle x\rangle}{dt} = \frac{j\hbar}{2m}\int_{-\infty}^{+\infty}\left(\Psi^* x\frac{\partial^2\Psi}{\partial x^2} - \frac{\partial^2\Psi^*}{\partial x^2}x\Psi\right)dx. \qquad (10.103)$$

To simplify this equation we need only consider the following:

$$\int_{-\infty}^{+\infty}\frac{\partial}{\partial x}\left(\Psi^* x\frac{\partial\Psi}{\partial x} - \frac{\partial\Psi^*}{\partial x}x\Psi\right)dx = \int_{-\infty}^{+\infty}\left(\frac{\partial\Psi^*}{\partial x}x\frac{\partial\Psi}{\partial x} - \frac{\partial\Psi^*}{\partial x}x\frac{\partial\Psi}{\partial x}\right)dx$$
$$+ \int_{-\infty}^{+\infty}\left(\Psi^*\frac{\partial\Psi}{\partial x} - \frac{\partial\Psi^*}{\partial x}\Psi\right)dx$$
$$+ \int_{-\infty}^{+\infty}\left(\Psi^* x\frac{\partial^2\Psi}{\partial x^2} - \frac{\partial^2\Psi^*}{\partial x^2}x\Psi\right)dx. \qquad (10.104)$$

The integral on the left-hand side of Equation 10.104 is just

$$\int_{-\infty}^{+\infty} d\left(\Psi^* x\frac{\partial\Psi}{\partial x} - \frac{\partial\Psi^*}{\partial x}x\Psi\right) = \left(\Psi^* x\frac{\partial\Psi}{\partial x} - \frac{\partial\Psi^*}{\partial x}x\Psi\right)\Bigg|_{-\infty}^{+\infty} = 0,$$

since the wave function solution and their derivatives must vanish at $x = \pm\infty$. Further, as the first integral on the right-hand side of Equation 10.104 is obviously zero, we have

$$\int_{-\infty}^{+\infty}\left(\Psi^* x\frac{\partial^2\Psi}{\partial x^2} - \frac{\partial^2\Psi^*}{\partial x^2}x\Psi\right)dx$$
$$= -\int_{-\infty}^{+\infty}\left(\Psi^*\frac{\partial\Psi}{\partial x} - \frac{\partial\Psi^*}{\partial x}\Psi\right)dx, \qquad (10.105)$$

which transforms Equation 10.103 into

$$\frac{d\langle x\rangle}{dt} = -\frac{j\hbar}{2m}\int_{-\infty}^{+\infty}\left(\Psi^*\frac{\partial\Psi}{\partial x} - \frac{\partial\Psi^*}{\partial x}\Psi\right)dx. \qquad (10.106)$$

Also, since

$$0 = \int_{-\infty}^{+\infty}\frac{\partial}{\partial x}(\Psi^*\Psi)\,dx$$
$$= \int_{-\infty}^{+\infty}\frac{\partial\Psi^*}{\partial x}\Psi dx + \int_{-\infty}^{+\infty}\Psi^*\frac{\partial\Psi}{\partial x}dx,$$

we have

$$-\int_{-\infty}^{+\infty} \frac{\partial \Psi^*}{\partial x} \Psi dx = \int_{-\infty}^{+\infty} \Psi^* \frac{\partial \Psi}{\partial x} dx \tag{10.107}$$

and Equation 10.106 becomes

$$\begin{aligned} \frac{d\langle x\rangle}{dt} &= -\frac{j\hbar}{2m} \int_{-\infty}^{+\infty} 2\Psi^* \frac{\partial \Psi}{\partial x} dx \\ &= -\frac{j\hbar}{m} \int_{-\infty}^{+\infty} \Psi^* \frac{\partial \Psi}{\partial x} dx. \end{aligned} \tag{10.108}$$

Now, substitution from Equation 10.101 gives

$$\frac{d\langle x\rangle}{dt} = \frac{\langle p\rangle}{m},$$

which is the desired result expressed in Equation 10.100. It is interesting to observe that Equation 10.106 can be expressed in terms of the one-dimensional *probability density flux* as

$$\frac{d\langle x\rangle}{dt} = \int_{-\infty}^{+\infty} S(x, t) dx, \tag{10.106}$$

and that Equation 10.107 can be expressed in terms of the momentum operator as (multiply by $-j\hbar$)

Hermitian Property

$$\int_{-\infty}^{+\infty} (\hat{p}\, {}^*\Psi^*)\, \Psi dx = \int_{-\infty}^{+\infty} \Psi^* \hat{p}\, \Psi dx. \tag{10.109}$$

Any operator behaving as $\hat{p}$ in this equation has what is known as an **Hermitian property.**

It is also straight forward to derive the quantum mechanical equivalent to Newton's second law in the form

$$\frac{d\langle p\rangle}{dt} = -\left\langle \frac{\partial}{\partial x} V(x, t) \right\rangle \tag{10.110}$$

by using the operator equivalence of momentum. That is,

$$\frac{d\langle p\rangle}{dt} = \frac{d}{dt}\int_{-\infty}^{+\infty} \Psi^* \hat{p}\, \Psi dx$$
$$= -j\hbar \int_{-\infty}^{+\infty} \frac{d}{dt}\left(\Psi^* \frac{\partial \Psi}{\partial x}\right) dx$$
$$= -j\hbar \int_{-\infty}^{+\infty} \left(\frac{\partial \Psi^*}{\partial t}\frac{\partial \Psi}{\partial x} + \Psi^* \frac{\partial}{\partial x}\frac{\partial \Psi}{\partial t}\right) dx. \quad (10.111)$$

Substitution from Schrödinger's equation and its complex conjugate (Equations 9.45a and 9.45b) and rearranging the terms yields

$$\frac{d\langle p\rangle}{dt} = -\int_{-\infty}^{+\infty} \Psi^* \frac{\partial V}{\partial x} \Psi dx - \frac{\hbar^2}{2m}\int_{-\infty}^{+\infty} \left(\frac{\partial^2 \Psi^*}{\partial x^2}\frac{\partial \Psi}{\partial x} - \Psi^* \frac{\partial^3 \Psi}{\partial x^3}\right) dx. \quad (10.112)$$

The second integral on the right-hand side of this equation is equivalent to

$$\int_{-\infty}^{+\infty} \frac{\partial}{\partial x}\left(\frac{\partial \Psi^*}{\partial x}\frac{\partial \Psi}{\partial x} - \Psi^* \frac{\partial^2 \Psi}{\partial x^2}\right) dx = 0, \quad (10.113)$$

from which we obtain

$$\frac{\partial^2 \Psi^*}{\partial x^2}\frac{\partial \Psi}{\partial x} = \Psi^* \frac{\partial^3 \Psi}{\partial x^3}. \quad (10.114)$$

Thus, the second integral on the right-hand side of Equation 10.112 reduces to zero and the equation reduces to Newton's second law in the form given by Equation 10.110. It is now trivial to demonstrate Newton's second law in the form

$$\frac{d\langle p\rangle}{dt} = m\frac{d^2\langle x\rangle}{dt^2}, \quad (10.115)$$

by taking a time differential of Equation 10.109, assuming $m \neq m(t)$, and equating the result to Equation 10.111.

## Operator Algebra

Although the results above were obtained by rather straight forward arguments involving differential calculus, the method illustrated is somewhat cumbersome. A far more elegant method is available by capitalizing on the properties of the position, momentum, and Hamiltonian operators. To generalize, imagine a quantum mechanical operator $Q$ to be an explicit function of position and time, $Q = Q(x, t)$. A general identity for the time derivative of the expectation value of $Q$ can be obtained from

$$\frac{d\langle Q\rangle}{dt} = \frac{d}{dt}\int_{-\infty}^{+\infty} \Psi^* Q \Psi dx$$
$$= \int_{-\infty}^{+\infty} \left( \frac{\partial \Psi^*}{\partial t} Q\Psi + \Psi^* \frac{\partial Q}{\partial t} \Psi + \Psi^* Q \frac{\partial \Psi}{\partial t} \right) dx,$$

by substitution of Schrödinger's equation in the form

$$\frac{\partial \Psi}{\partial t} = -\frac{j}{\hbar} H\Psi, \qquad (10.116a)$$

$$\frac{\partial \Psi^*}{\partial t} = +\frac{j}{\hbar} H^*\Psi^* = \frac{j}{\hbar} \Psi^* H, \qquad (10.116b)$$

to obtain

$$\frac{d\langle Q\rangle}{dt} = \int_{-\infty}^{+\infty} \left[ \Psi^* \frac{\partial Q}{\partial t} \Psi + \frac{j}{\hbar} (\Psi^* H Q \Psi - \Psi^* Q H \Psi) \right] dx. \qquad (10.117)$$

It should be noted that this result is also dependent on the *Hermitian property* of $H$, which was used in obtaining the second equality of Equation 10.116b. By using Equation 10.29 for the expectation value of $Q$ and the definition for the **commutator,**

Commutator

$$[H,Q] \equiv HQ - QH, \qquad (10.118)$$

($H$ and $Q$ could be any two operators in this defining equation) Equation 10.117 is transformed into

Generalized Operator Equation

$$\frac{d\langle Q\rangle}{dt} = \left\langle \frac{\partial Q}{\partial t} \right\rangle + \frac{j}{\hbar} \langle [H, Q] \rangle. \qquad (10.119)$$

It should be emphasized that $\langle Q\rangle$ represents an ensemble average of the results of a *single* measurement of $Q$ on each system of the ensemble. The derivative, $d\langle Q\rangle/dt$, is the time rate of change of this average, which is not the same as $\langle dQ/dt\rangle$. This is easily realized if we imagine $Q = x$, since then $\langle dx/dt\rangle$ represents the average of velocity measurements made on each system of the ensemble, but velocity operators simply do not occur in nonrelativistic quantum mechanics. Also, if the operator $Q$ is not an explicit function of time, the *generalized operator equation* (Equation 10.119) reduces to

$$\frac{d\langle Q\rangle}{dt} = \frac{j}{\hbar}\langle [H, Q]\rangle, \qquad Q \neq Q(t). \tag{10.120}$$

This equation is perfectly valid for $Q = x$ or $Q = p$, since in quantum mechanics position and momentum are time-independent operators.

As an example of the usefulness and ease of application of Equation 10.120 (or Equation 10.119), we will operationally derive Equation 10.100. That is, with $Q = x$, Equation 10.120 gives

$$\begin{aligned}\frac{d\langle x\rangle}{dt} &= \frac{j}{\hbar}\langle [H, x]\rangle \\ &= \frac{j}{\hbar}\left\langle \left[\left(\frac{p^2}{2m} + V\right), x\right]\right\rangle \\ &= \frac{j}{\hbar}\left\langle \left[\frac{p^2}{2m}, x\right] + [V, x]\right\rangle \\ &= \frac{j}{\hbar}\left\langle \left[\frac{1}{2m}\left(-j\hbar\frac{\partial}{\partial x}\right)^2, x\right]\right\rangle \\ &= -\frac{j\hbar}{2m}\left\langle \left[\frac{\partial}{\partial x}\frac{\partial}{\partial x}, x\right]\right\rangle.\end{aligned} \tag{10.121}$$

This equation can be reduced by realizing

$$[AB, C] = A[B, C] + [A, C]B \tag{10.122}$$

follows immediately from the definition of the *commutator* (Equation 10.118). Also, consider

$$\begin{aligned}\left[\frac{\partial}{\partial x}, x\right] A(x) &= \left(\frac{\partial}{\partial x} x - x\frac{\partial}{\partial x}\right) A \\ &= \frac{\partial}{\partial x} xA - x\frac{\partial A}{\partial x} \\ &= A + x\frac{\partial A}{\partial x} - x\frac{\partial A}{\partial x} \\ &= A(x),\end{aligned}$$

which means that

$$\left[\frac{\partial}{\partial x}, x\right] = 1. \tag{10.123}$$

Now, with Equations 10.122 and 10.123 we can transform Equation 10.121 into

$$\begin{aligned}\frac{d\langle x\rangle}{dt} &= -\frac{j\hbar}{2m}\left\langle\frac{\partial}{\partial x}\left[\frac{\partial}{\partial x}, x\right] + \left[\frac{\partial}{\partial x}, x\right]\frac{\partial}{\partial x}\right\rangle \\ &= -\frac{j\hbar}{2m}\left\langle 2\,\frac{\partial}{\partial x}\right\rangle \\ &= \frac{1}{m}\left\langle -j\hbar\,\frac{\partial}{\partial x}\right\rangle \\ &= \frac{\langle p\rangle}{m},\end{aligned}$$

which is identical to Equation 10.100. By using the above techniques, it is also easy to verify the relationship given in Equation 10.110, which is left as an exercise in the problem set. This problem is facilitated by realizing $[V, p] = -[p, V] \neq 0$ and taking note of the derivational procedure leading to Equation 10.123.

## 10.7 Free Particle in a Three-Dimensional Box

The development and application of Schrödinger's quantum mechanics has been primarily restricted to one-dimensional considerations for mathematical simplicity. Physical reality, however, requires an application of quantum theory in three dimensions, which necessitates a generalization of our previous discussions to include position variables in the $y$ and $z$ directions, as well as the $x$-direction. The general character of the eigenfunction solutions to Schrödinger's three-dimensional eigenvalue equation will be illustrated by considering a particle of mass $m$ confined to a three-dimensional *box*. This problem is but a generalization of the one-dimensional *free particle* in a *box* problem considered in Chapter 9, Section 9.7. Now, however, we consider the box to be a cube of edge $L$, with impenetrable walls parallel to the coordinate axes at $x = 0, L$; $y = 0, L$; $z = 0, L$. By analogy with Equation 10.99, Schrödinger's time-independent equation in three dimensions can be expressed in the form

$$\hat{H}\psi(x, y, z) = E\psi(x, y, z), \tag{10.124}$$

where the generalized *Hamiltonian operator*

$$\hat{H} = \frac{\hat{\mathbf{p}}^2}{2m} + V(x, y, z) \tag{10.125}$$

is given in terms of the **three-dimensional momentum operator**

Three-Dimensional Momentum Operator

$$\hat{\mathbf{p}} \equiv -j\hbar\left(\frac{\partial}{\partial x}\mathbf{i} + \frac{\partial}{\partial y}\mathbf{j} + \frac{\partial}{\partial z}\mathbf{k}\right). \tag{10.126}$$

Of course, $\hat{\mathbf{p}}^2$ in Equation 10.125 is easily obtained from the defining equation of $\hat{\mathbf{p}}$ by taking an inner product of $\hat{\mathbf{p}}$ with itself. That is,

$$\begin{aligned} \hat{\mathbf{p}}^2 &= \hat{\mathbf{p}} \cdot \hat{\mathbf{p}} \\ &= -\hbar^2\left(\frac{\partial^2}{\partial x^2} + \frac{\partial^2}{\partial y^2} + \frac{\partial^2}{\partial z^2}\right) \\ &= -\hbar^2\nabla^2, \end{aligned} \tag{10.127}$$

where the definition for the *Laplacian operator* $\nabla^2$ given in Equation 9.15 has been used in obtaining the second equality. Now, with this result substituted into the Hamiltonian (Equation 10.125) and that result substituted into Schrödinger's eigenvalue equation (Equation 10.124), we obtain

$$-\frac{\hbar^2}{2m}\nabla^2\psi(x, y, z) + V(x, y, z)\psi(x, y, z) = E\psi(x, y, z) \tag{10.128}$$

for the steady-state form of Schrödinger's equation in three dimensions. Although obtained here by operational algebra, this result is equivalent to that expressed in Equation 9.30. This equation simplifies for the free particle in a three-dimensional box to

$$-\frac{\hbar^2}{2m}\nabla^2\psi(x, y, z) = E\psi(x, y, z), \tag{10.129}$$

since the potential energy $V(x, y, z)$ is restricted to the values $V = 0$ inside and $V = \infty$ outside of the box. As a result of these values for $V$, the eigenfunction solution to Equation 10.129 has the boundary conditions $\psi(x, y, z) = 0$ at $x = 0$ or $L$, $y = 0$ or $L$, and $z = 0$ or $L$.

The form of the eigenfunction solution to Equation 10.129 is suggested by analogy with classical mechanics, where the position coordinates $x$, $y$, and $z$ for a free particle are considered to be independent variables. Accordingly, we consider a *separation of variables* and assume the eigenfunction solution to be of the form

$$\psi(x, y, z) = \psi_x(x)\psi_y(y)\psi_z(z). \tag{10.130}$$

Upon substitution of this assumed solution into the eigenvalue equation (Equation 10.129) and division by $\psi_x(x)\psi_y(y)\psi_z(z)$, we obtain

$$-\frac{\hbar^2}{2m}\left(\frac{1}{\psi_x}\frac{d^2\psi_x}{dx^2} + \frac{1}{\psi_y}\frac{d^2\psi_y}{dy^2} + \frac{1}{\psi_z}\frac{d^2\psi_z}{dz^2}\right) = E, \tag{10.131}$$

where the second-order partial derivatives become ordinary derivatives because of the separation of variables assumption. Since each differential term on the left-hand side of this equation is dependent on a different position variable and the right-hand side is a constant, each differential term must be set equal to a different constant. The identification of these constants is easily accomplished by realizing the total energy of the free particle is given by

$$\begin{aligned} E &= \frac{p^2}{2m} \\ &= \frac{p_x^2 + p_y^2 + p_z^2}{2m} \\ &= \frac{\hbar^2}{2m}(k_x^2 + k_y^2 + k_z^2), \end{aligned} \tag{10.132}$$

where de Broglie's *momentum postulate* has been employed in obtaining the final expression. Thus, a comparison of the last two equations results in

$$\frac{1}{\psi_x}\frac{d^2\psi_x}{dx^2} = -k_x^2, \tag{10.133a}$$

$$\frac{1}{\psi_y}\frac{d^2\psi_y}{dy^2} = -k_y^2, \tag{10.133b}$$

$$\frac{1}{\psi_z}\frac{d^2\psi_z}{dz^2} = -k_z^2, \tag{10.133c}$$

which are ordinary differential equations of identical form to that given by Equation 9.68 for the one-dimensional free particle in a box. Consequently, the *normalized* eigenfunction solutions to these equations can be chosen as (see Equation 9.85)

$$\psi_x(x) = \sqrt{\frac{2}{L}} \sin k_x x, \qquad (10.134a)$$

$$\psi_y(y) = \sqrt{\frac{2}{L}} \sin k_y y, \qquad (10.134b)$$

$$\psi_z(z) = \sqrt{\frac{2}{L}} \sin k_z z, \qquad (10.134c)$$

where the values of $k_x$, $k_y$, and $k_z$ (obtained from the boundary conditions) are *quantized* and given by the relations

$$k_x = \frac{n_x \pi}{L}, \qquad n_x = 1, 2, 3, \cdots, \qquad (10.135a)$$

$$k_y = \frac{n_y \pi}{L}, \qquad n_y = 1, 2, 3, \cdots, \qquad (10.135b)$$

$$k_z = \frac{n_z \pi}{L}, \qquad n_z = 1, 2, 3, \cdots. \qquad (10.135c)$$

The eigenfunction solutions associated with a particle in a cubical box are now obtained from Equation 10.130 by substitution of Equations 10.134a–c. That is,

Normalized Eigenfunctions

$$\psi_{n_x n_y n_z} = \sqrt{\frac{8}{V}} \sin k_x x \sin k_y y \sin k_z z, \qquad (10.136)$$

where $V = L^3$ and a subscript notation has been introduced on the eigenfunction to account for the dependence of $k_x$, $k_y$, and $k_z$ on the quantum numbers $n_x$, $n_y$, and $n_z$. For each eigenfunction represented in Equation 10.136 there exists an energy eigenvalue given by

Energy Eigenvalues

$$E_{n_x n_y n_z} = (n_x^2 + n_y^2 + n_z^2) \frac{\pi^2 \hbar^2}{2mL^2}, \qquad (10.137)$$

which is directly obtained by substitution of Equations 10.136 and 10.135a–c into Equation 10.129 or by substitution of Equations

10.135a–c into Equation 10.132. From this equation we note a general characteristic of three-dimensional problems, which is the requirement of *three principal quantum numbers* for the complete specification of each quantum state, as illustrated in Figure 10.1. Also, we note that Equation 10.137 predicts an *energy level degeneracy*, since, for example, the eigenfunctions $\psi_{112}$, $\psi_{121}$, and $\psi_{211}$ all describe quantum states with identical energy $E_{112} = E_{121} = E_{211}$. Although these quantum states have identical *energy eigenvalues,* they have physically different *momentum eigenvalues*.

## Free Electron Gas in Three Dimensions

An important application of the free particle in a box occurs in solid state physics and electrical engineering by applying the model to describe conduction electrons in a simple metal. As was pointed out in Chapter 9, Section 9.7, *conduction electrons* in condensed matter behave like a *gas* of noninteracting particles. As such, we consider a large number $N$ of conduction electrons in a three-dimensional crystal and ignore any interaction of the electrons with the *periodic* arrangement of the *ion cores*. This consideration is reasonable since the *matter* waves associated with a con-

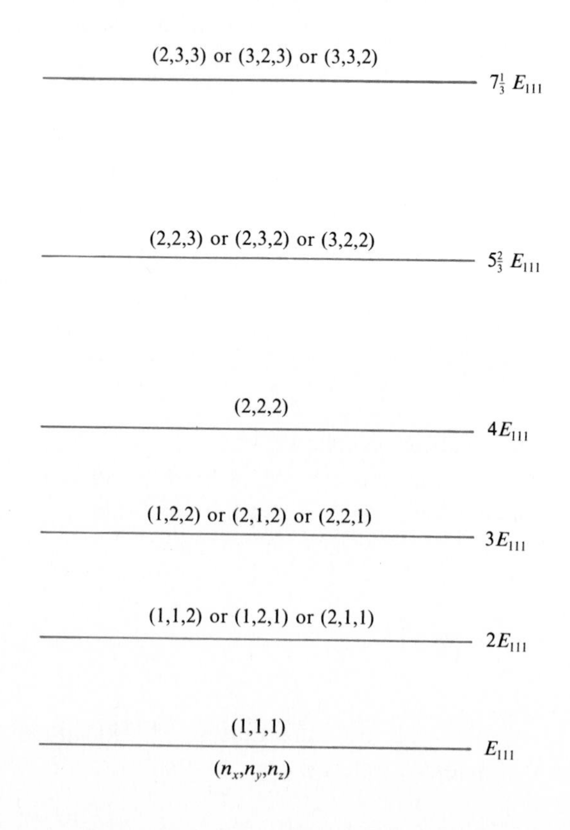


Figure 10.1
A few of the lower permitted energy levels for a free particle in a three dimensional box, with the corresponding level degeneracy indicated.

duction electron propagate freely in a *periodic* structure. Further, the medium can be thought of as *unbounded*, if we require the eigenfunctions to be *periodic* over a large distance $L$. This means that the particle in a box *boundary conditions* are replaced by **periodic boundary conditions,** which are given by

Periodic Boundary Conditions

$$\psi_x(x + L) = \psi_x(x), \tag{10.138a}$$
$$\psi_y(y + L) = \psi_y(y), \tag{10.138b}$$
$$\psi_z(z + L) = \psi_z(z). \tag{10.138c}$$

This method of *periodic boundary conditions* does not alter the physics of the problem in any essential respect, so long as the system contains a large number $N$ of particles. It allows for the enumeration of *quantum states* that is equally valid to those expressed in Equations 10.135a–c, although the results are slightly different. To see this, consider the eigenfunction solutions to Equation 10.129 to be of the traveling wave form

$$\begin{aligned}\psi(\mathbf{r}) &= \sqrt{\frac{1}{V}}\, e^{\frac{j}{\hbar}(\mathbf{p}\cdot\mathbf{r})} \\ &= \sqrt{\frac{1}{V}}\, e^{j\mathbf{k}\cdot\mathbf{r}},\end{aligned} \tag{10.139}$$

where $V = L^3$. The *normalization condition* is satisfied by this eigenfunction because

$$\begin{aligned}\int_0^V \psi^*(\mathbf{r})\psi(\mathbf{r})d\mathbf{r} &= \frac{1}{V}\int_0^V e^{-j\mathbf{k}\cdot\mathbf{r}} e^{j\mathbf{k}\cdot\mathbf{r}} d\mathbf{r} \\ &= \frac{1}{V}\int_0^L \int_0^L \int_0^L dxdydz \\ &= 1,\end{aligned} \tag{10.140}$$

and the eigenvalue equation is satisfied, since by direct substitution into Equation 10.129 we obtain $E = \hbar^2k^2/2m$. Further, the *separation of variables* eigenvalue equation expressed in Equation 10.131 is satisfied, as the eigenfunction of Equation 10.139 can be expressed as

$$\begin{aligned}\psi_n(\mathbf{r}) &= \psi_{n_x}(x)\psi_{n_y}(y)\psi_{n_z}(z) \\ &= \sqrt{\frac{1}{L}}\, e^{jk_x x} \sqrt{\frac{1}{L}}\, e^{jk_y y} \sqrt{\frac{1}{L}}\, e^{jk_z z}.\end{aligned} \tag{10.141}$$

Imposing the *periodic boundary condition* of Equation 10.138a on $\psi_{n_x}(x)$, we have

$$\sqrt{\frac{1}{L}}\, e^{jk_x(x+L)} = \sqrt{\frac{1}{L}}\, e^{jk_x x},$$

which results in

$$e^{jk_xL} = \cos k_xL + j\, sin\, k_xL = 1.$$

Clearly, this equation is satisfied by requiring

$$k_x = n_x \frac{2\pi}{L}, \qquad n_x = 0, 1, 2, \cdots, \tag{10.142}$$

and similar results are obtained for the $y$ and $z$ components. Thus, the free particle steady-state Schrödinger equation, the normalization condition, and the periodicity condition are satisfied by the assumed traveling wave eigenfunction (Equation 10.139 or Equation 10.141), provided the components of the *wave vector* **k** are given by

Wave Vector Quantization

$$\mathbf{k} = (k_{n_x}, k_{n_y}, k_{n_z}) = \left(\frac{2\pi n_x}{L}, \frac{2\pi n_y}{L}, \frac{2\pi n_z}{L}\right), \tag{10.143}$$

where $n_x$, $n_y$, and $n_z$ are integers.

With the above results, we are now prepared to derive the *density of states*, defined by

Density of States

$$D(E) \equiv \frac{d}{dE}\, \mathrm{N(E)}, \tag{9.89}$$

for the free electron gas (conduction electrons) in three dimensions. Taking the *Pauli exclusion principle* (see Chapter 9, Section 9.7) into account, the *quantum states* allowed for the $N$ conduction electrons are specified by the quantum numbers $n_x$, $n_y$, and $n_z$, along with the *spin* quantum number $m_s$. Thus, there are *two* allowed quantum states for a distinct triplet of quantum numbers $n_x$, $n_y$, and $n_z$, one with $m_s = -\frac{1}{2}$ and the other with $m_s = +\frac{1}{2}$, which can accommodate two conduction electrons. This means that in **k**-space a *volume element* $\Delta k$,

$$
\begin{aligned}
\Delta k &= \Delta k_{n_x} \Delta k_{n_y} \Delta k_{n_z} \\
&= \left(\frac{2\pi}{L}\right)^3 \Delta n_x \Delta n_y \Delta n_z,
\end{aligned} \tag{10.144}
$$

can accommodate

$$
\Delta N = 2\Delta n_x \Delta n_y \Delta n_z \tag{10.145}
$$

*quantum states* or electrons, where the factor 2 comes from the two allowed values of $m_s$. Now, the total number of *quantum states* or electrons contained within a sphere of radius $|\mathbf{k}|$ in **k**-space, is given by the product of the volume of that sphere and the number of states per unit volume. That is, in a sphere of volume (see Figure 10.2)

$$
V_k = \frac{4}{3}\pi k^3, \tag{10.146}
$$

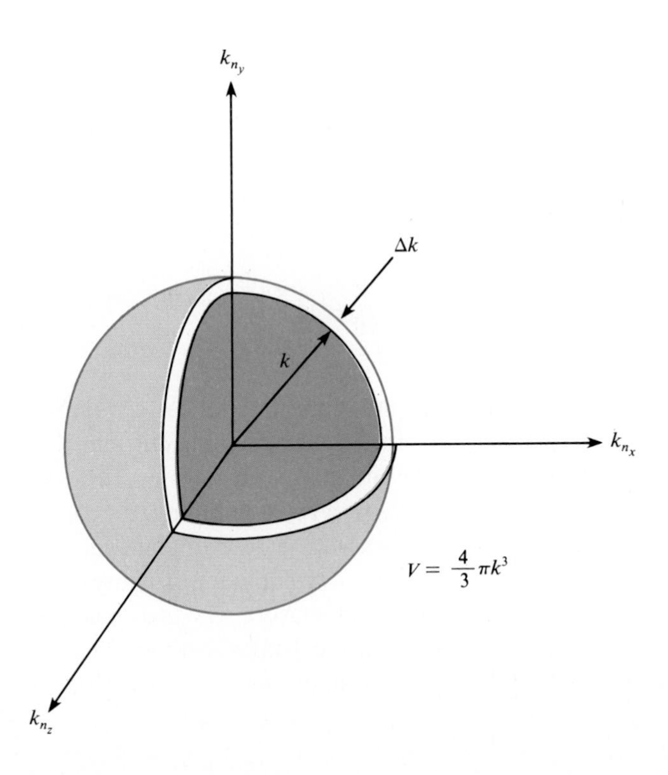


Figure 10.2
The volume of a spherical shell in **k**-space.

the total number of quantum states allowed is given by

$$
\begin{aligned}
N &= V_k \frac{\Delta N}{\Delta k} \\
&= \frac{4}{3}\pi k^3 \left[\frac{2}{(2\pi/L)^3}\right] \\
&= \frac{V}{3\pi^2}k^3,
\end{aligned}
\tag{10.147}
$$

where $V = L^3$ and the above relations (Equations 10.144 to 10.146) have been used. With $E = \hbar^2 k^2/2m$ solved for $k^3$,

$$
k^3 = \left(\frac{2mE}{\hbar^2}\right)^{3/2},
\tag{10.148}
$$

substituted into Equation 10.147, we obtain

$$
N(E) = \frac{V}{3\pi^2}\left(\frac{2m}{\hbar^2}\right)^{3/2} E^{3/2},
\tag{10.149}
$$

where the number of quantum states, equal to the number of electrons $N$, has been expressed as a function of the energy $E$. With this result, the *density of electronic states* is immediately obtained from the defining equation (Equation 9.89) above:

Density of Electronic States

$$
D(E) = \frac{V}{2\pi^2}\left(\frac{2m}{\hbar^2}\right)^{3/2} E^{1/2}.
\tag{10.150}
$$

This result is very important in the theoretical development and understanding of electrical phenomena in solids. Insight into these electrical properties of matter can not be attained with only our *introductory* treatment of quantum mechanics. However, combining the fundamentals of Schrödinger's quantum mechanics with a few fundamentals of *statistical mechanics* will allow for a rigorous development of many interesting electrical phenomena in solids. For this reason, we turn our attention to an introductory treatment of statistical mechanics, and in Chapter 12 we will return to the *free electron gas* problem and its application in solid state physics, material science, and electrical engineering.

## Review of Fundamental and Derived Equations

Below is a listing of the fundamental and derived equations of this chapter, along with newly introduced mathematical operators.

### FUNDAMENTAL EQUATIONS—QUANTUM MECHANICS

$$D(E) \equiv \frac{d}{dE}N(E)$$ Density of States

$$\rho(x, t) \equiv \frac{|\Psi|^2}{\int_{-\infty}^{+\infty} |\Psi|^2 dx}$$ Probability Density in Position-Space

$$\langle Q(x) \rangle \equiv \int_{-\infty}^{+\infty} Q(x)\rho(x, t)\, dx = \frac{\int_{-\infty}^{+\infty} \Psi^* Q \Psi dx}{\int_{-\infty}^{+\infty} |\Psi|^2\, dx}$$ Expectation Value in Position-Space

$$\rho(p, t) \equiv \frac{|\Phi|^2}{\int_{-\infty}^{+\infty} |\Phi|^2\, dp}$$ Probability Density in Momentum-Space

$$\langle Q(p) \rangle \equiv \int_{-\infty}^{+\infty} Q(p)\rho(p, t)\, dp = \frac{\int_{-\infty}^{+\infty} \Phi^* \mathrm{Q} \Phi dp}{\int_{-\infty}^{+\infty} |\Phi|^2\, dp}$$ Expectation Value in Momentum-Space

$$\hat{p}_x \equiv -j\hbar \frac{\partial}{\partial x}$$ Momentum Operator

$$-j\hbar \frac{d}{dx} \psi(x) = p_x \psi(x)$$ Momentum Eigenvalue Equation

$$\hat{\mathbf{p}} \equiv -j\hbar \nabla$$ Vector Momentum Operator

$$\hat{x} \equiv j\hbar \frac{\partial}{\partial p}$$ Position Operator

$$j\hbar \frac{d}{dp} \phi(p) = x\phi(p)$$ Position Eigenvalue Equation

$$\hat{E} \equiv +j\hbar \frac{\partial}{\partial t}$$ Energy Operator in Position-Space

$$\hat{E} \equiv -j\hbar \frac{\partial}{\partial t}$$ Energy Operator in Momentum-Space

$$\hat{H} \equiv \frac{\hat{p}^2}{2m} + \hat{V}$$ Hamiltonian Operator

$$\hat{T} \equiv \frac{\hat{p}^2}{2m}$$ Kinetic Energy Operator

$$\hat{H}\Psi(x, t) = \hat{E}\psi(x, t)$$ Schrödinger's Time-Dependent Equation

$$\hat{H}\psi(x) = E\psi(x)$$ Schrödinger's Steady-State Equation

$$\int_{-\infty}^{+\infty} \Psi^* Q\Psi dx = \int_{-\infty}^{+\infty} (\hat{Q}^*\Psi^*)\Psi dx$$ Operator—Hermitian Property

## GENERALIZED FREE PARTICLE WAVE FUNCTION

$$\Psi(x, t) = \sqrt{\frac{1}{2\pi\hbar}} \int_{-\infty}^{+\infty} \Phi(p, t) e^{+\frac{j}{\hbar}(px - Et)} dp = \psi(x) e^{-\frac{j}{\hbar}Et}$$ Position Representation

$$\Phi(p, t) = \sqrt{\frac{1}{2\pi\hbar}} \int_{-\infty}^{+\infty} \Psi(x, t) e^{-\frac{j}{\hbar}(px - Et)} dx = \phi(p) e^{+\frac{j}{\hbar}Et}$$ Momentum Representation

## MATHEMATICAL OPERATORS/RELATIONS

$$\nabla \equiv \frac{\partial}{\partial x}\mathbf{i} + \frac{\partial}{\partial y}\mathbf{j} + \frac{\partial}{\partial z}\mathbf{k}$$ Del Operator

$$\nabla^2 = \frac{\partial^2}{\partial x^2} + \frac{\partial^2}{\partial y^2} + \frac{\partial^2}{\partial z^2}$$ Laplacian Operator

$$[A, B] \equiv AB - BA$$ Commutator

$$\left.\begin{aligned} &[AB, C] = A[B, C] + [A, C]B \\ &\left[\frac{\partial}{\partial x}, x\right] = 1 \end{aligned}\right\}$$ Commutator Properties

$$\delta(k - k') \equiv \frac{1}{2\pi} \int_{-\infty}^{+\infty} e^{j(k-k')x} dx$$ Dirac Delta Function in Momentum-Space

$$\delta(x - x') \equiv \frac{1}{2\pi} \int_{-\infty}^{+\infty} e^{j(x-x')k} dk$$ Dirac Delta Function in Position-Space

$$\left.\begin{aligned}
&\delta(k - k') = 0,\ k \neq k' \\
&\delta(k - k') = \infty,\ k = k' \\
&\delta(k - k') = \delta(k' - k) \\
&f(k)\delta(k - k') = f(k')\delta(k - k') \\
&\delta\left[\frac{1}{a}(k - k')\right] = a\delta(k - k') \\
&\int_{-\infty}^{+\infty} \delta(k - k')\, dk = 1
\end{aligned}\right\} \quad \text{Dirac Delta Function Properties}$$

## FUNDAMENTAL DERIVATIONS

### *Dirac Delta Function*

$$\delta(x - x') = \frac{1}{2\pi\hbar} \int_{-\infty}^{+\infty} e^{\frac{j}{\hbar}(x-x')p} dp$$

$$\delta(p - p') = \frac{1}{2\pi\hbar} \int_{-\infty}^{+\infty} e^{\frac{j}{\hbar}(p-p')x} dx$$

$$\int_{-\infty}^{+\infty} f(k)\delta(k - k')\, dk = f(k')$$

### *Generalized Free Particle Wave Function*

$$\langle Q \rangle = \int_{-\infty}^{+\infty} \Psi^*(x, 0) Q \Psi(x, 0)\, dx$$

$$= \int_{-\infty}^{+\infty} \Phi^*(p, 0) Q \Phi(p, 0)\, dp \quad \text{Expectation Value}$$

$$\langle p \rangle = \int_{-\infty}^{+\infty} \Phi^*(p, 0)\, p\, \Phi(p, 0)\, dp$$

$$= \int_{-\infty}^{+\infty} \Psi^*(x, 0) \left(-j\hbar \frac{\partial}{\partial x}\right) \Psi(x, 0)\, dx \quad \text{Momentum Operator}$$

$$\langle x \rangle = \int_{-\infty}^{+\infty} \Psi^*(x, 0)\, x\, \Psi(x, 0)\, dx$$

$$= \int_{-\infty}^{+\infty} \Phi^*(p, 0) \left(+j\hbar \frac{\partial}{\partial p}\right) \Phi(p, 0)\, dp \quad \text{Position Operator}$$

$$\langle E \rangle = \int_{-\infty}^{+\infty} \Psi^*(x, 0) E \Psi(x, 0)\, dx$$

$$= \int_{-\infty}^{+\infty} \Psi^*(x, 0) \left(+j\hbar \frac{\partial}{\partial t}\right) \Psi(x, 0)\, dx \quad \text{Energy Operator in Position-Space}$$

$$\langle E \rangle = \int_{-\infty}^{+\infty} \Phi^*(p, 0) E \Phi(p, 0)\, dp$$

$$= \int_{-\infty}^{+\infty} \Phi^*(p, 0) \left(-j\hbar \frac{\partial}{\partial t}\right) \Phi(p, 0)\, dp \quad \text{Energy Operator in Momentum-Space}$$

### Free Particle in a One-Dimensional Box

$$\psi_n(x) = j\sqrt{\frac{2}{L}} \sin \frac{n\pi x}{L}$$ Energy Eigenfunctions

$$E_n = n^2\frac{\pi^2\hbar^2}{2mL^2}$$ Energy Eigenvalues

$$p_n^{\pm} = \pm\frac{n\pi\hbar}{L}$$ Momentum Eigenvalues

$$\psi_n^{\pm}(x) = \sqrt{\frac{1}{L}}\, e^{\pm jn\pi x/L}$$ Momentum Eigenfunctions

### Linear Harmonic Oscillator—Ground State

$$-\frac{\hbar^2}{2m}\frac{d^2\psi}{dx^2} + \tfrac{1}{2}kx^2\,\psi = E\psi$$ Eigenvalue Equation

$$\psi(x) = \left(\frac{d}{\pi^{1/2}}\right)^{1/2} e^{-d^2x^2/2}$$ Energy Eigenfunction

$$\phi(p) = \left(\frac{1}{d\hbar\pi^{1/2}}\right)^{1/2} e^{-p^2/2d^2\hbar^2}$$ Eigenfunction—Fourier Transform

$$E = \tfrac{1}{2}h\nu$$ Ground State Energy

$$E_n = (n + \tfrac{1}{2})h\nu,\ n = 0, 1, \cdots$$ Energy Eigenvalues

$$\langle x\rangle = \int_{-\infty}^{+\infty} \psi^*(x)x\psi(x)\,dx$$

$$= \int_{-\infty}^{+\infty} \phi^*(p)\hat{x}\phi(p)\,dp = 0$$ Position Expectation Value

$$\langle p\rangle = \int_{-\infty}^{+\infty} \phi^*(p)p\,\phi(p)\,dp$$

$$= \int_{-\infty}^{+\infty} \psi^*(x)\hat{p}\,\psi(x)\,dx = 0$$ Momentum Expectation Value

$$\langle x^2\rangle = \frac{1}{2d^2}$$

$$\langle p^2\rangle = \tfrac{1}{2}d^2\hbar^2$$

### Correspondence between Quantum and Classical Mechanics

$$\frac{d\langle x\rangle}{dt} = \frac{j\hbar}{2m}\int_{-\infty}^{+\infty}\left(\Psi^*x\,\frac{\partial^2\Psi}{\partial x^2} - \frac{\partial^2\Psi^*}{\partial x^2}\,x\Psi\right)dx$$

$$= -\frac{j\hbar}{2m}\int_{-\infty}^{+\infty}\left(\Psi^*\frac{\partial\Psi}{\partial x} - \frac{\partial\Psi^*}{\partial x}\Psi\right)dx$$

$$= -\frac{j\hbar}{m}\int_{-\infty}^{+\infty}\Psi^*\frac{\partial\Psi}{\partial x}\,dx$$

$$= \frac{\langle p\rangle}{m}$$ Linear Momentum

$$\frac{d\langle p\rangle}{dt} = -j\hbar\int_{-\infty}^{+\infty}\left(\frac{\partial\Psi^*}{\partial t}\frac{\partial\Psi}{\partial x} + \Psi^*\frac{\partial}{\partial x}\frac{\partial\Psi}{\partial t}\right)dx$$

$$= -\left\langle\frac{\partial V}{\partial x}\right\rangle$$ Newton's Second Law

$$\frac{d\langle Q\rangle}{dt} = \left\langle\frac{\partial Q}{\partial t}\right\rangle + \frac{j}{\hbar}\langle[H, Q]\rangle$$ Generalized Operator Equation

### Free Particle in a Three-Dimensional Box

$$\frac{\hat{p}^2}{2m}\psi(x, y, z) = E\psi(x, y, z)$$ Schrödinger's Steady-State Equation

$$\psi(x, y, z) = \psi_x(x)\psi_y(y)\psi_z(z)$$ Separation of Variables

$$\frac{1}{\psi_i}\frac{d^2\psi_i}{d_i^2} = -k_i^2,\ i = x, y, z$$ Eigenvalue Equations

$$\psi_i(i) = \sqrt{\frac{2}{L}}\sin k_i i,\ i = x, y, z$$ Normalized Eigenfunctions

$$k_i = \frac{n_i\pi}{L},\ i = x, y, z$$ Wave Vector Eigenvalues

$$E_{n_x n_y n_z} = \frac{\hbar^2 k^2}{2m} = \frac{\hbar^2}{2m}\sum_{i=x}^{z} k_i^2$$

$$= n^2\frac{\pi^2\hbar^2}{2mL^2}$$ Energy Eigenvalues

$$n^2 = n_x^2 + n_y^2 + n_z^2$$ Quantum Number

### Free Electron Gas in Three Dimensions

$$\psi_n(\mathrm{r}) = \psi_{n_x}(x)\psi_{n_y}(y)\psi_{n_z}(z)$$

$$= \sqrt{\frac{1}{V}}\,e^{j\mathbf{k}\cdot\mathbf{r}}$$ Normalized Traveling Waves

$$\psi_i(i + L) = \psi_i(i),\quad i = x, y, z$$ Periodic Boundary Conditions

$$\mathbf{k} = (k_{n_x}, k_{n_y}, k_{n_z}) = \frac{2\pi}{L}(n_x, n_y, n_z)$$ Wave Vector Quantization

$$N = V_k \frac{\Delta N}{\Delta k}$$

$$N = \frac{V}{3\pi^2} k^3$$ $N$-Particle Quantum States

$$D(E) = \frac{V}{2\pi^2}\left(\frac{2m}{\hbar^2}\right)^{3/2} E^{1/2}$$ Density of Electronic States

## Problems

**10.1** Starting with Equation 10.17, show that $\psi(x)$ in Equation 10.16 is the Fourier transform of $\phi(p)$.

*Solution:*
Multiplying the free particle eigenfunction in momentum-space, given in Equation 10.17 as

$$\phi(p) = \sqrt{\frac{1}{2\pi\hbar}} \int_{-\infty}^{+\infty} \psi(x) e^{-\frac{j}{\hbar}px} dx,$$

by $(1/2\pi\hbar)^{1/2} e^{(j/\hbar)px'} dp$ and integrating yields

$$\sqrt{\frac{1}{2\pi\hbar}} \int_{-\infty}^{+\infty} \phi(p) e^{\frac{j}{\hbar}px'} dp = \left(\frac{1}{2\pi\hbar}\right) \iint_{-\infty}^{+\infty} \psi(x) e^{\frac{j}{\hbar}(x'-x)} dx\,dp$$

$$= \int_{-\infty}^{+\infty} \psi(x) \left[\frac{1}{2\pi\hbar} \int_{-\infty}^{+\infty} e^{\frac{j}{\hbar}(x'-x)} dp\right] dx$$

$$= \int_{-\infty}^{+\infty} \psi(x) \delta(x' - x)\, dx$$

$$= \int_{-\infty}^{+\infty} \psi(x) \delta(x - x')\, dx$$

$$= \psi(x') \int_{-\infty}^{+\infty} \delta(x - x')\, dx$$

$$= \psi(x'),$$

where the properties of the Dirac delta function have been used. This result is identical to Equation 10.16, except for the *prime* associated with the position variable.

**10.2** Verify Equation 10.37 by using the one-dimensional free particle wave function of Equation 10.1 and the properties of the Dirac delta function.

*Answer:*

$$\int_{-\infty}^{+\infty} \Psi^*(x, t)\Psi(x, t)\, dx = \int_{-\infty}^{+\infty} \Phi^*(p, t)\Phi(p, t)\, dp$$

**10.3** Verify Equation 10.52 by using the free particle eigenfunction $\psi(x)$ given in Equation 10.16.

*Solution:*

With the free particle eigenfunction and its complex conjugate in position-space given by

$$\psi(x) = \sqrt{\frac{1}{2\pi\hbar}} \int_{-\infty}^{+\infty} \phi(p) e^{\frac{j}{\hbar}px}\, dp,$$

$$\psi^*(x) = \sqrt{\frac{1}{2\pi\hbar}} \int_{-\infty}^{+\infty} \phi^*(p') e^{-\frac{j}{\hbar}p'x}\, dp',$$

substituted into Equation 10.29, we obtain

$$\langle x \rangle = \int_{-\infty}^{+\infty} \Psi^*(x, 0)\, x\, \Psi(x, 0)\, dx$$

$$= \int_{-\infty}^{+\infty} \psi^*(x)\, x\, \psi(x)\, dx$$

$$= \frac{1}{2\pi\hbar} \iiint_{-\infty}^{+\infty} \phi^*(p') e^{-\frac{j}{\hbar}p'x}\, x\, \phi(p) e^{\frac{j}{\hbar}px}\, dp'\, dp\, dx,$$

where $Q \equiv x$ and $t \equiv 0$ in Equation 10.29. Since

$$\underbrace{\int_{-\infty}^{+\infty} \frac{d}{dp}\left(\phi(p) e^{\frac{j}{\hbar}px}\right) dp}_{0} = \int_{-\infty}^{+\infty} \frac{j}{\hbar}\, x\phi(p) e^{\frac{j}{\hbar}px}\, dp + \int_{-\infty}^{+\infty} e^{\frac{j}{\hbar}px} \frac{d}{dp}\phi(p)\, dp,$$

we obtain

$$x\phi(p) e^{\frac{j}{\hbar}px} = e^{\frac{j}{\hbar}px}\left(j\hbar \frac{d}{dp}\right)\phi(p).$$

Now, substitution of this result into the equation above for $\langle x \rangle$ yields

$$\langle x \rangle = \iint_{-\infty}^{+\infty} \phi^*(p')\left[\frac{1}{2\pi\hbar}\int_{-\infty}^{+\infty} e^{\frac{j}{\hbar}(p-p')x}\, dx\right]\left(j\hbar \frac{d}{dp}\right)\phi(p)\, dp'\, dp$$

$$= \iint_{-\infty}^{+\infty} \phi^*(p')\delta(p - p')\left(j\hbar \frac{d}{dp}\right)\phi(p)\, dp'\, dp$$

$$= \int_{-\infty}^{+\infty} \phi^*(p) \left[ \int_{-\infty}^{+\infty} \delta(p' - p)\, dp' \right] \left( j\hbar \frac{d}{dp} \right) \phi(p)\, dp$$

$$= \int_{-\infty}^{+\infty} \phi^*(p) \left( j\hbar \frac{d}{dp} \right) \phi(p)\, dp$$

$$= \int_{-\infty}^{+\infty} \Phi^*(p, 0) \left( j\hbar \frac{\partial}{\partial p} \right) \Phi(p, 0)\, dp.$$

**10.4** Find $\langle p \rangle$ for the free particle in a one-dimensional box, by using the normalized position-space eigenfunction given in Equation 9.85.

*Answer:* $\langle p \rangle = 0$

**10.5** Repeat Problem 10.4 for $\langle x \rangle$, using a table of indefinite integrals.

*Solution:*
With the normalized free particle eigenfunction

$$\psi^*(x) = \psi(x) = \sqrt{\frac{2}{L}} \sin \frac{n\pi x}{L}$$

for a free particle in a one-dimensional box, the expectation value of $x$ is simply

$$\langle x \rangle = \int_0^L \psi^* x \psi\, dx$$

$$= \frac{2}{L} \int_0^L x \sin^2 \frac{n\pi x}{L}\, dx$$

$$= \frac{2L}{n^2\pi^2} \int_0^{n\pi} u \sin^2 u\, du$$

$$= \frac{2L}{n^2\pi^2} \left[ \frac{u^2}{4} - \frac{u}{4} \sin 2u - \frac{1}{8} \cos 2u \right]_0^{n\pi}$$

$$= \frac{2L}{n^2\pi^2} \frac{n^2\pi^2}{4}$$

$$= \tfrac{1}{2}L,$$

where a change in the integration variable of

$$u = \frac{n\pi x}{L} \rightarrow du = \frac{n\pi}{L}\, dx$$

was performed in the 3rd equality.

**10.6** Using a table of indefinite integrals, repeat Problem 10.4 for $\langle x^2 \rangle$.

*Answer:* $\langle x^2 \rangle = L^2/3 - L^2/2n^2\pi^2$

**10.7** Starting with Schrödinger's one dimensional time-independent equation show that $\langle p^2 \rangle = 2m\langle [E - V(x)] \rangle$ in general. Using this result, find $\langle p^2 \rangle$ for the free particle in a one dimensional box.

*Solution:*
Schrödinger's one dimensional time-independent equation (Equation 9.29),

$$-\frac{\hbar^2}{2m}\frac{d^2\psi}{dx^2} + V(x)\psi(x) = E\psi(x),$$

can be rewritten in the form

$$\frac{1}{2m}\left(-j\hbar\frac{d}{dx}\right)\left(-j\hbar\frac{d}{dx}\right)\psi(x) = [E - V(x)]\psi(x),$$

which from Equation 10.46 can be expressed as

$$\frac{1}{2m}\hat{p}_x^2\psi(x) = [E - V(x)]\psi(x).$$

Now, multiplying this equation by $\psi^*(x)$ and integrating over the range of $x$ gives

$$\frac{1}{2m}\int_{-\infty}^{+\infty}\psi^*(x)\hat{p}_x^2\,\psi(x)dx = \int_{-\infty}^{+\infty}\psi^*(x)\,[E - V(x)]\psi(x)dx,$$

which defines the expectation value equation

$$\langle \hat{p}_x^2 \rangle = 2m\langle [E - V(x)] \rangle$$

or more simply

$$\langle p^2 \rangle = 2m\langle [E - V(x)] \rangle.$$

For a free particle in a one-dimensional box $V(x) \equiv 0$, so we have

$$\begin{aligned}\langle p^2 \rangle &= 2m\langle E \rangle \\ &= 2m\left\langle \frac{n^2\pi^2\hbar^2}{2mL^2} \right\rangle \\ &= n^2\frac{\pi^2\hbar^2}{L^2},\end{aligned}$$

where Equation 9.80 was used for the allowed energy eigenvalues corresponding to the values $n = 1, 2, 3, \cdots$ for the principal quantum number.

**10.8** Verify the result of Problem 10.7 by finding $\langle p^2 \rangle$ for the free particle in a one-dimensional box, using the normalized position-space eigenfunction given by Equation 9.85.

*Answer:* $\langle p^2 \rangle = n^2\pi^2\hbar^2/L^2$

**10.9** Find the *standard deviations* $\sigma_x \equiv [\langle x^2 \rangle - \langle x \rangle^2]^{1/2}$ and $\sigma_p \equiv [\langle p^2 \rangle$

$- \langle p\rangle^2]^{1/2}$ and their product $\sigma_x\sigma_p$ for the free particle in a one-dimensional box in the *ground state*.

*Solution:*

From Problems 10.5 and 10.6 we have

$$\langle x\rangle = \frac{L}{2},$$

$$\langle x^2\rangle = \frac{L^2}{3} - \frac{L^2}{2n^2\pi^2},$$

so $\sigma_x$ is simply

$$\begin{aligned}\sigma_x &\equiv [\langle x^2\rangle - \langle x\rangle^2]^{1/2} \\ &= \left(\frac{L^2}{3} - \frac{L^2}{2n^2\pi^2} - \frac{L^2}{4}\right)^{1/2} \\ &= \left(\frac{L^2}{12} - \frac{L^2}{2n^2\pi^2}\right)^{1/2} \\ &= L\left(\frac{1}{12} - \frac{1}{2n^2\pi^2}\right)^{1/2} \\ &= 0.181L.\end{aligned}$$

Likewise, from Problems 10.4 and 10.7 (or 10.8) we have

$$\langle p\rangle = 0,$$

$$\langle p^2\rangle = \frac{n^2\pi^2\hbar^2}{L^2},$$

thus $\sigma_p$ is

$$\begin{aligned}\sigma_p &\equiv [\langle p^2\rangle - \langle p\rangle^2]^{1/2} \\ &= \left(\frac{n^2\pi^2\hbar^2}{L^2} - 0^2\right)^{1/2} \\ &= \frac{n\pi\hbar}{L} \\ &= \frac{\pi\hbar}{L}\end{aligned}$$

for the ground state $n = 1$. The product of these standard deviations for position and momentum is

$$\sigma_x\sigma_p = (0.181L)\left(\frac{\pi\hbar}{L}\right) = 0.568\hbar.$$

**10.10** Find $\sigma_x$, $\sigma_p$, and $\sigma_x\sigma_p$ (defined in Problem 10.9) for the ground state of the linear harmonic oscillator.

*Answer:* $\sigma_x = (\hbar/2m\omega)^{1/2}$, $\sigma_p = \hbar(m\omega/2\hbar)^{1/2}$, $\sigma_x\sigma_p = \frac{1}{2}\hbar$

**10.11** Verify that the free particle in a box momentum eigenfunctions given in Equations 10.65a and 10.65b have associated momentum eigenvalues given by Equation 10.62.

*Solution:*
With the momentum eigenfunctions given in Equations 10.65a and 10.65b,

$$\psi_n^{\pm} = \frac{1}{2j}\sqrt{\frac{2}{L}}\, e^{\pm jn\pi x/L},$$

substituted into the momentum eigenvalue equation (Equation 10.61), we obtain

$$\begin{aligned} p_n^{\pm} &= -j\hbar \frac{d}{dx}\psi_n^{\pm} \\ &= (-j\hbar)\left(\pm\frac{jn\pi}{L}\right)\psi_n^{\pm} \\ &= \pm\frac{n\pi\hbar}{L}. \end{aligned}$$

**10.12** Find the expectation value of $x^2$ in the momentum representation for the linear harmonic oscillator, where the normalized ground state eigenfunction $\phi(p)$ is given by Equation 10.68.

*Answer:* $\langle x^2\rangle = 1/2d^2$

**10.13** Consider the linear harmonic oscillator and show exactly how Schrödinger's time-independent equation can be expressed in the form $d^2\psi/dx^2 - (m\omega/\hbar)^2x^2\psi = (-2mE/\hbar^2)\psi$.

*Solution:*
The potential energy for the linear harmonic oscillator is given by $V = \frac{1}{2}kx^2$. Since $V_{\max} = T_{\max} \rightarrow \frac{1}{2}kx^2 = \frac{1}{2}mv^2 = \frac{1}{2}m(x\omega)^2 = \frac{1}{2}mx^2\omega^2 \rightarrow k = m\omega^2$, then Schrödinger's equation (Equation 9.29)

$$-\frac{\hbar^2}{2m}\frac{d^2\psi}{dx^2} + V(x)\psi(x) = E\psi(x)$$

can be expressed as

$$-\frac{\hbar^2}{2m}\frac{d^2\psi}{dx^2} + \frac{1}{2}m\omega^2x^2\psi(x) = E\psi(x)$$

and rearranged in the form

$$\frac{d^2\psi}{dx^2} - \left(\frac{m\omega}{\hbar}\right)^2 x^2\psi(x) = -\frac{2mE}{\hbar^2}\psi(x).$$

**10.14** Consider the linear harmonic oscillator and its normalized ground state eigenfunction $\psi(x) = (d/\pi^{1/2})^{1/2}e^{-d^2x^2/2}$. Using the result of Problem 10.13, find the equivalence of $d^2$ in terms of $\omega$ and find the energy eigenvalue $E$.

*Answer:* $d^2 = m\omega/\hbar,\ E = \frac{1}{2}h\nu$

**10.15** By considering the free particle wave function $\Psi(x, t)$ representing a plane wave traveling in the positive $x$-direction (Equation 9.7), show how the momentum and energy operators in position-space are obtained.

*Solution:*
Taking a partial derivative of

$$\Psi(x, t) = Ae^{\frac{j}{\hbar}(px - Et)}$$

with respect to the position variable gives

$$\frac{\partial\Psi}{\partial x} = \frac{j}{\hbar}p\Psi \rightarrow p\Psi = \left(-j\hbar\frac{\partial}{\partial x}\right)\Psi,$$

while a first order partial with respect to time yields

$$\frac{\partial\Psi}{\partial t} = -\frac{j}{\hbar}E\Psi \rightarrow E\Psi = \left(j\hbar\frac{\partial}{\partial t}\right)\Psi.$$

**10.16** Find the position and energy operators in momentum-space, by considering the free particle wave function $\Phi(p, t)$ representing a plane wave traveling in the negative $x$-direction.

*Answer:* $x = j\hbar\frac{\partial}{\partial p},\ E = -j\hbar\frac{\partial}{\partial t}$

**10.17** Consider a one-dimensional free particle with $V = V(x) \equiv 0$ and show that the *Hamiltonian operator* is *Hermitian* in accordance with Equation 10.109.

*Solution:*
With the one-dimensional free particle Hamiltonian given by

$$\hat{H} = \frac{\hat{p}^2}{2m},$$

then the expectation value of $\hat{H}$ is

$$\begin{aligned}\langle \hat{H} \rangle &= \int_{-\infty}^{+\infty} \Psi^* \hat{H} \Psi \, dx \\ &= \frac{1}{2m} \int_{-\infty}^{+\infty} \Psi^* \hat{p}^2 \Psi \, dx \\ &= \frac{1}{2m} \int_{-\infty}^{+\infty} (\hat{p}^* \Psi^*) \hat{p} \Psi \, dx \\ &= \frac{1}{2m} \int_{-\infty}^{+\infty} (\hat{p}^* \hat{p}^* \Psi^*) \Psi \, dx \\ &= \int_{-\infty}^{+\infty} (\hat{H}^* \Psi^*) \Psi \, dx,\end{aligned}$$

where the first and last equality on the right-hand-side demonstrate the Hermitian property of $H$ in accordance with Equation 10.109. Clearly, the Hermitian property of $p$ was utilized in the intervening equalities.

**10.18** If the operators $\hat{g}$ and $\hat{q}$ are Hermitian, show that their linear combination $\hat{h} = a\hat{g} + b\hat{q}$ is also Hermitian, where $a$ and $b$ are constants.

*Answer:* $\int_{-\infty}^{+\infty} \Psi^* \hat{h} \Psi \, dx = \int_{-\infty}^{+\infty} (\hat{h}^* \Psi^*) \, \Psi \, dx$

**10.19** Show that the commutator $[\partial/\partial x, V]$ is equivalent to $\partial V/\partial x$, where $V = V(x, t)$ is the potential energy.

*Solution:*
This commutator identity is easily obtained by considering the product of the commutator and a function $F = F(x, t)$. That is,

$$\begin{aligned}\left[\frac{\partial}{\partial x}, V\right] F &= \left(\frac{\partial}{\partial x} V - V \frac{\partial}{\partial x}\right) F \\ &= \frac{\partial}{\partial x} (VF) - V \frac{\partial F}{\partial x} \\ &= \frac{\partial V}{\partial x} F + V \frac{\partial F}{\partial x} - V \frac{\partial F}{\partial x} \\ &= \frac{\partial V}{\partial x} F.\end{aligned}$$

Cancelling the common factor $F$ from the left-hand-side of the first equality and the right-hand-side of the last equality yields

$$\left[\frac{\partial}{\partial x}, V\right] = \frac{\partial V}{\partial x}.$$

**10.20** Verify Equation 10.110 by starting with Equation 10.119 for $Q = p$ and using the result of the last problem.

*Answer:* $\dfrac{d\langle p\rangle}{dt} = -\left\langle\dfrac{\partial V}{\partial x}\right\rangle$

**10.21** Show that the commutator $[\partial^2/\partial x^2, x)$ is equivalent to $2(\partial/\partial x)$, and that this result can be expressed in terms of the momentum operator as $[p^2, x] = -2j\hbar p$.

*Solution:*

Multiplying the commutator by a function $F = F(x, t)$ yields

$$\left[\frac{\partial^2}{\partial x^2}, x\right] F = \frac{\partial^2}{\partial x^2}(xF) - x\frac{\partial^2 F}{\partial x^2}$$

$$= \frac{\partial}{\partial x}\left(F + x\frac{\partial F}{\partial x}\right) - x\frac{\partial^2 F}{\partial x^2}$$

$$= \frac{\partial F}{\partial x} + \frac{\partial F}{\partial x} + x\frac{\partial^2 F}{\partial x^2} - x\frac{\partial^2 F}{\partial x^2}$$

$$= 2\frac{\partial F}{\partial x},$$

from which we obtain

$$\left[\frac{\partial^2}{\partial x^2}, x\right] = 2\frac{\partial}{\partial x}$$

by cancelling the common factor $F$ in the last equality. Now, multiplying both sides of this equation by $(-j\hbar)^2$ immediately yields

$$\left[(-j\hbar)^2\frac{\partial^2}{\partial x^2}, x\right] = -2j\hbar\left(-j\hbar\frac{\partial}{\partial x}\right),$$

which in view of the momentum operator equivalence gives

$$[p^2, x] = -2j\hbar p.$$

**10.22** By considering a free particle for which $V = V(x) = 0$ and a time derivative of Equation 10.119 for $Q = x$, verify the correspondence given in Equation 10.115 using *operator algebra*.

*Answer:* $\dfrac{d^2\langle x\rangle}{dt^2} = \dfrac{1}{m}\dfrac{d\langle p\rangle}{dt}$

**10.23** By considering a free particle for which $V = V(x) = 0$, verify the correspondence $d\langle x\rangle/dt = \langle p\rangle/m$ by using Equation 10.119 and the result of Problem 10.21.

*Solution:*
Allowing $Q = x$ in Equation 10.119,

$$\frac{d\langle Q\rangle}{dt} = \left\langle\frac{\partial Q}{\partial t}\right\rangle + \frac{j}{\hbar}\langle[H, Q]\rangle,$$

we obtain

$$\begin{aligned}
\frac{d\langle x\rangle}{dt} &= \cancelto{0}{\left\langle\frac{\partial x}{\partial t}\right\rangle} + \frac{j}{\hbar}\langle[H, x]\rangle \\
&= \frac{j}{\hbar}\left\langle\left[\left(\frac{p^2}{2m} + V\right), x\right]\right\rangle \\
&= \frac{j}{\hbar}\left\langle\left[\frac{p^2}{2m}, x\right]\right\rangle + \frac{j}{\hbar}\cancelto{0}{\langle[V, x]\rangle} \\
&= \frac{j}{\hbar}\frac{1}{2m}\langle[p^2, x]\rangle \\
&= \frac{j}{\hbar}\frac{1}{2m}(-2j\hbar)\langle p\rangle \\
&= \frac{\langle p\rangle}{m},
\end{aligned}$$

where the equality $[p^2, x] = -2j\hbar p$ of Problem 10.21 was substituted into the next-to-the-last equality.

**10.24** Consider an electron to be trapped in a three dimensional box of the size of a typical atomic diameter. Find the energy released when the electron makes a transition from the first excited state to the ground state.

*Answer:* $E_{112} - E_{111} = 112$ eV

**10.25** Consider an electron to be trapped in a three dimensional box of the size of a typical atom ($L = 1.0 \times 10^{-10}$ m). If the electron is in the ground state, calculate the *total probability* of finding the electron between $x = y = z = 0$ and $x = y = z = 0.50 \times 10^{-10}$ m.

*Solution:*
The normalized eigenfunction for the electron in terms of the quantum numbers $n_x$, $n_y$, and $n_z$ is given by combining Equations 10.135 and 10.136 to obtain

$$\psi_{n_x n_y n_z} = \sqrt{\frac{8}{L^3}}\sin\frac{n_x\pi x}{L}\sin\frac{n_y\pi y}{L}\sin\frac{n_z\pi z}{L}.$$

Using $n_x = n_y = n_z = 1$ for the ground state, we have

$$\psi_{111} = \sqrt{\frac{8}{L^3}} \sin\frac{\pi x}{L} \sin\frac{\pi y}{L} \sin\frac{\pi z}{L}$$

for the ground state eigenfunction. Now, substitution of $\psi_{111}$ into the three dimensional counterpart to Equation 9.43b,

$$P = \iiint_{-\infty}^{+\infty} \psi^*_{n_x n_y n_z} \psi_{n_x n_y n_z}\, dx\, dy\, dz,$$

gives the total probability as a triple integral over all of position-space, where each integral is of the same form. For example, the contribution to the probability by the x-component of the eigenfunction is just

$$\begin{aligned}
\int_{x_1}^{x_2} \sin^2\frac{\pi x}{L}\, dx &= \frac{L}{\pi}\int_{x_1}^{x_2} \sin^2 u\, du \\
&= \frac{L}{\pi}\left[\frac{u}{2} - \frac{1}{4}\sin 2u\right]_{x_1}^{x_2} \\
&= \frac{L}{\pi}\left[\frac{\pi x}{2L} - \frac{1}{4}\sin\frac{2\pi x}{L}\right]_{x_1}^{x_2} \\
&= \frac{L}{\pi}\left(\frac{\pi(0.50 \times 10^{-10}\text{ m})}{2(1.0 \times 10^{-10}\text{ m})} - 0\right) \\
&= 0.25L.
\end{aligned}$$

Identical results are obtained for the $y$ and $z$-components of the eigenfunction, so

$$P = \frac{8}{L^3}(0.25L)^3 = 0.125 = 12.5\%.$$

**10.26** Repeat Problem 10.25 for an electron in the third excited quantum state.

*Answer:* $P = 1.00 = 100\%$

CHAPTER 11

# Classical Statistical Mechanics

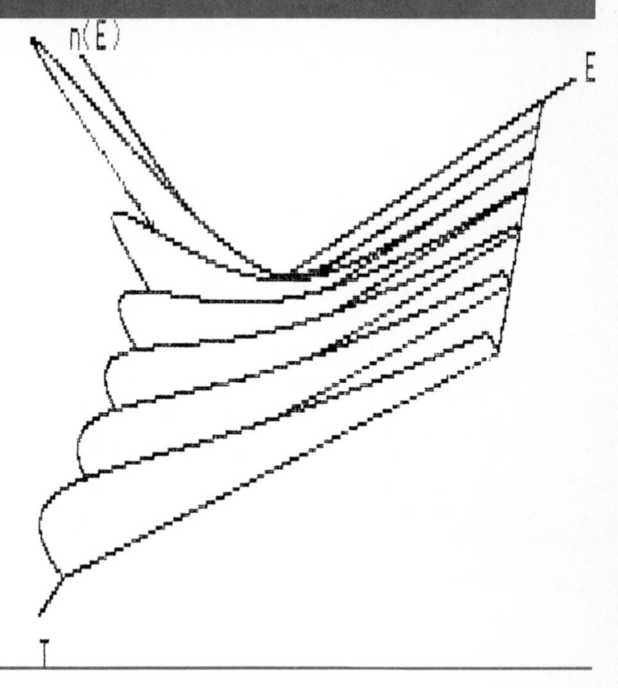


## Introduction

Statistical mechanics attempts to relate the *macroscopic* properties ($U$, $S$, $F$, $p$, etc.) of a many-bodied system to the *microscopic* properties of the system's particles. Because of its general formulation, statistical mechanics is equally applicable to problems of classical mechanics (e.g., molecules in a gas) and quantum mechanics (e.g., free electrons in a metal or semiconductor), where the classical equations of motion and Schrödinger's equation can *not* be solved exactly for a system consisting of a large number of particles. The *particles* of a statistical system can be representative of molecules, electrons, photons, or even wave functions. Statistical mechanics takes advantage of the fact that for a *statistical system* (one consisting of a very large number of particles) the *most probable* or, equivalently, the *ensemble average* properties of the system can be determined, even in the absence of any knowledge concerning the motions and interactions of the individual particles. Both equilibrium and nonequilibrium systems can be addressed by statistical mechanics, as illustrated in Figure 11.1. We will be totally concerned, however, with microscopic equilibrium systems and their associated macroscopic properties. In particular, statistical mechanics (classical in this chapter and quantum in the next) will be applied to microscopic variables of *position* and *momentum* to obtain (a) a *distribution* function for the *microscopic parameters* (e.g., molecular speeds and energies of particles in a gas) and (b) *macroscopic parameters* characterizing the system.

We begin with a discussion of a phase space description for the state of an isolated $N$ particle system at an instant in time, which culminates with a definition of the microcanonical ensemble. A simple example of five particles confined to a two cell phase space is then used to introduce the counting procedure of Maxwell-Boltzmann statistics and to distinguish macrostates from microstates. The counting procedure leads to the description of the *Maxwell-Boltzmann* (sometimes abbreviated as M-B) and *classical* thermodynamic probabilities for a system of $N$ identical particles that are considered to be *distinguishable* in M-B statistics and *indistinguishable* in classical statistics. From these results the definition of ensemble averaging becomes apparent and the concept of the most probable macrostate is introduced. The fundamental relation between thermodynamic probability of statistics and entropy of thermodynamics is then derived by considering two isolated systems in equilibrium. Next, the most probable distri-

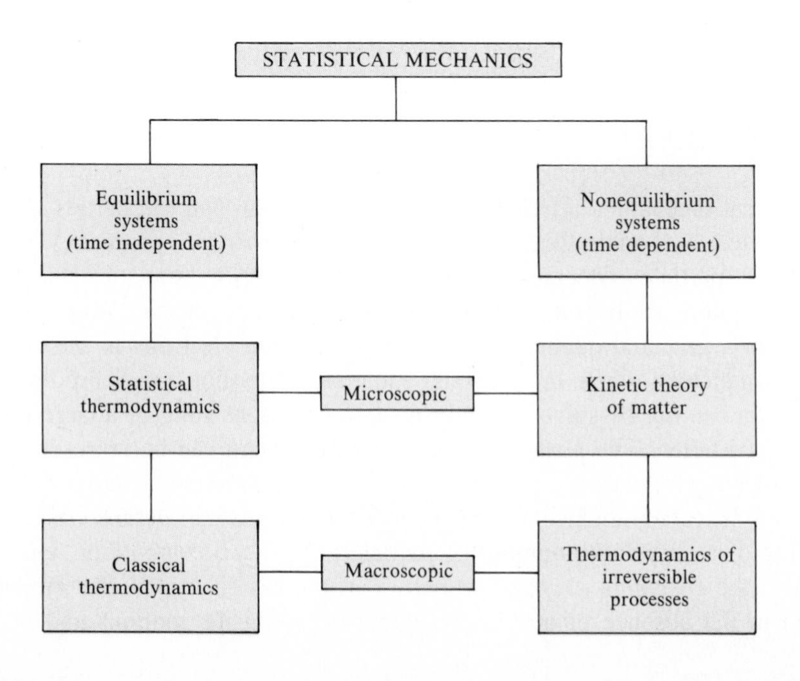


Figure 11.1
Theoretical subject areas of statistical mechanics.

bution for M-B statistics is derived by maximizing the M-B thermodynamic probability, using Stirling's approximation to evaluate logarithmic factorials and introducing the Lagrange multipliers $\alpha$ and $\beta$ to incorporate conservation of particles and energy requirements. The elimination of the undetermined multiplier $\alpha$, by imposing conservation of particles on the Maxwell-Boltzmann distribution, results in the well-known Boltzmann distribution and the definition of the partition function. Because the Boltzmann distribution and the partition function are both dependent on the Lagrange multiplier $\beta$, we turn our attention to a qualitative and quantitative identification of $\beta$. The fundamental significance of the partition function $Z$ in Maxwell-Boltzmann and classical statistics is then discussed, with the appropriate basic relations for total energy $E$, average energy $\bar{\epsilon}$, occupation number $n_j$, pressure $p$, entropy $S$, and the Helmholtz function $F$ being developed. Maxwell-Boltzmann statistics is then applied to the molecules of an ideal gas. After evaluating the degeneracy and partition function for a continuum of energy states, the average energy of an ideal gas molecule is determined, along with expressions for the pressure and entropy of the ideal gas. These considerations are followed by a derivation of distribution formulae for molecular momentum, energy, and speed. The Boltzmann distribution formulae are then used in fundamental applications to obtain expressions for average energy, average speed, root-mean-square speed, and the most probable speed of an ideal gas molecule. Our discussion of classical statistical mechanics concludes with the equipartition of energy principle being applied to the determination of the molal specific heat for a monatomic, diatomic, and polyatomic ideal gas.

## 11.1 Phase Space and the Microcanonical Ensemble

In classical mechanics the state of an $N$ particle system can be completely defined at a particular instant in time by enumerating the position $\mathbf{r}_i$ and momentum $\mathbf{p}_i$ of every particle in the system. Since we must know three position $(x_i, y_i, z_i)$ and three momentum $(p_{x_i}, p_{y_i}, p_{z_i})$ coordinates for each particle, the complete specification of the state of the system requires knowledge of $6N$ variables at an instant in time. It is convenient to consider a *six-dimensional* space, where each particle could be represented by a *point* having six coordinates $x$, $y$, $z$, $p_x$, $p_y$, and $p_z$. With this geometrical description, the state of a system of particles (usually molecules) is represented by a certain distribution of $N$ points in a six-dimensional **phase space** called **$\mu$-space** to suggest *molecular* space. Alternatively, a *system* of $N$ particles could be represented by *one point* in a $6N$-dimensional *phase space* called **$\Gamma$-space** to suggest *gas* space, where there would be dimensions for the six position and momentum coordinates and for the $N$ molecules. We will restrict the statistical description of a system to a representation in *μ-space* and frequently refer to it as simply *phase space*.

Consider partitioning μ-space into very small six-dimensional *cells* having sides of length $dx$, $dy$, $dz$, $dp_x$, $dp_y$, $dp_z$. The *volume of each cell*, as defined by

Cell Volume

$$\tau \equiv dx\,dy\,dz\,dp_x\,dp_y\,dp_z, \tag{11.1}$$

is very small compared to the spatial dimensions and range of momenta of the real system, but large enough such that *each cell* can contain a number of particles or representative *phase points*. Because of the *Heisenberg uncertainty principle* discussed in Chapter 8, Section 8.7, we have

$$dx\,dp_x = dy\,dp_y = dz\,dp_z = h. \tag{11.2}$$

Thus, quantum mechanically the position and momentum coordinates of a particle are restricted such that the representative phase point exists somewhere within an elemental volume of $h^3$ in μ-space. Consequently, the *minimum volume of a cell* in μ-space is given by

Minimum Volume Per Cell

$$\tau_0 = h^3. \tag{11.3}$$

Of course, the *cell volume* defined by Equation 11.1 is completely arbitrary, subject only to the restrictions $\tau_0 \leq \tau \ll V_\mu$, where $V_\mu$ represents the actual volume of the system in μ-space.

By numbering each cell in μ-space as $1, 2, 3, \cdots, i, \cdots$, *the number of phase points (particles) in each corresponding cell*, called the **occupation number,** can be denoted as $n_1, n_2, n_3, \cdots, n_i, \cdots$. With this definition of the *occupation number* it is apparent that the total number of particles in the system is given by

Conservation of Particles

$$N = \sum_i n_i, \tag{11.4}$$

where the summation extends over all cells in μ-space. Further, it should be noted that all particles in any one cell have exactly the same energy $\epsilon_1$ to within the limits of $\tau$. Consequently, the total energy of the particles in the $i$th cell is $n_i\,\epsilon_i$, and the total energy of the system is clearly

Conservation of Energy

$$E = \sum_i n_i\,\epsilon_i. \tag{11.5}$$

Frequently, the *thermodynamic internal energy* $U$ of a system is substituted for the total energy $E$ in this equation, which is completely valid for systems wherein the potential energy can be defined as zero.

Since an **assembly** is taken to be a *number of identical entities*, which may be particles or identical systems of particles, our system is an *assembly* of $N$ particles. If we constructed an assembly of replicas of a system, we would have an *assembly of assemblies*, which is commonly called an **ensemble** *of systems*. The conceptualization of an *ensemble* arises from the realization that initially a system of $N$ particles would have macroscopic properties that slowly change with the passing of time until equilibrium is established. After J. Willard Gibbs (recall Chapter 10, Section 10.2), we consider this intellectual construction of an *ensemble of systems* to simulate and represent at *one instant in time* the properties of the actual system, as would develop in the course of time. An *ensemble* is also considered to be suitably *randomized*, such that every configuration of position and momentum coordinates that are accessible to the actual system in the course of time is represented by one or more systems in the ensemble at one instant in time. Thus, Gibbs' scheme allows us to replace *time averages* over a single system by *ensemble averages* at a fixed time. The **Ergodic hypothesis** of statistical mechanics postulates the equivalence of *time averages* and *ensemble averages*, but such an equivalence has not been proven in general. It can be argued, however, that an *ensemble average* is more representative of the actual system than a *time average*, since we never know the initial conditions of a real system and, thus, cannot know exactly how to take a time average.

If our system of interest is in an external conservative force field (e.g., gravitational, electric, or magnetic), then from classical physics we know that the total energy of the system is a constant. An ensemble of such systems could be constructed such that the energy of every system is the same and independent of time. Such an ensemble is known as the *microcanonical ensemble*, and is appropriate for the discussion of an *isolated* system, since the system's total energy is necessarily a constant in time. Essentially, we can consider a system of a **microcanonical ensemble** as one wherein *particles are sufficiently independent of one another, such that an energy $\epsilon_i$ can be assigned to each, but sufficiently interactive with one another to establish thermodynamic equilibrium.*

## 11.2 System Configurations and Complexions: An Example

The **configuration** of a system is specified by enumerating the energies $(\epsilon_1, \epsilon_2, \epsilon_3, \cdots, \epsilon_i, \cdots)$ of all possible regions of equal phase volume in $\mu$-space and specifying the number of phase points $(n_1, n_2, n_3, \cdots, n_i, \cdots)$

in each cell. It should be emphasized that the energy associated with a cell in $\mu$-space is a constant of time for systems of a microcanonical ensemble. This results from the fact that such a system is composed of $N$ identical and essentially free particles, where the energy of a particle is dependent on only its momentum. That is, for a particle in the $i$th cell of $\mu$-space, its energy is given by

$$\epsilon_i = \frac{p_i^2}{2m}. \tag{11.6}$$

Thus, the system *configuration* consists of cells of equal extension (i.e., Equation 11.1) and constant energy (i.e., Equation 11.6), wherein particles (represented by *phase points*) are considered to be located. Certainly, the number of particles in a particular cell (the *occupation number*) will in general change with the passing of time, but at a particular instant in time this number associated with each cell in $\mu$-space is fixed. The specification of the *occupation numbers* $n_1, n_2, n_3, \cdots, n_i, \cdots$ associated with each cell in $\mu$-space at an instant in time is said to define a **macrostate** of the system. Thus, the *configuration* of a system at a particular instant, as illustrated in Table 11.1, represents *one* possible *macrostate* of the system. From this discussion, it should be clear that the *macroscopic* properties of our system will depend *only* on the *occupation numbers* of our system's *configuration*.

We are now confronted with the important problem of determining the number of different ways $n_1$, $n_2$, and so on particles can be selected from $N$ identical particles and placed in cells 1, 2, and so on for a specific macrostate. The total number of ways of making this selection is referred to as the number of **complexions** or **microstates** of the *macrostate*, which are allowed by the *nature* of the real system being considered. The counting procedure utilized in obtaining the number of allowed *complexions* for a system *configuration* will be illustrated by an example that should clarify the distinction between *macrostates* and *microstates* (complexions) and allow for generalizations. In particular, consider a system of five particles being represented in a two-cell $\mu$-space. As illustrated in Table 11.2, the number of possible *macrostates* of the system is six because *particle conservation* (Equation 11.4) allows for only six sets of values for the cell occupation numbers $n_1$ and $n_2$.

TABLE 11.1
A system configuration defining one particular macrostate.

| Cell Number | 1 | 2 | 3 | $\cdots$ | $i$ | $\cdots$ |
|---|---|---|---|---|---|---|
| Cell Energy | $\epsilon_1$ | $\epsilon_2$ | $\epsilon_3$ | $\cdots$ | $\epsilon_i$ | $\cdots$ |
| Occupation Number | $n_1$ | $n_2$ | $n_3$ | $\cdots$ | $n_i$ | $\cdots$ |

| $n_1$ | 5 | 4 | 3 | 2 | 1 | 0 |
|---|---|---|---|---|---|---|
| $n_2$ | 0 | 1 | 2 | 3 | 4 | 5 |

TABLE 11.2
The macrostates allowed for a five particle system in a 2-cell $\mu$-space.

The number of *complexions* or *microstates* associated with a particular *macrostate* is obtainable by enumerating the possible arrangements (or *permutations*) of the particles in the macrostate, exclusive of the combinations (*irrelevant permutations*) that merely interchange the particles within each phase cell. Before enumerating the *microstates* associated with each *macrostate* given in Table 11.2, we must realize that in **Maxwell-Boltzmann statistics** particles are considered to be *identical yet distinguishable*. That is, in principle it must be possible to label each particle as $a$, $b$, $c$, and so on, such that a microstate in which particle $a$ is in one cell and $b$ in another is regarded as distinct and different from the microstate in which they are reversed. Thus, for the macrostate corresponding to $n_1 = 2$ and $n_2 = 3$ of Table 11.2, the allowed microstates, exclusive of the irrelevant ones, are enumerated in Table 11.3. Clearly, the number of microstates in this case is ten for the $n_1 = 2$ and $n_2 = 3$ macrostate. The microstate *ba* in cell 1 and *cde* in cell 2 is equivalent to the first one tabulated in Table 11.3 and corresponds to an *irrelevant permutation* of the particles in cell 1. In defining the number of microstates, parentheses were used to indicate that it was equivalent to the *number of permutations* of the particles in the macrostate. The number of permutations of five particles is defined by $5! \equiv 1{\cdot}2{\cdot}3{\cdot}4{\cdot}5 = 120$, where 5!, read as "5 factorial," is an abbreviation for the product of the integers 1 through 5. This result of 120 is not equivalent to the number of microstates, however, because the *irrelevant permutations* within each cell have not been discarded. These irrelevant permutations correspond to $n_1! = 2! = 2$ for cell 1 and $n_2! = 3! = 6$ for cell 2. Thus, the total number of permutations (120) must be divided by the product of those that only permute particles

| Cell 1 | *ab* | *ac* | *ad* | *ae* | *bc* | *bd* | *be* | *cd* | *ce* | *de* |
|---|---|---|---|---|---|---|---|---|---|---|
| Cell 2 | *cde* | *bde* | *bce* | *bcd* | *ade* | *ace* | *acd* | *abe* | *abd* | *abc* |

TABLE 11.3
The number of microstates for the macrostate $n_1 = 2$, $n_2 = 3$ of Table 11.2, excluding the irrelevant permutations of particles in a cell.

within the individual cells, which gives the total number of microstates as $5!/2!3! = 120/2 \cdot 6 = 10$.

The result obtained above by considering *permutations* is certainly in agreement with the counting result of Table 11.3. This suggests a more generalized expression for the determination of the number of *microstates* (*complexions*) associated with a particular system *configuration* for the $k$th *macrostate* of

$$W_k = \frac{N!}{n_1!n_2!} \tag{11.7}$$

for the example considered. This result should be rather obvious from the above discussion, or we can consider the counting in a more general way for a system of $N$ particles in a two-cell $\mu$-space. For the first cell, the $n_1$ particles can be selected in

$$\begin{aligned} w_1 &= N(N-1)(N-2)\cdots\frac{(N-n_1+1)}{n_1!} \\ &= \frac{N!}{(N-n_1)!n_1!} \end{aligned} \tag{11.8}$$

ways, where the $n_1!$ in the denominator refers to the irrelevant permutations within the cell. Similarly, the $n_2$ particles for the second cell can be selected from the $N - n_1$ particles remaining in

$$\begin{aligned} w_2 &= (N-n_1)(N-n_1-1)(N-n_1-2)\cdots\frac{(N-n_1-n_2+1)}{n_2!} \\ &= \frac{(N-n_1)!}{(N-n_1-n_2)!n_2!} \end{aligned} \tag{11.9}$$

ways. Consequently, the total number of microstates corresponding to a particular $n_1$, $n_2$ macrostate is simply the product of $w_1$ and $w_2$. That is,

$$W_k = w_1 \cdot w_2 \tag{11.10}$$

$$= \frac{N!}{(N-n_1-n_2)!n_1!n_2!}$$

$$= \frac{N!}{n_1!n_2!} \tag{11.7}$$

represents the number of *microstates* (*complexions*) for a particular, say $k$th, *macrostate*. The result is equivalent to Equation 11.7, since from *particle conservation* $N - n_1 - n_2 = 0$ and $0! \equiv 1$.

Using the result of Equation 11.7, it is now easy to calculate the number of microstates corrresponding to each macrostate for the example illustrated in Table 11.2:

$$n_1 = 5, \qquad n_2 = 0 \rightarrow W = \frac{5!}{5!0!} = 1,$$

$$n_1 = 4, \qquad n_2 = 1 \rightarrow W = \frac{5!}{4!1!} = 5,$$

$$n_1 = 3, \qquad n_2 = 2 \rightarrow W = \frac{5!}{3!2!} = 10,$$

$$n_1 = 2, \qquad n_2 = 3 \rightarrow W = \frac{5!}{2!3!} = 10,$$

$$n_1 = 1, \qquad n_2 = 4 \rightarrow W = \frac{5!}{1!4!} = 5,$$

$$n_1 = 0, \qquad n_2 = 5 \rightarrow W = \frac{5!}{0!5!} = 1.$$

Altogether, there are 32 different microstates corresponding to the 6 different macrostates. In statistical mechanics the **principle of equal a priori probability** assumes that *each microstate occurs with equal probability*. From this fundamental postulate and our example, it is clear that each macrostate is not equally probable, since the number of microstates corresponding to different macrostates are in general different. We can, however, define the *probability of occurrence* of a particular macrostate as the ratio of its corresponding number of microstates to the total number of microstates. This means that the first and sixth macrostates in our example will be observed 1/32 of the time, the second and fifth will each occur 5/32 of the time, and the third and fourth will each be observed most frequently for 5/16 of the time. As a result of this interpretation of $W_k$, it is often referred to as the **thermodynamic probability** of the $k$th macrostate, which will be more completely discussed in the next section.

## 11.3 Thermodynamic Probability

Although Equation 11.7 was developed for an $N = 5$ particle system of *distinguishable* particles distributed among two cells, its generalization to a $\mu$-space configuration of many cells should clearly be

Total Microstates for a Macrostate

$$W_k = \equiv \Pi_i w_i, \tag{11.11}$$

where the symbol $\Pi_i$ defines a *product* of all terms that follow. Generalizing the $w_i'$s of Equations 11.8 and 11.9 and substituting into Equation 11.11 gives

$$W_k = \frac{N!}{\Pi_i n_i!} \tag{11.12}$$

for the number of microstates of the $k$th macrostate for $N$ identical but distinguishable particles in a $\mu$-space partitioned into $i$ cells. It should be understood that the *distinguishability* of $N$ identical particles is a feature of *classical* systems that has no validity in *quantum statistical mechanics*. Equation 11.12 represents the *thermodynamic probability* of a classical system for the configuration represented in Table 11.1.

The *thermodynamic probability* of Maxwell-Boltzmann statistics (Equation 11.12) is not completely general as it does not allow particles in different cells to possess the same energy. Such a *degeneracy* in a number of cells corresponding to the same energy must be allowed classically. This is easily understood by realizing that the energy of each cell in $\mu$-space is given by Equation 11.6, which is quadratic in the momentum coordinates associated with each cell. Thus, particles traveling with the same *speeds* in different directions would have *different momentum vectors* but *identical energies* because Equation 11.6 gives $\epsilon_i \propto \mathbf{p}_i \cdot \mathbf{p}_i$. For example, considering the two-dimensional momentum space illustrated by Figure 11.2, particles in the circular shell of radius $p = (p_x^2 + p_y^2)^{1/2}$ and thickness $dp$ possess momentum between $p$ and $p + dp$. Consequently, the phase points representing such particles would be in different cells of identical energy in $\mu$-space.

To understand how *cell-energy degeneracy* can be taken into account, imagine $N$ identical but distinguishable particles of which $n_1$ particles are to be placed into one or the other of two cells having the same cell-energy $\epsilon$. There are $N$ ways the first particle can be selected and put in the first cell and $N$ ways the same first particle can be put in the second cell. Because it cannot be placed in both cells simultaneously, the two events are *mutually exclusive*. Thus, the total number of ways of selecting the first particle and placing it in *either* cell is $2N$. For the second particle the number of ways is $2(N - 1)$, for the third particle $2(N - 2)$, and so forth for the $n_1$ particles. Clearly, from this discussion we have

$$\begin{aligned} w_1 &= (2N)[2(N-1)]\,[2(N-2)] \cdots \frac{[2(N-n_1+1)]}{n_1!} \\ &= N(N-1)\,(N-2) \cdots \frac{(N-n_1+1)2^{n_1}}{n_1!} \\ &= \frac{N!2^{n_1}}{(N-n_1)!n_1!}, \end{aligned} \tag{11.13a}$$

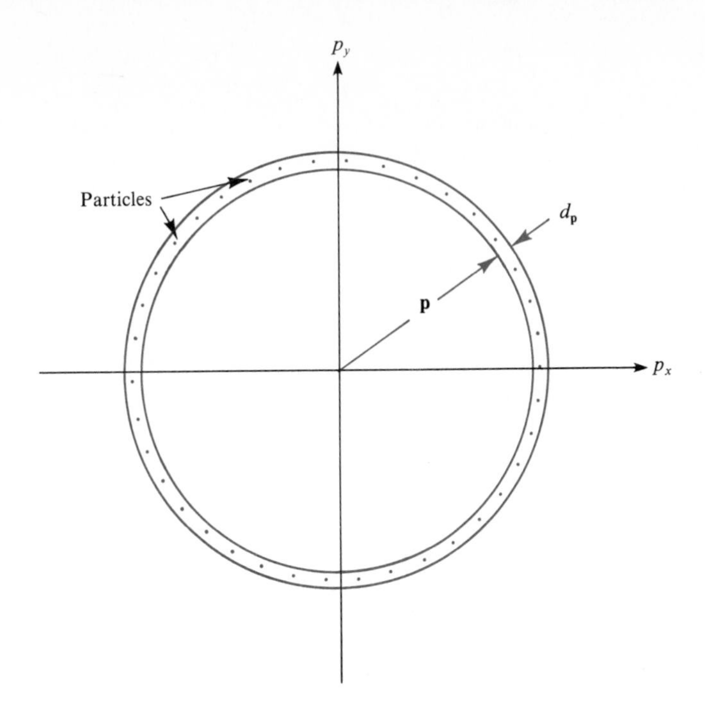


Figure 11.2
Particles in **p**-space having essentially the same energy.

which should be compared with Equation 11.8. If now there are $n_2$ particles to be distributed among three cells of identical energy $\epsilon_2$, then the number of ways would be

$$w_2 = (N - n_1)(N - n_1 - 1) \cdots \frac{(N - n_1 - n_2 + 1)3^{n_2}}{n_2!}$$

$$= \frac{(N - n_1)!3^{n_2}}{(N - n_1 - n_2)!n_2!}, \qquad (11.13b)$$

which is similar to Equation 11.9 except for the $3^{n_2}$ factor.

From the above discussion, it should be straight forward to generalize to a system having the configuration illustrated in Table 11.4 for $i$ cells in $\mu$-space, where the *cell-energy degeneracy* is denoted as $g_1, g_2, g_3, \cdots, g_i$. For this configuration of a system our interpretation of the occupation

TABLE 11.4
A system configuration for $i$ cells in $\mu$-space, including cell-energy degeneracy.

| Cell Number | 1 | 2 | 3 | ⋯ | $i$ |
|---|---|---|---|---|---|
| Cell Energy | $\epsilon_1$ | $\epsilon_2$ | $\epsilon_3$ | ⋯ | $\epsilon_i$ |
| Degeneracy | $g_1$ | $g_2$ | $g_3$ | ⋯ | $g_i$ |
| Occupation Number | $n_1$ | $n_2$ | $n_3$ | ⋯ | $n_i$ |

number is somewhat modified. For the $i$th cell of degeneracy $g_i$, the *occupation number* represents the *number of particles* distributed among the $g_i$ *cells* of *identical energy* $\epsilon_i$. We can think of the $g_i$ cells as different particle *energy states* at the same *energy level*, so a *macrostate* of the assembly is now defined by specifying the *number of particles* $n_i$ *in each energy level*. This definition of a *macrostate* for degenerate cells is equally valid for *distinguishable* or *indistinguishable* particles. If the particles are *distinguishable*, which is our present consideration for M-B statistics, a *microstate* of the assembly still corresponds to the specification of the particular *energy state* (i.e., the particular cell of the $g_i$ cells) of *each* particle. Thus, for the configuration of Table 11.4 a generalization of Equations 11.13a and 11.13b gives

$$w_1 = \frac{N!g_1^{n_1}}{(N - n_1)!n_1!} \tag{11.14a}$$

$$w_2 = \frac{(N - n_1)!g_2^{n_2}}{(N - n_1 - n_2)!n_2!} \tag{11.14b}$$

$$\vdots$$

$$w_i = \frac{(N - n_1 - \cdots - n_{i-1})!g_i^{n_i}}{(N - n_1 - \cdots - n_i)!n_i!}, \tag{11.14c}$$

which upon substitution into Equation 11.11 yields

Maxwell-Boltzmann Thermodynamic Probability

$$W_{\text{M-B}} = W_k = \Pi_i w_i = \frac{N!\Pi_i g_i^{n_i}}{\Pi_i n_i!}. \tag{11.15}$$

This result is known as the **Maxwell-Boltzmann thermodynamic probability** for the $k$th macrostate of a system of $N$ identical but *distinguishable* particles distributed among cells having an energy level degeneracy.

From the above discussion of M-B statistics for *distinguishable* particles, it is now relatively easy to count the number of microstates corresponding to a particular macrostate for identical and *indistinguishable* particles. In this case we do not need to specify the *energy state* of *each particle*, since the particles are *indistinguishable*. Instead, a *microstate* of the assembly corresponds to the specification of the *total number of particles* in *each energy level*. For the configuration of Table 11.4, the $i$th energy level contains $g_i$ cells of identical energy $\epsilon_i$. For the $n_i$ *indistinguishable* particles distributed among the $g_i$ cells, the first particle may be placed in any one of the $g_i$ cells. The second particle may also be placed in any one of the $g_i$ cells, since there is no limitation to the number of particles per cell. Similarly, there are $g_i$ ways of placing the third particle, the fourth particle, and so forth for each of the $n_i$ particles. Thus, the total number of possible distributions for the $n_i$ particles in the $g_i$ cells of energy level $\epsilon_i$ is simply

$$w_i = \frac{g_i^{n_i}}{n_i!},$$

where the $n_i!$ in the denominator takes into account the *irrelevant permutations* of the $n_i$ particles. Counting every possible distribution of particles in every *energy level* gives the so-called **classical thermodynamic probability**

Classical Thermodynamic Probability

$$W_C = W_k = \Pi_i w_i = \frac{\Pi_i g_i^{n_i}}{\Pi_i n_i!} \tag{11.16}$$

for the *k*th macrostate of a system of identical but *indistinguishable* particles.

## Ensemble Averaging

The *principle of equal a priori probability* of statistical mechanics has been interpreted to mean that every possible microstate of an isolated assembly is equally probable. The *thermodynamic probability* of the *k*th *macrostate*, denoted by $W_k$ and given by Equations 11.15 or 11.16 for classical statistical mechanics, represents the number of equally probable microstates corresponding to a *particular* macrostate. Thus, for a system of $N$ particles the *total number of equally probable microstates* $\Omega$ is defined by

Total Microstates

$$\Omega \equiv \sum_k W_k, \tag{11.17}$$

where the summation is over *all possible macrostates*. Frequently and for obvious reasons, $\Omega$ is referred to as the *thermodynamic probability of the assembly*. An alternative and sometimes more useful relation for the *total number of microstates* for M-B statistics is given by

Total M-B Microstates

$$\Omega_{\text{M-B}} = \left(\sum_i g_i\right)^N. \tag{11.18}$$

This relation can be argued for a system of $N$ distinguishable particles, since for the configuration defined in Table 11.4 we should in principle be able to specify which particles have energy $\epsilon_1$, which have energy $\epsilon_2$, and so forth for all $N$ particles. The total number of distinct cells in which a given particle can exist with energy $\epsilon_i$ is given by $\sum_i g_i$. Because each particle can exist in any one of the $\sum_i g_i$ possible cells, the total number of ways the $N$ particles can be distributed among the various cells is given by Equation 11.18. Hence, for the example of Section 11.2, where $N$ = 5 particles were distributed in a nondegenerate two-cell $\mu$-space, $g_1 = g_2 = 1$ and the total number of possible microstates is $\Omega_{\text{M-B}} = (1 + 1)^5 = 2^5 = 32$ in agreement with our previous result.

As defined previously, the *microcanonical ensemble* consists of a very large number of replicas of a given assembly of free particles having a constant total energy, where all *equally probable microstates* of the assembly are represented by one or more replicas in the *ensemble* at one instant in time. Thus, an *ensemble average* of any physical variable $A_{ik}$, giving its average *distribution* of values in the $i$-cells of the $k$th macrostate, can be obtained by multiplying $A_{ik}$ by the number of replicas in the $k$th macrostate, summing over all macrostates, and dividing by the total number of possible microstates. That is, the **ensemble average** of $A_{ik}$ is defined by

Ensemble Average

$$\bar{A}_i \equiv \frac{\sum_k A_{ik} W_k}{\Omega}, \qquad (11.19)$$

where the sum extends over all possible macrostates for which the conservation requirements of Equations 11.4 and 11.5 are valid. This definition for the *ensemble average* of a physical variable is completely general and valid for both *classical* and *quantum* statistical mechanics, and it should be compared with the quantum theory *expectation value* discussed in Chapter 10, Section 10.2.

The actual averaging of Equation 11.19 may be performed by several different general methods, including the Burns-Brown-Becker method (*Becker Averaging Technique for Obtaining Distribution Functions in Statistical Mechanics*, M. L. Burns, R. A. Brown, *Amer. J. Phys.* 39, no. 7 [1971]: 802–805). If the physical observable of interest is the occupation number (i.e., $A_i = n_i$), then Equation 11.19 would yield a *distribution* function for the *average* values of the occupation numbers. That is, a *distribution* function $\bar{n}_i = f(\epsilon_i)$ could be derived (see reference cited), which gives the average number of particles in the $i$th degenerate cell having an energy $\epsilon_i$. Allowing the $i$ subscript to take on all values for the cells in $\mu$-space, such a relation would describe the *average macrostate* for a system in thermodynamic equilibrium. For pedagogic reasons we derive in

Section 11.4 the *most probable* distribution $n_i = f(\epsilon_i)$, where the method detailed considers the *most probable* $W_k$ to be so much more probable than any other that all other macrostates can be ignored. This fundamental assumption of the *most probable distribution* means that the *average* value of any physical variable tends toward its *most probable* value. That is, from Equations 11.17 and 11.19 we have

$$\bar{A}_i \approx \frac{(A_i)_{\text{MP}}(W_k)_{\text{MP}}}{(W_k)_{\text{MP}}} = (A_i)_{\text{MP}} \equiv A_i, \tag{11.20}$$

where the second subscript MP denotes *most probable*. The assumption of this method is completely valid in the limit as the number of particles of the system goes to infinity. Indeed, we can recognize the merit of this result by applying Equation 11.19 to the example discussed in Section 11.2. For the case of $N = 5$ particles distributed between two nondegenerate cells, we have from Equation 11.19 for $i = 1$

$$\begin{aligned}\bar{n}_1 &= \frac{\sum_k n_{1k} W_k}{\Omega} \\ &= \frac{(5 \cdot 1 + 4 \cdot 5 + 3 \cdot 10 + 2 \cdot 10 + 1 \cdot 5 + 0 \cdot 1)}{32} \\ &= \frac{880}{32} = 2.5,\end{aligned}$$

where Table 11.2 and Equation 11.7 have been used. A similar calculation for $i = 2$ gives $\bar{n}_2 = 2.5$, such that the *average macrostate* is defined by $W(\bar{n}_1, \bar{n}_2) = W(2.5, 2.5)$. In Section 11.2 we found the *most probable macrostates* for this example to be $W(n_1, n_2) = W(3, 2) = W(2, 3)$, which makes us suspect that $\bar{n}_i$ trends toward $n_i$ for large values of $N$. Certainly, the results of this example demonstrate that the *average* of the two most probable macrostates is identical to $W(\bar{n}_1, \bar{n}_2)$.

## Entropy and Thermodynamic Probability

Before leaving our discussion of thermodynamic probabililty, we can capitalize on the concept of the *most probable distribution*. The assumption of the *most probable macrostate* can be interpreted as that state in which the system is most lilkely to exist. It is the macrostate toward which an isolated system would trend in attaining thermodynamic equilibrium and, consequently, that state with the maximum number of microstates. However, in classical thermodynamics the equilibrium state of an isolated system corresponds to the state of *maximum entropy*. Thus, we should expect some correlation between entropy $S$ and thermodynamic probability $W$, since they both have their maximum values in the equilibrium state of an isolated system. In this context $W$ without the $k$ subscript represents the *most probable* macrostate for *any kind of statistics*, including Maxwell-Boltzmann, classical, or the Bose-Einstein and Fermi-Dirac thermodynamic probabilities of Chapter 12.

Consider two isolated systems that are brought together and allowed to exchange energy but not particles through a *diathermic* partition, as illustrated in Figure 11.3. Since the two systems are allowed to exchange energy, thermodynamic equilibrium will be established after a sufficient contact time. In equilibrium the total entropy of the two systems is given by

$$S_T = S_1 + S_2, \tag{11.21}$$

while the total thermodynamic probability of the *most probable* macrostate is given by

$$W_T = W_1 W_2. \tag{11.22}$$

From the above discussion, it seems reasonable to assume entropy to be a function of thermodynamic probability. That is, we assume $S_T = S_T(W_T)$ and $S_i = S_i(W_i)$ for $i = 1$ or 2, which allows Equation 11.21 to be expressed as

$$S_T(W_T) = S_T(W_1, W_2) = S_1(W_1) + S_2(W_2). \tag{11.23}$$

This assumed functional dependence of entropy allows the *total derivative* of Equation 11.21,

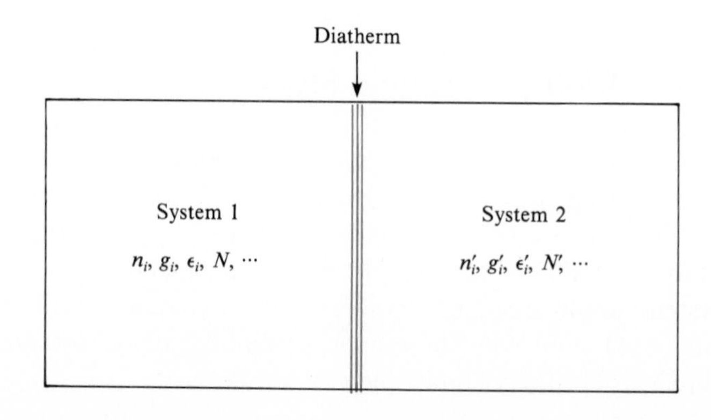


Figure 11.3
Two systems in thermodynamic equilibrium, separated by a diathermic partition.

$$dS_T = dS_1 + dS_2, \tag{11.24}$$

to be expressed as

$$\frac{\partial S_T}{\partial W_1} dW_1 + \frac{\partial S_T}{\partial W_2} dW_2 = \frac{dS_1}{dW_1} dW_1 + \frac{dS_2}{dW_2} dW_2.$$

For this equation to be valid, the coefficients of $dW_1$ must be equal and likewise for the coefficients of $dW_2$:

$$\frac{\partial S_T}{\partial W_1} = \frac{dS_1}{dW_1}, \tag{11.25a}$$

$$\frac{\partial S_T}{\partial W_2} = \frac{dS_2}{dW_2}. \tag{11.25b}$$

By employing the *chain rule* of differential calculus, the left-hand side of these equations can be rewritten as

$$\begin{aligned} \frac{\partial S_T}{\partial W_1} &= \frac{dS_T}{dW_T} \frac{\partial W_T}{\partial W_1} \\ &= W_2 \frac{dS_T}{dW_T}, \end{aligned} \tag{11.26a}$$

$$\begin{aligned} \frac{\partial S_T}{\partial W_2} &= \frac{dS_T}{dW_T} \frac{\partial W_T}{\partial W_2} \\ &= W_1 \frac{dS_T}{dW_T}, \end{aligned} \tag{11.26b}$$

where Equation 11.22 has been used in evaluating the partial derivatives of $W_T$. Thus, from Equations 11.25 (a – b) and 11.26 (a – b) we obtain the equations

$$W_2 \frac{dS_T}{dW_T} = \frac{dS_1}{dW_1}, \tag{11.27a}$$

$$W_1 \frac{dS_T}{dW_T} = \frac{dS_2}{dW_2}, \tag{11.27b}$$

which can be combined as

$$W_1 \frac{dS_1}{dW_1} = W_2 \frac{dS_2}{dW_2} \tag{11.28}$$

by multiplication of Equation 11.27a by $W_1$ and Equation 11.27b by $W_2$. Since $W_1$ and $W_2$ are independent, this equation is valid only if each side is equal to the same constant, say $k_B$. Thus, we have the results

$$W_1 \frac{dS_1}{dW_1} = k_B,$$
$$W_2 \frac{dS_2}{dW_2} = k_B,$$

which can be rearranged as

$$dS_1 = k_B \frac{dW_1}{W_1},$$
$$dS_2 = k_B \frac{dW_2}{W_2},$$

and integrated to obtain

$$S_1 = k_B \ln W_1, \tag{11.29a}$$
$$S_2 = k_B \ln W_2. \tag{11.29b}$$

These results suggest that

Entropy in Statistical Mechanics

$$S = k_B \ln W \tag{11.30}$$

for an isolated system in thermodynamic equilibrium, where $W$ represents the *most probable* or *maximum* macrostate.

The relation given by Equation 11.30 provides the fundamental connection between *classical thermodynamics* and *statistical mechanics*. Instead of being developed in terms of the thermodynamic probability of the *most probable macrostate* $W$, it could have been expressed in terms of the thermodynamic probability of the *assembly* $\Omega$, that is,

Entropy in Statistical Mechanics

$$S = k_B \ln \Omega, \tag{11.31}$$

by the same general arguments. Although the constant $k_B$ was arbitrarily chosen and undefined in our derivation, it must be selected such that thermodynamic and statistical values of a system's entropy are in agreement. Later in Section 11.7, $k_B$ will be shown to be the well-known Boltzmann constant. Further, Equations 11.30 and 11.31 are completely valid in both classical and quantum statistical mechanics.

## 11.4 Most Probable Distribution

The primary objective of this section is to derive an equation of the form $n_i = f(\epsilon_i)$ for the *most probable* distribution by determining the particular $W_k$ of Equation 11.15 that has the *largest* or *maximum* value. Since the *M-B thermodynamic probability* of *N distinguishable* particles (Equation 11.15) has products of cell-energy degeneracies and occupation numbers, it is simpler to work with $\ln W_{\text{M-B}}$ rather than $W_{\text{M-B}}$. This approach is completely justified since if $W_{\text{M-B}}$ is a *maximum*, then so is the natural logarithm of $W_{\text{M-B}}$. Hence, we consider

$$\begin{aligned} \ln W_{\text{M-B}} &= \ln N! + \sum_i \ln g_i^{n_i} - \sum_i \ln n_i! \\ &= \ln N! + \sum_i n_i \ln g_i - \sum_i \ln n_i!, \end{aligned} \tag{11.32}$$

where the product $\Pi$ has been replaced by the sum $\Sigma$, due to the properties of the logarithm, and Equation 11.15 has been used. The logarithm of the factorial of any number, say $A$, can be handled most easily by **Stirling's formula**

$$\ln A! \approx A \ln A - A, \qquad A >> 1. \tag{11.33}$$

Stirling's Formula

This formula is easily verified, since for $N = 60$ we have $\ln 60! \approx \ln (8.321 \times 10^{81}) \approx 188.6$ and $60 \ln 60 - 60 \approx 185.6$, which results in an error of only roughly 1.6 percent. The error is completely negligible in practical problems, where the number of particles is typically on the order of magnitude of Avogadro's number. With *Stirling's formula* Equation 11.32 becomes

$$\ln W_{\text{M-B}} = N \ln N - N + \sum_i n_i \ln g_i - \sum_i n_i \ln n_i + \sum_i n_i,$$

which immediately reduces to

$$\ln W_{\text{M-B}} = N \ln N + \sum_i n_i \ln \frac{g_i}{n_i}, \tag{11.34}$$

because the second and fifth terms cancel due to the conservation of particles requirement (Equation 11.4). The *classical* thermodynamic probability of Equation 11.16 can be handled similarly to obtain

$$\ln W_{\text{C}} = N + \sum_i n_i \ln \frac{g_i}{n_i}, \tag{11.35}$$

which is seen to differ from $\ln W_{\text{M-B}}$ in only the first term.

The requirement for $W_k$ or $\ln W_k$ to be the *most probable* is for its value to be unaffected by small changes in any of the occupation numbers. If the occupation numbers were *continuous* instead of *discrete*, this requirement could be expressed as

$$\frac{\partial W_k}{\partial n_i} = 0 \tag{11.36a}$$

or

$$\frac{\partial \ln W_k}{\partial n_i} = 0. \tag{11.36b}$$

For *discrete* occupation numbers, however, a change in $\ln W_k$ corresponding to a very small change in $n_i$ of $\delta n_i$ is denoted by $\delta \ln W_k$, which must be equal to zero for the most probable distribution. Thus, for M-B statistics and Equation 11.34 we have

$$\begin{aligned}\delta \ln W_{\text{M-B}} = N\delta \ln N + \ln N \delta N + \sum_i n_i \delta \ln g_i + \sum_i \ln g_i \, \delta n_i \\ - \sum_i n_i \delta \ln n_i - \sum_i \ln n_i \, \delta n_i = 0.\end{aligned} \tag{11.37}$$

Because $N$ and $g_i$ are constants of the system we have

$$\delta N = \delta g_i = 0, \tag{11.38}$$

and the first, second, third, and fifth terms vanish in Equation 11.37:

$$N\delta \ln N = N \frac{1}{N} \delta N = \delta N = 0, \tag{11.39a}$$

$$\ln N \delta N = 0, \tag{11.39b}$$

$$\sum_i n_i \delta \ln g_i = \sum_i n_i \frac{1}{g_i} \delta g_i = 0, \tag{11.39c}$$

$$\sum_i n_i \delta \ln n_i = \sum_i n_i \frac{1}{n_i} \delta n_i = \delta \sum_i n_i = \delta N = 0. \tag{11.39d}$$

Thus, the only terms remaining in Equation 11.37 are the fourth and sixth, which can be combined by a logarithm property in the form

$$\delta \ln W = \sum_i \ln \frac{g_i}{n_i} \delta n_i = 0. \tag{11.40}$$

Even though this equation has been derived for M-B statistics, the subscript M-B has been omitted from $W$, since an *identical* result is obtained by similar arguments for $\delta \ln W_C$ (see Problem 11.11).

Although Equation 11.40 must be satisfied by the *most probable* distribution, it does not specify such a distribution, since the $\delta n_i$'s are *not independent*. The $\delta n_i$'s must satisfy the conservation of particles and energy requirements

$$\delta N = 0 = \sum_i \delta n_i, \tag{11.41}$$

$$\delta E = 0 = \sum_i \epsilon_i \delta n_i. \tag{11.42}$$

In this last requirement the normal term $\sum_i n_i \delta\epsilon_i$ is omitted, since $\epsilon_i$ is constant for every cell in $\mu$-space and $\delta\epsilon_i = 0$. These two *conservation requirements* can be incorporated into Equation 11.40 by the **Lagrange method of undetermined multipliers.** The method consists of multiplying Equation 11.41 by $-\alpha$ and Equation 11.42 by $-\beta$, with the resulting expressions being added to Equation 11.40 to obtain

$$\sum_i \left( \ln \frac{g_i}{n_i} - \alpha - \beta\epsilon_i \right) \delta n_i = 0. \tag{11.43}$$

The *undetermined multipliers* $\alpha$ and $\beta$ are independent of the occupation numbers and the $\delta n_i$'s of this equation are effectively *independent*. Thus, Equation 11.43 is valid only under the condition that the coefficient of $\delta n_i$ vanishes for each and every value in the sum. That is,

$$\ln \frac{g_i}{n_i} - \alpha - \beta\epsilon_i = 0, \tag{11.44}$$

which by simple mathematics can be expressed as

Maxwell-Boltzmann Distribution

$$n_i = g_i e^{-\alpha} e^{-\beta\epsilon_i}. \tag{11.45}$$

This equation represents the **Maxwell-Boltzmann distribution** for the *most probable occupation number* and is a direct result of *maximizing* either $\ln W_{\text{M-B}}$ (Equation 11.34) or $\ln W_{\text{C}}$ (Equation 11.35).

The evaluation of the *undetermined multipliers* is the next important consideration. An expression involving $\alpha$ is easily obtained from the *conservation of particles* requirement. That is, substitution of $n_i$ from Equation 11.45 into Equation 11.4 yields

$$\begin{aligned} N &= \sum_i n_i \\ &= e^{-\alpha} \sum_i g_i e^{-\beta\epsilon_i}, \end{aligned}$$

which can easily be solved for $e^{-\alpha}$ in the form

$$e^{-\alpha} = \frac{N}{\sum_i g_i e^{-\beta\epsilon_i}}. \tag{11.46}$$

With this expression for $e^{-\alpha}$, Equation 11.45 becomes

Boltzmann Distribution

$$n_i = \frac{N}{Z} g_i e^{-\beta\epsilon_i}, \tag{11.47}$$

where we have made the symbolic definition

Partition Function

$$Z \equiv \sum_i g_i e^{-\beta\epsilon_i}. \tag{11.48}$$

The result given in Equation 11.47 is often referred to as the **Boltzmann distribution.** The *sum over states* in the denominator, represented by the letter $Z$ representing the German word *zustandssumme* is called the **partition function.** The *partition function*, defined in Equation 11.48, is very important in statistical mechanics as will be illustrated in Section 11.6. For now, however, we will accept it as a convenient symbolic simplification of the distribution function and turn our attention to the meaning of the second undetermined multiplier $\beta$.

## 11.5 Identification of β

In the previous section the derivation for the distribution law of classical statistical mechanics was initiated using the method of the most probable distribution and Stirling's formula to evaluate $\delta \ln W_k = 0$. The requirements for the conservation of total energy and total number of particles were accommodated using the Lagrange method of undetermined multipliers, where the parameter $\alpha$ was introduced for the latter and $\beta$ for the former conservation condition. This led to the well-known Maxwell-Boltzmann distribution law given by Equation 11.45, where the most probable occupation number is dependent on both undetermined multipliers $\alpha$ and $\beta$. The elimination of $\alpha$ from the distribution law was easily facilitated by considering conservation of particles, which immediately led to the Boltzmann distribution law (Equation 11.47) and the definition of the partition function (Equation 11.48). Since both of these equations have a dependence on the undetermined multiplier $\beta$, it is imperative that $\beta$ be evaluated. In the discussion that follows a qualitative interpretation of $\beta$ is developed by considering the *zeroth law of thermodynamics* and then a quantitative evaluation is accomplished using a fundamental relation between classical thermodynamics and statistical mechanics (i.e., Equation 11.30).

## β and the Zeroth Law of Thermodynamics

Consider the two systems in thermodynamic equilibrium of Figure 11.3, where the statistical quantities (energy, degeneracy, etc.) are *unprimed* in the first system and *primed* in the second system. Since the systems will exchange energy through the diathermic wall in attaining equilibrium, the conservation of energy requirement becomes

$$E_T = \sum_i n_i \epsilon_i + \sum_j n_j' \epsilon_j', \tag{11.49}$$

but the condition for the conservation of particles is just

$$N = \sum_i n_i \tag{11.50a}$$

and

$$N' = \sum_j n_j'. \tag{11.50b}$$

Now, however, the total number of microstates for a particular macrostate is given by the product

$$W_T = WW', \tag{11.51}$$

where for the sake of argument the M-B thermodynamic probability may be used for $W$ and $W'$, that is

$$W = \frac{N! \Pi_i g_i^{n_i}}{\Pi_i n_i!}, \tag{11.52a}$$

$$W' = \frac{N'! \Pi_j g_j'^{n'_j}}{\Pi_j n_j'!}. \tag{11.52b}$$

$W_T$ can be maximized by the method of the *most probable* distribution by considering

$$\delta \ln W_T = \delta \ln W + \delta \ln W' = 0. \tag{11.53}$$

By analogy with derivational steps leading from Equation 11.32 to Equation 11.40, the condition for the most probable distribution is

$$\delta \ln W_T = \sum_i \ln \frac{g_i}{n_i} \delta n_i + \sum_j \ln \frac{g'_j}{n'_j} \delta n'_j = 0. \tag{11.54}$$

Again, we can use *Lagrange's method of undetermined multipliers* to take into account the conservation requirements. That is, with

$$-\alpha \sum_i \delta n_i = 0, \tag{11.55a}$$

$$-\alpha' \sum_j \delta n'_j = 0 \tag{11.55b}$$

$$-\beta \sum_{i,j} (\epsilon_i \delta n_i + \epsilon'_j \delta n'_j) = 0 \tag{11.55c}$$

added to Equation 11.54, we obtain

$$\sum_i \left( \ln \frac{g_i}{n_i} - \alpha - \beta \epsilon_i \right) \delta n_i + \sum_j \left( \ln \frac{g'_j}{n'_j} - \alpha' - \beta \epsilon'_j \right) \delta n'_j = 0. \tag{11.56}$$

In this equation the $\delta n_i$'s and $\delta n'_j$'s are effectively independent, so their coefficients must be identically zero for all values of $i$ and $j$, respectively. Thus, after a little mathematical manipulation, we obtain

$$n_i = g_i e^{-\alpha} e^{-\beta \epsilon_i}, \tag{11.57a}$$

$$n'_j = g'_j e^{-\alpha'} e^{-\beta \epsilon'_j}. \tag{11.57b}$$

These two relations for the most probable distribution functions have only $\beta$ in *common*, while all other quantities are in general *different* for the two systems. But, the **zeroth law of thermodynamics** states that *temperature is the only property common to systems in thermodynamic equilibrium*. For this reason, $\beta$ is often referred to as the *empirical temperature* of a statistical system. We can not equate $\beta$ with the absolute temperature $T$, however, because our results (Equations 11.57a and 11.57b) and interpretation are still valid if we multiply $T$ by a constant or take its inverse. Fortunately, we are in a position to capitalize on previous results and derive an exact expression for $\beta$, which is the topic of our next discussion.

## Evaluation of $\beta$

Since $\ln W$ is dependent on $n_i$ (Equation 11.34 or Equation 11.35) and $n_i$ is dependent on $\beta$ (Equation 11.47), we propose the identification of $\beta$ by starting with the fundamental relation between thermodynamics and statistical mechanics given by Equation 11.30. This method of identifying $\beta$

will allow the entropy $S$ expressed by Equation 11.30 to be consistent with the entropy predicted by thermodynamics. Realizing that classical thermodynamics relates entropy to other physical properties of a system by *partial derivatives*, we begin by differentiating Equation 11.30 to obtain

$$dS = k_B\, d \ln W. \tag{11.58}$$

In view of Equation 11.40, $d \ln W$ must be

$$d \ln W = \sum_i \ln \frac{g_i}{n_i}\, dn_i,$$

where $g_i/n_i$ is obtainable from the Boltzmann distribution (Equation 11.47) as

$$\frac{g_i}{n_i} = \frac{Z}{N} e^{\beta\epsilon_i}.$$

Substitution of the last two equations into Equation 11.58 yields

$$\begin{aligned} dS &= k_B \sum_i \ln \left( \frac{Z}{N} e^{\beta\epsilon_i} \right) dn_i \\ &= k_B \sum_i (\ln Z - \ln N + \beta\epsilon_i)\, dn_i \\ &= k_B (\ln Z - \ln N) \sum_i dn_i + k_B\beta \sum_i \epsilon_i\, dn_i. \end{aligned}$$

Conservation of particles and energy require

$$dN = \sum_i dn_i,$$

$$dU = \sum_i \epsilon_i\, dn_i,$$

(see discussion of Equations 11.41 and 11.42) from which we obtain

$$dS = k_B \ln \frac{Z}{N}\, dN + k_B\beta\, dU. \tag{11.59}$$

Differential Entropy

Rearranging this equation to the form

$$dU = \frac{1}{k_B\beta}\, dS - \frac{1}{\beta} \ln \frac{Z}{N}\, dN \tag{11.60}$$

allows for a term-by-term comparison with the **combined first and second laws** of thermodynamics,

Combined First and Second Laws

$$dU = TdS - pdV + \mu dN, \tag{11.61}$$

under the assumption of *constant volume* (i.e., $dV = 0$),

$$dU = TdS + \mu dN. \tag{11.62}$$

This assumption of $dV = 0$ is completely valid in general, as a statistical system of an ideal gas can be thought of as isolated with a fixed volume $V$ (see Section 11.7). Hence, equating the coefficients of $dS$ from Equations 11.60 and 11.62 results in the identity of the Lagrange multiplier $\beta$ being

$$\beta = \frac{1}{k_B T}. \tag{11.63}$$

In essence this derivation has demonstrated that the *assumed* statistical relation for entropy, as given by Equation 11.30, yields an identical value for entropy as predicted by classical thermodynamics, *if* $\beta$ is *defined* by Equation 11.63. It needs to be emphasized that Equation 11.59 for $dS$ is valid for both *Maxwell-Boltzmann* and *classical statistics* under the Boltzmann distribution. The entropy $S$, however, is *not* the same in both cases, because $\ln W$ differs for each case (see Section 11.6).

It is also interesting to note that a comparison of the coefficients of $dN$ from Equations 11.60 and 11.62 gives an expression for the **classical chemical potential** $\mu_C$ as

Classical Chemical Potential

$$\mu_C = -k_B T \ln \frac{Z}{N}. \tag{11.64}$$

From this relation we immediately obtain

$$e^{\beta\mu_C} = \frac{N}{Z},$$

which can be substituted into the Boltzmann distribution to obtain the so-called **classical distribution**

Classical Distribution

$$n_i = g_i e^{\beta(\mu_C - \epsilon_i)}. \tag{11.65}$$

It needs to be emphasized that *classical statistics differs somewhat from*

*Maxwell-Boltzmann statistics* even though one can be derived from the other by an appropriate definition of the *partition function*. To be more specific, summing Equation 11.65 over $i$ gives

$$\sum_i n_i = \sum_i g_i e^{-\beta\epsilon_i} e^{\beta\mu_C} = e^{\beta\mu_C} \sum_i g_i e^{-\beta\epsilon_i},$$

which from the equations representing conservation of particles (Equation 11.4) and the partition function (Equation 11.48) becomes

$$N = Ze^{\beta\mu_C}.$$

Thus, substituting $N/Z$ for $e^{\beta\mu_C}$ in Equation 11.65 gives the Boltzmann distribution

$$n_i = \frac{N}{Z} g_i e^{-\beta\epsilon_i}.$$

As a last point, the *chemical potential* given by Equation 11.64 is *only* appropriate for *classical statistics*, where particles are considered to be *indistinguishable*. For a system of *distinguishable* particles obeying Maxwell-Boltzmann statistics, the chemical potential $\mu_{\text{M-B}}$ is given by Equation 11.87.

To illustrate the results of this and the previous section, consider the example described in Section 11.2 of $N = 5$ distinguishable particles distributed among two cells. Assuming $g_1 = g_2$, the most probable occupation number for each cell is given by the Boltzmann distribution (Equation 11.47) as

$$\begin{aligned} n_1 &= \frac{Ng_1 e^{-\beta\epsilon_1}}{g_1 e^{-\beta\epsilon_1} + g_2 e^{-\beta\epsilon_2}} \\ &= \frac{N}{1 + e^{\beta(\epsilon_1 - \epsilon_2)}}, \\ n_2 &= \frac{Ng_2 e^{-\beta\epsilon_2}}{g_1 e^{-\beta\epsilon_1} + g_2 e^{-\beta\epsilon_2}} \\ &= \frac{N}{1 + e^{\beta(\epsilon_2 - \epsilon_1)}}. \end{aligned}$$

Under the condition that any particle will have the same energy in either cell (i.e., $\epsilon_1 = \epsilon_2$), we have the particles being distributed equally among the cells (i.e., $n_1 = n_2 = N/2$). Of course, for our example where $N = 5$

this means that the most probable distribution will occur when $n_1 = 3$ and $n_2 = 2$ or vise versa, since we cannot have $n_1 = n_2 = N/2 = 2.5$ particles in each cell. Further, if we let $\epsilon_2 = 2\epsilon_1$ and $\theta \equiv \epsilon_1/k_B$, then the most probable distribution is given by

$$n_1 = \frac{N}{1 + e^{-\theta/T}},$$

$$n_2 = \frac{N}{1 + e^{+\theta/T}},$$

where the degeneracies $g_1$ and $g_2$ have cancelled because of our simplifying assumption. The quantity $\theta$ in these equations has the dimensions of temperature and is often referred to as the *characteristic temperature*. If $T$ is very small compared with $\theta$, then nearly all of the particles will be found in the first cell, as $n_1 \approx N$ and $n_2 \approx 0$. For $T = \theta$ we have $n_1 = 0.73N$ and $n_2 = 0.27N$, while for $T \gg \theta$ we have $n_1 = n_2 \approx N/2$.

## 11.6 Significance of the Partition Function

The partition function defined by Equation 11.48 is of fundamental importance in Maxwell-Boltzmann and classical statistics, because it can be easily related to the average particle energy, the occupation number, and the thermodynamic properties of a system. We can immediately obtain an expression for $\bar{\epsilon}$ and $n_j$ in terms of $Z$ by considering partial derivatives of $Z$ with respect to $\beta$ and with respect to $\epsilon_j$, respectively. For example, the partial derivative of $Z$,

$$Z \equiv \sum_i g_i e^{-\beta\epsilon_i},$$

with respect to $\beta$ gives

$$\frac{\partial Z}{\partial \beta} = -\sum_i \epsilon_i g_i e^{-\beta\epsilon_i}. \tag{11.66}$$

This result is very similar to that obtained from the conservation of energy equation,

$$E \equiv \sum_i \epsilon_i n_i,$$

after substitution of the Boltzmann distribution,

$$n_i = \frac{N}{Z} g_i e^{-\beta\epsilon_i},$$

except for a multiplicative factor of $N/Z$, that is,

$$\begin{aligned} E &= \sum_i \epsilon_i n_i \\ &= \sum_i \epsilon_i \left( \frac{N}{Z} g_i e^{-\beta\epsilon_i} \right) \\ &= \frac{N}{Z} \sum_i \epsilon_i g_i e^{-\beta\epsilon_i}. \end{aligned} \tag{11.67}$$

Comparing this equation with Equation 11.66 gives

$$E = -N \frac{1}{Z} \frac{\partial Z}{\partial \beta},$$

which can be expressed in a more compact form as

$$E = -N \frac{\partial \ln Z}{\partial \beta} \tag{11.68}$$

with the aid of differential calculus.

Also, from the identification of $\beta$ given by Equation 11.63 we have

$$d\beta = -\frac{1}{k_B T^2} dT, \tag{11.69}$$

and this allows the total energy $E$ (Equation 11.68) to be expressed in terms of the absolute temperature $T$ as

$$E = N k_B T^2 \frac{\partial \ln Z}{\partial T}. \tag{11.70}$$

Furthermore, since the average energy per particle in a system can be defined by the ratio of the total energy $E$ to the total number of particles $N$, we have

$$\bar{\epsilon} \equiv \frac{E}{N} = -\frac{\partial \ln Z}{\partial \beta} = k_B T^2 \frac{\partial \ln Z}{\partial T}, \tag{11.71}$$

where Equations 11.68 and 11.70 have been utilized in obtaining the last two equalities, respectively. Since the Boltzmann distribution was used in this derivation, the results for the total energy $E$ and the average energy $\bar{\epsilon}$ are valid for both Maxwell-Boltzmann and classical statistics (see Problem 11.12).

The relationship between the occupation number and the partition function is easily derived by considering the partial of $Z$ with respect to $\epsilon_j$,

$$\frac{\partial Z}{\partial \epsilon_j} = \frac{\partial}{\partial \epsilon_j} \sum_i g_i e^{-\beta\epsilon_i}.$$

In this consideration the partial derivative is zero for all terms in the summation, *except* for the *j*th *term*. Thus,

$$\frac{\partial Z}{\partial \epsilon_j} = -\beta g_j e^{-\beta\epsilon_j},$$

and multiplication by $-N/\beta Z$ yields

$$-\frac{N}{\beta Z}\frac{\partial Z}{\partial \epsilon_j} = \frac{N}{Z} g_j e^{-\beta\epsilon_j}.$$

A comparison of the right-hand side of this equation with the Boltzmann distribution (Equation 11.47) gives

$$n_j = -\frac{N}{\beta Z}\frac{\partial Z}{\partial \epsilon_j}, \tag{11.72}$$

which can clearly be expressed as

$$n_j = -\frac{N}{\beta}\frac{\partial \ln Z}{\partial \epsilon_j} = -Nk_B T\,\frac{\partial \ln Z}{\partial \epsilon_j}. \tag{11.73}$$

Again, this result is perfectly valid for both Maxwell-Boltzmann and classical statistics (see Problem 11.13).

Thermodynamic properties are also easily related to the partition function by use of Equation 11.30,

$$S = k_B \ln W,$$

where $W$ represents any thermodynamic probability $W_{M\text{-}B}$, $W_C$, and so on. Accordingly, for Maxwell-Boltzmann statistics and Equation 11.34,

$$\ln W_{\text{M-B}} = N \ln N - \sum_i n_i \ln \frac{n_i}{g_i},$$

we obtain

$$S_{\text{M-B}} = k_B \left( N \ln N - \sum_i n_i \ln \frac{n_i}{g_i} \right).$$

Because of the Boltzmann distribution, $n_i/g_i$ can be replaced by $(N/Z)e^{-\beta\epsilon_i}$ resulting in

$$\begin{aligned} S_{\text{M-B}} &= k_B \left[N \ln N - \sum_i n_i (\ln N - \ln Z - \beta\epsilon_i)\right] \\ &= k_B \left(N \ln N - \sum_i n_i \ln N + \sum_i n_i \ln Z + \beta \sum_i \epsilon_i n_i\right) \\ &= k_B (N \ln N - N \ln N + N \ln Z + \beta E) \\ &= k_B \beta E + N k_B \ln Z, \end{aligned} \tag{11.74}$$

where the *conservation requirements* (Equations 11.4 and 11.5) have been used in obtaining the third equality. Normally, this equation is expressed as

$$S_{\text{M-B}} = \frac{U}{T} + N k_B \ln Z, \tag{11.75}$$

M-B Entropy

where the *total particle energy E* has been replaced by the *total internal energy U* and Equation 11.63 has been used for $\beta$. Also, by combining Equations 11.74 and 11.70, $S_{\text{M-B}}$ can be expressed in terms of the partition function $Z$ as

$$S_{\text{M-B}} = N k_B T \frac{\partial \ln Z}{\partial T} + N k_B \ln Z. \tag{11.76}$$

M-B Entropy

These results for entropy are *only* valid for *Maxwell-Boltzmann statistics* and systems wherein the particles are considered *distinguishable*. For systems obeying *classical statistics*, where the particles are considered to be *indistinguishable*, we obtain

$$S_C = \frac{U}{T} + N k_B \left( \ln \frac{Z}{N} + 1 \right), \tag{11.77}$$

Classical Entropy

$$S_C = N k_B T \frac{\partial \ln Z}{\partial T} + N k_B \left( \ln \frac{Z}{N} + 1 \right) \tag{11.78}$$

Classical Entropy

by using arguments similar to those above (see Problem 11.14).

Other thermodynamic properties of a system can also be expressed in terms of the partition function. For example, from the defining equation of the **Helmholtz function,**

Helmholtz Function

$$F \equiv U - TS, \tag{11.79}$$

we immediately obtain

$$F_{\text{M-B}} = -Nk_BT \ln Z, \tag{11.80}$$

$$F_C = -Nk_BT \left( \ln \frac{Z}{N} + 1 \right) \tag{11.81}$$

by substitution from Equations 11.75 and 11.77, respectively. Also, we note that taking the total derivative of the Helmholtz function,

$$dF = dU - T\,dS - S\,dT,$$

and substituting for $dU$ from the combined first and second laws of thermodynamics (Equation 11.61) gives

$$dF = -pdV + \mu\,dN - S\,dT, \tag{11.82}$$

which immediately yields the general relations

Pressure

$$\left( \frac{\partial F}{\partial V} \right)_{N,T} = -p, \tag{11.83}$$

Chemical Potential

$$\left( \frac{\partial F}{\partial N} \right)_{V,T} = \mu, \tag{11.84}$$

Entropy

$$\left( \frac{\partial F}{\partial T} \right)_{V,N} = -S. \tag{11.85}$$

Thus, the thermodynamic *equation of state* for a statistical system can be expressed in terms of $Z$ from Equation 11.83 with $F$ being replaced by either $F_{\text{M-B}}$ (Equation 11.80) or $F_C$ (Equation 11.81). Surprisingly, in either case the result obtained is given by

Pressure

$$p = Nk_BT \left( \frac{\partial \ln Z}{\partial V} \right)_{N,T}, \tag{11.86}$$

and there is no need for a subscript (M-B or C) on $p$. It should be clear from Equation 11.84 that the chemical potential $\mu$ will be different for Maxwell-Boltzmann and classical statistics, since the Helmholtz function $F$ is different for the two cases. More specifically, for Maxwell-Boltzmann statistics the chemical potential is given by Equation 11.84,

$$\mu_{\text{M-B}} = \left(\frac{\partial F_{\text{M-B}}}{\partial N}\right)_{V,T},$$

upon substitution from Equation 11.80 and the assumption $Z \neq Z(N)$:

$$\mu_{\text{M-B}} = -k_B T \ln Z. \qquad (11.87)$$

Maxwell-Boltzmann Chemical Potential

The corresponding equation for classical statistics can be obtained in a similar manner, with the result being identical to that previously given by Equation 11.64,

$$\mu_C = -k_B T \ln \frac{Z}{N}. \qquad (11.64)$$

Classical Chemical Potential

Just as this expression for $\mu_C$ was used with the Boltzmann distribution to obtain the classical distribution, we can use the above relation for $\mu_{\text{M-B}}$ to obtain an alternative expression for the Boltzmann distribution. That is, from Equation 11.87 we have

$$Z = e^{-\beta\mu_{\text{M-B}}},$$

which can be substituted into the Boltzmann distribution (Equation 11.47) to obtain

$$n_i = N g_i e^{\beta(\mu_{\text{M-B}} - \epsilon_i)}. \qquad (11.88)$$

Boltzmann Distribution

The most probable occupation number for the Boltzmann distribution (Equation 11.88) is identical to that for the classical distribution (Equation 11.65), except for the multiplicative factor of $N$. As will be illustrated in Chapter 12, these particular distribution laws (Equations 11.65 and 11.88) are easily and directly compared with the Bose-Einstein and Fermi-Dirac distribution laws of quantum mechanics. Unlike quantum statistical mechanics, however, the thermodynamic properties of *internal energy* (Equations 11.68 and 11.70), *entropy* (Equations 11.75 to 11.78), the *Helmholtz function* (Equations 11.80 and 11.81), and *pressure* (Equation 11.86) are easily obtained, once the partition function of classical statistical mechanics has been evaluated. Further, we have demonstrated that the total energy, average energy per particle, occupation number, and pressure of classical statistical mechanics does *not* depend on whether the particles of a system are considered to be *distinguishable* or *indistinguishable*. Other thermodynamic properties, like the Helmholtz free energy and entropy, do depend on whether the system obeys *Maxwell-Boltzmann* or *classical* statistics.

## 11.7 Monatomic Ideal Gas

As a fundamental application of classical statistical mechanics (i.e., Maxwell-Boltzmann or classical statistics), consider an ideal gas consisting of $N$ identical but distinguishable (or indistinguishable) *monatomic* molecules of particle mass $m$ confined to a volume $V$. Since the particles of an ideal gas are essentially independent of one another, the potential energy between particles is effectively zero. Further, considering any gravitational potential energy associated with a particle as being insignificant, then the energy possessed by each particle in the system is all kinetic and given by Equation 11.6. As such, the total kinetic energy of the system is equal to the thermodynamic *internal energy* (i.e., $E = U$). In the discussions that follow, equations are derived for the internal energy $U$, the average energy per particle $\bar{\epsilon}$, entropy, pressure, and distribution laws for molecular energy, momentum, and speed. As will be illustrated in the problem section, the formulae developed in this section are of fundamental importance in the application of classical statistical mechanics.

### Energy, Entropy, and Pressure Formulae

The immediate objective is to find expressions for the internal energy $U$ and the average energy per particle $\bar{\epsilon}$, which can be accomplished most easily by the utilization of Equations 11.68 and 11.71, respectively. Since these equations have a strong dependence on the partition function $Z$, which itself depends on the *cell-energy degeneracy* $g_i$ (see Equation 11.48), we need to first determine an expression for *degeneracy* and then an expression for $Z$, before attempting an evaluation of $U$ and $\bar{\epsilon}$. Accordingly, recall that cell-energy degeneracy in classical statistical mechanics is defined as the *number of cells* in phase space corresponding to the same energy. To be consistent with a quantum mechanical interpretation of nature, this definition of degeneracy needs to be somewhat refined. In quantum mechanics, a *degeneracy of the energy level* occurs when more than one quantum state has the same energy. Hence, we need to consider $g_i$ in statistical mechanics as the number of possible and physically distinct particle states having a given energy $\epsilon_i$. With the *minimum* $\mu$-space volume of a particle state given by $\tau_0$ (Equation 11.3), then $g_i$ is essentially the *volume of cells in phase space of identical energy* divided by $\tau_0$ or

$$g_i = \frac{\text{Volume of } \mu\text{-space of Uniform Energy}}{\tau_0}. \tag{11.89}$$

To quantify the numerator of this expression, it is convenient to consider a *continuous* distribution of molecular energies, rather than the discrete values $\epsilon_1, \epsilon_2, \cdots, \epsilon_i$. Consequently, instead of Equation 11.6, we have the energy of a particle given by

$$\epsilon = \frac{p^2}{2m} = \frac{p_x^2 + p_y^2 + p_z^2}{2m}, \tag{11.90}$$

which is a continuous function of the particle's momentum **p.** For a continuous distribution of energy defined by this equation, momentum space must be considered as partitioned into *thin spherical shells* of essentially constant energy, as illustrated in Figure 11.4. That is, particles in the spherical shell of Figure 11.4 of radius **p** and thickness $d\mathbf{p}$ possess momentum between **p** and $\mathbf{p} + d\mathbf{p}$ and, consequently, have essentially the same energy according to Equation 11.90. Since the volume of a sphere in momentum space of radius $p$ is given by

$$V_p = \frac{4}{3}\pi p^3, \tag{11.91}$$

then the elemental volume

$$dV_p = 4\pi p^2 dp \tag{11.92}$$

represents the volume of a thin spherical shell of surface area $4\pi p^2$ and thickness $dp$. With this expression for a spherical shell of uniform energy in momentum space, the relation for *degeneracy* of Equation 11.89 can be expressed as

$$g(p)\,dp = \frac{1}{\tau_0}\iiint dx\,dy\,dz\,dV_p \tag{11.93}$$

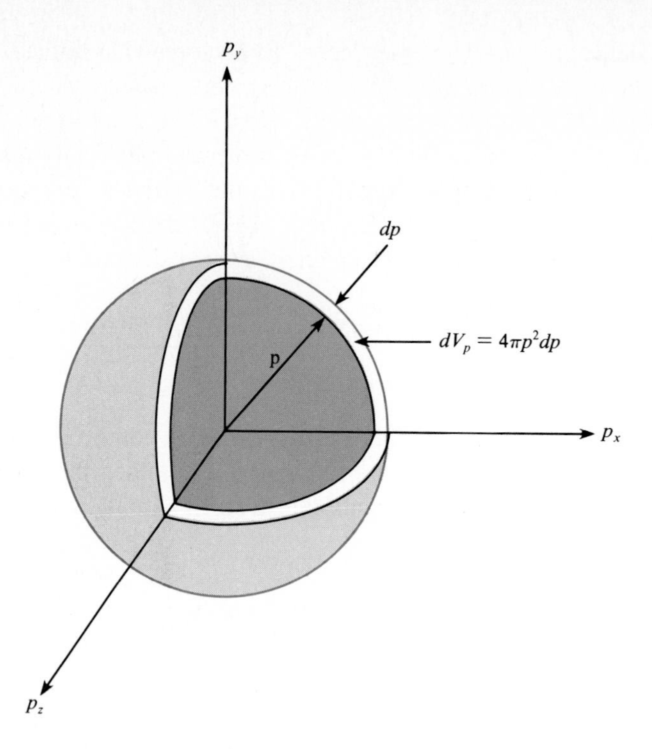


Figure 11.4
The volume of a thin spherical shell in **p**-space.

in terms of a continuous distribution of momenta. Since the volume occupied by the ideal gas in ordinary space is

$$V = \iiint dx\, dy\, dz, \tag{11.94}$$

then Equation 11.93 becomes

Degeneracy

$$g(p)\, dp = \frac{4\pi V}{h^3} p^2 dp, \tag{11.95}$$

where Equation 11.3 for $\tau_0$ and Equation 11.92 for $dV_p$ have been substituted. The degeneracy given by Equation 11.95 represents the distribution or number of *elemental cells* in $\mu$-space of, essentially, the same energy. Because all equal volumes of $\mu$-space are energetically accessible to a molecule and have equal *a priori probability*, $g(p)$ is often interpreted as the *a priori probability* that the momentum of an ideal gas molecule is between $p$ and $p + dp$.

In going from the *discrete* to a *continuous* distribution of molecular energies, the partition function of Equation 11.48,

$$Z \equiv \sum_i g_i e^{-\beta\epsilon_i},$$

can be expressed as

$$Z = \int_0^\infty g(p) e^{-\beta p^2/2m} dp, \tag{11.96}$$

where the summation is replaced by an integral, $g_i$ by $g(p)\,dp$, and $\epsilon_i$ by $\epsilon = p^2/2m$. Now, substituting from Equation 11.95 allows $Z$ to take the form

$$Z = \frac{4\pi V}{h^3} \int_0^\infty p^2 e^{-\beta p^2/2m} dp. \tag{11.97}$$

The integral in this expression is of the same form as that given by Equation 10.72 (also see Chapter 10, Section A.10), except here we must realize that

$$\int_{-\infty}^{+\infty} u^{2n} e^{-au^2} du = 2 \int_0^\infty u^{2n} e^{-au^2} du. \tag{11.98}$$

Therefore, the generalized integral is of the form

$$\int_0^\infty u^{2n} e^{-au^2} du = \frac{1}{2}\sqrt{\frac{\pi}{a}} \left[\frac{1}{2} \cdot \frac{3}{2} \cdots \frac{2n-1}{2}\right] a^{-n} \tag{11.99}$$

and Equation 11.97 becomes

$$Z = \frac{4\pi V}{h^3} \left(\frac{1}{2}\right) \left(\frac{2\pi m}{\beta}\right)^{1/2} \left(\frac{1}{2}\right) \left(\frac{\beta}{2m}\right)^{-1},$$

which simplifies to

$$Z = V \left(\frac{2\pi m}{h^2}\right)^{3/2} \beta^{-3/2}. \tag{11.100}$$

Partition Function

Having evaluated the partition function, it is now relatively easy to obtain the thermodynamic properties of the monatomic ideal gas. In particular, the thermodynamic internal energy $U$ can be evaluated using Equation 11.68,

$$U = -N\frac{\partial \ln Z}{\partial \beta} = -\frac{N}{Z}\frac{\partial Z}{\partial \beta},$$

upon substitution from Equation 11.100 to obtain

$$U = -\frac{N}{Z}V\left(\frac{2\pi m}{h^2}\right)^{3/2}\left(-\frac{3}{2}\right)\beta^{-5/2}$$

$$= \frac{3}{2}\frac{N}{Z}Z\beta^{-1},$$

which simplifies to

Internal Energy

$$U = \frac{3}{2}\frac{N}{\beta} = \frac{3}{2}Nk_{\rm B}T. \tag{11.101}$$

With this result for the total internal energy of the monatomic ideal gas, the average energy per molecule $\bar{\epsilon}$ is simply given by $U$ divided by the total number of particles $N$ as

Average Particle Energy

$$\bar{\epsilon} = \frac{3}{2}\frac{1}{\beta} = \frac{3}{2}k_{\rm B}T, \tag{11.102}$$

where Equation 11.63 for $\beta$ has been used. This result can also be directly obtained without knowledge of $U$, by substituting the partition function of Equation 11.100 into the expression for the average particle energy given by Equation 11.71,

$$\bar{\epsilon} \equiv \frac{E}{N} = -\frac{\partial \ln Z}{\partial \beta},$$

and performing the indicated partial derivative. Actually, we can obtain the result for $\bar{\epsilon}$ (Equation 11.102) without first evaluating $Z$ and $U$. That is, since

$$\bar{\epsilon} = \frac{U}{N}$$

$$= \frac{\sum_i \epsilon_i n_i}{N}$$

$$= \frac{\sum_i \epsilon_i g_i e^{-\beta\epsilon_i}}{\sum_i g_i e^{-\beta\epsilon_i}},$$

then for a continuous distribution of energies the summations are replaced by integrals, $\epsilon_i$ by $\epsilon$, and $g_i$ by $g(\epsilon)\, d\epsilon$ to obtain

$$\bar{\epsilon} = \frac{\int_0^\infty \epsilon g(\epsilon) e^{-\beta\epsilon} d\epsilon}{\int_0^\infty g(\epsilon) e^{-\beta\epsilon} d\epsilon}.$$

Now, substitution of Equations 11.6 and 11.95, along with the equality

$$d\epsilon = \frac{p}{m} dp,$$

into the above equation for $\bar{\epsilon}$ yields

$$\bar{\epsilon} = \frac{1}{2m} \frac{\int_0^\infty p^4 e^{-\beta p^2/2m} dp}{\int_0^\infty p^2 e^{-\beta p^2/2m} dp} \tag{11.103}$$

for the average energy per molecule in terms of a continuous distribution of momentum. Obviously, there is no advantage in using this approach, since the partition function integral (Equation 11.97) must be evaluated along with another integral.

Having evaluated an expression for the total internal energy (Equation 11.101), it is relatively easy to obtain expressions for the *thermal capacity* and entropy. From general physics *thermal capacity* for an isometric process is given by the partial derivative of $U$ with respect to $T$ (see also Equations 11.146 to 11.149),

$$C_V = \left(\frac{\partial U}{\partial T}\right)_V,$$

so we immediately obtain

Thermal Capacity

$$C_V = \frac{3}{2} N k_B \tag{11.104}$$

for the *thermal capacity* of a monatomic ideal gas. Also, the entropy can be evaluated using Equation 11.75,

$$S_{\text{M-B}} = \frac{U}{T} + N k_B \ln Z,$$

upon substitution of the expressions for $U$ from Equation 11.101 and $Z$ from Equation 11.100. That is,

$$S_{\text{M-B}} = \frac{(3/2)Nk_{\text{B}}T}{T} + Nk_{\text{B}} \ln \left[ V \left( \frac{2\pi m}{h^2} \right)^{3/2} \beta^{-3/2} \right],$$

which becomes

$$S_{\text{M-B}} = Nk_{\text{B}} \left\{ \frac{3}{2} + \ln \left[ V \left( \frac{2\pi m k_{\text{B}} T}{h^2} \right)^{3/2} \right] \right\} \tag{11.105}$$

after the obvious cancellation of $T$ and substitution for $\beta$. If the ideal gas is fairly dense, where the molecules must be considered as essentially *indistinguishable*, then Equation 11.77 must be used to evaluate the entropy $S_{\text{C}}$. In either case, it is interesting to note that in classical thermodynamics we obtain *differences* in entropy, while relations for entropy contain an undetermined constant. When statistical mechanics is applied to different problems, however, we obtain exact expressions for entropy without any undetermined constants.

The central role of the statistical partition function in analyzing the monatomic ideal gas should be obvious from the above evaluations of energy and entropy formulae. It can also be employed to obtain the *equation of state* for the ideal gas molecules, by direct substitution into Equation 11.86. That is, with

$$p = Nk_{\text{B}}T \left( \frac{\partial \ln Z}{\partial V} \right)_{N,T}$$

from Equation 11.86 and

$$Z = V \left( \frac{2\pi m}{h^2} \right)^{3/2} \beta^{-3/2}$$

from Equation 11.100, we immediately obtain

$$p = \frac{Nk_{\text{B}}T}{V}$$

or, as it is customarily written,

Ideal Gas
Equation of State

$$pV = Nk_{\text{B}}T. \tag{11.106}$$

An interesting and fundamental relationship can now be realized by comparing this result with that predicted by general classical physics. Elementary classical thermodynamics gives the **ideal gas equation of state** in the form

Ideal Gas Equation of State

$$pV = nRT, \tag{11.107a}$$

where $n$ represents the number of moles and $R$ is the **universal gas constant.** Since we can define the number of moles by the ratio of the number of particles $N$ to Avogadro's number $N_o$,

Number of Moles

$$n \equiv \frac{N}{N_o}, \tag{5.69}$$

then Equation 11.107a can be expressed as

$$pV = N\frac{R}{N_o}T. \tag{11.107b}$$

A comparison of this equation from classical thermodynamics with Equation 11.106 from statistical mechanics gives

Boltzmann Constant

$$k_B \equiv \frac{R}{N_o} = 1.38066 \times 10^{-23}\ \frac{\text{J}}{\text{K}}. \tag{11.108}$$

This result can also be obtained by comparing our result for $U$ (see Equation 11.101) with that predicted by *kinetic theory*, that is, $U = (3/2)nRT = (3/2)N(R/N_o)T$. The important point, however, is that although $k_B$ in statistics was simply an unknown constant for Equation 11.28, it is now recognized as *necessarily* being identical to $R/N_o$ and called the **Boltzmann constant,** if classical statistical mechanics is to agree with classical thermodynamics and kinetic theory.

## Energy, Momentum, and Speed Distribution Formulae

Now that the cell-energy degeneracy and partition function have been evaluated for the monatomic ideal gas, we can obtain an expression for the Boltzmann distribution,

$$n_i = \frac{N}{Z} g_i e^{-\beta\epsilon_i},$$

in terms of a continuous distribution of molecular momenta. In this case $n_i$ is replaced by $n(p)\,dp$, $g_i$ by $g(p)\,dp$, and $\epsilon_i$ by $\epsilon = p^2/2m$ to obtain

$$n(p)\,dp = \frac{N}{Z}\,g(p)\,e^{-\beta p^2/2m}\,dp. \tag{11.109}$$

Substitution for $g(p)\,dp$ and $Z$ from Equations 11.95 and 11.100 gives

$$n(p)\,dp = \frac{N(4\pi V/h^3)p^2 e^{-\beta p^2/2m}\,dp}{V(2\pi m/h^2)^{3/2}\beta^{-3/2}},$$

which with the algebra of exponents reduces to

Boltzman Distribution of Momenta

$$n(p)\,dp = 4\pi N\left(\frac{\beta}{2\pi m}\right)^{3/2} p^2 e^{-\beta p^2/2m}\,dp. \tag{11.110}$$

This result, called the **Boltzmann distribution of momenta,** represents the number of ideal gas molecules having momenta between $p$ and $p + dp$, where $\beta$ is related to the absolute temperature of the system by Equation 11.63. The Boltzmann distribution for an ideal gas can also be expressed in terms of molecular energies and speeds by fundamental substitutions into Equation 11.110. That is, with

$$p = (2m\epsilon)^{1/2}, \tag{11.111}$$

$$dp = m(2m\epsilon)^{-1/2}\,d\epsilon, \tag{11.112}$$

Equation 11.110 becomes

$$n(\epsilon)\,d\epsilon = 4\pi N\left(\frac{\beta}{2\pi m}\right)^{3/2} 2m\epsilon e^{-\beta\epsilon} m(2m\epsilon)^{-1/2}\,d\epsilon,$$

which with a little effort reduces to the **Boltzmann distribution of energies**

Boltzman Distribution of Energies

$$n(\epsilon)\,d\epsilon = 2\pi N\left(\frac{\beta}{\pi}\right)^{3/2} \epsilon^{1/2} e^{-\beta\epsilon}\,d\epsilon \tag{11.113}$$

for the number of gas molecules having energies between $\epsilon$ and $\epsilon + d\epsilon$. In a similar manner, substituting

$$p = mv,$$
$$dp = mdv$$

into the right-hand side of Equation 11.110 yields

$$n(v)dv = 4\pi N \left(\frac{\beta m}{2\pi}\right)^{3/2} v^2 e^{-\beta m v^2/2} dv \qquad (11.114)$$

Boltzman Distribution of Speeds

for the **Boltzmann distribution of speeds.** This result, originally obtained by Maxwell in 1859, represents the distribution of speeds between $v$ and $v + dv$ for the molecules of an ideal gas. Although it was directly obtained from the *Boltzmann distribution of momenta* (Equation 11.110), it can also be easily derived from the *energy distribution* formula given by Equation 11.113 (see Problem 11.17).

It should be obvious from the above discussion that starting with any one of the three distribution formulae (Equations 11.110, 11.113, or 11.114) allows for the direct derivation of the other two distribution functions. In this sense the distribution formulae are redundant, even though each one has its particular usefulness in statistical applications for an ideal gas. For example, by analogy with the definitions for the *ensemble average* (Equation 11.19) and the quantum mechanical *expectation value* (Equation 10.25), it seems appropriate to define the **statistical average** of any physical variable $A$ by the general equation

$$\bar{A} \equiv \frac{\displaystyle\int_0^\infty A n(q)\, dq}{\displaystyle\int_0^\infty n(q)\, dq}, \qquad (11.115)$$

Statistical Average

where $q$ is a generalized coordinate that can be replaced by $p$, $v$, $\epsilon$, and so forth. Clearly, the *distribution formulae* are redundant, since the average value of any physical variable (e.g., $\bar{A} \rightarrow \bar{v}, \bar{p}, \bar{\epsilon}$, etc.) can be obtained in any one of three different ways corresponding to $q = v$, $q = p$, or $q = \epsilon$. It is also interesting to note that unlike the quantum mechanical *expectation value* defined by Equation 10.29, the integral in the denominator of Equation 11.115 is not *normalized* to unity. Instead, *conservation of particles* must be required in statistical mechanics, which means that

$$\int_0^\infty n(q)\, dq = N, \qquad q = p, v, \epsilon, \cdots. \qquad (11.116)$$

Conservation of Particles

As an example of the validity of this equation, for an ideal gas with $q = p$ we have

$$\int_0^\infty n(p)dp = 4\pi N\left(\frac{\beta}{2\pi m}\right)^{3/2}\int_0^\infty p^2 e^{-\beta p^2/2m}dp$$
$$= 4\pi N\left(\frac{\beta}{2\pi m}\right)^{3/2}\left[\frac{1}{4}\left(\frac{2\pi m}{\beta}\right)^{1/2}\left(\frac{2m}{\beta}\right)\right]$$
$$= N, \tag{11.117}$$

where the integral was evaluated using Equation 11.99. This same result is certainly obtained for an ideal gas with $q = v$ and $q = \epsilon$ (see Problems 11.18 and 11.19). Hence, the generalized equation defining a *statistical average* for a *continuous distribution* can be expressed as

Statistical Average

$$\bar{A} = \frac{1}{N}\int_0^\infty An(q)\,dq, \qquad q = p, v, \epsilon, \cdots . \tag{11.118}$$

Further, it should be clear that in general the number of *particles* with values between $q$ and $q + dq$ can always be expressed in the form

$$n(q)\,dq \equiv dn_q, \qquad q = p, v, \epsilon, \cdots . \tag{11.119}$$

With this relation, the result expressed by Equation 11.117 is obvious for $q = p$ and likewise for $q = v$ or $q = \epsilon$.

The defining equation for the *statistical average* of a physical variable is extremely useful in a number of physical applications. For example, the average energy of an ideal gas molecule given by Equation 11.102 can be obtained using Equation 11.118 by replacing $A$ with $\epsilon$ and $q$ with $\epsilon$. Accordingly,

$$\bar{\epsilon} = \frac{1}{N}\int_0^\infty \epsilon n(\epsilon)\,d\epsilon$$
$$= 2\pi\left(\frac{\beta}{\pi}\right)^{3/2}\int_0^\infty \epsilon^{3/2}e^{-\beta\epsilon}d\epsilon, \tag{11.120}$$

where the *energy distribution* of Equation 11.113 has been used in obtaining the second equality. With a new variable of integration defined by

$$\alpha = \epsilon^{1/2}, \tag{11.121a}$$
$$d\epsilon = 2\alpha\,d\alpha, \tag{11.121b}$$

Equation 11.120 can be expressed in a form that is amenable to integration using Equation 11.99. That is, combining the last three equations results in

$$\bar{\epsilon} = 2\pi \left(\frac{\beta}{\pi}\right)^{3/2} \left(2 \int_0^\infty \alpha^4 e^{-\beta\alpha^2} d\alpha\right) \qquad (11.122)$$

$$= 2\pi \left(\frac{\beta}{\pi}\right)^{3/2} \left[2 \frac{1}{2} \left(\frac{\pi}{\beta}\right)^{1/2} \frac{1}{2} \frac{3}{2} \left(\frac{1}{\beta}\right)^2\right]$$

$$= \frac{3}{2\beta} = \frac{3}{2} k_B T, \qquad (11.102)$$

which is identical to our previously derived result of Equation 11.102. Of course, because of the redundancy of the *distribution formulae*, this result could be obtained from Equation 11.118 for the *momentum distribution* (Equation 11.110) or *speed distribution* (Equation 11.114), that is,

$$\bar{\epsilon} = \frac{1}{N} \int_0^\infty \epsilon n(p)\, dp, \qquad (11.123)$$

$$\bar{\epsilon} = \frac{1}{N} \int_0^\infty \epsilon n(v)\, dv, \qquad (11.124)$$

by using $\epsilon = p^2/2m$ and $\epsilon = \frac{1}{2}mv^2$, respectively, in the integrals. The verification of these two equations for an ideal gas molecule is left as an exercise in the problem set.

By capitalizing on the result for the average molecular energy $\bar{\epsilon}$, it is simple to obtain an expression for the average speed squared $\bar{v}^2$ and the *root-mean-square speed* $v_{\text{rms}}$. That is, since $\epsilon = \frac{1}{2}mv^2$ we have

$$\bar{\epsilon} = \tfrac{1}{2}m\bar{v}^2, \qquad (11.125)$$

which when compared with Equation 11.102 gives

$$\frac{3}{2} k_B T = \tfrac{1}{2}m\bar{v}^2.$$

Thus, $\bar{v}^2$ is given by

$$\bar{v}^2 = \frac{3k_B T}{m} \qquad (11.126)$$

from which the **root-mean-square speed,** defined by

Root-Mean-Square Speed

$$v_{\rm rms} \equiv \sqrt{\bar{v}^2}, \tag{11.127}$$

is simply given by

Root-Mean-Square Speed

$$v_{\rm rms} = \sqrt{\frac{3k_{\rm B}T}{m}}. \tag{11.128}$$

Had we not first evaluated $\bar{\epsilon}$, the result for $\bar{v}^2$ could be easily derived from Equation 11.118. The important point above, however, is that once an expression for $\bar{v}^2$ is obtained, the result for $v_{\rm rms}$ is immediate from its defining equation. It is also important to emphasize that $v_{\rm rms}$ is not the same as $\bar{v}$. For an ideal gas obeying the Boltzmann distribution, $\bar{v}$ is given by (see Problem 11.23)

Average Speed

$$\bar{v} = 2\sqrt{\frac{2k_{\rm B}T}{\pi m}}, \tag{11.129}$$

so the relationship between $v_{\rm rms}$ and $\bar{v}$ is

$v_{\rm rms}$ Versus $\bar{v}$

$$v_{\rm rms} = \sqrt{\frac{3\pi}{8}}\,\bar{v}. \tag{11.130}$$

The results for $v_{\rm rms}$ and $\bar{v}$ can also be compared with the *most probable speed* $v$, where $v$ is obtained by *maximizing* $n(v)$, that is,

Maximizing $n(v)$

$$\frac{dn(v)}{dv} = 0, \tag{11.131}$$

and solving the resulting equation for $v$. For the *Boltzmann distribution of speeds* (Equation 11.114) substituted into this equation (see Problem 11.25) we obtain

Most Probable Speed

$$v = \sqrt{\frac{2k_{\rm B}T}{m}}. \tag{11.132}$$

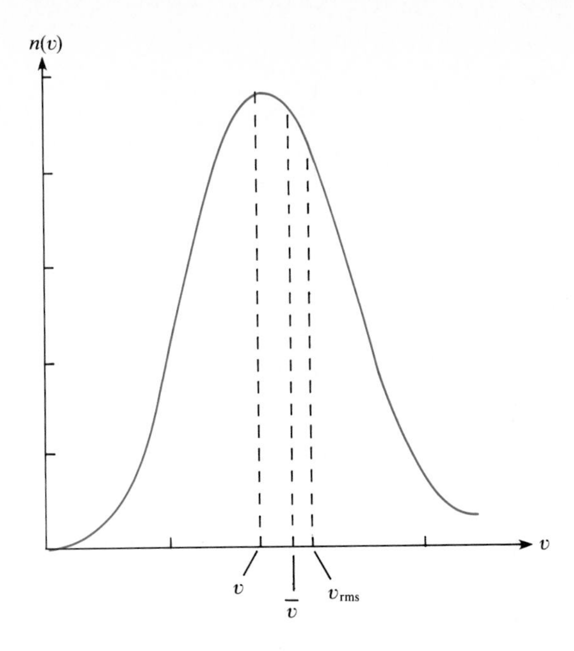


Figure 11.5
The Boltzman distribution of speeds $n(v)$ with the most probable speed $v$, the average speed $\bar{v}$, and the root-mean-square speed $v_{rms}$ indicated.

Clearly, from the above results we have

$v$ versus $\bar{v}$ and $v_{rms}$

$$v = \sqrt{\frac{\pi}{4}}\,\bar{v} = \sqrt{\frac{2}{3}}\,v_{rms} \tag{11.133}$$

and $v < \bar{v} < v_{rms}$. This relationship between $v$, $\bar{v}$, and $v_{rms}$ is indicated in Figure 11.5, where the Boltzmann distribution of speeds (Equation 11.114) is plotted.

## 11.8 Equipartition of Energy

Consider the energy of a molecule to be expressed in terms of generalized parameters in the form $\epsilon = \epsilon(x_1, x_2, \cdots, x\text{'s})$, where the $x$'s represent the various position, linear momenta, and angular momenta coordinates of the molecule. If this energy can be written in terms of a quadratic term as

$$\epsilon = ax_1^2 + \epsilon_0(x_2, x_3, \cdots, x_n) \equiv \epsilon_q + \epsilon_0, \tag{11.134}$$

then the **mean energy** associated with the quadratic term is

Mean Energy

$$\bar{\epsilon}_q = \frac{\sum_i (\epsilon_q)_i n_i}{\sum_i n_i} \tag{11.135}$$

$$= \frac{\sum_i (\epsilon_q)_i g_i e^{-\beta\epsilon_i}}{\sum_i g_i e^{-\beta\epsilon_i}}. \tag{11.136}$$

In the limit of a continuous distribution of energy states, this equation becomes

$$\bar{\epsilon}_q = \frac{\int_{-\infty}^{+\infty} \epsilon_q e^{-\beta\epsilon} dx_1 \cdots dx_n}{\int_{-\infty}^{+\infty} e^{-\beta\epsilon} dx_1 \cdots dx_n}, \tag{11.137}$$

where $\epsilon = \epsilon(x_1, x_2, \cdots, x_n)$ and the degeneracy has been thought of as (see Equations 11.89 and 11.93)

$$g = \frac{\int_{-\infty}^{+\infty} dx_1 dx_2 \cdots dx_n}{\tau_0}. \tag{11.138}$$

Expressing the exponentials in Equation 11.137 as (see Equation 11.134)

$$e^{-\beta\epsilon(x_1,x_2,\cdots,x_n)} = e^{-\beta\epsilon_q} e^{-\beta\epsilon_0(x_2,x_3,\cdots,x_n)},$$

the expression for $\bar{\epsilon}_q$ simplifies to

$$\bar{\epsilon}_q = \frac{\int_{-\infty}^{+\infty} a x_1^2 e^{-\beta a x_1^2} dx_1}{\int_{-\infty}^{+\infty} e^{-\beta a x_1^2} dx_1}, \tag{11.139}$$

as the coefficient

$$\int_{-\infty}^{+\infty} e^{-\beta\epsilon_0(x_2,x_3,\cdots,x_n)} dx_2 dx_3 \cdots dx^n$$

cancels from the numerator and denominator. Now, employing the transformation

$$y = ax_1^2 \rightarrow dy = 2ax_1 dx_1, \tag{11.140}$$

Equation 11.139 becomes

$$\bar{\epsilon}_q = \frac{\int_0^{\infty} y^2 e^{-\beta y^2} dy}{\int_0^{\infty} e^{-\beta y^2} dy}, \tag{11.141}$$

where Equation 11.98 has been used. The integrals of this equation are evaluated using Equation 11.99 for the numerator and Equation 10.67 for the denominator (see also Appendix A, Section A.10) to obtain

$$\bar{\epsilon}_q = \frac{(1/2)(\pi/\beta)^{1/2}(1/2)(1/\beta)}{(1/2)(\pi/\beta)^{1/2}},$$

which reduces to

$$\bar{\epsilon}_q = \frac{1}{2\beta} = \tfrac{1}{2}k_B T. \tag{11.142}$$

Energy Per Degree of Freedom

This result shows that each *quadratic term* in an expression for the total energy of a particle has associated with it a *mean energy* of $\frac{1}{2}k_B T$. We can think of *each parameter* associated with the energy of a particle as representing a *degree of freedom*. Consequently, for each *degree of freedom* that is consistent with the above requirements, its contribution to the total energy of a particle in an assembly in thermodynamic equilibrium is given by Equation 11.142. From these observations, we can state that *the energy of a particle depends only on temperature and is equally distributed among its independent degrees of freedom*. This statement represents the **equipartition of energy principle** of thermal physics.

The quadratic conditions discussed above are certainly fulfilled for an *essentially free* particle of an ideal gas, since the energy is the sum of quadratic momenta terms $p_x^2/2m$, $p_y^2/2m$, and $p_z^2/2m$. They are also satisfied by a linear harmonic oscillator, since its maximum potential energy given by $\frac{1}{2}kx_m^2$, in terms of its maximum displacement $x_m$, is equivalent to its maximum kinetic energy $\frac{1}{2}mv_m^2$ (see Equation 7.98). The conditions are not satisfied for particles in a gravitational field, however, because the gravitational potential energy given by $mgy$ (Equation 1.23) is not quadratic in the $y$-coordinate. Further, it should be clear that the principle is not valid for the quantized energies predicted by quantum mechanics, as the discrete energy values cannot be expressed as a *continuous* function of coordinates. Consequently, for classical systems obeying the *equipartition of energy principle*, we can consider the total energy per particle to be given by

Equipartition of Particle Energy

$$\bar{\epsilon} = \frac{N_f}{2} k_B T, \tag{11.143}$$

where $N_f$ represents the total number of degrees of freedom. Thus, for a system of $N$ particles, the *total internal energy* is

Equipartition of Internal Energy

$$U = \frac{N_f}{2} N k_B T, \tag{11.144}$$

which can be expressed as

Equipartition of Internal Energy

$$U = \frac{N_f}{2} nRT, \tag{11.145}$$

because of Equations 11.108 and 5.69.

## Classical Specific Heat

The above results are particularly useful in applications to *specific heat* problems. To fully appreciate this application, we will derive an expression for *specific heat* in terms of a particle's number of degrees of freedom $N_f$. To begin, **thermal capacity,** often called *specific heat*, is defined in thermodynamics as

Thermal Capacity

$$C \equiv \frac{dQ}{dT}, \tag{11.146}$$

where $Q$ is the heat energy and $T$ is the absolute temperature. From the **first law of thermodynamics,**

First Law of Thermodynamics

$$dU = dQ - pdV, \tag{11.147}$$

we have for an *isometric* process

$$\left(\frac{\partial U}{\partial T}\right)_V = \left(\frac{\partial Q}{\partial T}\right)_V. \tag{11.148}$$

Thus, *thermal capacity* at constant volume is obtained from Equations 11.146 and 11.148 as

$$C_V = \left(\frac{\partial U}{\partial T}\right)_V. \tag{11.149}$$

Dividing both sides of this equation by the number of moles $n$ of a system gives

$$c_v = \left(\frac{\partial u}{\partial T}\right)_v, \tag{11.150}$$

which identifies the **molal specific thermal capacity,**

$$c_v \equiv \frac{C_V}{n}, \tag{11.151}$$

Molal Specific Thermal Capacity

as equal to the partial derivative of the **molal specific internal energy,**

$$u \equiv \frac{U}{n}, \tag{11.152}$$

Molal Specific Internal Energy

with respect to absolute temperature. The subscript denotes that the **molal specific volume,**

$$v \equiv \frac{V}{n}, \tag{11.153}$$

Molal Specific Volume

is constant with respect to the derivative. In terms of the *molal specific internal energy* (defined by Equation 11.152), the total internal energy of a system obeying the *equipartition of energy principle*, as given by Equation 11.145, can be expressed as

$$u = \frac{N_f}{2} RT. \tag{11.154}$$

Now, substituting this equation into Equation 11.150 and performing the partial differential gives

$$c_v = \frac{N_f}{2} R \tag{11.155}$$

Molal Specific Heat

for the *molal specific thermal capacity* or *molal specific heat* in terms of the number of degrees of freedom $N_f$.

As a particular example of the usefulness of Equation 11.155, consider a **monatomic** ideal gas, where the molecules are thought of as being spherical in shape. The number of degrees of freedom $N_f$ for each molecule is *three*, arising from each of the three *translational* terms in Equation 11.90. Thus, Equation 11.155 gives

Monatomic Ideal Gas

$$\begin{aligned} c_v &= \frac{3}{2} R \\ &\approx 3 \frac{\text{cal}}{\text{mole} \cdot \text{K}}. \end{aligned} \tag{11.156}$$

This result is also predicted by Equation 11.104 when the definition for the Boltzmann constant $k_B$ (Equation 11.108) and the number of moles $n$ (Equation 5.69) are taken into account.

For a **diatomic** ideal gas, we may imagine the molecules as dumbell-like consisting of two spatially separated atoms aligned on the $X$-axis. There are still *three translational* degrees of freedom and, in addition, *two rotational* degrees of freedom exist. The latter originate from a rotation of the dumbell-like molecule in the $x$-$y$ plane and a rotation in the $x$-$z$ plane. Rotation in the $y$-$z$ plane, which is rotation about the imaginary line connecting the two atoms, is considered to be insignificant by comparison with the other two rotation modes. Thus, there are a total of *five* degrees of freedom, $N_f = 5$, and the *molal specific heat* from Equation 11.155 is

Diatomic Ideal Gas

$$\begin{aligned} c_v &= \frac{5}{2} R \\ &\approx 5 \frac{\text{cal}}{\text{mole} \cdot \text{K}}. \end{aligned} \tag{11.157}$$

For a **polyatomic** ideal gas, each molecule consists of three or more atoms having a finite separation. Consequently, there are *three translational* and *three rotational* degrees of freedom, $N_f = 6$, giving rise to a *molal specific heat* of

Polyatomic Ideal Gas

$$\begin{aligned} c_v &= 3R \\ &\approx 6 \frac{\text{cal}}{\text{mole} \cdot \text{K}}. \end{aligned} \tag{11.158}$$

Coincidentally, this result for polyatomic molecules is identical to the **Dulong-Petit law** (see Problem 11.30) of solid state physics, where the atoms in a simple cubic lattice of a solid are considered to be a series of identical harmonic oscillators.

## Review of Fundamental and Derived Equations

A listing of the fundamental and derived equations of this chapter is presented below. The derivations of classical statistical mechanics are presented in a logical listing that parallels their development in each section of this chapter.

### FUNDAMENTAL EQUATIONS—CLASSICAL PHYSICS

| | |
|---|---|
| $\epsilon = \frac{1}{2}mv^2 = \frac{p^2}{2m}$ | Free Particle Kinetic Energy |
| $\bar{\epsilon} \equiv \frac{E}{N}$ | Average Particle Energy |
| $n \equiv \frac{N}{N_o}$ | Number of Moles |
| $v \equiv \frac{V}{n}$ | Molal Specific Volume |
| $u \equiv \frac{U}{n}$ | Molal Specific Internal Energy |
| $C \equiv \frac{dQ}{dT}$ | Thermal Capacity |
| $c \equiv \frac{C}{n}$ | Molal Specific Thermal Capacity |
| $dU = dQ - p\,dV + \mu\,dN$ | First Law of Thermodynamics |
| $dQ = TdS$ | Second Law of Thermodynamics |
| $F \equiv U - TS$ | Helmholtz Function |

### BASIC EQUATIONS—CLASSICAL STATISTICAL MECHANICS

| | |
|---|---|
| $\tau \equiv dxdydzdp_xdp_ydp_z$ | Cell Volume in μ-Space |
| $\tau_0 = h^3$ | Minimum Volume Per Cell |
| $N = \sum_i n_i$ | Conservation of Particles |
| $E = \sum_i n_i\epsilon_i,\ E \equiv U$ for $V = 0$ | Conservation of Energy |

| | |
|---|---|
| $W_k \equiv \Pi_i w_i$ | Total Microstates for $k$th Macrostate |
| $W_{\text{M-B}} = W_k = \dfrac{N!\Pi_i g_i^{n_i}}{\Pi_i n_i!}$ | M-B Thermodynamic Probability |
| $W_{\text{C}} = W_k = \dfrac{\Pi_i g_i^{n_i}}{\Pi_i n_i!}$ | Classical Thermodynamic Probability |
| $\Omega \equiv \sum_k W_k$ | Total Microstates |
| $\Omega_{\text{M-B}} = \left(\sum_i g_i\right)^N$ | Total Microstates for M-B Statistics |
| $\bar{A}_i \equiv \dfrac{\sum_k A_{ik} W_k}{\Omega}$ | Ensemble Average |
| $\bar{A}_i = \dfrac{1}{N} \sum_i A_i n_i$ | Statistical Average—Discrete Case |
| $\bar{A} = \dfrac{1}{N} \int_0^\infty A n(q)\, dq,\ q = v, p, \epsilon$ | Statistical Average—Continuous Case |
| $Z \equiv \sum_i g_i e^{-\beta\epsilon_i}$ | Partition Function |

## DERIVED EQUATIONS

### *Entropy and Thermodynamic Probability—Two Systems in Equilibrium*

| | |
|---|---|
| $W_T = W_1 W_2$ | Total Thermodynamic Probability |
| $S_T(W_1, W_2) = S_1(W_1) + S_2(W_2)$ | Total Entropy |
| $S = k_{\text{B}} \ln W$ | Entropy in Statistical Mechanics |

### *Most Probable Distribution*

| | |
|---|---|
| $\ln A! \approx A \ln A - A,\ A \gg 1$ | Stirling's Formula |
| $\ln W_{\text{M-B}} = N \ln N + \sum_i n_i \ln \dfrac{g_i}{n_i}$ | |
| $\ln W_{\text{C}} = N + \sum_i n_i \ln \dfrac{g_i}{n_i}$ | |
| $\delta \ln W_{\text{M-B}} = \delta \ln W_{\text{C}} = 0$ | Condition for Most Probable Distribution |
| $\delta \ln W = \sum_i \ln \dfrac{g_i}{n_i}\, \delta n_i$ | |
| $n_i = g_i e^{-\alpha} e^{-\beta\epsilon_i}$ | Maxwell-Boltzmann Distribution |
| $n_i = \dfrac{N}{Z} g_i e^{-\beta\epsilon_i}$ | Boltzmann Distribution |

*Evaluation of β*

$$dS = k_B \, d \ln W$$

$$= k_B \ln \frac{Z}{N} dN + k_B \beta dU \quad \text{Differential Entropy}$$

$$dU = T\, dS - p\, dV + \mu\, dN \quad \text{Combined First and Second Laws of Thermodynamics}$$

$$\beta = \frac{1}{k_B T} \quad \text{Lagrange Multiplier}$$

$$\mu_C = -k_B T \ln \frac{Z}{N} \quad \text{Classical Chemical Potential}$$

$$n_i = g_i e^{\beta(\mu_C - \epsilon_i)} \quad \text{Classical Distribution}$$

*Significance of the Partition Function*

$$E = -N \frac{\partial \ln Z}{\partial \beta}$$

$$= N k_B T^2 \frac{\partial \ln Z}{\partial T} \quad \text{Total Particle Energy}$$

$$n_j = -\frac{N}{\beta} \frac{\partial \ln Z}{\partial \epsilon_j}$$

$$= -N k_B T \frac{\partial \ln Z}{\partial \epsilon_j} \quad \text{Occupation Number}$$

$$S_{\text{M-B}} = \frac{U}{T} + N k_B \ln Z \quad \text{M-B Entropy}$$

$$S_C = \frac{U}{T} + N k_B \left( \ln \frac{Z}{N} + 1 \right) \quad \text{Classical Entropy}$$

$$F_{\text{M-B}} = -N k_B T \ln Z \quad \text{M-B Helmholtz Function}$$

$$F_C = -N k_B T \left( \ln \frac{Z}{N} + 1 \right) \quad \text{Classical Helmholtz Function}$$

$$dF = dU - T dS - S dT \quad \text{Differential Helmholtz Function}$$

$$\left( \frac{\partial F}{\partial V} \right)_{N,T} = -p \quad \text{Pressure}$$

$$\left( \frac{\partial F}{\partial N} \right)_{V,T} = \mu \quad \text{Chemical Potential}$$

$$\left( \frac{\partial F}{\partial T} \right)_{V,N} = -S \quad \text{Entropy}$$

$$p = N k_B T \left( \frac{\partial \ln Z}{\partial V} \right)_{N,T} \quad \text{Thermodynamic Pressure}$$

$$\mu_{\text{M-B}} = -k_B T \ln Z$$ M-B Chemical Potential

$$\mu_C = -k_B T \ln \frac{Z}{N}$$ Classical Chemical Potential

$$n_i = N g_i e^{\beta(\mu_{\text{M-B}} - \epsilon_i)}$$ Boltzmann Distribution

### *Monatomic Ideal Gas*

$$g(p)\,dp = \frac{1}{\tau_0} \iiint dx\,dy\,dz\,dV_p$$

$$= \frac{4\pi V}{h^3} p^2\,dp$$ Degeneracy—Continuous Case

$$Z = \int_0^\infty g(p) e^{-\beta p^2/2m}\,dp$$

$$= V \left(\frac{2\pi m}{h^2}\right)^{3/2} \beta^{-3/2}$$ Partition Function—Continuous Case

$$U = -N \frac{\partial \ln Z}{\partial \beta}$$

$$= \frac{3}{2} N k_B T$$ Internal Energy

$$C_V = \frac{3}{2} N k_B$$ Thermal Capacity

$$p = N k_B T \left(\frac{\partial \ln Z}{\partial V}\right)_{N,T}$$

$$= \frac{N k_B T}{V}$$ Pressure—Equation of State

$$k_B = \frac{R}{N_o}$$ Boltzmann Constant

### *Boltzmann Distribution Formulae*

$$n(p)\,dp = 4\pi N \left(\frac{\beta}{2\pi m}\right)^{3/2} p^2 e^{-\beta p^2/2m}\,dp$$ Distribution of Momenta

$$n(\epsilon)\,d\epsilon = 2\pi N \left(\frac{\beta}{\pi}\right)^{3/2} \epsilon^{1/2} e^{-\beta\epsilon}\,d\epsilon$$ Distribution of Energies

$$n(v)\,dv = 4\pi N \left(\frac{\beta m}{2\pi}\right)^{3/2} v^2 e^{-\beta m v^2/2}\,dv$$ Distribution of Speeds

$$v_{\text{rms}} \equiv \sqrt{\bar{v}^2}$$

$$= \sqrt{\frac{3k_BT}{m}} \qquad \text{Root-Mean-Square Speed}$$

$$\bar{v} = 2\sqrt{\frac{2k_BT}{\pi m}} \qquad \text{Average Speed}$$

$$v = \sqrt{\frac{2k_BT}{m}} \qquad \text{Most Probable Speed}$$

### *Equipartition of Energy*

$$\epsilon = ax_1^2 + \epsilon_0(x_2, x_3, \cdots, x_n)$$

$$\equiv \epsilon_q + \epsilon_0$$

$$\bar{\epsilon}_q = \frac{\sum (\epsilon_q)_i n_i}{\sum n_i}$$

$$= \frac{\int_{-\infty}^{+\infty} \epsilon_q e^{-\beta\epsilon}\, dx_1 \cdots dx_n}{\int_{-\infty}^{+\infty} e^{-\beta\epsilon}\, dx_1 \cdots dx_n}$$

$$= \tfrac{1}{2}k_BT \qquad \text{Energy Per Degree of Freedom}$$

$$U = \frac{N_f}{2}nRT \qquad \text{Equipartition of Internal Energy}$$

$$c_v = \frac{N_f}{2}R \qquad \text{Molal Specific Thermal Capacity}$$

## Problems

**11.1** Consider a system of six particles being represented in a two-cell $\mu$-space. Enumerate the number of macrostates associated with this system and calculate the number of microstates corresponding to each macrostate.

*Solution:*
The distribution of six particles in two cells can be accomplished in *seven* different ways, as given by

$$(n_1, n_2) \rightarrow (6, 0), (5, 1), (4, 2), (3, 3), (2, 4), (1, 5), (0, 6).$$

The number of microstates corresponding to each of these macrostates is given by Equation 11.7,

$$W_k = \frac{N!}{n_1!n_2!},$$

to be

$$W(6, 0) = \frac{6!}{6!0!} = 1,$$
$$W(5, 1) = \frac{6!}{5!1!} = 6,$$
$$W(4, 2) = \frac{6!}{4!2!} = 15,$$
$$W(3, 3) = \frac{6!}{3!3!} = 20,$$
$$W(2, 4) = \frac{6!}{2!4!} = 15,$$
$$W(1, 5) = \frac{6!}{1!5!} = 6,$$
$$W(0, 6) = \frac{6!}{0!6!} = 1.$$

**11.2** Consider a system of four particles being represented in a three-cell μ-space. Find (a) the number of different macrostates, (b) the total number of microstates, and (c) the most probable macrostates.

*Answer:* (a) 15, (b) 81, (c) $W(2, 1, 1,) = W(1, 2, 1) = W(1, 1, 2)$

**11.3** Find the total number of microstates and the most probable distribution for $N = 5$ distinguishable particles distributed among three degenerate cells having $g_1 = g_2 = g_3 = 2$.

*Solution:*
The total number of microstates is easily obtained from Equation 11.18 as

$$\Omega_{\text{M-B}} = \left(\sum_i g_i\right)^N,$$
$$= (g_1 + g_2 + g_3)^N$$
$$= (2 + 2 + 2)^5 = 7776.$$

The most probable distribution of $n_1$, $n_2$, and $n_3$ is that for which Equation 11.15,

$$W_{\text{M-B}} = W_k = \frac{N!\Pi_i g_i^{n_i}}{\Pi_i n_i!},$$

is a maximum. Because the interchange of particles in any two cells results in the same value for $W_k$, there is only the need to calculate $W_k$ for the cases where $n_1 \geq n_2 \geq n_3$. Thus, with $W_k$ denoted as $W(n_1, n_2, n_3)$, we have

$$W(5, 0, 0) = \frac{5!2^5 2^0 2^0}{5!0!0!} = 32,$$

$$W(4, 1, 0) = \frac{5!2^4 2^1 2^0}{4!1!0!} = 160,$$

$$W(3, 2, 0) = \frac{5!2^3 2^2 2^0}{3!2!0!} = 320,$$

$$W(3, 1, 1) = \frac{5!2^3 2^1 2^1}{3!1!1!} = 640,$$

$$W(2, 2, 1) = \frac{5!2^2 2^2 2^1}{2!2!1!} = 960,$$

which shows the most probable distribution is given by

$$W(n_1, n_2, n_3) = W(2, 2, 1) = W(2, 1, 2) = W(1, 2, 2).$$

**11.4** Use the basic equation for an *ensemble average* defined by Equation 11.19 to calculate $\bar{n}_i$ for $i = 1, 2, 3$ for the distribution of particles described in Problem 11.2.

*Answer:* $\frac{4}{3}$

**11.5** Find the most probable distribution for a system of $N = 10^4$ distinguishable particles distributed among three degenerate cells having $g_1 = 1$, $g_2 = 2$, and $g_3 = 3$. Assume the system is at a temperature such that $\epsilon_2 - \epsilon_1 = \epsilon_3 - \epsilon_2 = k_B T$.

*Solution:*
From the description of this system the choice of $\epsilon_1$ is arbitrary. Thus, with $\epsilon_1 \equiv 0$, the partition function (Equation 11.48) is easily evaluated for $i = 1, 2, 3$ as

$$\begin{aligned} Z &\equiv \sum_i g_i e^{-\beta \epsilon_i} \\ &= g_1 e^{-\beta \epsilon_1} + g_2 e^{-\beta \epsilon_2} + g_3 e^{-\beta \epsilon_3} \\ &= 1 \cdot e^0 + 2 \cdot e^{-1} + 3 \cdot e^{-2} \\ &= 1.0000 + 0.7358 + 0.2707 = 2.0065. \end{aligned}$$

Therefore, the most probable set of occupation numbers is given by the Boltzmann distribution (Equation 11.47) as

$$n_1 = \frac{N g_1 e^{-\beta \epsilon_1}}{Z} = \frac{10^4 \cdot 1 \cdot e^{-0}}{2.0065} = 4984,$$

$$n_2 = \frac{Ng_2e^{-\beta\epsilon_2}}{Z} = \frac{10^4 \cdot 2 \cdot e^{-1}}{2.0065} = 3667,$$

$$n_3 = \frac{Ng_3e^{-\beta\epsilon_3}}{Z} = \frac{10^4 \cdot 3 \cdot e^{-2}}{2.0065} = 2023.$$

**11.6** Find the most probable set of occupation numbers for a system of $N = 10^5$ distinguishable particles distributed among three cells having degeneracies $g_1 = 1$, $g_2 = 2$, and $g_3 = 3$. Assume the system is at a temperature such that $\epsilon_2 - \epsilon_1 = 2/\beta$ and $\epsilon_3 - \epsilon_2 = 1/\beta$.

*Answer:* $n_1 = 70421$, $n_2 = 19061$, $n_3 = 10518$

**11.7** Verify that the total number of microstates $W(n_1, n_2, n_3)$ is a maximum for the set of $n_i$ obtained in Problem 11.5, by varying the set of occupation numbers such that the conservation requirements (*energy* and *particles*) are maintained.

*Solution:*
For $W(n_1, n_2, n_3)$ to be a maximum for the set of $n_i$ calculated in Problem 11.5, any slight variation in the values of $n_1$, $n_2$, $n_3$ should result in a new set $n_1'$, $n_2'$, $n_3'$ such that $W'(n_1', n_2', n_3') < W(n_1, n_2, n_3)$. Consequently, the ratio of

$$W = \frac{N!g_1^{n_1}g_2^{n_2}g_3^{n_3}}{n_1!n_2!n_3!}$$

for the most probable set of $n_i$ to

$$W' = \frac{N!g_1^{n_1'}g_2^{n_2'}g_3^{n_3'}}{n_1'!n_2'!n_3'!}$$

for the slightly varied set of $n_i'$, as given by

$$\frac{W}{W'} = g_1^{n_1-n_1'}g_2^{n_2-n_2'}g_3^{n_3-n_3'}\,\frac{n_1'!}{n_1!}\,\frac{n_2'!}{n_2!}\,\frac{n_3'!}{n_3!},$$

should be slightly greater than one. This observation can be tested by allowing $\delta n_1$ to vary by $+1$ or $-1$ and generating values for $\delta n_2$ and $\delta n_3$ such that the conservation requirements are maintained. For $\delta n_1 = +1$ the conservation of total energy requires that $\delta n_3 = +1$ and $\delta n_2 = -2$, since then $\delta E = 1 \cdot \epsilon_1 - 2 \cdot \epsilon_2 + 1 \cdot \epsilon_3 = 1 \cdot 0 - 2 \cdot k_BT + 1 \cdot 2k_BT = 0$. Maintaining the conservation of total energy in this case also results in the conservation of particles being preserved (i.e., $\delta N = \delta n_1 + \delta n_2 + \delta n_3 = 1 - 2 + 1 = 0$), and the result can be generalized to $\delta n_1 = \delta n_3 = -\delta n_2/2$. Consequently, the new set of occupation numbers for the case $\delta n_1 = 1$ is given by

$$\delta n_1 = +1 \rightarrow n_1' = n_1 + 1, \qquad n_2' = n_2 - 2, \qquad n_3' = n_3 + 1,$$

from which we obtain

$$\frac{W}{W'} = g_1^{-1} g_2^2 g_3^{-1} (n_1 + 1) \left[ \frac{1}{(n_2 - 1)\, n_2} \right] (n_3 + 1)$$

$$= 1^{-1} 2^2 3^{-1} (4985) \left[ \frac{1}{(3666)(3667)} \right] (2024)$$

$$= 1.0007.$$

Applying similar arguments for $\delta n_1 = -1$ gives

$$\delta n_1 = -1 \rightarrow n_1' = n_1 - 1, \qquad n_2' = n_2 + 2, \qquad n_3' = n_3 - 1,$$

with the result for $W/W'$ being

$$\frac{W}{W'} = g_1^1 g_2^{-2} g_3^1 \left( \frac{1}{n_1} \right) [n_2 + 1)(n_2 + 2)] \left( \frac{1}{n_3} \right)$$

$$= 1^1 2^{-2} 3^1 \left( \frac{1}{4984} \right) [(3668)(3669)] \left( \frac{1}{2023} \right)$$

$$= 1.0011.$$

Since $W/W' > 1$ for an increase or decrease in $n_1$, $W'$ is smaller than $W$ and the original distribution for $n_1$, $n_2$, and $n_3$ must be the most probable.

**11.8** While maintaining conservation requirements consider $\delta n_1 = \pm 1$ and generate the set of $n_i'$ for *each case* from the most probable set of $n_i$ calculated in Problem 11.6. Now, verify that $W(n_i)$ is a maximum by showing that $W(n_i)/W'(n_i') > 1$.

*Answer:* $\delta n_1 = +1 \rightarrow \dfrac{W}{W'} = 1.0004, \qquad \delta n_1 = -1 \rightarrow \dfrac{W}{W'} = 1.0005$

**11.9** An ideal gas consisting of atomic hydrogen is at a temperature of 300 K. Find (a) the ratio of the number of atoms in the first excited state to the number in the ground state and (b) the temperature at which 20.0 percent of the atoms are in the first excited state.

*Solution:*
Using the Boltzmann distribution (Equation 11.47)

$$n_i = \frac{N}{Z} g_i e^{-\beta \epsilon_i},$$

we obtain

$$\frac{n_2}{n_1} = \frac{g_2}{g_1} e^{\beta(\epsilon_1 - \epsilon_2)}$$

for the relative population of the first excited state to that of the ground state. Recalling that the maximum number of electrons allowed in a shell is given in terms of the principal quantum number $n$ by Equation 7.104,

$$N_n = 2n^2,$$

then the degeneracy of the ground state ($n = 1$) is $g_1 = 2$ and the first excited state ($n = 2$) has a degeneracy of $g_2 = 8$. Of course the energies associated with these electron states are $\epsilon_1 = -13.6$ eV and $\epsilon_2 = -3.4$ eV, which are obvious from Equations 7.39 and 7.41, so the expression for the relative population becomes

$$\frac{n_2}{n_1} = 4e^{\beta(-10.2\text{ eV})}.$$

With $\beta = 1/k_B T$, $k_B = 1.38 \times 10^{-23}$ J/K (Equation 11.108), and $T = 300$ K, we obtain

$$\beta = \frac{1}{(1.38 \times 10^{-23}\text{ J/K})(300\text{ K})}$$

$$= \frac{1}{4.14 \times 10^{-21}\text{ J}} = \frac{1}{0.0259\text{ eV}}$$

and

$$\frac{n_2}{n_1} = 4e^{-(10.2\text{ eV})/(0.0259\text{ eV})}$$

$$= 4e^{-394} = \frac{4}{1.29 \times 10^{171}} = 3.1 \times 10^{-171}.$$

We see that few atoms are in the first excited state, because the energy difference $\epsilon_1 - \epsilon_2$ is so large compared with $k_B T$. In fact, for one atom to be in the first excited state, the gas would have to contain approximately $0.32 \times 10^{171}$ atoms of hydrogen, which is physically impossible, since the mass of the gas would exceed the mass of the universe.

To find the temperature at which 20.0 percent of the atoms are in the first excited state, we require $n_2/n_1 = 0.200/0.800 = 0.250$ and solve for $T$. That is,

$$0.250 = 4e^{(\beta - 10.2\text{ eV})} \rightarrow \frac{10.2\text{ eV}}{k_B T} = \ln 16,$$

from which we obtain

$$T = \frac{10.2\text{ eV}}{(2.77)(1.38 \times 10^{-23}\text{ J/K})}$$

$$= \frac{5.89 \times 10^{-19}\text{ J}}{1.38 \times 10^{-23}\text{ J/K}}$$

$$= 4.27 \times 10^4\text{ K}.$$

**11.10** Find the average energy per molecule of polarized HCL molecules, if they are in a uniform external electric field of $E = 1.38 \times$

$10^7$ N/C at a temperature of 344 K. Imagine each molecule to have a charge distribution of $+q$ separated from a charge of $-q$ by a distance $r$, such that its *dipole moment* is given by $\mu = qr = 3.44 \times 10^{-30}$ C · m and its potential energy is given by $-\mu E$ if aligned parallel to the **E** field or $+\mu E$ if aligned antiparallel.

*Answer:* $\bar{E} = -4.75 \times 10^{-25}$ J

**11.11** Starting with the expression of Equation 11.16 for $W_C$, derive an equation for $\delta \ln W_C$ using Stirling's formula and conservation of particles and energy requirements.

*Solution:*
Taking the logarithm of Equation 11.16,

$$W_C = \Pi_i \frac{g_i^{n_i}}{n_i!},$$

immediately yields

$$\begin{aligned}
\ln W_C &= \sum_i \ln \frac{g_i^{n_i}}{n_i!} \\
&= \sum_i (\ln g_i^{n_i} - \ln n_i!) \\
&= \sum_i (n_i \ln g_i - n_i \ln n_i + n_i) \\
&= N + \sum_i n_i \ln g_i - \sum_i n_i \ln n_i,
\end{aligned}$$

where Stirling's approximation and conservation of particles have been used. Now, $\delta \ln W_C$ is given by

$$\begin{aligned}
\delta \ln W_C &= \delta N + \sum_i \left( \ln g_i \, \delta n_i + n_i \frac{1}{g_i} \overset{0}{\cancel{\delta g_i}} \right) \\
&\quad - \sum_i \left( \ln n_i \, \delta n_i + n_i \frac{1}{n_i} \delta n_i \right) \\
&= \delta N + \sum_i \ln \frac{g_i}{n_i} \delta n_i - \sum_i \delta n_i \\
&= \delta N + \sum_i \ln \frac{g_i}{n_i} \delta n_i - \delta \sum_i n_i \\
&= \delta N + \sum_i \ln \frac{g_i}{n_i} \delta n_i - \delta N \\
&= \sum_i \ln \frac{g_i}{n_i} \delta n_i,
\end{aligned}$$

which is identical to the expression obtained for $\delta \ln W_{\text{M-B}}$ (see Equation 11.40).

**11.12** Starting with the fundamental equation defining the average energy per particle (i.e., $\bar{\epsilon} \equiv E/N$) and using the equations for conservation of particles and energy, derive an equation of the form $\bar{\epsilon} = \bar{\epsilon}(Z)$ for (a) the Maxwell-Boltzmann distribution, and (b) the classical distribution.

*Answer:* $\bar{\epsilon} = -\dfrac{\partial \ln Z}{\partial \beta}$

**11.13** By considering the partial derivative of $Z$ with respect to $\epsilon_j$, derive Equation 11.73 for (a) the Maxwell-Boltzmann distribution and (b) the classical distribution.

*Solution:*
Performing the indicated partial derivative on $Z$ gives

$$\frac{\partial Z}{\partial \epsilon_j} = \frac{\partial}{\partial \epsilon_j} \sum_i g_i e^{-\beta \epsilon_i}$$
$$= -\beta g_j e^{-\beta \epsilon_j},$$

which can be rewritten in the form

$$g_j e^{-\beta \epsilon_j} = -\frac{1}{\beta}\frac{\partial Z}{\partial \epsilon_j}.$$

If we multiply this equation by $e^{-\alpha}$, the left-hand side becomes identical to the Maxwell-Boltzmann distribution (Equation 11.45) in the form

$$n_j = g_j e^{-\alpha} e^{-\beta \epsilon_j}.$$

Thus, we obtain

$$n_j = -e^{-\alpha}\frac{1}{\beta}\frac{\partial Z}{\partial \epsilon_j}.$$

But, from Equation 11.46

$$e^{-\alpha} = \frac{N}{Z},$$

so our equation for $n_j$ becomes

$$n_j = -\frac{N}{Z}\frac{1}{\beta}\frac{\partial Z}{\partial \epsilon_j}$$
$$= -\frac{N}{\beta}\frac{\partial \ln Z}{\partial \epsilon_j}.$$

Similarly, after multiplying

$$g_j e^{-\beta\epsilon_j} = -\frac{1}{\beta}\frac{\partial Z}{\partial \epsilon_j}$$

by $e^{\beta\mu_C}$, the left-hand side is recognized as the classical distribution

$$n_j = g_j e^{\beta\mu_C} e^{-\beta\epsilon_j} = -e^{\beta\mu_C}\frac{1}{\beta}\frac{\partial Z}{\partial \epsilon_j}.$$

But, from Equation 11.64 we have

$$e^{\beta\mu_C} = \frac{N}{Z},$$

so again we obtain

$$n_j = -\frac{N}{\beta}\frac{\partial \ln Z}{\partial \epsilon_j}.$$

**11.14** Starting with $S = k_B \ln W$ and using $\ln W_C$ given by Equation 11.35, derive an expression for entropy $S_C$ that is appropriate for *classical* statistical mechanics.

*Answer:* $S_C = \frac{U}{T} + Nk_B\left(\ln\frac{Z}{N} + 1\right)$

**11.15** By starting with the equation $p = -(\partial F/\partial V)_{N,T}$, derive an equation for pressure in terms of the partition function for (a) Maxwell-Boltzmann statistics and (b) classical statistics.

*Solution:*
Substituting the Helmholtz function $F_{\text{M-B}}$ from Equation 11.80,

$$F_{\text{M-B}} = -Nk_BT \ln Z,$$

into the above expression for pressure gives

$$p = -\left(\frac{\partial[-Nk_BT \ln Z]}{\partial V}\right)_{N,T}$$

$$= Nk_BT\left(\frac{\partial \ln Z}{\partial V}\right)_{N,T}.$$

Similarly, using

$$F_C = -Nk_BT\left(\ln\frac{Z}{N} + 1\right)$$

for classical statistics, we obtain

$$p = -\left(\frac{\partial F_C}{\partial V}\right)_{N,T}$$

$$= Nk_B T \left(\frac{N}{Z}\right)\left(\frac{1}{N}\right)\left(\frac{\partial Z}{\partial V}\right)_{N,T}$$

$$= Nk_B T \left(\frac{\partial \ln Z}{\partial V}\right)_{N,T} .$$

**11.16** Starting with the equation $\mu = (\partial F/\partial N)_{V,T}$, derive an equation for the classical chemical potential $\mu_C$, in terms of the partition function.

*Answer:* $\mu_C = -k_B T \ln \frac{Z}{N}$

**11.17** Starting with the ideal gas *distribution of energies* formula (Equation 11.113), derive Equation 11.114 for the *Boltzmann distribution of speeds*.

*Solution:*
Starting with the Boltzmann distribution of energies,

$$n(\epsilon)\, d\epsilon = 2\pi N \left(\frac{\beta}{\pi}\right)^{3/2} \epsilon^{1/2} e^{-\beta\epsilon}\, d\epsilon,$$

and substituting

$$\epsilon = \tfrac{1}{2}mv^2,$$

$$d\epsilon = mv\, dv,$$

yields

$$n(v)\, dv = 2\pi N \left(\frac{\beta}{\pi}\right)^{3/2} \left(\frac{m}{2}\right)^{1/2} v\, e^{-\beta m v^2/2} mv\, dv,$$

which reduces to the Boltzmann distribution of speeds in the form

$$n(v)\, dv = 4\pi N \left(\frac{\beta m}{2\pi}\right)^{3/2} v^2 e^{-\beta m v^2/2}\, dv.$$

**11.18** Verify the *normalization condition* of statistical mechanics (Equation 11.116) for an ideal gas with $q = v$.

*Answer:* $\int_0^\infty n(v)\, dv = N$

**11.19** Verify that the *Boltzmann distribution of energies* is normalized to $N$ for an ideal gas.

*Solution:*
From the *normalization condition* (Equation 11.116) with $q = \epsilon$ and the *energy distribution* for an ideal gas (Equation 11.113) we have

$$\int_0^\infty n(\epsilon)\, d\epsilon = 2\pi N \left(\frac{\beta}{\pi}\right)^{3/2} \int_0^\infty \epsilon^{1/2} e^{-\beta\epsilon}\, d\epsilon.$$

By changing the variable of integration, the integral on the right-hand side can be evaluated using Equation 11.99. That is, with

$$\epsilon^{1/2} = \alpha,$$

$$d\epsilon = 2\alpha\, d\alpha,$$

we have

$$\int_0^\infty n(\epsilon)\, d\epsilon = 2\pi N \left(\frac{\beta}{\pi}\right)^{3/2} \left(2 \int_0^\infty \alpha^2 e^{-\beta\alpha^2} d\alpha\right)$$

$$= 2\pi N \left(\frac{\beta}{\pi}\right)^{3/2} \left[2\,\frac{1}{4}\left(\frac{\pi}{\beta}\right)^{1/2}\left(\frac{1}{\beta}\right)\right]$$

$$= N.$$

**11.20** Starting with Equation 11.123, find the average energy $\bar{\epsilon}$ of an ideal molecule predicted by Maxwell-Boltzmann statistics.

*Answer:* $\bar{\epsilon} = \dfrac{3}{2} k_B T$

**11.21** Starting with Equation 11.124, show that $\bar{\epsilon} = (3/2)k_B T$ for an ideal gas molecule obeying Maxwell-Boltzmann statistics.

*Solution:*
Substituting the Boltzmann distribution of speeds (Equation 11.114) and $\epsilon = \frac{1}{2}mv^2$ into Equation 11.124 gives

$$\bar{\epsilon} = 4\pi \left(\frac{\beta m}{2\pi}\right)^{3/2} \int_0^\infty \tfrac{1}{2}mv^4\, e^{-\beta m v^2/2}\, dv$$

$$= 4\pi \left(\frac{\beta m}{2\pi}\right)^{3/2} \left(\frac{m}{2}\right) \left[\frac{1}{2}\left(\frac{2\pi}{\beta m}\right)^{1/2} \frac{1}{2}\frac{3}{2}\left(\frac{2}{\beta m}\right)^2\right]$$

$$= \left(\frac{m}{2}\right)\left(\frac{3}{2}\right)\left(\frac{2}{\beta m}\right)$$

$$= \frac{3}{2} k_B T,$$

where the integral has been evaluated using Equation 11.99.

**11.22** Starting with Equation 11.118 and using the Boltzmann distribution of speeds, find the average speed squared for a monatomic ideal gas molecule.

*Answer:* $\bar{v}^2 = \dfrac{3k_BT}{m}$.

**11.23** For an ideal gas governed by Maxwell-Boltzmann statistics, find an expression for the average molecular speed $\bar{v}$.

*Solution:*
From Equation 11.118, with $A = v$ and $q = v$, and the Boltzmann distribution of speeds (Equation 11.114), we have

$$\bar{v} = 4\pi \left(\frac{\beta m}{2\pi}\right)^{3/2} \int_0^\infty v^3 e^{-\beta m v^2/2}\, dv.$$

The integral is of the form

$$I_n(a) \equiv \int_0^\infty u^n e^{-au^2} du$$

given in Appendix A, Section A.10. With $n = 3$ and $a = \beta m/2$, we have

$$\int_0^\infty v^3 e^{-av^2} dv = \frac{1}{2\,a^2}.$$

Thus, the equation for $\bar{v}$ reduces to

$$\begin{aligned}\bar{v} &= 4\pi \left(\frac{\beta m}{2\pi}\right)^{3/2} \left[\frac{1}{2}\left(\frac{2}{\beta m}\right)^2\right] \\ &= 2\,\frac{1}{\pi}\left(\frac{2\pi}{\beta m}\right)^{1/2} \\ &= 2\sqrt{\frac{2k_BT}{\pi m}}.\end{aligned}$$

**11.24** Starting with Equation 11.118, find an expression for the average momentum of an ideal gas molecule obeying Maxwell-Boltzmann statistics.

*Answer:* $\bar{p} = 2\sqrt{\dfrac{2mk_BT}{\pi}}$

**11.25** Derive an expression for the *most probable speed* of an ideal gas molecule for a system obeying Maxwell-Boltzmann statistics.

*Solution:*
Maximizing the *Boltzmann distribution of speeds* given by Equation 11.114,

$$n(v)\,dv = 4\pi N\left(\frac{\beta m}{2\pi}\right)^{3/2} v^2 e^{-\beta m v^2/2}\,dv,$$

requires the first order derivative of $n(v)$ with respect to $v$ to vanish. Thus, with $A = 4\pi N(\beta m/2\pi)^{3/2}$, we have

$$0 = A(2v - \beta m v^3)e^{-\beta m v^2/2},$$

which immediately reduces to

$$2v - \beta m v^3 = 0.$$

Solving this result for $v$ and using $\beta = 1/k_B T$, gives

$$v = \sqrt{\frac{2k_B T}{m}}.$$

**11.26** Derive an expression for the *most probable energy* of an ideal gas molecule obeying Maxwell-Boltzmann statistics.

*Answer:* $\epsilon = \frac{1}{2}k_B T$

**11.27** The theoretical expression for the speed of sound in a monatomic ideal gas is given in terms of presssure $p$ and mass density $\rho$ by $(5p/3\rho)^{1/2}$. By what factor does this differ from the root-mean-square speed $v_{rms}$ and the most probable speed $v$?

*Solution:*
With the theoretical speed of sound represented by $v_s$, the ideal gas equation of state (Equation 11.107a), and the defining equation for mass density (Equation 5.33), we have

$$v_s = \left(\frac{5p}{3\rho}\right)^{1/2}$$

$$= \left[\frac{5(nRT/V)}{3(M/V)}\right]^{1/2}$$

$$= \left[\frac{5(N/N_o)\,RT}{3M}\right]^{1/2}$$

$$= \left(\frac{5Nk_B T}{3M}\right)^{1/2}$$

$$= \sqrt{\frac{5k_B T}{3m}},$$

where the defining equation $k_B \equiv R/N_o$ for the Boltzmann constant and $m = M/N$ have been used in obtaining the last two equalities, respectively. Now, a comparison of this expression for $v_s$ with that of

$$v_{\rm rms} = \sqrt{\frac{3k_BT}{m}}$$

given by Equation 11.128 yields

$$\begin{aligned} v_s &= \sqrt{\frac{5k_BT}{3m}} \\ &= \left(\frac{5}{3}\right)^{1/2} 3^{-1/2}\left(\frac{3k_BT}{m}\right)^{1/2} \\ &= \sqrt{\frac{5}{9}}\, v_{\rm rms}. \end{aligned}$$

Likewise, for the *most probable* speed

$$v = \sqrt{\frac{2k_BT}{m}},$$

we immediately obtain

$$\begin{aligned} v_s &= \sqrt{\frac{5k_BT}{3m}} \\ &= \left(\frac{5}{3}\right)^{1/2} 2^{-1/2}\, v \\ &= \sqrt{\frac{5}{6}}\, v \end{aligned}$$

**11.28** Find (a) $v_{\rm rms}$ of a nitrogen molecule at a temperature of $T = 273$ K, and (b) its translational kinetic energy according to Maxwell-Boltzmann statistics.

*Answer:* (a) $v_{\rm rms} = 493\ \frac{\rm m}{\rm s}$, (b) $\epsilon = 3.77 \times 10^{-21}$ J

**11.29** Show that the *Boltzmann distribution of speeds* can be expressed as $n(v)\,dv = (4N/\pi^{1/2}v_{\rm mp})(v/v_{\rm mp})^2 e^{-(v/v_{\rm mp})^2}dv$, where $v_{\rm mp}$ represents the *most probable speed*. Now, by considering a system of $N_o$ (Avogadro's number) molecules and approximating $dv$ by $\Delta v = 0.01v_{\rm mp}$, find the number of molecules $\Delta n$ with speeds in $dv$ at $v = 0$, $v = v_{\rm mp}$, $v = 2v_{\rm mp}$, $v = 3v_{\rm mp}$, $v = 4v_{\rm mp}$, $v = 5v_{\rm mp}$, $v = 6v_{\rm mp}$, $v = 7v_{\rm mp}$, and $v = 8v_{\rm mp}$.

*Solution:*
From Equations 11.114 and 11.119 we have

$$dn_v = 4\pi N\left(\frac{\beta m}{2\pi}\right)^{3/2} v^2 e^{-\beta m v^2/2}\, dv$$

$$= \frac{4N}{\pi^{1/2}}\left(\frac{m}{2k_BT}\right)^{3/2} v^2 e^{-v^2(m/2k_BT)}\, dv$$

$$= \frac{4N}{\pi^{1/2}}\frac{1}{v_{mp}^3} v^2 e^{-(v/v_{mp})^2}\, dv$$

$$= \frac{4N}{\pi^{1/2} v_{mp}}\left(\frac{v}{v_{mp}}\right)^2 e^{-(v/v_{mp})^2}\, dv,$$

where Equation 11.132 has been used in obtaining the third equality. With $dn_v \rightarrow \Delta n_v$ and $dv \rightarrow \Delta v = 0.01 v_{mp}$, we obtain

$$\Delta n_v = \frac{4N_o}{\pi^{1/2} v_{mp}}\left(\frac{v}{v_{mp}}\right)^2 e^{-(v/v_{mp})^2}(0.01 v_{mp})$$

$$= 1.36 \times 10^{22}\left(\frac{v}{v_{mp}}\right)^2 e^{-(v/v_{mp})^2}.$$

Thus, we obtain the following:

$v = 0 \quad\rightarrow \Delta n_v = 1.36 \times 10^{22}\,(0)^2 e^{-0} = 0$

$v = v_{mp} \quad\rightarrow \Delta n_v = 1.36 \times 10^{22}\,(1)^2 e^{-1} = 5.00 \times 10^{21}$

$v = 2v_{mp} \rightarrow \Delta n_v = 1.36 \times 10^{22}\,(2)^2 e^{-4} = 9.96 \times 10^{20}$

$v = 3v_{mp} \rightarrow \Delta n_v = 1.36 \times 10^{22}\,(3)^2 e^{-9} = 1.51 \times 10^{19}$

$v = 4v_{mp} \rightarrow \Delta n_v = 1.36 \times 10^{22}\,4^2 e^{-16} = 2.45 \times 10^{16}$

$v = 5v_{mp} \rightarrow \Delta n_v = 1.36 \times 10^{22}\,5^2 e^{-25} = 4.72 \times 10^{12}$

$v = 6v_{mp} \rightarrow \Delta n_v = 1.36 \times 10^{22}\,6^2 e^{-36} = 1.14 \times 10^{8}$

$v = 7v_{mp} \rightarrow \Delta n_v = 1.36 \times 10^{22}\,7^2 e^{-49} = 3.49 \times 10^{2}$

$v = 8v_{mp} \rightarrow \Delta n_v = 1.36 \times 10^{22}\,8^2 e^{-64} = 1.40 \times 10^{-4}$

**11.30** Consider a solid to consist of regularly spaced atoms on a simple cubic lattice connected by springs. Using the *equipartition of energy principle*, find the *molal specific thermal capacity* of the solid.

*Answer:* $c_v = 3R$

CHAPTER 12

# Quantum Statistical Mechanics

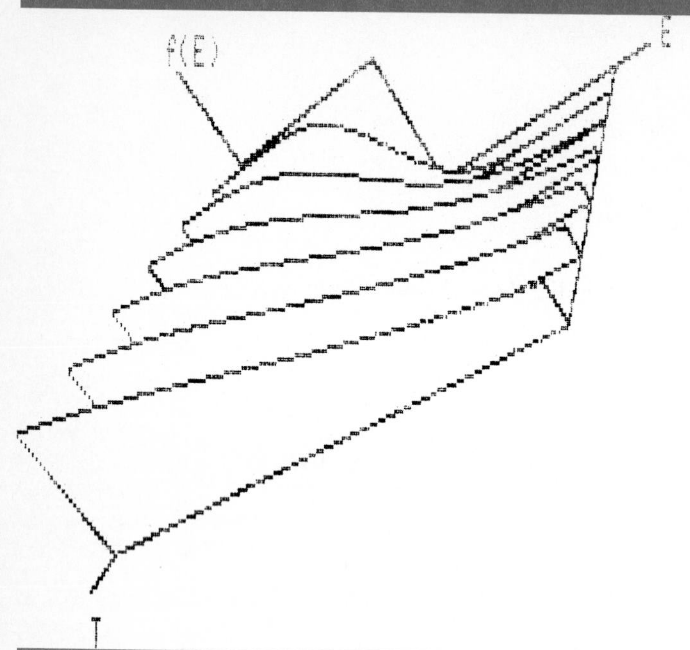


## Introduction

In the preceding chapter the development of Maxwell-Boltzmann statistics involved an *essentially* classical counting procedure, wherein identical particles were explicitly considered to be *distinguishable*. This assumed distinguishability of particles is applicable in classical physics, especially for the molecules of an ideal *rarefied* gas or the molecules of a crystal lattice, but it is not valid in quantum mechanics for, say, the description of inherently *indistinguishable* conduction electrons in a metal. If free conduction electrons are considered as a monatomic ideal gas, classical statistical mechanics predicts a thermal capacity of $(3/2)nR$ according to Equation 11.104. But, the observed heat capacity of a metal at *high temperatures* is given by the Dulong and Petit law as $3nR$ (recall Equation 11.158 and Problem 11.30), which is due solely to the metal lattice. Since the thermal capacity predicted by classical statistical mechanics is valid for both Maxwell-Boltzmann (distinguishable particles) and classical (indistinguishable particles) statistics, the difference between classical and quantum statistical mechanics is more than just the *indistinguishability* of particles. The correct quantum mechanical distribution law for electrons differs significantly from that of classical statistical mechanics in two important, and related, respects. First, in quantum theory electrons are *quantized* and second, they obey the *Pauli exclusion principle* where no *quantum state* can be occupied by more than one electron.

We have also seen (Chapter 6) how electromagnetic radiation is *quantized* in nature, exhibiting particle-like behavior in the photoelectric and the Compton effects. If we consider the classical statistical mechanics of an ideal gas consisting of indistinguishable *photons*, the classical energy distribution formula obtained also predicts the distribution in frequency (or wavelength) because the energy of a photon is directly proportional to its frequency (or inversely proportional to its wavelength). In this case, classical statistical mechanics predicts a frequency dependence for the radiation energy density (energy per unit volume) that is inconsistent with observation and the well-known Planck radiation law. Even though photons do not obey the Pauli exclusion principle, the associated quantum statistics differs significantly from that of classical statistical mechanics in that photons are *quantized*. This means that the concept of representative phase points being continuously distributed in phase space is no longer valid in quantum theory because the energy of a *particle* (e.g., electrons, photons, etc.) is restricted to a discrete set of *quantized* energy states and cannot change in a continuous manner or assume an arbitrary value.

The difficulties arising from employing classical statistical mechanics to describe electrons and photons is completely resolved by quantum statistics. The discussion of this chapter is still restricted to the statistics of a *microcanonical ensemble* of systems consisting of, essentially, *free* particles. However, each system will now be treated in a completely quantum mechanical manner, where identical particles are considered to be *indistinguishable* and *quantized*. Certainly, because of these requirements, quantum statistics will differ from the classical theory in the counting of the number of microstates associated with a particular macrostate. This difference is illustrated by a simple example where the formulation of quantum statistics is carefully discussed. From this understanding, the thermodynamic probabilities for Bose-Einstein and Fermi-Dirac statistics are developed, which apply, respectively, to quantized indistinguishable particles to which the Pauli exclusion principle is not applicable (e.g., photons) and to particles that obey the principle (e.g., electrons). Next, the quantum distribution functions for Bose-Einstein and Fermi-Dirac statistics are derived, using the *method of the most probable distribution* that was detailed for classical statistical mechanics. Then useful insights are obtained by comparing all four of the distribution functions with one another. Further, after evaluating the Lagrange undetermined multiplier $\beta$ for quantum statistics, the distribution functions for Maxwell-Boltzmann, Bose-Einstein, and Fermi-Dirac statistics are applied to different quantum mechanical examples. In particular, the specific heat of a solid is obtained by applying (a) Maxwell-Boltzmann statistics to quantized harmonic oscillators and (b) Bose-Einstein statistics to *quanta of vibrational energy*, called *phonons*, in the respective development of the Einstein theory and the Debye theory. Bose-Einstein statistics is also applied to an ideal gas consisting of *photons* for the analysis of radiation in a cavity, where the Planck radiation formula, Wien's displacement law, and the Stefan-Boltzmann law are derived. Finally, Fermi-Dirac statistics is applied to the conduction electrons in a metal for the determination of the electronic density of states and specific heat.

## 12.1 Formulation of Quantum Statistics

Because classical statistical mechanics is not in agreement with observed physical phenomenon involving quantized particles (e.g., electrons, photons, etc.), there must be a fundamental conceptual difficulty with the counting procedure used in obtaining Maxwell-Boltzmann and classical statistics. This is especially true for the discussion presented in Chapter 11, Sections 11.1 and 11.2, where cells in $\mu$-space were considered to be very small compared to the spatial dimensions and range of momenta of the real system, yet large enough to accommodate a number of representative phase points. Usually, the number of particles (or representative phase points) per cell is quite large for a real system described by classical statistical mechanics, as was suggested by several problems at the end of Chapter 11. It is noteworthy, however, that the introduction of the Heisenberg uncertainty principle, predicting a minimum volume per cell of

Minimum Volume Per Cell

$$\tau_0 = h^3 \approx 3 \times 10^{-104}\ (\mathrm{J \cdot s})^3, \tag{12.1}$$

suggests that there are considerably more *elemental cells* in accessible $\mu$-space than particles. Consequently, only a small fraction of the elemental cells in $\mu$-space have a nonzero occupation number. This statement is easily verified for Maxwell-Boltzmann statistics by considering the quantity

Occupation Index

$$f(\epsilon_i) \equiv \frac{n_i}{g_i}, \tag{12.2}$$

called the **occupation index** of a cell of energy $\epsilon_i$. It represents the average number of particles in each of the $g_i$ cells of identical energy $\epsilon_i$. Using the Boltzmann distribution (Equation 11.47) in this equation gives

$$f(\epsilon_i) = \frac{N}{Z} e^{-\beta\epsilon_i},$$

from which it is obvious that $f(\epsilon_i)$ is a maximum for cells where $\epsilon_i = 0$. Thus, for a monatomic gas with

$$Z = \frac{V}{h^3}(2\pi m k_B T)^{3/2} \qquad (11.100)$$

we have

$$[f(\epsilon_i)]_{\max} = \frac{Nh^3}{V(2\pi m k_B T)^{3/2}}. \qquad (12.3)$$

This equation can be evaluated for an ideal monatomic gas by solving the *ideal gas equation of state* for $V$ (see Equations 11.107a to 11.108),

$$V = \frac{nRT}{p} = \frac{N}{N_o}\frac{RT}{p} = \frac{Nk_BT}{p},$$

and substituting to obtain

$$[f(\epsilon_i)]_{\max} = \frac{ph^3}{k_BT(2\pi m k_B T)^{3/2}}. \qquad (12.4)$$

Considering helium, as the lightest monatomic gas, at **standard conditions** (i.e., STP $\rightarrow T = 273$ K, $p = 1.013 \times 10^5$ N/m$^2$), we have, without the inclusion of units,

$$\begin{aligned}(2\pi m k_B T)^{3/2} &= [2\pi(6.6466 \times 10^{-27})(1.3807 \times 10^{-23})(273)]^{3/2} \\ &= (1.5742 \times 10^{-46})^{3/2} = 1.975 \times 10^{-69}\end{aligned}$$

and Equation 12.4 gives

$$\begin{aligned}[f(\epsilon_i)]_{\max} &= \frac{(1.013 \times 10^5)(6.626 \times 10^{-34})^3}{(1.381 \times 10^{-23})(273)(1.975 \times 10^{-69})} \\ &= 3.958 \times 10^{-6} \approx \frac{1}{252{,}700}.\end{aligned}$$

This result means that there is only one particle to every 252,700 cells in the *most densely occupied region* of $\mu$-space. The *occupation index* is even smaller for heavier gases (i.e., $m > m_{\mathrm{He}}$) and for real gases, where the kinetic energy per particle is nonzero (i.e., $\epsilon_i > 0$). Clearly, the number of elemental cells in $\mu$-space is prodigiously larger than the number of particles in a real gas and, consequently, only a very small fraction of the cells are occupied.

From the above discussion, it is clear that introducing the fundamental quantum mechanical concept of the Heisenberg uncertainty principle necessitates a fundamental revision of the statistical argument. This revision was, essentially, addressed in Chapter 11, Section 11.3 with the introduction of energy-cell degeneracy and the system configuration given in Table 11.4. We need, however, to restate the statistical argument using purely quantum mechanical concepts and terminology. In quantum mechanics the energy of a *particle* (e.g., electron, photon, etc.) is *quantized* to a set of discrete values that are allowed by the *principal quantum number*. But, because of the existence of three other quantum numbers (i.e., orbital quantum number, magnetic quantum number, and spin quantum number), we have *in general* the existence of different *quantum states* having the *same energy*. Thus, a *degeneracy* in the *energy level*, defined by the principle quantum number $n$, is fundamental in quantum theory. For bound electrons, as an example, a unique *quantum state* is completely defined by specifying all four quantum numbers. The *quantum state* for a *free* electron, however, is completely defined by specifying its principal and spin quantum numbers $n$ and $m_s$. Again, we have an obvious *energy level degeneracy*, because of the two values $(-\frac{1}{2}, +\frac{1}{2})$ allowed for $m_s$. Further, there is an additional degeneracy resulting from the requirement of *three principal quantum numbers* $(n_x, n_y, n_z)$ for the complete specification of each *energy level*, as was discussed in connection with Equation 10.137. Thus, instead of thinking of cell-energy degeneracy, we need to consider *energy level degeneracy* in quantum statistics. Accordingly, the **occupation number** must now be thought of as the number of particles distributed among the allowed *quantum states* of a degenerate *energy level*, and a **macrostate** is specified by enumerating the occupation number $n_i$ in each quantum mechanically allowed *energy level* $\epsilon_i$. With this, the *configuration* of a quantum mechanical system in $\mu$-space can be represented by that illustrated in Table 12.1, where the **degeneracy** $g_i$ refers to the number of unique and allowed *quantum states* for the energy level $\epsilon_i$. The energy level $\epsilon_0$, corresponding to the principal quantum number $n = 0$, has been included in this configuration to accommodate the **zero point energy** of quantum mechanics (e.g., recall the quantum mechanical harmonic oscillator of Chapter 10, Section 10.4).

It is also important to note that although the energy levels in quantum

| Principal Quantum Number | 0 | 1 | 2 | 3 | $\cdots$ | $i$ | $\cdots$ |
|---|---|---|---|---|---|---|---|
| Energy Level | $\epsilon_0$ | $\epsilon_1$ | $\epsilon_2$ | $\epsilon_3$ | $\cdots$ | $\epsilon_i$ | $\cdots$ |
| Degeneracy | $g_0$ | $g_1$ | $g_2$ | $g_3$ | $\cdots$ | $g_i$ | $\cdots$ |
| Occupation Number | $n_0$ | $n_1$ | $n_2$ | $n_3$ | $\cdots$ | $n_i$ | $\cdots$ |

TABLE 12.1
A quantum system configuration for the allowed degenerate energy levels in $\mu$-space.

theory are discrete and *not continuous*, there are a number of cases in quantum statistics where the energy levels for free *particles* (e.g., electrons, phonons, and photons) are so closely spaced that they form a *continuum*. In these cases the same arguments that led to Equation 11.95 may be used to obtain a *general* expression for the degeneracy in the form

$$g(p)\,dp = 2\,\frac{4\pi V}{h^3}\,p^2 dp. \tag{12.5}$$

Degeneracy

This expression differs from Equation 11.95 by only the multiplicative factor 2, which takes into account the two directions of polarization for photons or the two directions of spin for electrons (see Sections 12.6 and 12.7, respectively). This equation gives the *total number of quantum states* in $\mu$-space that have a momentum between $p$ and $p + dp$ and, as will be later illustrated, can be easily transformed into a degeneracy in terms of the energy of the quantum states allowed for free electrons, photons, and even phonons (see Section 12.5).

The above discussion for the *configuration* of a system in quantum statistics is basically the same as that following Table 11.4, where we modified, somewhat, our interpretation of the classical counting procedure. For an isolated system, we still have the conservation requirements for the total number of particles,

$$N = \sum_i n_i, \tag{11.4}$$

Conservation of Particles

and the total energy,

$$E = \sum_i n_i \epsilon_i, \tag{11.5}$$

Conservation of Energy

of a system, where now the summation extends over all allowed *energy levels* in $\mu$-space. To be consistent with the notion of energy level degeneracy, we may now imagine an energy level $\epsilon_i$ of $\mu$-space to be *partitioned* into the appropriate number of *quantum states* that will exactly accommodate the degeneracy $g_i$, wherein the $n_i$ particles may be distributed. An

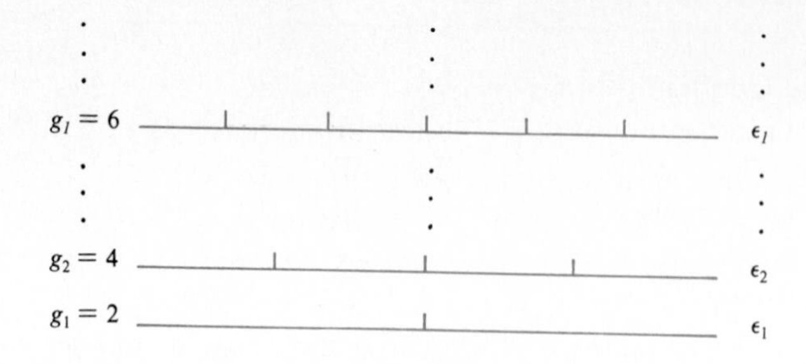


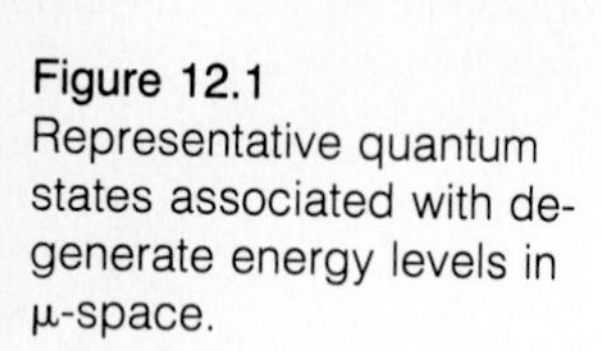
**Figure 12.1** Representative quantum states associated with degenerate energy levels in μ-space.

example of this scheme is depicted in Figure 12.1, where the energy level $\epsilon_1$ has been subdivided by one partition to accommodate $g_1 = 2$ quantum states, $\epsilon_2$ has three partitions to subdivide the energy level in $g_2 = 4$ quantum states, and $\epsilon_i$ has five partitions for $g_i = 6$. Of course, the counting of the total number of different distributions of $n_i$ particles in each of the $\epsilon_i$ energy levels must be done in a manner to ensure the *indistinguishability* requirement that is fundamental in quantum statistics. This requirement and the appropriate counting procedure for Bose-Einstein and Fermi-Dirac statistics will be discussed in some detail in the next section.

## 12.2 Thermodynamic Probabilities in Quantum Statistics

The method of determining the thermodynamic probability in quantum statistics is rather similar to that presented for Maxwell-Boltzmann and classical statistics. That is, the number of *microstates* $W_k$ corresponding to a particular macrostate requires an assumed set of occupation numbers,

$$n_0, n_1, n_2, \cdots, n_i, \cdots,$$

for a *set of different energy levels*,

$$\epsilon_0, \epsilon_1, \epsilon_2, \cdots, \epsilon_i, \cdots.$$

The total number of *microstates* $W_k$ for the $k$th macrostate is then obtained by enumerating or calculating the allowed permutations $w$ for the *quantities* occupying *each* separate energy level,

$$w_0, w_1, w_2, \cdots, w_i, \cdots,$$

and multiplying the results, that is,

$$W_k \equiv \Pi_i w_i. \qquad (12.6)$$

Thermodynamic Probability

The product of the $w_i$'s in this equation should be reasonably obvious, since for the permutations $w_0$ allowed for the *quantities* of energy level $\epsilon_0$, there are $w_1$ independent permutations allowed for $\epsilon_1$, $w_2$ for $\epsilon_2$, and so forth. The actual counting for the various statistics is easily understood by a simple example, like the one discussed in Chapter 11, Section 11.2. More specifically, however, we will consider a very simple example of distributing $N = 2$ particles in a $\mu$-space consisting of only one energy level $\epsilon_1$ having a degeneracy $g_1 = 4$. Before considering the enumeration of the microstates for Bose-Einstein and Fermi-Dirac statistics, the Maxwell-Boltzmann case will be discussed for the purposes of comparison.

## Maxwell-Boltzmann Statistics Revisited

In this example of two particles occupying one energy level, there is only one possible macrostate corresponding to $n_1 = 2$. The allowed microstates associated with this macrostate for Maxwell-Boltzmann statistics are illustrated in Figure 12.2, where two conditions on the distribution of the par-

**Maxwell-Boltzmann Statistics**

| Quantum States of $\epsilon_1$ | | | |
|---|---|---|---|
| $QS_1$ | $QS_2$ | $QS_3$ | $QS_4$ |
| *ab* | | | |
| | *ab* | | |
| | | *ab* | |
| | | | *ab* |
| *a* | *b* | | |
| *a* | | *b* | |
| *a* | | | *b* |
| | *a* | *b* | |
| | *a* | | *b* |
| | | *a* | *b* |
| *b* | *a* | | |
| *b* | | *a* | |
| *b* | | | *a* |
| | *b* | *a* | |
| | *b* | | *a* |
| | | *b* | *a* |

Figure 12.2
The M-B microstates associated with two particles restricted to a $\mu$-space of one quadruply degenerate energy level.

ticles have been observed. First, the particles are treated as *distinguishable* by labeling them as $a$ and $b$ and, second, there are no restrictions on the number of particles that can occupy a given quantum state.

The sixteen microstates illustrated in Figure 12.2 can be predicted by the statistical counting method discussed previously in Chapter 11, Sections 11.2 and 11.3. Generalizing and recapitulating in the terminology of quantum statistics, the first particle can be chosen in $N$ ways and placed in any one of the $g_1$ quantum states in energy level $\epsilon_1$. So the total number of ways of selecting the first particle and placing it somewhere is $Ng_1$. For the second particle the number of ways is $(N - 1)g_1$, for the third particle its $(N - 2)g_1$, and so forth for the $n_1$ particles occupying energy level $\epsilon_1$. Since the last particle for $\epsilon_1$ can be chosen in $(N - n_1 + 1)g_1$ ways, then the total number of ways of selecting $n_1$ particles to occupy $g_1$ quantum states of energy level $\epsilon_1$ is simply

$$\begin{aligned} w_1 &= \frac{Ng_1(N - 1)g_1 \cdots (N - n_1 + 1)g_1}{n_1!} \\ &= \frac{N! g_1{}^{n_1}}{(N - n_1)! n_1!}, \end{aligned}$$

where the $n_1!$ in the denominator eliminates the *irrelevant permutations*. This same reasoning is used if there is a second energy level $\epsilon_2$ having a degeneracy $g_2$ only, now, the first particle may be selected in $(N - n_1)g_2$ ways, the second in $(N - n_1 - 1)g_2$ ways, and so forth, until the last of the $n_2$ particles is selected in $(N - n_1 - n_2 + 1)g_2$ ways. Thus, eliminating the $n_2!$ irrelevant permutations we have

$$\begin{aligned} w_2 &= \frac{(N - n_1)g_2 \, (N - n_1 - 1)g_2 \cdots (N - n_1 - n_2 + 1)g_2}{n_2!} \\ &= \frac{(N - n_1)! g_2{}^{n_2}}{(N - n_1 - n_2)! n_2!}, \end{aligned}$$

and generalizing to the $i$th energy level gives

$$w_i = \frac{(N - n_1 \cdots - n_{i-1})! g_i{}^{n_i}}{(N - n_1 \cdots - n_i)! n_i!}.$$

In this recapitulation we have reproduced Equations 11.14a through 11.14c, which upon substitution into Equation 12.6 yields the familiar **Maxwell-Boltzmann thermodynamic probability,**

$$W_{\text{M-B}} = \frac{N!\Pi_i\, g_i^{n_i}}{\Pi_i n_i!}. \tag{12.7}$$

Maxwell-Boltzmann Thermodynamic Probability

This result is clearly capable of predicting the number of microstates listed in Figure 12.2, since for $N = 2$, $i = 1$, $g_i = 4$, and $n_i = 2$ we have

$$W_{\text{M-B}} = \frac{2!4^2}{2!} = 16.$$

An interesting difference between Maxwell-Boltzmann and quantum statistics can be obtained from this example and the definition

$$\mathcal{N} \equiv \frac{\text{probability of all particles occupying the same quantum state}}{\text{probability of particles occupying different quantum states}} \tag{12.8}$$

For our simple example

$$\mathcal{N}_{\text{M-B}} = \frac{4}{12} = \frac{1}{3},$$

which will be interesting when compared with the results predicted by quantum statistics.

## Bose-Einstein Statistics

In quantum statistics particles are considered to be inherently *indistinguishable*. But, in Bose-Einstein (frequently abbreviated as B-E) statistics the Pauli exclusion principle is *not obeyed*, which means that *any number* of *particles* can occupy *any one quantum state*. Particles with this behavior are collectively called **bosons** (see Table 12.2 for listing) and represent

| *Bosons* | *Spin* | *Fermions* | *Spin* |
|---|---|---|---|
| $\alpha$ particle | 0 | Electron | $\frac{1}{2}$ |
| He atom | 0 | Neutron | $\frac{1}{2}$ |
| $\pi$-meson (pion) | 0 | Proton | $\frac{1}{2}$ |
| Phonon | 1 | Positron | $\frac{1}{2}$ |
| Photon | 1 | $\mu$-meson (muon) | $\frac{1}{2}$ |
| Deuteron | 1 | Neutrino | $\frac{1}{2}$ |

TABLE 12.2
A partial listing of particles classified as bosons and fermions.

**Bose-Einstein Statistics**

| Quantum States of $\epsilon_1$ | | | |
|---|---|---|---|
| $QS_1$ | $QS_2$ | $QS_3$ | $QS_4$ |
| *aa* | | | |
| | *aa* | | |
| | | *aa* | |
| | | | *aa* |
| *a* | *a* | | |
| *a* | | *a* | |
| *a* | | | *a* |
| | *a* | *a* | |
| | *a* | | *a* |
| | | *a* | *a* |

Figure 12.3
The microstates allowed by B-E statistics for two particles restricted to a μ-space of a quadruply degenerate energy level.

those particles in nature that have an integral (0, 1, 2, ⋯) total spin angular momentum measured in units of $\hbar$. Since the particles are *indistinguishable*, both particles in our simple example must be labeled the same, say $a$, and the allowed microstates may be enumerated as illustrated in Figure 12.3. There are $g_1 = 4$ ways of placing all particles in the same quantum state and six ways of placing them in different states, so the total number of microstates is ten. It should be noted that the number of microstates allowed by B-E statistics is less than those allowed by M-B statistics by exactly the number of microstates corresponding to an interchange of particles (i.e., distinguishability) between different quantum states.

To quantify the counting procedure that is appropriate for the B-E distribution of particles, observe that *in general* $g_i - 1$ partitions or vertical lines are required to subdivide an energy level $\epsilon_i$ into $g_i$ quantum states. We have $n_i$ particles to place in $g_i$ quantum states that are defined by $g_i - 1$ lines. Thus, there are $n_i + g_i - 1$ *quantities* to be distributed, and any order in which these quantities are placed will represent a microstate of the *i*th energy level. The total number of possible ways of distributing $n_i + g_i - 1$ quantities is given by $(n_i + g_i - 1)!$. But, of these *possible* permutations there are $n_i!$ and $(g_i - 1)!$ *irrelevant permutations* of the particles and lines, respectively. Consequently, there are

$$w_i = \frac{(n_i + g_i - 1)!}{(g_i - 1)!n_i!} \tag{12.9}$$

allowed and *unique* distributions of the $n_i$ *indistinguishable* particles among the $g_i$ quantum states of the $\epsilon_i$ energy level. Clearly, a similar result may be argued for any energy level, so the **Bose-Einstein thermodynamic probability** for a *particular macrostate* is obtained from Equations 12.6 and 12.9 as

Bose-Einstein Thermodynamic Probability

$$W_{\text{B-E}} = \Pi_i \frac{(n_i + g_i - 1)!}{(g_i - 1)!n_i!}, \tag{12.10}$$

where the product extends over all possible *energy levels*.

It is easy to verify that ten microstates are allowed for our simple example illustrated in Figure 12.3 by direct substitution of $i = 1$, $n_i = 2$, and $g_i = 4$ into Equation 12.10, that is,

$$\begin{aligned} W_{\text{B-E}} &= \frac{(2 + 4 - 1)!}{(4 - 1)!2!} \\ &= \frac{5!}{3!2!} = 10. \end{aligned}$$

Also, since there are only four ways of placing all the particles in any one quantum state, Equation 12.8 gives

$$\mathcal{N}_{\text{B-E}} = \frac{4}{6} = \frac{2}{3}.$$

Because this result is larger than that obtained for $\mathcal{N}_{\text{M-B}}$, there is a greater relative tendency for *bosons* to *bunch together* than particles obeying M-B statistics.

## Fermi-Dirac Statistics

In Fermi-Dirac (frequently abbreviated as F-D) statistics particles are considered to be *indistinguishable* and to *obey the Pauli exclusion principle*. Such particles are collectively called **fermions** (see listing in Table 12.2) and have a total spin angular momentum (measured in units of $\hbar$) that is half-integral (1/2, 3/2, 5/2, ···). The enumeration of allowed microstates in this case for our simple example of two particles distributed among four quantum states is illustrated in Figure 12.4. Because of the *Pauli exclusion principle*, there are only six distinguishably different ways of placing the two particles in different quantum states.

From the above example we can generalize and obtain the appropriate quantitative expression for the F-D distribution of *indistinguishable* particles. This situation differs from the B-E case in that now a quantum state

**Fermi-Dirac Statistics**

| Quantum States of $\epsilon_1$ | | | |
|---|---|---|---|
| $QS_1$ | $QS_2$ | $QS_3$ | $QS_4$ |
| *a* | *a* | | |
| *a* | | *a* | |
| *a* | | | *a* |
| | *a* | *a* | |
| | *a* | | *a* |
| | | *a* | *a* |

Figure 12.4
The microstates allowed by F-D statistics for two particles restricted to a μ-space of a quadruply degenerate energy level.

can either be vacant or occupied by only *one* particle. In general then, $g_i - n_i$ quantum states are vacant and $n_i$ *quantum states* are filled. Thus, there are $g_i$ ways of selecting the first quantum state to be occupied by any one of the $n_i$ particles, $g_i - 1$ ways of selecting the second quantum state to be populated, and so forth, and $g_i - n_i + 1$ ways of selecting the $n_i$th quantum state to be occupied with the last of the $n_i$ particles. Thus, there are $g_i \cdot (g_i - 1) \cdots (g_i - n_i + 1)$ ways of selecting $n_i$ of the $g_i$ quantum states to be populated by the $n_i$ indistinguishable particles, where the order in which the $n_i$ quantum states were selected has been counted. But this order should not be counted, so we must divide by the irrelevant permutations $n_i!$ of the filled quantum states. Consequently, there are

$$w_i = \frac{g_i\,(g_i - 1)(g_i - 2) \cdots (g_i - n_i + 1)}{n_i!}$$

$$= \frac{g_i!}{(g_i - n_i)!n_i!} \tag{12.11}$$

allowed and *unique* distributions of $n_i$ *indistinguishable* particles among $g_i$ quantum states of equal energy $\epsilon_i$. Since this result was argued in general for any degenerate energy level being populated by indistinguishable particles obeying the Pauli exclusion principle, then the **Fermi-Dirac thermodynamic probability** for a particular macrostate of a system is immediately obtained from Equations 12.6 and 12.11 in the form

Fermi-Dirac Thermodynamic Probability

$$W_{\text{F-D}} = \Pi_i \frac{g_i!}{(g_i - n_i)!n_i!}. \tag{12.12}$$

As before in the previous cases, this distribution law can be used to predict the number of allowed microstates illustrated in Figure 12.4 for our simple example, that is,

$$W_{\text{F-D}} = \frac{4!}{(4-2)!2!} = \frac{4!}{2!2!} = 6.$$

Further, since (see Equation 12.8 and Figure 12.4)

$$\mathcal{N}_{\text{F-D}} = \frac{0}{6} = 0$$

is less than $\mathcal{N}_{\text{M-B}} = 1/3$ or $\mathcal{N}_{\text{B-E}} = 2/3$, there is a greater relative tendency for *fermions* to be *separated* in different states than that allowed by either M-B or B-E statistics.

## 12.3 Most Probable Distribution

The method of the *most probable* distribution, which was detailed in Chapter 11, Section 11.4 for classical statistical mechanics, will now be employed to find a distribution law of the form $n_i = f(\epsilon_i)$ that gives the most probable occupation number for B-E and F-D statistics. The recipe for these derivations is to first take the appropriate thermodynamic probability $W_k$ for B-E or F-E statistics and find $\ln W_k$, using Stirling's formula,

Stirling's Formula

$$\ln A! \approx A \ln A - A, \qquad A >> 1 \tag{11.33}$$

to evaluate logarithmic factorials. Second, the relation for $\ln W_k$ is *maximized* by considering $\delta \ln W_k = 0$, which gives a relation of the form

$$\sum_i f(n_i, g_i)\, \delta n_i = 0 \tag{12.13}$$

after judicious application of the requirements $\delta N = \delta g_i = 0$. Since the $\delta n_i$'s of this equation are *not independent* then, third, the *Lagrange method of undetermined multipliers* is employed to incorporate the *conservation requirement* by adding

$$-\alpha\delta N = 0 = -\alpha \sum_i \delta n_i, \tag{12.14}$$

$$-\beta\delta E = 0 = -\beta \sum_i \epsilon_i \delta n_i \tag{12.15}$$

to Equation 12.13 to obtain

$$\sum_i [f(n_i, g_i) - \alpha - \beta\epsilon_i]\, \delta n_i = 0. \tag{12.16}$$

Since undetermined multipliers $\alpha$ and $\beta$ are independent of the occupation numbers $n_i$, this equation is in terms of effectively *independent* $\delta n_i$'s. Thus, the coefficient in the brackets must vanish for every value in the sum and we have the relation

$$f(n_i, g_i) - \alpha - \beta\epsilon_i = 0, \tag{12.17}$$

which can be solved algebraically for the most probable occupation number $n_i$ in terms of the energy level $\epsilon_i$. This three part recipe will now be employed to find the appropriate distribution laws for B-E and F-D statistics.

## Bose-Einstein Distribution

Before taking the logarithm of the Bose-Einstein thermodynamic probability,

$$W_{\text{B-E}} = \Pi_i \frac{(n_i + g_i - 1)!}{(g_i - 1)!n_i!},$$

let us assume

$$g_i >> 1,$$

so that the numerator and denominator of $W_{\text{B-E}}$ simplifies and

$$W_{\text{B-E}} = \Pi_i \frac{(n_i + g_i)!}{g_i!n_i!}. \tag{12.18}$$

The logarithm of $W_{\text{B-E}}$ is easily obtained by considering

$$\begin{aligned} \ln W_{\text{B-E}} &= \sum_i [\ln (n_i + g_i)! - \ln g_i! - \ln n_i!] \\ &= \sum_i [(n_i + g_i) \ln (n_i + g_i) - g_i \ln g_i - n_i \ln n_i], \end{aligned} \tag{12.19}$$

where Stirling's formula has been used. Realizing that $\delta g_i = 0$, $\delta \ln W_{\text{B-E}}$ is immediately determined in the form

$$\delta \ln W_{\text{B-E}} = \sum_i [\delta n_i \ln (n_i + g_i) + \delta n_i - \delta n_i \ln n_i - \delta n_i],$$

which obviously reduces to

$$\delta \ln W_{\text{B-E}} = \sum_i [\ln (n_i + g_i) - \ln n_i] \, \delta n_i = 0. \qquad (12.20)$$

Introducing the Lagrange multipliers by way of the conservation requirements, as given by Equations 12.14 and 12.15, we obtain

$$\delta \ln W_{\text{B-E}} = 0 = \sum_i [\ln (n_i + g_i) - \ln n_i - \alpha - \beta\epsilon_i] \, \delta n_i.$$

It is interesting to note that this same result is obtained under the assumption $n_i + g_i \gg 1$, which reduces only the numerator of $W_{\text{B-E}}$ (see Problems 12.3 and 12.4). Since the $\delta n_i$'s are independent in this expression, the coefficient in the brackets must vanish for each and every value in the sum. Thus,

$$\ln (n_i + g_i) - \ln n_i - \alpha - \beta\epsilon_i = 0,$$

which can be rearranged as

$$\frac{n_i + g_i}{n_i} = e^{\alpha + \beta\epsilon_i}$$

and solved for $n_i$ in the form

$$n_i = \frac{g_i}{e^{\alpha} e^{\beta\epsilon_i} - 1}. \qquad (12.21)$$

Bose-Einstein Distribution

This result represents the **Bose-Einstein distribution** for the *most probable occupation number* for a system consisting of *bosons* (see Table 12.2).

In this derivation the conservation of particles was explicitly assumed, when Equation 12.14 involving the Lagrange multiplier $\alpha$ was added to Equation 12.20. Such a requirement is not applicable to certain bosons like *photons* and *phonons*. For example, the photons of an *ideal photon gas* enclosed in a container of volume $V$ will be annihilated (absorbed) and created (emitted) by the container walls. For this particular class of bosons, $\alpha$ is not introduced in the above derivation (i.e., particles are not conserved), so the *Bose-Einstein distribution* (Equation 12.21) reduces to

$$n_i = \frac{g_i}{e^{\beta\epsilon_i} - 1} \qquad (12.22)$$

Photon/Phonon Statistics

for the special case of *photon (or phonon) statistics*. The undetermined multiplier $\beta$ in this equation, as well as in Equation 12.21, is related to the absolute temperature of the system and, as before, is given by $\beta = 1/k_BT$. The derivation of this identity, however, is postponed to Section 12.4, where it will be demonstrated to be the same for B-E and F-D statistics.

## Fermi-Dirac Distribution

An expression for the most probable occupation number for Fermi-Dirac statistics is easily obtained from the expression for $W_{\text{F-D}}$,

$$W_{\text{F-D}} = \Pi_i \frac{g_i!}{(g_i - n_i)!n_i!},$$

given by Equation 12.12. In this case there is no need for any assumption (recall that $g_i \gg 1$ was assumed for B-E statistics) before evaluating the logarithm of $W_{\text{F-D}}$. Using the properties of the logarithm and Stirling's formula, we immediately obtain

$$\ln W_{\text{F-D}} = \sum_i [-(g_i - n_i) \ln (g_i - n_i) + g_i \ln g_i - n_i \ln n_i], \quad (12.23)$$

from which

$$\delta \ln W_{\text{F-D}} = \sum_i [\ln (g_i - n_i) - \ln n_i]\, \delta n_i = 0. \quad (12.24)$$

It should be noted that these last two equations are rather similar to the corresponding ones obtained for B-E statistics (i.e., Equations 12.19 and 12.20, respectively). Now, however, adding the conservation requirements with Lagrange multipliers (Equations 12.14 and 12.15) to the expression for $\delta \ln W_{\text{F-D}}$ gives

$$\ln (g_i - n_i) - \ln n_i - \alpha - \beta\epsilon_i = 0,$$

from which we obtain

Fermi-Dirac Distribution

$$n_i = \frac{g_i}{e^{\alpha}e^{\beta\epsilon_i} + 1}. \quad (12.25)$$

This equation represents the **Fermic-Dirac distribution,** which gives the *most probable occupation number* for a system containing *fermions* (see

Table 12.2). It differs only slightly from the B-E distribution, in that the sign in the denominator is *positive* instead of *negative*. One of the most important applications of this distribution law is in the description of the free conduction electrons of a metal, which will be addressed in some detail in Section 12.7.

## Classical Limit of Quantum Distributions

The form of the Maxwell-Boltzmann distribution given by Equation 11.45 lends itself to a comparison with the Bose-Einstein and Fermi-Dirac distributions. The three distributions,

$$n_i = \frac{g_i}{e^{\alpha}e^{\beta\epsilon_i}}, \qquad (11.45)$$

Maxwell-Boltzmann

$$n_i = \frac{g_i}{e^{\alpha}e^{\beta\epsilon_i} - 1}, \qquad (12.21)$$

Bose-Einstein

$$n_i = \frac{g_i}{e^{\alpha}e^{\beta\epsilon_i} + 1}, \qquad (12.25)$$

Fermi-Dirac

can all be represented by the generalized equation

$$n_i = \frac{g_i}{e^{\beta(\epsilon_i - \mu)} + d}, \qquad \begin{aligned} d &= \phantom{+}0 \rightarrow \text{M-B}, \\ d &= -1 \rightarrow \text{B-E}, \\ d &= +1 \rightarrow \text{F-D}, \end{aligned} \qquad (12.26)$$

Generalized Distribution

where $d = 0, -1, +1$ corresponds to the M-B, B-E, and F-D distributions, respectively. In writing this general expression, the parameter $\alpha$ was replaced by $-\beta\mu$, that is,

$$\alpha = -\beta\mu, \qquad (12.27)$$

because of a comparison of the *classical distribution*,

$$n_i = g_i e^{\beta(\mu_C - \epsilon_i)}, \qquad (11.65)$$

Classical Distribution

with the Maxwell-Boltzmann distribution (see also Section 12.4). Since the classical distribution (indistinguishable particles) is derivable from the Maxwell-Boltzmann distribution (distinguishable particles) or vice versa, then the distribution given by Equation 12.26 is perfectly general for classical and quantum statistical mechanics. Further, if the system of interest contains a fixed number of particles $N$, the quantity $\mu$ (or $\alpha$) is in principle always determined by the conservation of particles condition

$$N = \sum_i n_i = \sum_i \frac{g_i}{e^{\beta(\epsilon_i - \mu)} + d}, \tag{12.28}$$

where the sum extends over all *energy levels* of the system.

From the generalized form of Equation 12.26, it is obvious that B-E and F-D statistics ($d = -1, +1$) tend toward M-B statistics ($d = 0$) under the condition

Classical Limit

$$e^{\beta(\epsilon_i - \mu)} >> 1, \tag{12.29}$$

which will be referred to as the **classical limit.** Alternatively, the condition for the *classical limit* may be expressed in terms of the *occupation index* (see Equation 12.2) as

Classical Limit

$$f(\epsilon_i) = \frac{n_i}{g_i} << 1, \tag{12.30}$$

since for $n_i/g_i \ll 1$ the relation given by Equation 12.29 is necessarily required by both Equations 12.26 and 12.28. Thus, the *classical limit* corresponds to a small number of particles for the available quantum states, which is in total agreement with the discussion and example of Section 12.1.

Another interesting interpretation of the classical limit is directly obtainable from Equation 12.29 for *bosons* that do not obey the conservation of particles requirement. This case corresponds to *photon* or *phonon* statistics, where $\alpha = 0$ and, consequently, $\mu = 0$ from Equation 12.27. Accordingly, the *classical limit* expressed by Equation 12.29 becomes

$$e^{\beta\epsilon_i} = e^{\epsilon_i/k_B T} >> 1, \tag{12.31}$$

which is clearly valid at energies $\epsilon_i$ that are large compared to $k_B T$. To be more specific, consider a photon gas in the visible portion of the electromagnetic spectrum, where the wavelength of photons (see Table 6.1) is roughly on the order of $5 \times 10^{-7}$ m. With the energy of the photons given by

$$\begin{aligned}\epsilon_i &= h\nu = \frac{hc}{\lambda} \\ &= \frac{(6.6 \times 10^{-34}\ \mathrm{J \cdot s})\,(3 \times 10^8\ \mathrm{m/s})}{5 \times 10^{-7}\ \mathrm{m}} \\ &\approx 4 \times 10^{-19}\ \mathrm{J}\end{aligned}$$

and $\beta$ evaluated as

$$\begin{aligned}\beta &= \frac{1}{k_B T} \\ &= \frac{1}{(1.4 \times 10^{-23}\ \mathrm{J/K})T} \\ &\approx \frac{7 \times 10^{22}\ \mathrm{K/J}}{T},\end{aligned}$$

the classical limit expressed by Equation 12.31 becomes

$$e^{\epsilon_i/k_B T} \approx e^{(3 \times 10^4\ \mathrm{K})/T} >> 1. \tag{12.32}$$

This result for photon statistics is certainly valid at low and intermediate temperatures. From this example it should be rather obvious that at low temperatures Equation 12.31 is easily satisfied for photons ranging from gamma to infrared radiation.

Additional insight on the essential features of the statistical distributions is obtainable from the *occupation index*. For example, at all temperatures when $\epsilon_i = \mu$ the occupation index (Equation 12.2) can be obtained from Equation 12.26 as

$$\epsilon_i = \mu \rightarrow f(\epsilon) = \frac{1}{1 + d} = \begin{matrix} 1 & \text{for M-B,} \\ \infty & \text{for B-E,} \\ \frac{1}{2} & \text{for F-D,} \end{matrix} \tag{12.33}$$

while at sufficiently low temperatures (in the limit as $T = 0$ K) we have results

$$\epsilon_i < \mu \rightarrow f(\epsilon) = \frac{1}{d} = \begin{matrix} \infty & \text{for M-B,} \\ -1 & \text{for B-E,} \\ 1 & \text{for F-D,} \end{matrix} \tag{12.34}$$

$$\epsilon_i > \mu \rightarrow f(\epsilon) = 0 \qquad \text{for M-B, B-E, F-D.} \tag{12.35}$$

Thus, for $\epsilon_i = \mu$ in B-E statistics, the occupation index $f(\epsilon)$ becomes infinite, while it is zero for energy levels greater than $\mu$ ($\epsilon_i > \mu$) and meaningless for energy levels smaller than $\mu$ ($\epsilon_i < \mu$). Consequently, *bosons* tend to concentrate in energy levels $\epsilon_i$ that are only slightly greater than $\mu$. *Fermions*, on the other hand, tend to populate levels that are *equal to and less* than $\mu$, with the lower energy levels being fully populated with one particle for every allowed $g_i$ quantum state. The nature of the F-D distribution is more fully discussed in Section 12.7, where the free electron theory of metals is considered.

## 12.4 Identification of the Lagrange Multipliers

Before the B-E and F-D distribution laws are applied to different quantum mechanical problems, it is convenient to evaluate the Lagrange multiplier $\beta$ in quantum statistics and to show in general that it is related to the undetermined multiplier $\alpha$ by Equation 12.27. The identification of $\beta$ is easily obtained by considering arguments similar to those presented in Chapter 11, Section 11.5, where $\beta$ was found to be the inverse of the product of the Boltzmann constant $k_B$ and the absolute temperature $T$. This same result will be obtained in quantum statistical mechanics for both B-E and F-D statistics by evaluating, as before,

$$dS = k_B \, d \ln W \tag{12.36}$$

for both cases and comparing the results with the combined first and second laws of thermodynamics. Since the derivational method and the result are not new, this section provides a slight respite from our fundamental considerations of quantum statistical mechanics. Many of the basic relations of quantum statistics will be reiterated in our evaluation of $\beta$, and some interesting results of entropy $S$ and differential entropy $dS$ will be noted for B-E and F-D statistics.

From the above overview of the derivational requirements, the evaluation of Equation 12.36 immediately requires a determination of $d \ln W$ for both B-E and F-D statistics, where the respective thermodynamic probabilities are given by

$$W_{\text{B-E}} = \Pi_i \frac{(n_i + g_i)!}{g_i! n_i!}, \qquad g_i >> 1, \tag{12.18}$$

$$W_{\text{F-D}} = \Pi_i \frac{g_i!}{(g_i - n_i)! n_i!}. \tag{12.12}$$

Differentiating the logarithm of these expressions (see Equations 12.19 and 12.23),

$$\ln W_{\text{B-E}} = \sum_i \left[ n_i \ln \left( \frac{g_i}{n_i} + 1 \right) + g_i \ln \left( \frac{n_i}{g_i} + 1 \right) \right], \tag{12.37}$$

$$\ln W_{\text{F-D}} = \sum_i \left[ n_i \ln \left( \frac{g_i}{n_i} - 1 \right) - g_i \ln \left( -\frac{n_i}{g_i} + 1 \right) \right], \tag{12.38}$$

gives results that are essentially identical to the previously derived Equations 12.20 and 12.24, respectively. That is, from Equation 12.20 we obtain

$$d \ln W_{\text{B-E}} = \sum_i \ln \left( \frac{g_i}{n_i} + 1 \right) dn_i, \tag{12.39}$$

whereas Equation 12.24 gives

$$d \ln W_{\text{F-D}} = \sum_i \ln \left( \frac{g_i}{n_i} - 1 \right) dn_i, \tag{12.40}$$

which are also obvious from Equations 12.37 and 12.38 by realizing that $dg_i = 0$ and $d \ln (g_i/n_i) = -d \ln (n_i/g_i)$. These two equations can be further reduced by employing, respectively, the B-E and F-D distribution laws,

$$n_i = \frac{g_i}{e^{\alpha + \beta\epsilon_i} \pm 1}, \qquad \begin{array}{l} + \rightarrow \text{F-D}, \\ - \rightarrow \text{B-E}, \end{array} \tag{12.41}$$

in the form

$$\frac{g_i}{n_i} \pm 1 = e^{\alpha + \beta\epsilon_i}, \qquad \begin{array}{l} + \rightarrow \text{B-E}, \\ - \rightarrow \text{F-D}. \end{array} \tag{12.42}$$

Thus, the argument of the logarithm in Equation 12.39 is identical to that in Equation 12.40, and the two equations can be expressed as

$$\begin{aligned} d \ln W &= \sum_i (\alpha + \beta\epsilon_i)\, dn_i \\ &= \alpha \sum_i dn_i + \beta \sum_i \epsilon_i\, dn_i, \end{aligned} \tag{12.43}$$

upon substitution from Equation 12.42. Using the conservation of particles and energy requirements in the form

$$dN = \sum_i dn_i,$$

$$dU = \sum_i \epsilon_i \, dn_i,$$

(see discussion of Equations 11.41 and 11.42) allows Equation 12.43 to be expressed as

$$d \ln W = \alpha \, dN + \beta \, dU. \qquad (12.44)$$

This result is perfectly general and valid for both B-E and F-D statistics, hence the inclusion of B-E and F-D subscripts on $W$ is unnecessary. With this determination of $d \ln W$ in quantum statistics, substitution into Equation 12.36 immediately yields

Differential Entropy

$$dS = k_B\alpha \, dN + k_B\beta \, dU \qquad (12.45)$$

for the **differential entropy.** A comparison of this result with Equation 11.59 shows that differential entropy has a common term of $k_B\beta \, dU$ in classical and quantum statistical mechanics. Further, although the relation for differential entropy $dS$ is common in B-E and F-D statistics, the entropy $S$ is *not* the same in both cases, because $\ln W$ differs for each case (see Equations 12.37 and 12.38).

Having obtained an expression for the differential entropy in quantum statistics, we rewrite the result in the form

$$dU = \frac{1}{k_B\beta} \, dS - \frac{\alpha}{\beta} \, dN, \qquad (12.46)$$

which is nicely amenable to a term-by-term comparison with the combined first and second laws of thermodynamics,

$$dU = T \, dS - p \, dV + \mu \, dN. \qquad (11.61)$$

Taking the partial derivative of internal energy $U$ with respect to entropy $S$ at constant $V$ and $N$, these two equations give

$$\left(\frac{\partial U}{\partial S}\right)_{V,\,N} = \frac{1}{k_B\beta} = T,$$

from which we obtain the identity of β in quantum statistics as

$$\beta = \frac{1}{k_B T}. \tag{12.47}$$

Hence, the Lagrange multiplier β has an identical form in both classical (see Equation 11.63) and quantum statistical mechanics. It is also rather interesting to note that from Equations 12.46 and 11.61 a comparison of the coefficients of $dN$ gives the general relation

$$\alpha = -\beta\mu \tag{12.27}$$

that was previously assumed in writing Equation 12.26.

## 12.5 Specific Heat of a Solid

Over a century ago the molal specific heat of a solid was experimentally observed by P. L. Dulong and A. L. Petit to be very nearly the same for all solids, approximately 6 cal/mole · K. More specifically, the amount of heat energy required to raise the temperature of one mole of a substance by 1° C at constant volume, called the *molal specific thermal capacity* $c_v$ (see Equations 11.146 and 11.151), is essentially independent of the chemical composition of a solid. This result was easily understood from the ideas of classical statistical mechanics for the *equipartition of energy* (see Equation 11.158). Regarding each atom of a solid as executing simple harmonic motion about its lattice point, then each atom in a *one-dimensional* solid could be represented by a linear harmonic oscillator. In this case the total energy of an oscillator consists of two quadratic terms, owing to its kinetic ($p_x^2/2m$) and elastic potential ($\frac{1}{2}kx^2$) energies (recall Equation 7.93), resulting in the linear oscillator having *two degrees* of freedom. Hence, the *average energy per oscillator* is given by Equation 11.143 as

$$\bar{\epsilon}_o = \frac{N_f}{2} k_B T$$
$$= k_B T, \tag{12.48}$$

Average Energy of Classical Linear Oscillator

since $N_f = 2$. In *three dimensions* each atom of a solid may be represented by three harmonic oscillators, so the average energy per atom is simply $\bar{\epsilon}_a = 3\bar{\epsilon}_o$. Thus, the total *internal energy* $U$ for a system of $N$ particles is

$$\begin{aligned} U &= N\bar{\epsilon}_a = N(3\bar{\epsilon}_o) \\ &= 3Nk_BT \\ &= 3N\frac{R}{N_o}T \\ &= 3nRT, \end{aligned} \tag{12.49}$$

where the fundamental relations given by Equations 11.108 and 5.69 have been used. From this result, the *molal specific internal energy* (recall Equation 11.152) is

$$u = 3RT, \tag{12.50}$$

and the *molal specific thermal capacity* (recall Equation 11.150) is given by

Dulong and Petit Law

$$\begin{aligned} c_v &= \left(\frac{\partial u}{\partial T}\right)_v = 3R \\ &= 5.97\frac{\text{cal}}{\text{mole}\cdot\text{K}}. \end{aligned} \tag{12.51}$$

This result for a classsical solid, which we originally derived in Problem 11.30, is known as the **law of Dulong and Petit.** It is rather closely obeyed by solids at high temperatures. As temperatures are lowered toward room temperature, however, serious discrepancies are observed especially for the less massive elements (e.g., $(c_v)_{Be} = 3.85$ cal/mole · K for beryllium and $(c_v)_B = 3.34$ cal/mole · K for boron). At lower temperatures, observed specific heats of *all* solids depart even more dramatically from the Dulong and Petit law, which suggests that the classical analysis is *fundamentally* in error.

## Einstein Theory (M-B Statistics)

At the turn of this century, it was well recognized experimentally that the specific heat of any solid (a) tends to obey the Dulong and Petit law at a high temperature, (b) tends toward zero as the temperature decreases, and (c) varies as $T^3$ nears absolute zero. It was Einstein in 1907 who first recognized that the basic error in the classical analysis resulted from the equipartition factor of $k_B T$ (Equation 12.48) for the average energy *per oscillator* in a solid. This factor had to be replaced by one that properly accounts for the *energy quantization* of a linear harmonic oscillator, as the energy spectrum is not continuous but, rather, discrete in multiples of *quantized energy* $h\nu$. More specifically, the *quantum mechanical energy eigenvalues* for a *linear harmonic oscillator* are given by (recall Equation 10.84 with $E$ replaced by $\epsilon$)

Linear Oscillator Eigenvalues

$$\epsilon_n = (n + \tfrac{1}{2})h\nu, \tag{12.52}$$

where $\nu$ is the fundamental frequency of oscillation and $n = 0, 1, 2, \cdots, \infty$. With this fundamental correction to the classical analysis, we need to first find the average energy per oscillator $\bar{\epsilon}_o$, from which the total internal energy $U$ and molal specific thermal capacity $c_v$ are readily obtainable by arguments similar to those above (i.e., Equations 12.49 to 12.51). Accordingly, the *average energy per oscillator* is given by

$$\begin{aligned}
\bar{\epsilon}_o &\equiv \frac{U}{N} \\
&= \frac{1}{N}\sum_i \epsilon_i n_i \\
&= \frac{1}{N}\sum_i \epsilon_i \frac{N}{Z_o} g_i e^{-\beta\epsilon_i} \\
&= \frac{\sum_i \epsilon_i g_i e^{-\beta\epsilon_i}}{\sum_i g_i e^{-\beta\epsilon_i}},
\end{aligned} \tag{12.53}$$

where the Boltzmann distribution (Equation 11.47) and the defining equation for the classical partition function $Z$ (Equation 11.48) have been used. The summations in Equation 12.53 are over all energy levels, which are defined for the quantized linear oscillator by Equation 12.52. Since there is, obviously, no degeneracy in the energy levels, $g_i = 1$ in the above equation. Further, we need to replace the $i$-subscript with an $n$-subscript to particularize Equation 12.53 for the *quantized oscillator* problem. Thus Equation 12.53 becomes

$$\bar{\epsilon}_o = \frac{\sum_n \epsilon_n e^{-\beta\epsilon_n}}{\sum_n e^{-\beta\epsilon_n}}$$

$$= \frac{\sum_n (n + \frac{1}{2})\, h\nu e^{-\beta\epsilon_n}}{\sum_n e^{-\beta\epsilon_n}}$$

$$= \frac{1}{2}h\nu + \frac{h\nu \sum_n n e^{-\beta\epsilon_n}}{\sum_n e^{-\beta\epsilon_n}}, \tag{12.54}$$

where Equation 12.52 has been utilized in the expansion of the first equality. The summations in the second term of Equation 12.54 can be evaluated by expansion. For example, expanding the denominator we have

$$\sum_{n=0}^{\infty} e^{-\beta(n+\frac{1}{2})h\nu} = e^{-\frac{1}{2}\beta h\nu} + e^{-(3/2)\beta h\nu} + e^{-(5/2)\beta h\nu} + \cdots$$

$$= e^{-\frac{1}{2}\beta h\nu}(1 + e^{-\beta h\nu} + e^{-2\beta h\nu} + \cdots). \tag{12.55}$$

The geometric series in parenthesis is well known and can be obtained from the **binomial expansion,**

$$(x + y)^n = x^n + nx^{n-1}y + \frac{n(n-1)}{2!}x^{n-2}y^2 + \cdots, \tag{12.56}$$

given in Appendix A, Section A.7. That is, with $x = 1$, $y = -z$, and $n = -1$, the *binomial expansion* gives (see Problem A.7)

$$(1-z)^{-1} = 1^{-1} + (-1)1^{-2}(-z) + \frac{(-1)(-2)}{2!}1^{-3}(-z)^2 + \cdots$$

$$= 1 + z + z^2 + z^3 + \cdots. \tag{12.57}$$

Thus, Equation 12.55 can be represented in the reduced form

$$\sum_{n=0}^{\infty} e^{-\beta(n+\frac{1}{2})h\nu} = \frac{e^{-\frac{1}{2}\beta h\nu}}{1 - e^{-\beta h\nu}}. \tag{12.58}$$

In a similar manner the summation in the numerator of the second term of Equation 12.54 can be expressed as

$$\sum_{n=0}^{\infty} ne^{-\beta\epsilon_n} = e^{-(3/2)\beta h\nu} + 2e^{-(5/2)\beta h\nu} + 3e^{-(7/2)\beta h\nu} + \cdots$$
$$= e^{-(3/2)\beta h\nu}(1 + 2e^{-\beta h\nu} + 3e^{-2\beta h\nu} + \cdots)$$
$$= \frac{e^{-(3/2)\beta h\nu}}{(1 - e^{-\beta h\nu})^2}, \tag{12.59}$$

where the geometric series in the second equality is of the form

$$(1 - z)^{-2} = 1 + 2z + 3z^2 + 4z^3 + \cdots . \tag{12.60}$$

Now, substitution of Equations 12.58 and 12.59 into Equation 12.54 immediately yields

$$\bar{\epsilon}_o = \tfrac{1}{2}h\nu + \frac{h\nu e^{-\beta h\nu}}{1 - e^{-\beta h\nu}},$$

which can be reduced to the form

$$\bar{\epsilon}_o = \tfrac{1}{2}h\nu + \frac{h\nu}{e^{\beta h\nu} - 1}. \tag{12.61}$$

Average Energy of Quantized Oscillator

This equation represents the *average energy of a quantized linear harmonic oscillator*. Since $\beta = 1/k_BT$, only the second term is temperature dependent. Consequently, the first term gives the ground-state energy of the oscillator, called the *zero-point energy*, while the second term gives the *energy of thermal excitation*. Further, with the Einstein *characteristic temperature*, represented by $\theta_E$ (recall the illustrative considerations of Chapter 11, Section 11.5), being that temperature at which $k_BT = h\nu$, then

$$\theta_E \equiv \frac{h\nu}{k_B} = \beta h\nu T, \tag{12.62}$$

Einstein Temperature

and we can express Equation 12.61 as

$$\bar{\epsilon}_o = \tfrac{1}{2}h\nu + \frac{h\nu}{e^{\theta_E/T} - 1}. \tag{12.63}$$

Average Energy of Quantized Oscillator

In this form, we can see that $\bar{\epsilon}_o$ depends on the ratio of the characteristic temperature to the actual temperature. Since the greater the *natural fre-*

*quency* $\nu$ of the assembly of oscillators the higher the characteristic temperature, $\theta_E$ provides a reference temperature for the assembly. For example, if the natural frequency is in the infrared region of the electromagnetic spectrum (see Table 6.1) with say $\lambda = 1.44 \times 10^{-5}$ m, then

$$\begin{aligned}\theta_E &= \frac{h\nu}{k_B}\\ &= \frac{hc}{\lambda k_B}\\ &= \frac{(6.63 \times 10^{-34})(3 \times 10^{8})}{(1.44 \times 10^{-5})(1.38 \times 10^{-23})}\\ &= 1000 \text{ K.}\end{aligned}$$

Thus, an actual temperature of $T = 100$ K is equivalent to $\theta_E/10$, while a temperature of $T = 5000$ K is equal to $5\theta_E$.

With $\bar{\epsilon}_o$ being the average energy per *linear oscillator*, then the total internal energy $U$ for a three-dimensional solid consisting of $N$ atoms is

$$\begin{aligned}U &= N(3\bar{\epsilon}_o)\\ &= 3\frac{N}{N_o}N_o\bar{\epsilon}_o\\ &= 3\,nN_o\bar{\epsilon}_o.\end{aligned} \tag{12.64}$$

Thus, the *molal specific internal energy* of the solid is

$$\begin{aligned}u &= 3N_o\bar{\epsilon}_o\\ &= 3N_oh\nu[\tfrac{1}{2} + (e^{\theta_E/T} - 1)^{-1}],\end{aligned} \tag{12.65}$$

and its *molal specific thermal capacity* is

$$\begin{aligned}c_v &= \left(\frac{\partial u}{\partial T}\right)_v\\ &= \frac{3N_oh\nu\theta_E T^{-2}e^{\theta_E/T}}{(e^{\theta_E/T} - 1)^2}.\end{aligned} \tag{12.66}$$

With $N_ok_B = R$ and $h\nu/k_B = \theta_E$, this equation reduces to

Einstein's Specific Heat Formula

$$c_v = \frac{3R(\theta_E/T)^2e^{\theta_E/T}}{(e^{\theta_E/T} - 1)^2}, \tag{12.67}$$

which is known as the **Einstein specific heat formula.** It is also interesting to note that the zero-point energy $\frac{1}{2}h\nu$ does not contribute to $c_v$, since it is independent of temperature and vanishes when the partial derivative $(\partial u/\partial T)_v$ is performed. Also of importance is that Einstein's formula immediately reduces to the Dulong and Petit law at high temperatures and trends toward zero at low temperatures. That is, at high temperatures, $T \gg \theta_E$ and

$$
\begin{aligned}
e^{\theta_E/T} &= 1 + \frac{\theta_E}{T} + \frac{(\theta_E/T)^2}{2!} + \cdots \\
&\approx 1 + \frac{\theta_E}{T}, \qquad T \gg \theta_E, \qquad (12.68)
\end{aligned}
$$

so Equation 12.67 reduces to

$$
\begin{aligned}
c_v &\approx \frac{3R(\theta_E/T)^2 e^{\theta_E/T}}{(1 + \theta_E/T - 1)^2} \\
&= 3Re^{\theta_E/T}, \qquad T \gg \theta_E. \qquad (12.69)
\end{aligned}
$$

Thus, for $T \gg \theta_E$ the exponential goes to one and the law of Dulong and Petit is obtained. On the other hand, as $T$ approaches zero so does $c_v$, since $\theta_E \gg T$, $e^{\theta_E/T} \gg 1$, and Equation 12.67 reduces to

$$
\begin{aligned}
c_v &\approx \frac{3R\,(\theta_E/T)^2 e^{\theta_E/T}}{e^{2\theta_E/T}} \\
&= 3R\left(\frac{\theta_E}{T}\right)^2 e^{-\theta_E/T}, \qquad \theta_E \gg T. \qquad (12.70)
\end{aligned}
$$

It is rather interesting to note that virtually all of the oscillators are found in the four lowest energy levels, when the condition $\theta_E \geq T$ is valid. This is illustrated in Problems 12.9 and 12.10 by evaluating the fractional number of oscillators in the $i$th level $n_i/N$ for the cases $\theta_E = T$ and $\theta_E = 2T$, respectively. Although Einstein's model of a solid is in good agreement with observed $c_v$ data over a wide range of temperatures, (see Figure 12.5) its inability to predict $c_v \propto T^3$ near absolute zero requires that we search further for a complete theoretical explanation of the specific heats of solids.

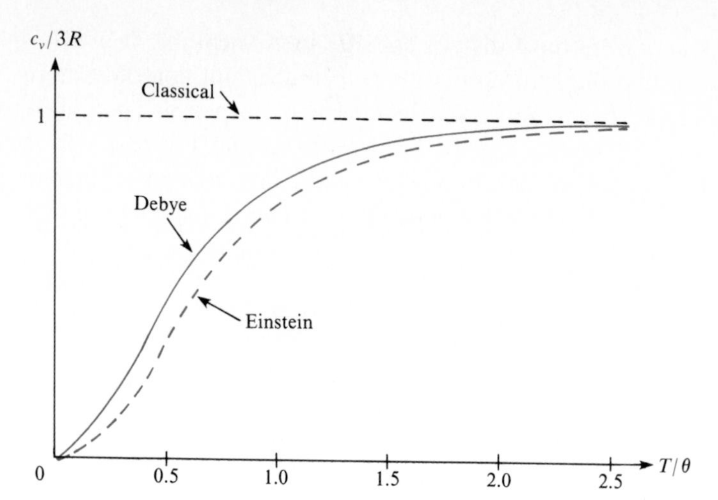


Figure 12.5
The general characteristics of the molal specific thermal capacity by the classical, Einstein, and Debye theories.

## Debye Theory (Phonon Statistics)

The principal deficiency of the Einstein theory lies in considering the atoms of a solid as oscillating independently at the same fundamental frequency. Although efforts to rectify this deficiency were made by Max Born, Theodor von Karman, and others, it was not until 1912 that a satisfactory theory was proposed by Peter Debye. He considered the atoms of a solid to represent a system of coupled oscillators having a continuous range of frequencies. Although the statistics of noninteracting or free particles is inappropriate for such a model, Debye further assumed a solid to be a *continuous elastic body*. This assumption allows that the frequencies of *thermal excitation* of the atoms of a solid are equivalent to the frequencies of the possible standing *accoustic* (or *elastic*) *waves* of an elastic solid. Thus, the $N$ atoms of a solid, modeled as a three-dimensional array of particles connected by springs, are replaced by $3N$ elastic modes of vibration. By analogy with the vibrating string analysis presented in Chapter 8, Section 8.2, we immediately realize that in the Debye model the elastic modes of vibration are independent and noninteracting. Furthermore, each vibrating *mode* in a three-dimensional model would be characterized by a *unique set* of $n_x$, $n_y$, and $n_z$ numbers, whereas for the vibrating string only $n_x$ (see Equation 8.23) was required. Hence, the Debye model allows for the *distinguishability* of the vibration modes, so Maxwell-Boltzmann statistics is applicable.

Instead of using M-B statistics to develop the Debye theory of the specific heats of solids, we will employ B-E statistics by considering each elastic wave to be a *particle* called a *phonon*. A **phonon** is simply defined as a *quanta of vibrational energy* in a solid. *Phonons* are rather similar to *photons* in that they are *indistinguishable* entities having *quantized energy* given by the equation

$$\epsilon = h\nu, \tag{12.71}$$

Phonon/Photon Energy Quantization

where $h$ is Planck's constant. Unlike photons, which travel at the speed of light, *phonons* propagate through a solid with the speed of sound obeying the wave equation

$$v_s = \lambda\nu. \tag{12.72}$$

Phonon Speed

Further, *phonons* obey the de Broglie relation (see Equation 8.53)

$$p = \frac{h}{\lambda} = \frac{h\nu}{v_s}. \tag{12.73}$$

Phonon Momentum

Hence, we propose replacing the assembly of elastic waves in the Debye model with an *ideal phonon gas*, where the allowable wave frequencies and energies are given by the above relations. Since *phonons* are indistinguishable particles that do not obey the Pauli exclusion principle, they are classified as *bosons* and obey Bose-Einstein statistics or, more specifically, *photon/phonon statistics* (see Equation 12.22).

Before the specific heat of a solid can be determined, we need to first find the total internal energy $U$ of the ideal gas. Even though *particles* are not conserved in a *phonon gas*, energy is conserved. From this most fundamental conservation principle, then, we have

$$U = \sum_i \epsilon_i n_i$$

$$= \sum_i \frac{\epsilon_i g_i}{e^{\beta\epsilon_i} - 1}, \tag{12.74}$$

where Equation 12.22 has been used for the phonon distribution law. For a continuous distribution of frequencies in the solid, this equation becomes

$$U = \int_0^{\nu_{\mathrm{m}}} \frac{h\nu g(\nu)\, d\nu}{e^{\beta h\nu} - 1}, \tag{12.75}$$

where Equation 12.71 has been used and $\nu_{\mathrm{m}}$ represents a *maximum* or *cutoff* frequency that limits the internal energy to a finite value. The degeneracy $g(\nu)$ represents the number of states having the same frequency (or energy since $\epsilon = h\nu$) in $\mu$-space for the *ideal phonon gas*. Accordingly, we can use arguments similar to those that led to Equation 11.95 and write

$$g(p)\,dp = \left(\frac{4\pi V}{h^3}\right) p^2 dp, \tag{11.95}$$

which from de Broglie's relation (Equation 12.73) becomes

$$g(\nu)\,d\nu = \left(\frac{4\pi V}{v_s^3}\right) \nu^2 d\nu. \tag{12.76}$$

This equation can be interpreted as the maximum possible number of states (vibration modes) having a frequency between $\nu$ and $\nu + d\nu$. The constant factor in parenthesis can be expressed in terms of the cutoff frequency $\nu_{\mathrm{m}}$ by properly normalizing Equation 12.76. That is, for a solid consisting of $N$ atoms, we assume there to be $3N$ vibration modes or phonons with allowed frequencies varying from zero to $\nu_{\mathrm{m}}$. Hence, the total number of phonon states is limited to

$$\begin{aligned} 3N &= \int_0^{\nu_{\mathrm{m}}} g(\nu)\,d\nu = \frac{4\pi V}{v_s^3} \int_0^{\nu_{\mathrm{m}}} \nu^2\,d\nu \\ &= \frac{4\pi V}{v_s^3} \frac{\nu_{\mathrm{m}}^3}{3}, \end{aligned} \tag{12.77}$$

from which we obtain

$$\frac{4\pi V}{v_s^3} = \frac{9N}{\nu_{\mathrm{m}}^3}. \tag{12.78}$$

It should be mentioned that for any direction of propagation in an elastic solid there are three types of elastic waves. These correspond to a *longitudinal wave* propagating with a speed $v_l$ and two mutually perpendicular *transverse waves* propagating with a speed of $v_t$. Accordingly, we have

$$\frac{1}{v_s^3} = \frac{1}{v_l^3} + \frac{2}{v_t^3}, \tag{12.79}$$

and the right-hand side of this relation is *normally* substituted in Equations 12.76 to 12.78 for $1/v_s^3$. Irrespectively, however, from Equations 12.78 and 12.76 we obtain

$$g(\nu)\,d\nu = \left(\frac{9N}{\nu_{\mathrm{m}}^3}\right)\nu^2\,d\nu, \tag{12.80}$$

and substitution into Equation 12.75 gives

$$U = \frac{9N}{\nu_{\mathrm{m}}^3}\int_0^{\nu_{\mathrm{m}}}\frac{h\nu^3 d\nu}{e^{\beta h\nu} - 1}. \tag{12.81}$$

Letting the integration variable be the dimensionless quantity

$$x = \frac{h\nu}{k_{\mathrm{B}}T} = \beta h\nu, \tag{12.82}$$

from which

$$d\nu = \frac{dx}{\beta h} \tag{12.83}$$

and

$$x_{\mathrm{m}} = \beta h\nu_{\mathrm{m}}, \tag{12.84}$$

the total energy can be expressed in a more compact form as

$$\begin{aligned} U &= \frac{9N}{\beta x_{\mathrm{m}}^3}\int_0^{x_{\mathrm{m}}}\frac{x^3 dx}{e^x - 1} \\ &= \frac{9nRT}{x_{\mathrm{m}}^3}\int_0^{x_{\mathrm{m}}}\frac{x^3 dx}{e^x - 1}, \end{aligned} \tag{12.85}$$

where $N/\beta = k_{\mathrm{B}}NT = (R/N_o)NT = nRT$ has been used in obtaining the second equality.

The specific heat of a solid for the Debye model is now easily determined from the total internal energy relation given by Equation 12.85. First, however, defining the **Debye characteristic temperature** $\theta_D$ by

$$\theta_{\mathrm{D}} \equiv \frac{h\nu_{\mathrm{m}}}{k_{\mathrm{B}}}, \tag{12.86}$$

Debye Temperature

from which

$$\frac{\theta_{\mathrm{D}}}{T} = \beta h\nu_{\mathrm{m}} = x_{\mathrm{m}}, \tag{12.87}$$

Equation 12.85 becomes

$$U = \frac{9nRT^4}{\theta_D^3} \int_0^{\theta_D/T} \frac{x^3 dx}{e^x - 1} \tag{12.88}$$

for the *total internal energy*. Dividing this equation by the number of moles $n$ and taking the partial derivative with respect to temperature yields

Debye Specific Heat Formula

$$c_v = 9R \left[ 4\left(\frac{T}{\theta_D}\right)^3 \int_0^{\theta_D/T} \frac{x^3 dx}{e^x - 1} - \frac{\theta_D/T}{e^{\theta_D/T} - 1} \right] \tag{12.89}$$

for the *molal specific thermal capacity*. This relation is known as the **Debye specific heat formula** and can be compared (as illustrated in Figure 12.5) with the Einstein formula given by Equation 12.67. As Problems 12.13 to 12.16 illustrate, the Debye temperature is independent of specific heat measurements and can be obtained directly from the elastic properties of a solid. Using such theoretically determined values in Debye's formula yields $c_v$ values that are generally in good to excellent agreement with experimental observations at all values of the absolute temperature $T$.

We can immediately see that this formula is in agreement with the Dulong and Petit law at high temperatures and that $c_v$ is proportional to $T^3$ at very low temperatures. That is, at high temperatures ($T \gg \theta_D$) the integral of Equation 12.89 becomes

$$\begin{aligned} \int_0^{\theta_D/T} \frac{x^3 dx}{e^x - 1} &\approx \int_0^{\theta_D/T} \frac{x^3 dx}{1 + x - 1} \\ &= \int_0^{\theta_D/T} x^2 dx \\ &= \frac{1}{3}\left(\frac{\theta_D}{T}\right)^3 \end{aligned} \tag{12.90}$$

and the second term reduces to

$$\frac{\theta_D/T}{e^{\theta_D/T} - 1} \approx \frac{\theta_D/T}{1 + \theta_D/T - 1} = 1, \tag{12.91}$$

where an approximation similar to that of Equation 12.68 has been used in both cases. Hence, Equation 12.89 reduces to

$$c_v \approx 9R\left[4\left(\frac{T}{\theta_D}\right)^3 \frac{1}{3}\left(\frac{\theta_D}{T}\right)^3 - 1\right]$$
$$= 9R\left(\frac{4}{3} - 1\right)$$
$$= 3R, \qquad T >> \theta_D. \tag{12.92}$$

On the other hand, at very low temperatures $\theta_D \gg T$ and $\theta_D/T \rightarrow \infty$. Consequently, the second term of Equation 12.89 becomes insignificant and the first term contains the definite integral

$$\int_0^\infty \frac{x^3 dx}{e^x - 1} = \frac{\pi^4}{15}, \tag{12.93}$$

so Equation 12.89 reduces to

$$c_v \approx 9R\left[4\left(\frac{T}{\theta_D}\right)^3 \frac{\pi^4}{15} + 0\right]$$
$$= \frac{12\pi^4 R}{5}\left(\frac{T}{\theta_D}\right)^3, \qquad \theta_D >> T. \tag{12.94}$$

Debye $T^3$ Law

This relation is frequently referred to as the **Debye $T^3$ law.** Although the Debye theory is remarkably good in predicting observed $c_v$ values, it is somewhat limited and not universally applicable to all solids because the actual frequency spectrum of elastic waves depends on the particular crystalline structure of a solid. Further, it considers only the contributions to specific heat from atomic vibrations of individual atoms that are assumed to occupy the crystal lattice points and must be modified for molecular solids. Also, there are contributions to the specific heat from free conduction electrons in a solid, which will be addressed for metals in Section 12.7.

## 12.6 Blackbody Radiation (Photon Statistics)

Before the development of quantum statistics, it was well known that the electromagnetic radiation absorbed and emitted by every substance was dependent on the nature and absolute temperature of the substance. By considering this phenomenon, Max Planck formulated his revolutionary *quantum hypothesis* (see Equation 6.50) by developing a theoretical explanation for the spectral emission of electromagnetic radiation from an *ideal emitter*. It should be mentioned that the spectral energy distribution from an *ideal emitter* is experimentally found to be independent of its material composition and dependent only on its absolute temperature. Such an emitter is commonly referred to as an *ideal blackbody*, because a good emitter is a good absorber of radiation and, at nonluminous temperatures, any body that absorbs all radiation incident upon it appears black in color. An *ideal blackbody*, or simply a *blackbody*, can be closely approximated in the laboratory by a small hole in a hollow enclosure composed of almost any heat resisting material. Since virtually all of the radiation entering the opening is reflected within the cavity and eventually absorbed, the radiator is an *ideal absorber*. Further, if the enclosure is at a uniform temperature, the radiation in thermal equilibrium with its surroundings has the property of emitting radiation at the same rate as it absorbs energy and is, hence, an *ideal emitter*. Our primary objective is to apply the methods of statistics to such a radiator and obtain the well known Planck formula for observed blackbody radiation, which is graphically illustrated in Figure 12.6.

The first theoretical attempts at explaining the spectral energy distribution of a blackbody was made by Lord J. W. S. Rayleigh and later modified by Sir James H. Jeans. Their classical formulation of the problem combined kinetic theory and the classical theory of electromagnetic radiation, by considering the radiation in a blackbody cavity as a series of standing electromagnetic waves. We can easily obtain an expression of their formula by considering the electromagnetic waves as classical oscil-

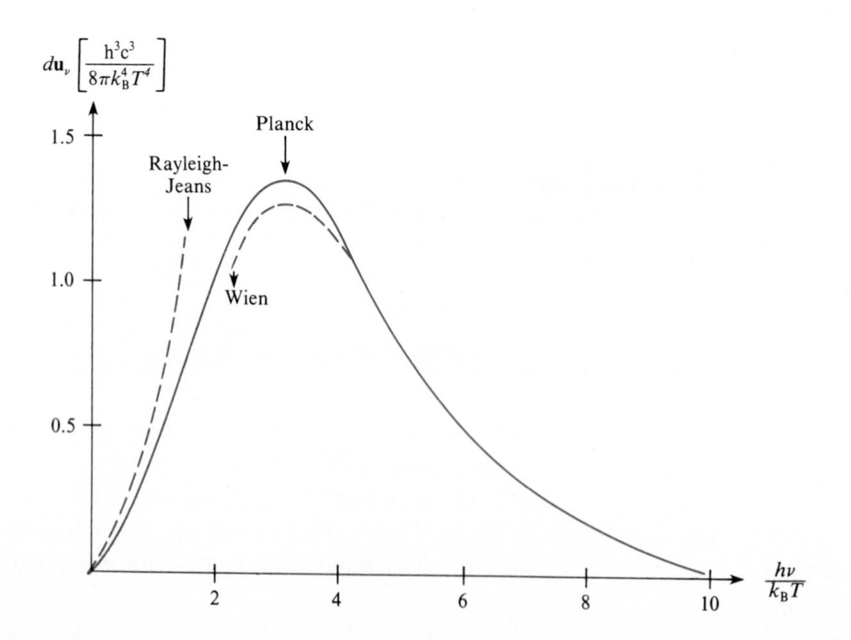


Figure 12.6
A comparison of the Rayleigh-Jeans, Wien, and Planck formulae for blackbody spectral emission of electromagnetic radiation.

lators and calculating the radiant energy per unit volume *within* the black-body enclosure. With $U$ representing the total energy of the electromagnetic radiation confined to a cavity of volume $V$, the *total radiant* **energy density** is defined by

$$\mathbf{u} \equiv \frac{U}{V}, \tag{12.95}$$

Total Energy Density

where bold type has been used here so as not to confuse energy density with the molal specific internal energy. Since the total radiant energy within the cavity remains constant, this relation may be expressed as

$$\mathbf{u} = \frac{1}{V} \sum_i \epsilon_i n_i \tag{12.96}$$

for *quanta* of electromagnetic energy. Considering the cavity to contain a large number of waves of *all possible frequencies from zero to infinity*, then in the limit of a continuous distribution of frequency Equation 12.96 becomes

$$\mathbf{u} = \frac{1}{V} \int_0^\infty \epsilon n(\nu)\, d\nu. \tag{12.97}$$

Total Energy Density

From this relation we can immediately write down an expression for the **spectral energy density** as (see also Equation 11.119)

$$d\mathbf{u}_\nu = \mathbf{u}(\nu)\, d\nu = \frac{\epsilon n(\nu)\, d\nu}{V}, \tag{12.98}$$

Spectral Energy Density

which is the radiation energy per unit volume having frequencies between $\nu$ and $\nu + d\nu$. In this relation, $n(\nu)\, d\nu$ is the *number of waves* having a frequency between $\nu$ and $\nu + d\nu$. Let us assume that $n(\nu)\, d\nu$ is *identical* to the available number of states with frequencies between $\nu$ and $\nu + d\nu$, where the latter is given by the arguments that led to Equation 11.95. Accordingly, assuming *all possible energy states* to be *occupied*, we have

$$n(p)\, dp \rightarrow g(p)\, dp = 2\frac{4\pi V}{h^3} p^2 dp, \tag{12.99}$$

which when combined with de Broglie's relation $p = h/\lambda$ and the wave equation $c = \lambda\nu$ yields

$$n(\nu)\,d\nu \rightarrow g(\nu)\,d\nu = \frac{8\pi V}{c^3}\,\nu^2 d\nu. \tag{12.100}$$

The multiplicative factor of 2 in Equation 12.99 takes into account the two possible polarization directions associated with transverse electromagnetic waves. Incidentally, this corrective factor is attributed to Jeans in his modification of Rayleigh's original radiation formula. Now, if we combine Equation 12.100 with Equation 12.98 and take the classical value of $k_B T$ (see Equation 12.48) as the energy $\epsilon$ associated with each quanta of electromagnetic energy, then the energy distribution is given by

Rayleigh-Jeans Formula

$$d\mathbf{u}_\nu = \mathbf{u}(\nu)\,d\nu = \frac{8\pi}{c^3}\,k_B T \nu^2 d\nu. \tag{12.101}$$

This equation is known as the **Rayleigh-Jeans formula** for blackbody radiation and is illustrated graphically in Figure 12.6. It agrees reasonably well with experimental data at low frequencies, but it is absurdly in error at predicting the radiant energy density at high frequencies. In fact, for frequencies in the ultraviolet region of the electromagnetic spectrum, the total energy density (see Equation 12.97) goes to infinity. This critical flaw in the classical theory became known as the **ultraviolet catastrophe.**

The correct formula for the *spectral energy density* in a blackbody cavity was advanced by Max Planck in 1900. His formula can be easily derived by considering the electromagnetic waves to be represented by *particles*, called *photons*, of an assembly. Since *photons* are indistinguishable particles that do not obey the Pauli exclusion principle, they are classified as *bosons* and Bose-Einstein statistics is applicable to the assembly. More specifically, we are considering the electromagnetic radiation in the cavity to be an *ideal photon gas* that obeys the *photon distribution* given by Equation 12.22,

$$n_i = \frac{g_i}{e^{\beta\epsilon_i} - 1}, \tag{12.22}$$

which for a continuous distribution of frequencies becomes

$$dn_\nu = n(\nu)\,d\nu = \frac{g(\nu)\,d\nu}{e^{\beta\epsilon} - 1}. \tag{12.102}$$

The problem of applying this distribution law is very similar to that encountered for the *ideal phonon gas* of the Debye theory. Here, however, $g(\nu)\,d\nu$ is given by the arguments that led to Equation 12.100 (recall that the multiplicative factor of 2 accounts for the two allowed states associated with polarization) and the energy of $\epsilon$ of a photon is given by $\epsilon = h\nu = hc/\lambda$. Hence, for the radiation in a cavity of volume $V$ at the absolute temperature $T$, we have (from Equations 12.100 and 12.102) the relation

$$dn_\nu = n(\nu)\, d\nu = \frac{8\pi V}{c^3} \frac{\nu^2 d\nu}{e^{\beta h\nu} - 1} \tag{12.103}$$

representing the number of photons having frequencies between $\nu$ and $\nu + d\nu$. Substitution of this equation and $\epsilon = h\nu$ into the *spectral energy density* relation (Equation 12.98) immediately yields the well known **Planck radiation formula,**

Planck Radiation Formula

$$\mathbf{u}(\nu)\, d\nu = \frac{8\pi h}{c^3} \frac{\nu^3 d\nu}{e^{\beta h\nu} - 1}, \tag{12.104}$$

which very nicely predicts observed spectral emission data (see Figure 12.6).

The Planck radiation formula can be seen to immediately reduce to the Rayleigh-Jeans formula at low frequencies for which $h\nu << k_B T$, since

$$\begin{aligned} e^{\beta h\nu} - 1 &\approx 1 + \beta h\nu - 1 \\ &= \frac{h\nu}{k_B T}, \qquad h\nu << k_B T. \end{aligned} \tag{12.105}$$

At high frequencies for which $h\nu >> k_B T$, however,

$$e^{\beta h\nu} - 1 \approx e^{\beta h\nu}, \qquad h\nu >> k_B T, \tag{12.106}$$

and Equation 12.104 reduces to

Wien Formula

$$\mathbf{u}(\nu)\, d\nu = \frac{8\pi h}{c^3} \nu^3 e^{-\beta h\nu} d\nu. \tag{12.107}$$

This relation is called the **Wien formula** for blackbody radiation and is illustrated graphically in Figure 12.6. It was originally advanced as an empirical relation by Wilhelm Wien, shortly after the development of the Rayleigh-Jeans formula, to predict the spectral emission at *high frequencies*. Incidentally, Planck's formula was originally empirically developed to agree with the Wien formula at high frequencies and the Rayleigh-Jeans formula at low frequencies. Planck's efforts at deriving the equation from fundamental physical principles necessitated his advancing the *quantum hypothesis* of atomic oscillators.

Another interesting result can be obtained by integrating the Planck radiation formula over the allowed frequency range. Clearly, from Equation 12.98 the *total energy density* is given by

$$\mathbf{u} = \int_0^\infty d\mathbf{u}_\nu = \int_0^\infty u(\nu)\, d\nu, \tag{12.108}$$

which upon substitution from Equation 12.104 gives

$$\mathbf{u} = \frac{8\pi h}{c^3}\int_0^\infty \frac{\nu^3 d\nu}{e^{\beta h\nu} - 1}. \tag{12.109}$$

Using the dimensionless variable defined by Equation 12.82 (i.e., $x = \beta h\nu$), this equation takes the form

$$\begin{aligned}\mathbf{u} &= \frac{8\pi(k_B T)^4}{(hc)^3}\int_0^\infty \frac{x^3 dx}{e^x - 1} \\ &= \frac{8\pi(k_B T)^4}{(hc)^3}\frac{\pi^4}{15} \\ &= \frac{8\pi^5 k_B^4}{15h^3c^3}T^4,\end{aligned} \tag{12.110}$$

where Equation 12.93 has been used to evaluate the definite integral. Using the symbolic definition

$$\sigma \equiv \frac{8\pi^5 k_B^4}{15h^3c^3}, \tag{12.111}$$

we obtain the well known **Stefan-Boltzmann law**

Stefan-Boltzmann Law

$$\mathbf{u} = \sigma T^4. \tag{12.112}$$

Thus, the *total energy density* is dependent on only the absolute temperature raised to the fourth power. Incidentally, when Equation 12.111 is used to calculate $\sigma$, the value obtained agrees perfectly with that determined experimentally.

## 12.7 Free Electron Theory of Metals (F-D Statistics)

As a last consideration illustrating the methods of statistical mechanics, we will apply Fermi-Dirac statistics to the free electrons in a metal. We already know from Chapter 9, Section 9.7 that conduction electrons are the *liberated* outer valence electrons associated with the atoms of a metal. Since they are assumed to be, essentially, noninteracting with each other and the positively charged ion cores, conduction electrons move freely throughout a metal behaving like an *ideal electron gas*. Further, since free electrons obey the Pauli exclusion principle, they are characterized as *fermions* and the assembly obeys the Fermi-Dirac statistics of noninteracting particles.

The quantum mechanics of a *free electron gas* has already been detailed in Chapter 10, Section 10.7, where the relation for the *density of electronic states* was derived (see Equation 10.150). It is interesting to note that the *density of states* introduced in quantum mechanics (Equation 9.89) and defined by (replacing $E$ with $\epsilon$)

$$D(\epsilon) \equiv \frac{dN(\epsilon)}{d\epsilon} \tag{12.113}$$

Density of States

is equivalent to our interpretation of degeneracy $g(\epsilon)$ in statistical mechanics. This is perhaps more evident from the interpretation of $dN(\epsilon)$ as *the number of quantum states available to electrons having energies between* $\epsilon$ *and* $\epsilon + d\epsilon$. Hence, from this qualitative interpretation and the above equation

$$\begin{aligned} dN(\epsilon) &= D(\epsilon)\, d\epsilon \\ &= g(\epsilon)\, d\epsilon, \end{aligned} \tag{12.114}$$

and the *general* equivalence between $D(\epsilon)$ and $g(\epsilon)$ is obvious. To see this equivalence quantitatively, for the *free electron gas*, we can use the arguments that led to Equation 11.95 and write

$$g(p)\, dp = 2\frac{4\pi V}{h^3} p^2 dp \tag{12.115}$$

for the degeneracy of quantum states in terms of momentum. The multiplicative factor 2 in this relation takes into account the two spin states

($m_s = -\frac{1}{2}, +\frac{1}{2}$) allowed for free electrons. This relation can be expressed in terms of energy by using the classical (nonrelativistic) free particle energy relations

$$p = (2m\epsilon)^{1/2}, \tag{11.111}$$

$$dp = m(2m\epsilon)^{-1/2}\, d\epsilon, \tag{11.112}$$

with the result

$$\begin{aligned} g(\epsilon)\, d\epsilon &= \frac{8\pi V}{h^3}(2m\epsilon)m(2m\epsilon)^{-1/2}\, d\epsilon \\ &= \frac{8\pi V}{h^3} 2^{1/2} m^{3/2} \epsilon^{1/2}\, d\epsilon. \end{aligned}$$

Multiplying and dividing this result by $2\pi^2$ yields

Density of Electron States

$$g(\epsilon)\, d\epsilon = \frac{V}{2\pi^2}\left(\frac{2m}{\hbar^2}\right)^{3/2} \epsilon^{1/2}\, d\epsilon, \tag{12.116}$$

which should be compared with the *density of electronic states* given by Equation 10.150. Clearly, statistical mechanics offers an alternative and perhaps simpler method for the derivation of the *density of states* allowed quantum mechanically for conduction electrons. Of course, with this relation and the fundamentals of statistical mechanics, it is relatively easy to obtain expressions for the energy distribution, internal energy, and specific heat of conduction electrons.

## Fermi Energy

Before derivations for energy and specific heat are attempted, however, recall that in the *Debye theory* a maximum or *cutoff frequency* was required to limit the internal energy of the *phonon gas* to a finite value. For the same reason, we assume the *free electron gas* to be in a *ground state* or minimum energy configuration. This means that in a metal of $N$ atoms contributing $N$ conduction electrons, the available quantum states will be populated by one electron each, because of the Pauli exclusion principle, from the lowest energy state $\epsilon = 0$ to the highest state, say $\epsilon = \epsilon_F$. It is customary to refer to *the energy of the topmost filled quantum state* as the **Fermi energy** and denote it symbolically as $\epsilon_F$. The *Fermi energy* for the assembly in its ground state can be determined by the same reasoning employed in the Debye theory for $\nu_m$. That is, we normalize the available quantum states $g(\epsilon)$ of Equation 12.116 to the number of conduction electrons $N$ by evaluating the integral relation

$$N = \int_0^{\epsilon_F} g(\epsilon)\, d\epsilon. \qquad (12.117)$$

Normalization of Quantum States

Substitution from Equation 12.116 for $g(\epsilon)\, d\epsilon$ immediately yields

$$N = \frac{V}{2\pi^2}\left(\frac{2m}{\hbar^2}\right)^{3/2} \int_0^{\epsilon_F} \epsilon^{1/2} d\epsilon$$

$$= \frac{2}{3}\frac{V}{2\pi^2}\left(\frac{2m}{\hbar^2}\right)^{3/2} \epsilon_F^{3/2}, \qquad (12.118)$$

from which we obtain

$$\epsilon_F^{-3/2} = \frac{2}{3N}\frac{V}{2\pi^2}\left(\frac{2m}{\hbar^2}\right)^{3/2} \qquad (12.119)$$

or in simpler form

$$\epsilon_F = \frac{\hbar^2}{2m}\left(\frac{3\pi^2 N}{V}\right)^{2/3}. \qquad (12.120)$$

Fermi Energy

The first equation for $\epsilon_F^{-3/2}$ is given because a comparison of it with Equation 12.116 immediately allows $g(\epsilon)\, d\epsilon$ to be expressed as

$$g(\epsilon)\, d\epsilon = \frac{3}{2} N\, \epsilon_F^{-3/2} \epsilon^{1/2}\, d\epsilon, \qquad (12.121)$$

Density of States

while the second equation for $\epsilon_F$ in reduced form is more amenable to computational problems.

The expression for the *Fermi energy* (Equation 12.120) can be expressed in terms of more fundamental quantities by recognizing that the **electron density,** defined by

$$\eta \equiv \frac{N}{V}, \qquad (12.122)$$

Electron Density

can be represented as

$$\eta \equiv \frac{N}{V}\frac{M}{M} = \frac{N\rho}{M} = \frac{N_o\rho}{\mathcal{M}}, \tag{12.123}$$

where the definitions for *mass density*,

Mass Density

$$\rho \equiv \frac{M}{V} \tag{5.33}$$

and the number of moles,

Number of Moles

$$n \equiv \frac{N}{N_o} = \frac{M}{\mathcal{M}}, \tag{5.76}$$

have been employed. Hence, a more amenable form of the Fermi energy relation for computational purposes is given by

Fermi Energy

$$\begin{aligned} \epsilon_F &= \frac{\hbar^2}{2m}(3\pi^2\eta)^{2/3} \\ &= \frac{\hbar^2}{2m}\left(\frac{3\pi^2 N_o\rho}{\mathcal{M}}\right)^{2/3}. \end{aligned} \tag{12.124}$$

For example, potassium has a *ground state electron configuration* given by (see Chapter 7, Section 7.7)

$${}^{39}_{19}\mathrm{K}{:}\ 1s^2 2s^2 2p^6 3s^2 3p^6 4s^1,$$

so each atom contributes a single $4s$ electron to the electron gas. With mass density $\rho$ and molal atomic mass $\mathcal{M}$ (using the chemical atomic weight instead of the relative atomic mass) given by $\rho_K = 0.86\ \mathrm{g/cm^3}$, $\mathcal{M}_K = 39.1$ g, Equation 12.123 gives the *electron density* as

$$\begin{aligned} \eta &= \frac{N_o\rho}{\mathcal{M}} \\ &= \frac{(6.02 \times 10^{23})\,(0.86\ \mathrm{g/cm^3})}{39.1\ \mathrm{g}} \\ &= 1.32 \times 10^{22}\ \frac{\text{electrons}}{\mathrm{cm^3}} \end{aligned}$$

$$= 1.32 \times 10^{28} \frac{\text{electrons}}{\text{m}^3},$$

and the Fermi energy is obtained from Equation 12.124 as

$$\begin{aligned}
\epsilon_F &= \frac{\hbar^2}{2m}(3\pi^2\eta)^{2/3} \\
&= \frac{(1.05 \times 10^{-34}\ \text{J} \cdot \text{s})^2}{2(9.11 \times 10^{-31}\ \text{kg})}[3\pi^2(1.32 \times 10^{28}/\text{m}^3)]^{2/3} \\
&= (6.05 \times 10^{-39}\ \text{J} \cdot \text{m}^2)(3.90 \times 10^{29}/\text{m}^3)^{2/3} \\
&= (6.05 \times 10^{-39}\ \text{J} \cdot \text{m}^2)(5.34 \times 10^{19}/\text{m}^2) \\
&= 3.23 \times 10^{-19}\ \text{J} = 2.02\ \text{eV}.
\end{aligned}$$

Thus, in the ground state energy configuration, conduction electrons in potassium would have energies from zero up to 2.02 eV. Electrons at the Fermi energy are said to have a classical **Fermi velocity** defined by

$$v_F \equiv \left(\frac{2\epsilon_F}{m_e}\right)^{1/2} \qquad (12.125)$$

Fermi Velocity

and a **Fermi temperature** defined by

$$T_F \equiv \frac{\epsilon_F}{k_B}, \qquad (12.126)$$

Fermi Temperature

The *Fermi temperature* corresponds to the approximate temperature of a metal described by classical theory at which the electron would have an energy $\epsilon_F$. A more complete discussion and example of this point is presented following Equation 12.135.

There is another very interesting point concerning the Fermi energy and its interpretation with the Fermi-Dirac distribution,

$$n_i = \frac{g_i}{e^{\beta(\epsilon_i - \mu)} + 1}. \qquad (12.26)$$

Recall that the *occupation index* for this distribution at *absolute zero* was given by (see Equations 12.34 and 12.35)

$$\begin{aligned}
f(\epsilon) &= 1, \qquad \epsilon < \mu, \\
&= 0, \qquad \epsilon > \mu,
\end{aligned} \qquad (12.127)$$

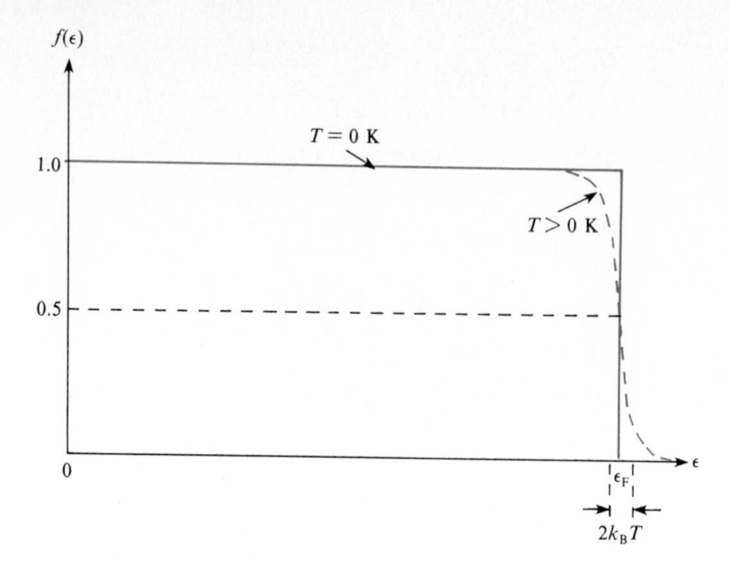


Figure 12.7
A generalized illustration of the Fermi-Dirac occupation index for $T = 0$ K (solid line) and $T > 0$ K (dashed line).

which means that electrons will tend to populate all energy states that are less than $\mu$ up to and including those that are equal to $\mu$. Consequently, from our definition of the Fermi energy, we set

$$\mu = \epsilon_F. \tag{12.128}$$

For a continuous distribution of electron energies, the occupation index is now expressed by

$$\begin{aligned} f(\epsilon) &= \frac{n(\epsilon)}{g(\epsilon)} \\ &= \frac{1}{e^{\beta(\epsilon - \epsilon_F)} + 1} \end{aligned} \tag{12.129}$$

and illustrated graphically in Figure 12.7 for $T = 0$ K and $T > 0$ K. From this interpretation of $\mu$ and $\epsilon_F$, the most interesting observation is that *even at absolute zero* conduction electrons have energies from zero up to the Fermi energy $\epsilon_F$. This is decidedly a nonclassical behavior, since particles in a *classical ideal gas* would have zero energy (recall $\epsilon = (3/2)k_B T$ from Equation 11.102) at a temperature of absolute zero.

## Electronic Energy and Specific Heat Formulae

It is now straightforward to obtain an expression for the total internal energy of the free electron gas. Assuming a continuous distribution of electron energies, the Fermi-Dirac distribution law of Equation 12.129 gives

$$dn_\epsilon = n(\epsilon)\, d\epsilon = \frac{g(\epsilon)\, d\epsilon}{e^{\beta(\epsilon - \epsilon_F)} + 1}, \tag{12.130}$$

which upon substitution from Equation 12.116 for $g(\epsilon)\, d\epsilon$ becomes

$$dn_\epsilon = n(\epsilon)\, d\epsilon = \frac{(V/2\pi^2)(2m/\hbar^2)^{3/2}\epsilon^{1/2}\, d\epsilon}{e^{\beta(\epsilon - \epsilon_F)} + 1}. \tag{12.131}$$

In terms of the Fermi energy $\epsilon_F$, the number of electrons having energies between $\epsilon$ and $\epsilon + d\epsilon$ is given by Equations 12.121 and 12.130 as

$$dn_\epsilon = n(\epsilon)\, d\epsilon = \frac{(3/2)N\epsilon_F^{-3/2}\, \epsilon^{1/2}\, d\epsilon}{e^{\beta(\epsilon - \epsilon_F)} + 1}. \tag{12.132}$$

Thus, for the assembly of free conduction electrons the *total internal energy* is given by the integral expression

$$\begin{aligned} U_e &= \int_0^\infty \epsilon n(\epsilon)\, d\epsilon \\ &= \frac{3}{2} N\epsilon_F^{-3/2} \int_0^\infty \frac{\epsilon^{3/2}\, d\epsilon}{e^{\beta(\epsilon - \epsilon_F)} + 1}. \end{aligned} \tag{12.133}$$

The integral of this expression cannot be evaluated in closed form but must be expressed as an infinite series.

Although the reduction of Equation 12.133 requires mathematics that is beyond the scope of this textbook, we can evaluate $U$ at absolute zero. That is, assuming the assembly to be in the *ground state* with the highest energy state being $\epsilon_F$, then for $T = 0$ K the integral expression for $U$ reduces to

$$\begin{aligned} U_e &= \frac{3}{2} N\epsilon_F^{-3/2} \int_0^{\epsilon_F} \epsilon^{3/2}\, d\epsilon \\ &= \frac{3}{5} N\epsilon_F, \qquad T = 0 \text{ K}, \end{aligned} \tag{12.134}$$

from which the *average energy per electron* becomes

$$\bar{\epsilon}_e \equiv \frac{U_e}{N} = \frac{3}{5}\,\epsilon_{\mathrm{F}}. \tag{12.135}$$

This result is considerably different than that predicted by the classical theory for a monatomic ideal gas,

$$\bar{\epsilon}_e = \frac{3}{2}\,k_{\mathrm{B}}T, \tag{11.102}$$

since at $T = 0$ K the average particle energy is zero. According to the classical theory of Maxwell-Boltzmann statistics, a sample of potassium would have to be at a temperature of (see previous example for $\epsilon_{\mathrm{F}}$ value)

$$\begin{aligned} T &= \frac{2}{3}\frac{\bar{\epsilon}_e}{k_{\mathrm{B}}} \\ &= \frac{2}{3}\left(\frac{3}{5}\right)\frac{\epsilon_{\mathrm{F}}}{k_{\mathrm{B}}} \\ &= \frac{2}{5}\frac{\epsilon_{\mathrm{F}}}{\mathrm{k_B}} \\ &= \frac{2}{5}\frac{3.23 \times 10^{-19}\ \mathrm{J}}{1.38 \times 10^{-23}\ \mathrm{J/K}} \\ &= 9.36 \times 10^{3}\ \mathrm{K} \end{aligned} \tag{12.136}$$

for its electrons to be at the same average energy as that predicted by Fermi-Dirac statistics (Equation 12.135) at 0 K.

A more detailed evaluation of Equation 12.133 gives

$$U_e = \frac{3}{5}N\epsilon_{\mathrm{F}}\left[1 + \frac{5}{3}\left(\frac{\pi k_{\mathrm{B}}T}{2\epsilon_{\mathrm{F}}}\right)^2 - \left(\frac{\pi k_{\mathrm{B}}T}{2\epsilon_{\mathrm{F}}}\right)^4 + \cdots\right] \tag{12.137}$$

for the total internal energy, which obviously reduces to Equation 12.134 at $T = 0$ K. With this relation for $U$, it is straightforward to determine the electronic specific heat. That is, the *thermal capacity* at constant volume is given by

$$C_{V-e} = \left(\frac{\partial U}{\partial T}\right)_V$$

$$= \frac{\pi^2}{2} Nk_B \frac{k_B T}{\epsilon_F}\left[1 - \frac{6}{5}\left(\frac{\pi k_B T}{2\epsilon_F}\right)^2 + \cdots\right], \quad (12.138)$$

from which the electronic *molal specific thermal capacity* becomes (note that $Nk_B = nR$)

$$c_{v-e} = \frac{\pi^2}{2}\frac{k_B T}{\epsilon_F} R\left[1 - \frac{6}{5}\left(\frac{\pi k_B T}{2\epsilon_F}\right)^2 + \cdots\right]. \quad (12.139)$$

Usually, only the first term in this expression is retained, as the second and higher order terms in $T$ are rather small compared to one. For example, in the case of potassium at $T = 500$ K, we have the second term in the brackets of Equation 12.139 given by $1.35 \times 10^{-3}$. Thus, at relatively low temperatures, the *molal specific thermal capacity* of the free electrons in a metal is given by

$$c_{v-e} = \frac{\pi^2}{2}\frac{k_B T}{\epsilon_F} R. \quad (12.140)$$

Electronic Specific Heat

Since the coefficient of $R$ is quite small for metals over a broad range of temperatures (e.g., for potassium at 300 K, $c_{v-e} = 6.32 \times 10^{-2} R$), the electronic specific heat $c_{v-e}$ does not appreciably contribute to the specific heat of metals. The lattice specific heat $c_v$, given by Equation 12.89, dominates $c_{v-e}$ at all but very low temperatures, where $c_v \propto T^3$ and $c_{v-e} \propto T$.

It is also interesting to note that the result for $c_{v-e}$ given by Equation 12.140 is essentially consistent with allowing only electrons within about $k_B T$ of the Fermi energy (see Figure 12.7) to absorb energy as a solid is heated. The *effective* number of electrons in this region is approximately given by

$$N_{\text{eff}} \approx g(\epsilon_F)\, k_B T, \quad (12.141)$$

which from Equation 12.121 becomes

$$N_{\text{eff}} \approx \frac{3}{2} N\epsilon_F^{-1}\, k_B T. \quad (12.142)$$

Assuming each of these electrons within $k_B T$ of $\epsilon_F$ to acquire (3/2) $k_B T$ of energy, the total electronic contribution to the internal energy of the metal is

$$U_e = N_{eff} \frac{3}{2} k_B T, \tag{12.143}$$

which becomes

$$\begin{aligned} U_e &= \frac{9}{4} N k_B \frac{k_B}{\epsilon_F} T^2 \\ &= \frac{9}{4} nR \frac{k_B}{\epsilon_F} T^2. \end{aligned} \tag{12.144}$$

Thus, the *electronic molal specific thermal capacity* is

$$\begin{aligned} c_{v-e} &= \frac{1}{n}\left(\frac{\partial U_e}{\partial T}\right)_v \\ &= \frac{9}{2}\frac{k_B T}{\epsilon_F} R, \end{aligned} \tag{12.145}$$

which is only slightly different from the more exact result given by Equation 12.140. In either case, however, it should be clear that the electrons contribution to the specific heat of a solid is essentially negligible.

## Review of Fundamental and Derived Equations

A listing of the fundamental and derived equations of this chapter is presented below. The derivations and applications of statistical mechanics are presented in a logical listing that parallels their development in each section of the chapter.

### FUNDAMENTAL EQUATIONS—CLASSICAL PHYSICS

$$\epsilon = \tfrac{1}{2} m v^2 = \frac{p^2}{2m}$$ Free Particle Kinetic Energy

$$\bar{\epsilon} \equiv \frac{E}{N}$$ Average Particle Energy

$\bar{\epsilon} = \dfrac{N_f}{2} k_B T$ — Free Particle Average Energy

$\rho \equiv \dfrac{M}{V}$ — Mass Density

$\mathbf{u} \equiv \dfrac{U}{V}$ — Total Energy Density

$\eta \equiv \dfrac{N}{V}$ — Particle Density

$n \equiv \dfrac{N}{N_o}$ — Number of Moles

$pV = nRT = Nk_B T$ — Ideal Gas Equation of State

$dU = T\,dS - p\,dV + \mu\,dN$ — 1st & 2nd Laws of Thermodynamics

## FUNDAMENTAL EQUATIONS—STATISTICAL MECHANICS

### *Discrete Distribution of Particle Energies*

$N = \sum_i n_i$ — Conservation of Particles

$E = \sum_i n_i \epsilon_i$ — Conservation of Energy

$\mathbf{u} = \dfrac{1}{V} \sum_i n_i \epsilon_i$ — Total Energy Density

$W_k = \Pi_i w_i$ — Thermodynamic Probability

$S = k_B \ln W$ — Entropy in Statistical Mechanics

$W_{\text{M-B}} = N! \Pi_i \dfrac{g_i^{n_i}}{n_i!}$ — M-B Thermodynamic Probability

$W_{\text{B-E}} = \Pi_i \dfrac{(n_i + g_i - 1)!}{(g_i - 1)! n_i!}$ — B-E Thermodynamic Probability

$W_{\text{F-D}} = \Pi_i \dfrac{g_i!}{(g_i - n_i)! n_i!}$ — F-D Thermodynamic Probability

$$n_i = \frac{g_i}{e^{\alpha} e^{\beta \epsilon_i} + d}, \qquad \begin{array}{l} d = 0 \rightarrow \text{M-B} \\ d = -1 \rightarrow \text{B-E} \\ d = +1 \rightarrow \text{F-D} \end{array}$$

Distribution Laws

$\beta = \dfrac{1}{k_B T}, \qquad \alpha = -\beta\mu$ — Lagrange Multipliers

$f(\epsilon_i) \equiv \dfrac{n_i}{g_i}$ — Occupation Index

*Continuous Distribution of Particle Energies—Generalized Equations*

$$dn_q = n(q)\,dq$$
$$= f(q)g(q)\,dq, \qquad q = \epsilon, \nu, \lambda, p$$ Distribution of Particles

$$g(p)\,dp = [1 \text{ or } 2]\,\frac{4\pi V}{h^3}\,p^2\,dp$$ Distribution of Energy States

$$N = \int_0^\infty n(q)\,dq, \qquad q = \epsilon, \nu, \lambda, p$$ Conservation of Particles

$$E = \int_0^\infty \epsilon n(q)\,dq, \qquad q = \epsilon, \nu, \lambda, p$$ Conservation of Energy

$$d\mathbf{u}_q = \mathbf{u}(q)\,dq = \frac{\epsilon n(q)\,dq}{V}$$ Spectral Energy Density

$$\mathbf{u} = \frac{1}{V}\int_0^\infty \epsilon n(q)\,dq$$ Total Energy Density

## DERIVED EQUATIONS

*Most Probable Distribution*

$$\ln A! \approx A\ln A - A, \qquad A \gg 1$$ Stirling's Formula

$$\delta \ln W_{\text{B-E}} = \sum_i \ln\left(\frac{g_i}{n_i} + 1\right)\delta n_i$$

$$n_i = \frac{g_i}{e^{\alpha}e^{\beta\epsilon_i} - 1}$$ Bose-Einstein Distribution

$$\delta \ln W_{\text{F-D}} = \sum_i \ln\left(\frac{g_i}{n_i} - 1\right)\delta n_i$$

$$n_i = \frac{g_i}{e^{\alpha}e^{\beta\epsilon_i} + 1}$$ Fermi-Dirac Distribution

$$e^{\beta(\epsilon_i - \mu)} \gg 1$$ Classical Limit

$$f(\epsilon_i) \equiv \frac{n_i}{g_i} \ll 1$$ Classical Limit

*Identification of Lagrange Multipliers*

$$dS = k_B\,d\ln W$$
$$= k_B\sum_i(\alpha + \beta\epsilon_i)\,dn_i$$
$$= k_B\alpha\,dN + k_B\beta\,dU$$ Differential Entropy

$$\beta = \frac{1}{k_B T}, \qquad \alpha = -\beta\mu$$ Lagrange Multipliers

### Specific Heat of a Solid—Classical Theory

$\bar{\epsilon}_o = k_B T$ Average Energy of Linear Oscillator

$U = N\bar{\epsilon}_a = N(3\bar{\epsilon}_o) = 3nRT$ Internal Energy for $N$-Atoms

$c_v = 3R$ Law of Dulong and Petit

### Specific Heat of a Solid—Einstein Theory (M-B Statistics)

$\epsilon_n = (n + \frac{1}{2})h\nu$ Linear Oscillator Eigenvalues

$$\bar{\epsilon}_0 = \tfrac{1}{2}h\nu + \frac{h\nu}{e^{\beta h\nu} - 1}$$ Average Energy—Quantized Oscillator

$$\theta_E \equiv \frac{h\nu}{k_B} = \beta h\nu T$$ Einstein's Characteristic Temperature

$u = 3N_o h\nu\,[\frac{1}{2} + (e^{\theta_E/T} - 1)^{-1}]$ Molal Specific Internal Energy

$$c_v = \frac{3R\,(\theta_E/T)^2 e^{\theta_E/T}}{(e^{\theta_E/T} - 1)^2}$$ Einstein's Specific Heat Formula

### Specific Heat of a Solid—Debye Theory (Phonon Statistics)

$\epsilon = h\nu$ Phonon/Photon Energy Quantization

$v_s = \lambda\nu$ Phonon Speed

$$p = \frac{h}{\lambda} = \frac{h\nu}{v_s}$$ Phonon Momentum

$$\frac{1}{v_s^3} = \frac{1}{v_l^3} + \frac{2}{v_t^3}$$ Phonon Speed—Elastic Waves

$$3N = \int_0^{\nu_m} g(\nu)\,d\nu$$ Normalization of Phonon States

$\theta_D \equiv \beta h\nu_m T = x_m T$ Debye Characteristic Temperature

$$U = \int_0^{\nu_m} \frac{h\nu g(\nu)\,d\nu}{e^{\beta h\nu} - 1}$$

$$= \frac{9nRT^4}{\theta_D^3}\int_0^{\theta_D/T} \frac{x^3\,dx}{e^x - 1}$$ Internal Energy

$$c_v = 9R\left[4\left(\frac{T}{\theta_D}\right)^3 \int_0^{\theta_D/T} \frac{x^3\,dx}{e^x - 1} - \frac{\theta_D/T}{e^{\theta_D/T} - 1}\right]$$

$$c_v = 3R, \quad T \gg \theta_D \qquad \text{High Temperature Limit}$$

$$c_v = \frac{12\pi^4 R}{5}\left(\frac{T}{\theta_D}\right)^3, T \ll \theta_D \qquad \text{Low Temperature Limit}$$

### Blackbody Radiation (Photon Statistics)

$$d\mathbf{u}_\nu = \mathbf{u}(\nu)\,d\nu = \frac{\epsilon n(\nu)\,d\nu}{V}$$

$$= \frac{\epsilon g(\nu)\,d\nu}{V\,(e^{\beta\epsilon} - 1)} \qquad \text{Spectral Energy Density}$$

$$\mathbf{u}(\nu)\,d\nu = \frac{8\pi h}{c^3}\frac{\nu^3\,d\nu}{e^{\beta h\nu} - 1} \qquad \text{Planck Radiation Formula}$$

$$\mathbf{u}(\nu)\,d\nu = \frac{8\pi}{c^3}\,k_B T\nu^2\,d\nu \qquad \text{Rayleigh-Jeans Formula}$$

$$\mathbf{u}(\nu)\,d\nu = \frac{8\pi h}{c^3}\,\nu^3 e^{-\beta h\nu}\,d\nu \qquad \text{Wien Formula}$$

$$\mathbf{u} = \frac{8\pi^5 k_B^4}{15h^3c^3}\,T^4 = \sigma T^4 \qquad \text{Stefan-Boltzmann Law}$$

### Free Electron Theory of Metals (F-D Statistics)

$$g(\epsilon)\,d\epsilon = D(\epsilon)\,d\epsilon$$

$$= \frac{V}{2\pi^2}(2m\hbar^2)^{3/2}\epsilon^{1/2}\,d\epsilon$$

$$= \frac{3}{2}N\,\epsilon_F{}^{-3/2}\epsilon^{1/2}\,d\epsilon \qquad \text{Degeneracy/Density of States}$$

$$N = \int_0^{\epsilon_F} g(\epsilon)\,d\epsilon \qquad \text{Normalization of Quantum States}$$

$$\epsilon_F = \frac{\hbar^2}{2m}\,(3\pi^2\eta)^{2/3}$$

$$= \frac{\hbar^2}{2m}\left(\frac{3\pi^2 N_o\rho}{\mathcal{M}}\right)^{2/3} \qquad \text{Fermi Energy}$$

$$v_F \equiv \left(\frac{2\epsilon_F}{m_e}\right)^{1/2}$$ Fermi Velocity

$$T_F \equiv \frac{\epsilon_F}{k_B}$$ Fermi Temperature

$$n(\epsilon)\, d\epsilon = \frac{(3/2)N\epsilon_F^{-3/2}\, \epsilon^{1/2}\, d\epsilon}{e^{\beta(\epsilon - \epsilon_F)} + 1}$$ Distribution of Electron Energies

$$U_e = \frac{3}{2} N\epsilon_F^{-3/2} \int_0^\infty \frac{\epsilon^{3/2}\, d\epsilon}{e^{\beta(\epsilon - \epsilon_F)} + 1}$$ Electronic Internal Energy

$$U_e = \frac{3}{5} N\epsilon_F, \qquad T = 0 \text{ K}$$ Internal Energy—Low Temperature

$$U_e = N_{\text{eff}} \frac{3}{2} k_B T$$

$$= g(\epsilon_F)\, k_B T \frac{3}{2} k_B T$$

$$= \frac{9}{4} nR \frac{k_B}{\epsilon_F} T^2$$ Effective Internal Energy

$$c_{v-e} = \frac{9}{2} \frac{k_B T}{\epsilon_F} R$$ Effective Molal Specific Thermal Capacity

## Problems

**12.1** Consider a system of $N = 4$ particles in a $\mu$-space consisting of one energy level $\epsilon_1$ having a degeneracy $g_1 = 6$. Calculate the *thermodynamic probability* for the one allowed macrostate for M-B, B-E, and F-D statistics and determine the value for $N$ of Equation 12.8 for each case.

*Solution:*

With $N = 4$, $g_1 = 6$, and $n_1 = 4$, the M-B, B-E, and F-D thermodynamic probabilities are given by

$$W_{\text{M-B}} = N!\Pi_i \frac{g_i^{n_i}}{n_i!} = \frac{4!6^4}{4!} = 1296,$$

$$W_{\text{B-E}} = \Pi_i \frac{(n_i + g_i - 1)!}{(g_i - 1)!n_i!} = \frac{(4 + 6 - 1)!}{(6 - 1)!4!}$$

$$= \frac{9!}{5!4!}$$
$$= 126,$$

$$W_{\text{F-D}} = \Pi_i \frac{g_i!}{(g_i - n_i)!n_i!}$$
$$= \frac{6!}{(6 - 4)!4!}$$
$$= \frac{6!}{2!4!}$$
$$= 15.$$

Since there are only $g_1 = 6$ ways of placing all the particles in one quantum state for M-B and B-E statistics (recall that Pauli exclusion principle is not applicable in these cases), then from the above results and Equation 12.8 we obtain

$$\mathcal{N}_{\text{M-B}} = \frac{6/1296}{1290/1296} = \frac{1}{215},$$

$$\mathcal{N}_{\text{B-E}} = \frac{6/126}{120/126} = \frac{1}{20}.$$

Of course, for F-D statistics we have

$$\mathcal{N}_{\text{F-D}} = \frac{0/15}{15/15} = 0.$$

**12.2** Consider a system of $N = 3$ particles in a $\mu$-space consisting of two energy levels having degeneracies given by $g_1 = 2$, $g_2 = 4$. Determine the number of microstates associated with each *allowed* macrostate for M-B, B-E, and F-D statistics.

*Answer:* 8, 48, 96, 64; 4, 12, 20, 20; 4, 12

**12.3** Assuming $n_i + g_i \gg 1$, derive an expression for the logarithm of the Bose-Einstein thermodynamic probability.

*Solution:*
Assuming $n_i + g_i \gg 1$, the B-E thermodynamic probability of Equation 12.10 reduces to

$$W_{\text{B-E}} = \Pi_i \frac{(n_i + g_i)!}{(g_i - 1)!n_i!},$$

from which $\ln W_{\text{B-E}}$ is easily determined using Stirling's approximation, that is,

$$\ln W_{\text{B-E}} = \sum_i [\ln (n_i + g_i)! - \ln (g_i - 1)! - \ln n_i!]$$

$$= \sum_i [(n_i + g_i) \ln (n_i + g_i) - (n_i + g_i) - (g_i - 1) \ln (g_i - 1) + (g_i - 1) - n_i \ln n_i + n_i]$$

$$= \sum_i [(n_i + g_i) \ln (n_i + g_i) - (g_i - 1) \ln (g_i - 1) - 1 - n_i \ln n_i].$$

This result should be compared with that given by Equation 12.19.

**12.4** Using the result of Problem 12.3, derive an expression for $\delta \ln W_{\text{B-E}}$.

*Answer:* $\delta \ln W_{\text{B-E}} = \sum_i [\ln (n_i + g_i) - \ln n_i] \, \delta n_i$

**12.5** Starting with Equation 12.36 and using Equations 12.13 and 12.16, show that for any statistics $\beta = 1/k_B T$ and $\alpha = -\beta\mu$.

*Solution:*

Since the method of the most probable distribution results in a relation of the form (Equation 12.13)

$$d \ln W = 0 = \sum_i f(n_i, g_i) \, dn_i,$$

then Equation 12.36,

$$dS = k_B d \ln W,$$

can be expressed as

$$dS = k_B \sum_i f(n_i, g_i) \, dn_i.$$

But, from Equation 12.16,

$$\sum_i [f(n_i, g_i) - \alpha - \beta\epsilon_i] \, dn_i = 0,$$

we obtain

$$\sum_i f(n_i, g_i) \, dn_i = \sum_i (\alpha + \beta\epsilon_i) \, dn_i = \alpha \, dN + \beta \, dU,$$

and the differential entropy becomes

$$dS = k_B \beta \, dU + k_B \alpha \, dN.$$

A comparison of this result with the combined first and second laws of thermodynamics in the form

$$dS = \frac{1}{T} dU + \frac{p}{T} dV - \frac{\mu}{T} dN$$

immediately yields

$$\beta = \frac{1}{k_B T},$$

$$\alpha = -\frac{\mu}{k_B T} = -\beta\mu.$$

**12.6** Using Maxwell-Boltzmann statistics, expand the partition function $Z_o$ for the quantized linear harmonic oscillator and simplify it using the geometric series of Equation 12.57.

*Answer:* $Z_o = \frac{e^{-\frac{1}{2}\beta h\nu}}{1 - e^{-\beta h\nu}}$

**12.7** Derive the equation for the average energy of a quantized linear harmonic oscillator, using the result of Problem 12.6 and the equality $\bar{\epsilon}_o = -\partial \ln Z_o/\partial\beta$.

*Solution:*
With the expression of the partition function for the quantized linear harmonic oscillator given by

$$Z_o = \frac{e^{-\frac{1}{2}\beta h\nu}}{1 - e^{-\beta h\nu}},$$

the average oscillator energy is determined by

$$\begin{aligned}
\bar{\epsilon}_o &= -\frac{\partial}{\partial\beta} \ln Z_o \\
&= -\frac{\partial}{\partial\beta}[-\tfrac{1}{2}\beta h\nu - \ln(1 - e^{-\beta h\nu})] \\
&= \tfrac{1}{2}h\nu + \frac{h\nu e^{-\beta h\nu}}{1 - e^{-\beta h\nu}} \\
&= \tfrac{1}{2}h\nu + \frac{h\nu}{e^{\beta h\nu} - 1}.
\end{aligned}$$

**12.8** Using the Boltzmann distribution and the partition function $Z_o$ of Problem 12.6, derive a relation for the *fractional number of quantized oscillators* in the *i*th energy level, $n_i/N$, in terms of the Einstein characteristic temperature $\theta_E$.

*Answer:* $\frac{n_i}{N} = (1 - e^{-\theta_E/T})e^{-n\theta_E/T}$

**12.9** Using the result of Problem 12.8, determine $n_i/N$ for the quantized linear oscillator for the four lowest energy levels when $\theta_E = T$.

*Solution:*
With the fractional number of oscillators in the *i*th level expressed by

$$\frac{n_i}{N} = (1 - e^{-\theta_E/T})e^{-n\theta_E/T},$$

then for $\theta_E = T$ we have

$$\frac{n_i}{N} = (1 - e^{-1})\, e^{-n}$$
$$= (0.6321)\, e^{-n}.$$

Consequently, $n_i/N$ for the four lowest energy levels ($n = 0, 1, 2, 3$) is

$$\frac{n_0}{N} = (0.6321)\, e^{-0} = 0.6321,$$
$$\frac{n_1}{N} = (0.6321)\, e^{-1} = 0.2325,$$
$$\frac{n_2}{N} = (0.6321)\, e^{-2} = 0.0855,$$
$$\frac{n_3}{N} = (0.6321)\, e^{-3} = 0.0315.$$

Thus, about 63 percent of the oscillators are in the lowest energy level when $\theta_E = T$.

**12.10** Allowing $\theta_E = 2T$, evaluate $n_i/N$ for the quantized linear oscillator for the four lowest energy levels.

*Answer:* 0.8647, 0.1170, 0.0158, 0.0021

**12.11** Consider each atom of a solid to be represented by three quantized linear harmonic oscillators, and derive an expression for the entropy $S_{\text{M-B}}$ of the solid in terms of the Einstein characteristic temperature.

*Solution:*
For Maxwell-Boltzmann statistics the entropy is given by Equation 11.75,

$$S_{\text{M-B}} = \frac{U}{T} + Nk_B \ln Z.$$

Expressing the internal energy by

$$U = N\bar{\epsilon}_a = 3N\bar{\epsilon}_o$$

and using

$$\frac{1}{T} = k_B\beta,$$

the expression for entropy becomes

$$S_{\text{M-B}} = 3Nk_B\,(\beta\bar{\epsilon}_o + \ln Z_o).$$

It should be noted that we have used

$$\ln Z = 3 \ln Z_o.$$

since

$$\bar{\epsilon} = -\frac{\partial \ln Z}{\partial \beta}$$

suggests that

$$3\bar{\epsilon}_o = -\frac{\partial}{\partial \beta}(3 \ln Z_o).$$

Now, with $\bar{\epsilon}$ given by Equation 12.61,

$$\bar{\epsilon}_o = \tfrac{1}{2}h\nu + \frac{h\nu}{e^{\beta h\nu} - 1},$$

and $\ln Z_o$ given by (see result of Problem 12.6)

$$\ln Z_o = \ln\left(\frac{e^{-\frac{1}{2}\beta h\nu}}{1 - e^{-\beta h\nu}}\right)$$
$$= -\tfrac{1}{2}\beta h\nu - \ln(1 - e^{-\beta h\nu}),$$

the expression for entropy becomes

$$S_{\text{M-B}} = 3Nk_B\left[\frac{\beta h\nu}{e^{\beta h\nu} - 1} - \ln(1 - e^{-\beta h\nu})\right].$$

In terms of the Einstein characteristic temperature this expression becomes

$$S_{\text{M-B}} = 3Nk_B\left[\frac{\theta_E/T}{e^{\theta_E/T} - 1} - \ln(1 - e^{-\theta_E/T})\right].$$

**12.12** What does the entropy for Einstein oscillators (see Problem 12.11) approach at high temperatures ($T \gg \theta_E$) and at low temperatures ($T \ll \theta_E$)?

*Answer:* $S = 3Nk_B\left(1 + \ln\frac{T}{\theta_E}\right)$, $S = 0$

**12.13** Express the maximum frequency $\nu_m$ and characteristic temperature $\theta_D$ of the Debye theory in terms of the mass density $\rho$ of a solid.

*Solution:*

From the Debye theory of the specific heat of a solid the cutoff frequency is obtainable from Equation 12.78 in the form

$$\nu_m = v_s\left(\frac{9N}{4\pi V}\right)^{1/3},$$

where $N/V$ represents the **atomic density** (*number of atoms per unit volume*). From the definitions of mass density and number of moles (see development of Equation 12.123), the *atomic density* can be expresssed as

$$\frac{N}{V} = \frac{N}{V}\frac{M}{M}$$

$$= \rho \frac{N}{M}$$

$$= \rho \left(\frac{N}{N\mathcal{M}/N_o}\right)$$

$$= \frac{\rho N_o}{\mathcal{M}},$$

with which $\nu_{\mathrm{m}}$ becomes

$$\nu_{\mathrm{m}} = v_s \left(\frac{9\rho N_o}{4\pi \mathcal{M}}\right)^{1/3}$$

and the Debye temperature defined by Equation 12.86 becomes

$$\theta_D \equiv \frac{h\nu_{\mathrm{m}}}{k_{\mathrm{B}}}$$

$$= \frac{hv_s}{k_B}\left(\frac{9\rho N_o}{4\pi \mathcal{M}}\right)^{1/3}$$

**12.14** Using Equation 12.79 for $v_s$ and the results of Problem 12.13, calculate $\nu_{\mathrm{m}}$ and $\theta_{\mathrm{D}}$ for aluminum, where $v_l = 6420$ m/s, $v_t = 3040$ m/s, $\rho_{\mathrm{Al}} = 2.70 \times 10^3$ kg/m$^3$, and $\mathcal{M}_{\mathrm{Al}} = 26.98$ g. Note that specific heat measurements of aluminum give $\theta_{\mathrm{D}} = 398$ K.

*Answer:* $\nu_{\mathrm{m}} = 8.32 \times 10^{12}\ \mathrm{s}^{-1}$, $\theta_{\mathrm{D}} = 400$ K

**12.15** Calculate $v_s$, $\nu_{\mathrm{m}}$, and $\theta_{\mathrm{D}}$ for silver, using $v_l = 3650$ m/s, $v_t = 1610$ m/s, $\rho_{\mathrm{Ag}} = 10.5 \times 10^3$ kg/m$^3$, and $\mathcal{M} = 107.87$ g. How does the calculated value of $\theta_{\mathrm{D}}$ compare with the value obtained from specific heat measurements of $\theta_{\mathrm{D}} = 215$ K?

*Solution:*
Using the values of $v_l = 3650$ m/s and $v_t = 1610$ m/s, the speed of sound in silver is given by (see Equation 12.79)

$$v_s = \left(\frac{v_t^3 v_l^3}{v_t^3 + 2v_l^3}\right)^{1/3}$$

$$= \left(\frac{2.03 \times 10^{20}}{1.01 \times 10^{11}}\frac{\mathrm{m}^3}{\mathrm{s}^3}\right)^{1/3}$$

$$= 1.26 \times 10^3\ \frac{\mathrm{m}}{\mathrm{s}}.$$

With the above values for $v_s$, $\rho_{\mathrm{Ag}}$, and $\mathcal{M}_{\mathrm{Ag}}$, the maximum frequency $\nu_{\mathrm{m}}$ can be calculated using (see Problem 12.13)

$$\nu_m = v_s \left(\frac{9\rho_{Ag}N_o}{4\pi\mathcal{M}_{Ag}}\right)^{1/3}.$$

Hence, dropping units we obtain

$$\nu_m = v_s \left[\frac{9(10.5 \times 10^3)(6.02 \times 10^{23})}{4(3.14)(0.10787)}\right]^{1/3}$$

$$= \left(1.26 \times 10^3 \frac{\text{m}}{\text{s}}\right)(4.16 \times 10^{28} \text{ m}^3)^{1/3}$$

$$= 4.37 \times 10^{12} \text{ s}^{-1},$$

and the Debye characteristic temperature is given by

$$\theta_D = \frac{h\nu_m}{k_B}$$

$$= \frac{(6.63 \times 10^{-34} \text{ J} \cdot \text{s})(4.37 \times 10^{12} \text{ s}^{-1})}{1.38 \times 10^{-23} \text{ J/K}}$$

$$= 210 \text{ K}.$$

This calculated value compares favorably with the experimentally determined value of $\theta_D = 215$ K, being in error by only about 2 percent.

**12.16** At 10 K the measured specific thermal capacity of copper is 0.860 J/kg · K. Find $\theta_D$ using the low temperature approximation given by Equation 12.94 and $\mathcal{M}_{Cu} = 63.55$ g.

*Answer:* $\theta_D = 329$ K

**12.17** Calculate $v_s$ and $\nu_m$ for copper, using $v_l = 4560$ m/s, $v_t = 2250$ m/s, $\rho_{Cu} = 8.96$ g/cm$^3$, and $\mathcal{M}_{Cu} = 63.55$ g. Compare the calculated value of $\nu_m$ with that predicted by experiment from Problem 12.16.

*Solution:*

With the data given and Equation 12.79 we have

$$v_s = \left(\frac{v_t^3 v_l^3}{v_t^3 + 2v_l^3}\right)^{1/3}$$

$$= 1751 \frac{\text{m}}{\text{s}}$$

and from Problem 12.13 we have

$$\nu_m = v_s \left(\frac{9\rho_{Cu}N_o}{4\pi\mathcal{M}_{Cu}}\right)^{1/3}$$

$$= (1751 \text{ m/s})(3.93 \times 10^9 \text{ m}^{-1})$$

$$= 6.88 \times 10^{12} \text{ s}^{-1}.$$

This value of $\nu_m$ obtained from measurements of the velocity of sound can be compared with that obtained from specific heat measurements by using the value of $\theta_D$ obtained in Problem 12.16. That is, with $(\theta_D)_{SH} = 329$ K, then from Equation 12.86 we have

$$(\nu_m)_{SH} = \frac{k_B(\theta_D)_{SH}}{h} = 6.85 \times 10^{12}\ s^{-1}.$$

The value obtained using velocity of sound data differs from that obtained from specific heat measurements by only

$$\frac{\nu_m - (\nu_m)_{SH}}{(\nu_m)_{SH}} = 0.00438.$$

**12.18** Find the *density of photon states* in the energy interval between $\epsilon$ and $\epsilon + d\epsilon$ for blackbody radiation.

*Answer:*

$$g(\epsilon)\ d\epsilon = \frac{8\pi V}{h^3 c^3}\epsilon^2\ d\epsilon$$

**12.19** Express the *spectral energy density* for blackbody radiation in terms of wavelength.

*Solution:*

To express the Planck radiation formula,

$$d\mathbf{u}_\nu = \mathbf{u}(\nu)\ d\nu = \frac{8\pi h}{c^3}\frac{\nu^3\ d\nu}{e^{\beta h\nu} - 1},$$

in terms of wavelength $\lambda$ rather than frequency $\nu$, note that

$$\nu = \frac{c}{\lambda} \rightarrow d\nu = -\frac{c}{\lambda^2}\ d\lambda \cdot$$

Further, since an increase in wavelength corresponds to a decrease in frequency,

$$\mathbf{u}(\nu)\ d\nu = -\mathbf{u}(\lambda)\ d\lambda,$$

then direct substitution of the above three relations into the equation for $\mathbf{u}(\nu)\ d\nu$ gives

$$d\mathbf{u}_\lambda = \mathbf{u}(\lambda)\ d\lambda = \frac{(8\pi hc\lambda^5)d\lambda}{e^{\beta hc/\lambda} - 1}.$$

**12.20** Using the result of Problem 12.19, find the value of the wavelength for which the spectral energy density is a maximum. That is, consider $d\mathbf{u}(\lambda)/d\lambda = 0$ and solve for $\lambda \equiv \lambda_{max}$ to obtain an expression of the form $\lambda_{max} T$ = CONSTANT, which is known as the **Wien displacement law.**

*Answer:* $\lambda_{max} T = 2.8977 \times 10^{-3}\ \text{m} \cdot \text{K}$

**12.21** Consider a *boson gas* of $N$ particles confined to a volume $V$. Show that the Lagrange multiplier $\alpha$ is an increasing function of temperature $T$, by normalizing the system.

*Solution:*

The *normalization condition* in statistical mechanics is given by (see Equation 11.116 and Problem 11.18)

$$N = \int_0^{\infty} dn_{\epsilon},$$

where for a *boson gas* the Bose-Einstein distribution (Equation 12.21) gives

$$dn_{\epsilon} = n(\epsilon)\, d\epsilon = \frac{g(\epsilon)\, d\epsilon}{e^{\alpha}e^{\beta\epsilon} - 1}$$

for a continuous distribution of energy. The degeneracy of quantum states can be obtained by using the arguments that led to Equation 11.95,

$$g(p)\, dp = \frac{4\pi V}{h^3} p^2\, dp.$$

This expression for degeneracy can be transformed in terms of energy by using (see discussion following Equation 12.115)

$$p = (2m\epsilon)^{1/2} \rightarrow dp = m(2m\epsilon)^{-1/2}\, d\epsilon$$

to obtain

$$\begin{aligned} g(\epsilon)\, d\epsilon &= 2\,\frac{2\pi V}{h^3}\, 2m\epsilon m 2^{-1/2} m^{-1/2} \epsilon^{-1/2}\, d\epsilon \\ &= \frac{2\pi V}{h^3}(2m)^{3/2}\epsilon^{1/2}\, d\epsilon. \end{aligned}$$

Now, substitution of $dn_{\epsilon}$ and $g(\epsilon)\, d\epsilon$ into the normalization integral gives

$$N = \frac{2\pi V}{h^3}(2m)^{3/2}\int_0^{\infty} \frac{\epsilon^{1/2}\, d\epsilon}{e^{\alpha}e^{\beta\epsilon} - 1}.$$

Letting the variable of integration be the quantity

$$x = \beta\epsilon \rightarrow dx = \beta d\epsilon,$$

we obtain

$$N = \frac{2\pi V}{h^3}(2mk_B T)^{3/2}\int_0^{\infty} \frac{x^{1/2}\, dx}{e^{\alpha}e^{x} - 1}.$$

This equation implicitly defines $\alpha$ as a function of $T$. Since $N$ is finite, as $T$ increases the integral must correspondingly decrease, which means that $\alpha$ must increase. Further, the integral must always converge for $N$ to be finite, so $\alpha$ can never be a negative quantity.

**12.22** Find the lowest possible temperature $T_0$ of the boson gas of Problem 12.21 that is consistent with Bose-Einstein statistics. Express $T_0$ in terms of the particle density $\eta \equiv N/V$ and the Einstein energy-mass relation $\epsilon = mc^2$.

*Answer:* $T_0 = (2.42 \times 10^{-28}\ \text{JKm}^2)\ \eta^{2/3}\epsilon^{-1}$

**12.23** Consider a *fermion gas* of $N$ spin-$\frac{1}{2}$ particles confined to a volume $V$. Show that the Lagrange multiplier $\alpha$ is an increasing function of temperature $T$.

*Solution:*

This problem is similar to that of Problem 12.21, except now the Fermi-Dirac distribution (Equation 12.25)

$$dn_\epsilon = \frac{g(\epsilon)\ d\epsilon}{e^{\alpha}e^{\beta\epsilon} + 1}$$

must be used for a continuous distribution of energy. The degeneracy $g(\epsilon)\ d\epsilon$ must be multiplied by a factor of two in this case for spin-$\frac{1}{2}$ fermions (see Equation 12.115), so we have

$$g(\epsilon)\ d\epsilon = \frac{4\pi V}{h^3}\,(2m)^{3/2}\epsilon^{1/2}d\epsilon$$

by analogy with Problem 12.21. Thus, the *normalization condition* is

$$N = \int_0^\infty dn_\epsilon = \frac{4\pi V}{h^3}\,(2m)^{3/2}\int_0^\infty \frac{\epsilon^{1/2}\ d\epsilon}{e^{\alpha}e^{\beta\epsilon} + 1}$$

$$= \frac{4\pi V}{h^3}\,(2mk_BT)^{3/2}\int_0^\infty \frac{x^{1/2}\ dx}{e^{\alpha}e^{x} + 1},$$

where $x = \beta\epsilon \rightarrow dx = \beta d\epsilon$ was used in obtaining the final expression. That $\alpha$ is an increasing function of $T$ follows by the arguments of Problem 12.21.

**12.24** Derive an equation for the Fermi energy of conduction electrons in terms of the electron density $\eta$, and find $\epsilon_F$ for sodium using the data from Appendix B.

*Answer:* $\epsilon_F = \dfrac{\hbar^2}{2m}\,(3\pi^2\eta)^{2/3} = 3.13\ \text{eV}$

**12.25** Find the Fermi energy, velocity, and temperature for electrons in copper using the data of Appendix B.

*Solution:*

With $\rho_{Cu} = 8.96\ \text{g/cm}^3$, $\mathcal{M}_{Cu} = 63.546$ g, and

$$_{29}\text{Cu: } 1s^2 2s^2 2p^6 3s^2 3p^6 4s^2 3d^{10} 4p^1,$$

Equation 12.123 gives the *electron density* as

$$\begin{aligned}\eta &= \frac{N_o\rho}{\mathcal{M}} \\ &= (6.02 \times 10^{23}\ \text{electrons/mole})\,\frac{(8.96\ \text{g/cm}^3)}{(63.546\ \text{g/mole})} \\ &= 8.49 \times 10^{22}\ \frac{\text{electrons}}{\text{cm}^2} \\ &= 8.49 \times 10^{28}\ \frac{\text{electrons}}{\text{m}^3},\end{aligned}$$

from which the Fermi energy is given by Equation 12.124 as

$$\begin{aligned}\epsilon_F &= \frac{\hbar^2}{2m}(3\pi^2\eta)^{2/3} \\ &= (6.05 \times 10^{-39}\ \text{J}\cdot\text{m}^2)\,(2.51 \times 10^{30}\ \text{m}^{-3})^{2/3} \\ &= (6.05 \times 10^{-39}\ \text{J}\cdot\text{m}^2)\,(1.85 \times 10^{20}\ \text{m}^{-2}) \\ &= 1.12 \times 10^{-18}\ \text{J} = 7.00\ \text{eV}.\end{aligned}$$

Now, from Equation 12.125 we obtain

$$\begin{aligned}v_F &= \left(\frac{2\epsilon_F}{m_e}\right)^{1/2} \\ &= \left[\frac{2\,(1.12 \times 10^{-18}\ \text{J})}{9.11 \times 10^{-31}\ \text{kg}}\right]^{1/2} \\ &= 1.57 \times 10^6\ \frac{\text{m}}{\text{s}}\end{aligned}$$

for the Fermi velocity, and from Equation 12.126 we have

$$\begin{aligned}T_F &= \frac{\epsilon_F}{k_B} \\ &= \frac{1.12 \times 10^{-18}\ \text{J}}{1.38 \times 10^{-23}\ \text{J/K}} \\ &= 8.12 \times 10^4\ \text{K}\end{aligned}$$

for the Fermi temperature.

**12.26** Find the Fermi energy and temperature for electrons in silver, using the data of Appendix B. Consider the *effective mass* of an electron in silver to be $0.99m_e$.

*Answer:* $\epsilon_F = 5.50$ eV, $T_F = 6.38 \times 10^4$ K

**12.27** The *effective mass* of an electron in zinc is $0.85m_e$. Find the Fermi energy, the average energy per electron at absolute zero, and the temperature necessary for an ideal gas molecule (classical theory) to have the energy $\epsilon_F$.

*Solution:*
With the ground state electronic configuration of zinc given by

$$_{30}\text{Zn}: 1s^2 2s^2 2p^6 3s^2 3p^6 4s^2 3d^{10} 4p^2,$$

we consider there to be two $4p$ electrons liberated as conduction electrons. Thus, with $\rho_{Zn} = 7.13$ g/cm$^3$ and $\mathcal{M}_{Zn} = 65.38$ g from Appendix B, the *electron density* (Equation 12.123 multiplied by two) is

$$\eta = \frac{2N_o\rho}{\mathcal{M}}$$
$$= 1.31 \times 10^{29} \frac{\text{electrons}}{\text{m}^3}.$$

Taking the *effective mass* into account, Equation 12.124 for the Fermi energy becomes

$$\epsilon_F = \frac{\hbar^2}{2m_{\text{eff}}}(3\pi^2\eta)^{2/3}$$
$$= \frac{6.05 \times 10^{-39}\ \text{J} \cdot \text{m}^2}{0.85}(3.87 \times 10^{30}\ \text{m}^{-3})^{2/3}$$
$$= (7.12 \times 10^{-39}\ \text{J} \cdot \text{m}^2)(2.46 \times 10^{20}\ \text{m}^{-2})$$
$$= 1.75 \times 10^{-18}\ \text{J}$$
$$= 10.9\ \text{eV}.$$

Using this value for the Fermi energy, the average energy per electron is given by (see Equation 12.135)

$$\bar{\epsilon}_e = \frac{3}{5}\epsilon_F$$
$$= 1.05 \times 10^{-18}\ \text{J}$$
$$= 6.56\ \text{eV}.$$

For a classical ideal gas molecule to have this same energy, the gas would have to be at a temperature of (also see Equation 12.136)

$$T = \frac{2}{3}\frac{\bar{\epsilon}_e}{k_B}$$
$$= \frac{2}{3}\frac{1.05 \times 10^{-18}\ \text{J}}{1.38 \times 10^{-23}\ \text{J/K}}$$
$$= 5.07 \times 10^4\ \text{K}.$$

**12.28** Find the Fermi energy and electronic specific heat of aluminum at 25° C. Consider all $M$-shell electrons in an aluminum atom to contribute to the conduction electrons and take the electron effective mass to be $0.97m_e$.

*Answer:* $\epsilon_F = 11.9$ eV, $c_{v-e} = (1.06 \times 10^{-2})\,R$

**12.29** In a semiconductor the Fermi energy is in the energy gap $\epsilon_g$ between the valence and conduction bands. Assuming $\epsilon - \epsilon_F = \frac{1}{2}\epsilon_g$, $\epsilon_g = 1$ eV, and $T = 25°$ C, show that the Fermi-Dirac distribution for conduction electrons reduces to the Maxwell-Boltzmann distribution.

*Solution:*
With $\epsilon = 0$ being the lowest energy state in the conduction band, the Fermi-Dirac distribution of Equation 12.129 becomes

$$\begin{aligned} n(\epsilon) &= \frac{g(\epsilon)}{e^{\beta(\epsilon-\epsilon_F)} + 1} \\ &= \frac{g(\epsilon)}{e^{\frac{1}{2}\epsilon_g/k_BT} + 1} \\ &= \frac{g(\epsilon)}{e^{(0.5\text{ eV})/(1.38\times10^{-23}\text{ J/K})(298\text{ K})} + 1} \\ &= \frac{g(\epsilon)}{e^{(0.5\text{ eV})/(2.57\times10^{-2}\text{ eV})} + 1} \\ &= \frac{g(\epsilon)}{e^{19.5} + 1} \\ &\approx \frac{g(\epsilon)}{e^{19.5}}. \end{aligned}$$

Since the exponential $e^{19.5} = 2.94 \times 10^8$ is much greater than 1, the result using the Fermi-Dirac distribution reduces to that predicted by the Maxwell-Boltzmann distribution of Equation 11.45,

$$n(\epsilon) = \frac{g(\epsilon)}{e^{\alpha}e^{\beta\epsilon}},$$

with $\alpha$ replaced by $-\beta\epsilon_F$.

**12.30** Compute the *occupation index* for conduction electrons in germanium, assuming $\epsilon_F = \frac{1}{2}\epsilon_g$, $\epsilon = \epsilon_g$, $\epsilon_g = 0.7$ eV, and $T = 25°$ C.

*Answer:* $f(\epsilon) = 1.24 \times 10^{-6}$

APPENDIX A

# Basic Mathematics

The mathematical symbols and formulas presented here are all that is required for a successful study of this textbook. Some of the formulas will not be explicitly utilized in this text, but are presented for completeness in the mathematics review of algebra, trigonometry, and introductory calculus. The theme throughout this mathematical review is for commonly used formulas to be initially presented and discussed, with other infrequently used mathematical relationships being derived from knowledge of these basic identities.

## A.1 Mathematical Symbols

$\equiv$ defined by
$=$ equal to
$\neq$ not equal to
$\approx$ approximately equal to
$\propto$ proportional to
$>$ greater than
$\gg$ much greater than
$<$ less than
$\ll$ much less than
$\geq$ greater than or equal to
$\leq$ less than or equal to
$\rightarrow$ implying, yielding, approaching
$\Delta$ change in
$\infty$ infinity

## A.2 Exponential Operations

The quantity $x^n$ is referred to as an algebraic **exponential,** where $n$ is the *exponent* and the *base* $x$ has been raised to the $n$th power. The most trivial exponential involves any base $x$ raised to the *zeroth* power, as given by

$$x^0 \equiv 1.$$

Arithmetic and algebraic operations involving exponentials are most easily accomplished by employing one or more of the basic principles defined below.

*Multiplication* of exponential numbers is generalized by the formula

Multiplication

$$x^n x^m = x^{n+m},$$

where a number $x$ raised to the $n$th power is multiplied by the *same number* or *base* raised to the $m$th power by simply adding the exponents. The exponentials are additive for the multiplication of numbers having the *same base*, but there is no way of simplifying an expression like

$$x^n y^m = x^n y^m,$$

since the bases are different. If the expression contained dissimilar bases but *identical exponents*, then the rule is

$$x^n y^n = (xy)^n.$$

Further, an exponential number can also be raised to the $m$th power, according to the rule

$$(x^n)^m = x^{nm},$$

where $nm$ represents an *implied product* of the two exponents.

Changing the sign of the exponent on an exponential number in the numerator of an expression allows it to be represented in the denominator or vice versa, as given by the rule

Negative Exponents

$$x^{-n} = \frac{1}{x^n}$$

or

$$\frac{1}{x^{-n}} = x^n.$$

*Negative exponents* are primarily used to indicate division, since by combining this definition with the multiplication principle the rule for *division of exponents* is obviously given by

Division

$$\frac{x^n}{x^m} = x^{n-m}.$$

Another commonly used property of exponential numbers employs *fractional exponents* to indicate roots. By definition

Fractional Exponents

$$x^{1/n} = \sqrt[n]{x},$$

which means that $x$ to the fractional exponent of $1/n$ is equal to the $n$th root of $x$.

## A.3 Logarithmic Operations

For an *exponential equation* of the form

$$y = a^x,$$

Exponential Equation

the value of the exponent is obtained by taking the *logarithm to the base a of y*, as given by

$$x = \log_a y.$$

This equation represents a *common* logarithm for $a = 10$ and a *natural* logarithm for $a = e$, where $e = 2.71828$. Normally the subscript is omitted by adopting the *convention*

$$x = \log y$$

Common Logarithm

to represent a *common* logarithm (log) and

$$x' = \ln y$$

Natural Logarithm

to mean the *natural* logarithm (ln), where of course

$$\ln e = 1, \qquad \log 10 = 1$$

and

$$\ln 1 = \log 1 = 0.$$

Since logarithms are exponents, exponential properties are also properties of logarithms. There are essentially three basic principles, which are stated in general below, with the *base* of the logarithm being deleted.

1. The logarithm of the product of two numbers equals the sum of their logarithms,

$$\log xy = \log x + \log y.$$

2. The logarithm of the ratio (quotient) of two numbers is equal to the difference of the logarithm of the numerator and the logarithm of the denominator,

$$\log \frac{x}{y} = \log x - \log y.$$

3. The logarithm of an exponential number equals the product of the exponent and the logarithm of the base,

$$\log x^n = n \log x.$$

From this last property it should be obvious that the logarithm of the $n$th root of a number is just equal to the ratio of the logarithm of the number to $n$, as given by

$$\log x^{1/n} = \frac{\log x}{n}.$$

Frequently it is desirable to effect a *change in base* of a logarithm, such that, for example, a *common* logarithm can be expressed as its equivalent *natural* logarithm or vice versa. It is rather straightforward to derive a general equation to see how a *change in base* can be effected. Taking the logarithm to the base $b$ of the equation $y = a^x$ gives

$$\begin{aligned} \log_b y &= \log_b a^x \\ &= x \log_b a, \end{aligned}$$

where the exponential principle above has been utilized. Since $x = \log_a y$, direct substitution gives

$$\log_b y = \log_a y \log_b a,$$

which can be rewritten in the form

Base Conversion

$$\log_a y = \frac{\log_b y}{\log_b a}.$$

This general equation can be employed in a change of base from $b$ to $a$ for any values of $b$ and $a$. Consequently, in terms of the *common* and *natural* logarithms and the convention adopted above, the base conversion formula yields

$$\log y = \frac{\ln y}{\ln 10},$$

where $(\ln 10)^{-1} = 0.43429$, and

$$\ln y = \frac{\log y}{\log e},$$

where $(\log e)^{-1} = 2.3026$.

## A.4 Scientific Notation and Useful Metric Prefixes

Although the rules governing the algebra of exponents apply to any base, the *decimal* system in general and the *metric* systems of units in particular place major emphasis on *powers of ten*. The important **metric prefixes,** their abbreviations, and their equivalence expressed in **scientific notation** as powers of ten are listed below, with the most commonly used prefixes indicated in bold type.

| PREFIX | ABBREVIATION | VALUE |
|---|---|---|
| tera | T | $10^{12}$ |
| giga | G | $10^{9}$ |

| | | |
|---|---|---|
| **mega** | **M** | $\mathbf{10^6}$ |
| **kilo** | **k** | $\mathbf{10^3}$ |
| hecto | h | $10^2$ |
| deka | da | $10^1$ |
| deci | d | $10^{-1}$ |
| **centi** | **c** | $\mathbf{10^{-2}}$ |
| **milli** | **m** | $\mathbf{10^{-3}}$ |
| **micro** | **μ** | $\mathbf{10^{-6}}$ |
| nano | n | $10^{-9}$ |
| pico | p | $10^{-12}$ |

The primary advantage of knowing the metric prefixes is the ease with which such knowledge allows for the conversion of units from one metric system to the other. For example the number of centimeters in a meter is easily found, since the prefix *centi* can be replaced with its equivalence $10^{-2}$ (i.e., cm $= 10^{-2}$ m). From that relation it follows that 1 m $=$ cm/$10^{-2}$ $= 10^{+2}$ cm. Another way of utilizing a metric prefix is to realize that a prefix divided by its power of ten equivalence defines 1 exactly. Thus, for example, 1 m $=$ (milli/$10^{-3}$)m $= 10^3$ mm, and we realize there are one thousand ($10^3$) millimeters in a meter. Since the metric prefixes are so commonly used in science and engineering, the table of prefixes given above is reproduced on the inside cover of this textbook for ease of reference.

## A.5 Quadratic Equations

Frequently, physical equations representing fundamental laws of nature are linear, since the physical variables (unknowns) are raised only to the first power. Occasionally, there is the need to solve a more complex equation known as a **quadratic equation,** where the physical variable is raised to the second power as well as to the first power. The general quadratic formula, expressed by

$$ax^2 + bx + c = 0,$$

Quadratic Formula

has *solutions* given by

$$x = \frac{-b \pm \sqrt{b^2 - 4ac}}{2a}.$$

From the form of the solution equation, it is obvious that in general *two* solutions are obtainable for a quadratic equation. If $b^2 > 4ac$, the solutions are *real*; whereas, if $b^2 < 4ac$, the solutions are *complex* or *imaginary*. For the special case where $b^2 = 4ac$, the two solutions coincide.

## A.6 Trigonometry

The use of the *sine*, *cosine*, and *tangent* functions of a variable are often essential in the formulation of a physical problem. These functions can be defined for acute angles in terms of the sides of a right triangle. Consider the right triangle given below, where side $c$ is the

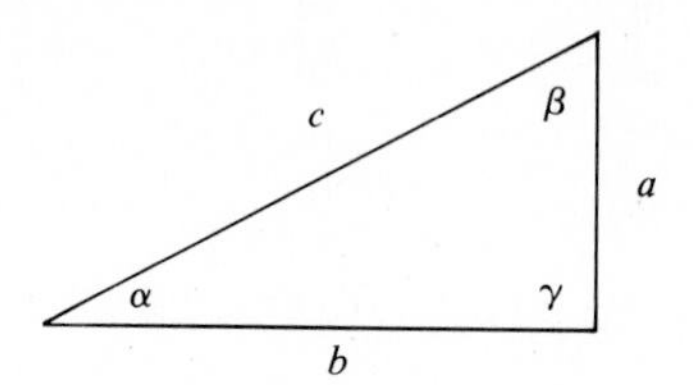


*hypothenuse* and sides $a$ and $b$ are the *legs*. In terms of the sides of this right triangle the well known **Pythagorean theorem** is given by

Pythagorean Theorem

$$c^2 = a^2 + b^2.$$

Since for any triangle the sum of the interior angles always equals 180° ($\pi$),

$$\alpha + \beta + \gamma = \pi,$$

then for $\gamma = 90°$,

Complementary Angles

$$\alpha + \beta = \frac{\pi}{2}.$$

The interior angles $\alpha$ and $\beta$ are called **complementary angles,** since their sum is equal to 90° ($\pi/2$). The *sine*, *cosine*, and *tangent* functions of these complementary angles are related to the sides of the right triangle by the relations

$$\sin\alpha = \frac{a}{c} \qquad \sin\beta = \frac{b}{c}$$

$$\cos\alpha = \frac{b}{c} \qquad \cos\beta = \frac{a}{c}$$

$$\tan\alpha = \frac{a}{b} \qquad \tan\beta = \frac{b}{a}.$$

Defining the common trigonometric functions with respect to a right triangle is advantageous as several other basic relationships are immediately suggested. For example, the *sine* of any angle $\theta$ is known to be equal to the *cosine* of the complement of $\theta$ or vice versa, as given by

$$\sin\theta = \cos(90° - \theta)$$

or

$$\cos\theta = \sin(90° - \theta),$$

which is suggested above by $\sin \alpha = \cos \beta$ and $\cos \alpha = \sin \beta$. Also, it is obvious from the above relations that $\tan \alpha = \sin \alpha / \cos \alpha$ and $\tan \beta = \sin \beta / \cos \beta$. These observations suggest

$$\tan \theta = \frac{\sin \theta}{\cos \theta}$$

is valid in general for any angle $\theta$. Further, the familiar identity

$$\sin^2 \theta + \cos^2 \theta = 1$$

is easily verified for angles $\alpha$ and $\beta$ for our right triangle. That is,

$$\begin{aligned} \sin^2 \alpha + \cos^2 \alpha &= \frac{a^2}{c^2} + \frac{b^2}{c^2} \\ &= \frac{a^2 + b^2}{c^2} \\ &= 1, \end{aligned}$$

where the Pythagorean theorem has been utilized in obtaining the last equality.

In addition to the above trigonometric formulas involving the sine, cosine, and tangent functions, there are two (actually four) other very useful formulas that are well worth remembering. They are called **addition formulas** and in terms of *any* two angles $\alpha$ and $\beta$ are given by

Addition Formulas

$$\sin (\alpha \pm \beta) = \sin \alpha \cos \beta \pm \sin \beta \cos \alpha,$$

$$\cos (\alpha \pm \beta) = \cos \alpha \cos \beta \mp \sin \alpha \sin \beta.$$

Since $\tan (\alpha \pm \beta) = \sin (\alpha \pm \beta)/\cos (\alpha \pm \beta)$, then a direct derivation using the above addition formulas immediately yields

$$\tan (\alpha \pm \beta) = \frac{\tan \alpha \pm \tan \beta}{1 \mp \tan \alpha \tan \beta}.$$

Likewise, the following trigonometric identities are easily derived by employing the addition formulas and the previously given basic relationships between the sine, cosine, and tangent functions:

$$\left.\begin{aligned} \sin (-\alpha) &= -\sin \alpha \\ \cos (-\alpha) &= \cos \alpha \end{aligned}\right\} \quad \rightarrow \quad \tan (-\alpha) = -\tan \alpha$$

$$\sin \left(\alpha \pm \frac{\pi}{2}\right) = \pm \cos \alpha$$

$$\cos \left(\alpha \pm \frac{\pi}{2}\right) = \mp \sin \alpha \quad \rightarrow \quad \tan \left(\alpha \pm \frac{\pi}{2}\right) = -\frac{1}{\tan \alpha}$$

$$\left.\begin{aligned} \sin (\alpha \pm \pi) &= -\sin \alpha \\ \cos (\alpha \pm \pi) &= -\cos \alpha \end{aligned}\right\} \quad \rightarrow \quad \tan (\alpha + \pi) = \tan \alpha$$

$$\sin 2\alpha = 2 \sin \alpha \cos \alpha$$
$$\cos 2\alpha = 1 - 2 \sin^2 \alpha = 2 \cos^2 \alpha - 1$$
$$\tan 2\alpha = \frac{2 \tan \alpha}{1 - \tan^2 \alpha}$$

Normally, knowledge of the basic relationships between the sine, cosine, and tangent functions along with the sine and cosine addition formulas are more than adequate for the formulation of physical problems. Occasionally, the less common functions of *cotangent* (cot), *secant* (sec), and *cosecant* (csc) are utilized but are easily handled by employing the relations

$$\cot \theta = \frac{1}{\tan \theta},$$

$$\sec \theta = \frac{1}{\cos \theta},$$

$$\csc \theta = \frac{1}{\sin \theta}.$$

With these relationships and the basic identities for the common functions, it is easy to verify that

$$\sec^2 \theta - \tan^2 \theta = 1$$

and

$$\csc^2 \theta - \cot^2 \theta = 1.$$

In addition to the basic relationships given above, there are two fundamental identities that are very useful in finding unknown sides and angles of any triangle. With respect to our right triangle, the **law of cosines** is defined by

Law of Cosines

$$c^2 = a^2 + b^2 - 2ab \cos \gamma$$

and the **law of sines** by

Law of Sines

$$\frac{a}{\sin \alpha} = \frac{b}{\sin \beta} = \frac{c}{\sin \gamma}.$$

These identities are completely general for any triangle having sides $a$, $b$, and $c$ and interior angles $\alpha$, $\beta$, and $\gamma$. Since $\gamma = 90°$ in our triangle, the *law of cosines* immediately yields the equation representing the Pythagorean theorem, as $\cos 90° = 0$. Further, a general interpretation is applicable for the equation representing the *law of cosines*, as the unknown side $c$ is given in terms of two known sides ($a$ and $b$) and the interior angle $\gamma$ between the known sides. Consequently, we could also represent the *law of cosines* by

$$b^2 = a^2 + c^2 - 2ac \cos \beta$$

or

$$a^2 = b^2 + c^2 - 2bc \cos \alpha.$$

In general physics the addition of two *vectors* (velocities, acceleration, forces, etc.) is one important application of the *laws of sines and cosines*.

## A.7 Algebraic Series

Sometimes in a theoretical derivation of a physical problem it is necessary to represent a function or expression by a *converging series*. The **binomial expansion** is most common and can be defined in general by the equation

$$(x + y)^n = x^n + nx^{n-1}y + \frac{n(n-1)}{2!}x^{n-2}y^2 + \cdots,$$

Binomial Expansion

where $2! = 1 \cdot 2,$

and is called a **factorial** as defined in general by

$$m! \equiv 1 \cdot 2 \cdot 3 \cdots m$$

for $m$ a positive integer. The **Taylor series,** represented by

$$\begin{aligned} f(x + h) &= f(x) + hf'(x) + \frac{h^2}{2!}f''(x) + \cdots \\ &= f(h) + xf'(h) + \frac{x^2}{2!}f''(h) + \cdots \end{aligned}$$

Taylor Series

and the **MacLaurin series**

$$f(x) = f(0) + xf'(0) + \frac{x^2}{2!}f''(0) + \cdots$$

MacLaurin Series

are also very useful. In the above $f'$ represents the first derivative of $f$ and $f''$ is the second order derivative of $f$.

It is instructive to utilize the MacLaurin series and expand $e^{jx}$, $\sin x$, and $\cos x$ to obtain

$$\begin{aligned} e^{jx} &= \left(1 - \frac{x^2}{2!} + \frac{x^4}{4!} - \cdots\right) + j\left(x - \frac{x^3}{3!} + \frac{x^5}{5!} - \cdots\right), \\ \sin x &= x - \frac{x^3}{3!} + \frac{x^5}{5!} - \frac{x^7}{7!} + \cdots, \\ \cos x &= 1 - \frac{x^2}{2!} + \frac{x^4}{4!} - \frac{x^6}{6!} + \cdots, \end{aligned}$$

where

$$j \equiv \sqrt{-1}$$

Imaginary Unit

is the so-called **imaginary unit** for which

$$j^2 = -1.$$

Clearly, from the above expansions the *complex exponential* $e^{jx}$ is related to the sine and cosine functions by the simple expression

Euler's Relation

$$e^{jx} = \cos x + j \sin x,$$

which is known as **Euler's relation.** It can also be shown that

Euler's Relation

$$e^{-jx} = \cos x - j \sin x$$

from which it immediately follows that

$$\cos x = \tfrac{1}{2}\,(e^{jx} + e^{-jx})$$

and

$$\sin x = \frac{1}{2j}\,(e^{jx} - e^{-jx}).$$

It is most beneficial to verify these examples, as the mathematical techniques involved will prove useful in many physical derivations.

## A.8 Basic Calculus

Derivations of some very common mathematical functions along with formulas illustrating some fundamental properties of the derivative are listed below, where $u$ and $v$ are arbitrary functions of $x$ $[u = u(x),\ v = v(x)]$, $f$ is a function of $u$ $[f = f(u)]$, and $n$ is a constant.

$$\frac{dx}{dx} = 1$$

$$\frac{d}{dx}\,nu = n\,\frac{du}{dx}$$

$$\frac{d}{dx}\,(u \pm v) = \frac{du}{dx} \pm \frac{dv}{dx}$$

$$\frac{d}{dx}\,uv = \frac{du}{dx}\,v + u\,\frac{dv}{dx}$$

$$\frac{df}{dx} = \frac{df}{du}\,\frac{du}{dx}$$

$$\frac{d}{dx}\,u^n = nu^{n-1}\,\frac{du}{dx}$$

$$\frac{d}{dx}\,e^u = e^u\,\frac{du}{dx}$$

$$\frac{d}{dx}\,\ln |u| = \frac{1}{u}\,\frac{du}{dx}$$

$$\frac{d}{dx}\sin u = \cos u \frac{du}{dx}$$

$$\frac{d}{dx}\cos u = -\sin u \frac{du}{dx}$$

$$\frac{d}{dx}\tan u = \sec^2 u \frac{du}{dx}$$

Knowledge of these *common derivatives* is completely adequate for the development of other infrequently used derivatives, which are easily obtained by repetitive application of one or more of the above identities. For example, with the notation $D \equiv d/dx$ representing a first order **differential operator,** that operates on anything following it, we have

$$\begin{aligned} D\frac{u}{v} &= D\,(uv^{-1}) \\ &= v^{-1}Du + uDv^{-1} \\ &= v^{-1}Du + u\,(-1)\,v^{-2}Dv \\ &= v\,(v^{-2})\,Du - uv^{-2}\,Dv \\ &= \frac{1}{v^2}\,(vDu - uDv). \end{aligned}$$

Likewise, the derivatives for cotangent, secant, and cosecant, given by

$$\frac{d}{dx}\cot u = -\csc^2 u \frac{du}{dx},$$

$$\frac{d}{dx}\sec u = \sec u \tan u \frac{du}{dx},$$

$$\frac{d}{dx}\csc u = -\csc u \cot u \frac{du}{dx},$$

are directly obtained from basic trigonometric identities and the above common derivatives. The point of this observation is that other identities involving derivatives need not be memorized, as they can be easily derived from knowledge of the more basic identities. As a last example of this derivational approach, consider the following:

$$\begin{aligned} D\cot u &= D\tan^{-1} u = (-1)\tan^{-2} u\, D\tan u \\ &= -\tan^{-2} u \sec^2 u \frac{du}{dx} \\ &= -\frac{\cos^2 u}{\sin^2 u}\frac{1}{\cos^2 u}\frac{du}{dx} \\ &= -\frac{1}{\sin^2 u}\frac{du}{dx} \\ &= -\csc^2 u \frac{du}{dx}. \end{aligned}$$

Our discussion of derivatives has been limited to *total derivatives* of a function depending on only *one* variable. If a function $w$ depends on more than one variable, say

$$w = w(x, y, z),$$

and the variables $x$, $y$, and $z$ are differentiable functions of time $t$, then the **total derivative** of $w$ is given by the well known *chain rule*

Total Derivative

$$\frac{dw}{dt} \equiv \left(\frac{\partial w}{\partial x}\right)\frac{dx}{dt} + \left(\frac{\partial w}{\partial y}\right)\frac{dy}{dt} + \left(\frac{\partial w}{\partial z}\right)\frac{dz}{dt}.$$

This identity is normally condensed to the form

Total Derivative

$$dw = \left(\frac{\partial w}{\partial x}\right) dx + \left(\frac{\partial w}{\partial y}\right) dy + \left(\frac{\partial w}{\partial z}\right) dz,$$

where the factors in parenthesis are called **partial derivatives.** That is, the *curly dees* in the expression $\partial w/\partial x$ is interpreted as the *partial derivative of $w$ with respect to $x$*, where $y$ and $z$ are understood to be held constant. The *total derivative* defined above will prove useful in a number of physical applications throughout this textbook, beginning with Einsteinian relativity.

A few commonly used *indefinite integrals* are listed below, where $u$ and $v$ are functions of $x$ and $n$ is a constant. Unless the *limits of integration* are known, a constant of integration should be added to the right-hand side of each identity.

$$\int du = u$$

$$\int nu\, dx = n \int u\, dx$$

$$\int (u + v)\, dx = \int u\, dx + \int v\, dx$$

$$\int u\, dv = uv - \int v\, du$$

$$\int x^n\, dx = \frac{x^{n+1}}{(n + 1)}, \qquad (n \neq -1)$$

$$\int \frac{1}{x}\, dx = \ln |x|$$

$$\int e^x\, dx = e^x$$

$$\int \ln x\, dx = x \ln x - x$$

$$\int \sin x\, dx = -\cos x$$

$$\int \cos x\, dx = \sin x$$

$$\int \tan x\, dx = \ln |\sec x|$$

Integrals can be thought of as *antiderivatives* and are easily derived by making use of the properties of derivatives. For example, the identity for $\int \sin x \, dx$ is easily obtained by imagining the function which upon differentiation would yield $\sin x$. Clearly, we can use the cosine function and differentiate to obtain

$$\frac{d}{dx} \cos x = -\sin x.$$

Multiplying both sides of this equation by $dx$ and integrating gives

$$\int d(\cos x) = -\int \sin x \, dx,$$

which immediately yields

$$-\cos x = \int \sin x \, dx,$$

since $\int d\,(\cos x)$ is of the form $\int du = u$. Although this approach requires some mathematical imagination at times, it is an easy way of obtaining the more common integrals from a knowledge of derivatives. As another example, consider taking the derivative of $x^{n+1}$ to obtain

$$\begin{aligned} \frac{d}{dx} x^{n+1} &= (n+1)\, x^n \frac{dx}{dx} \\ &= (n+1)\, x^n. \end{aligned}$$

Again, multiplying both sides of the equation by $dx$ and integrating gives

$$\int d\,(x^{n+1}) = (n+1) \int x^n \, dx,$$

where $n + 1$ is a constant and is placed outside of the integral. Integrating the left-hand side of this equation and simplifying gives the identity

$$\int x^n \, dx = \frac{x^{n+1}}{n+1}.$$

The advantage of this operational approach in calculus is that very few basic identities need to be remembered. However, an operational facility with these identities and the ability to make derivations is required or must be developed for physics and engineering.

## A.9 Vector Calculus

To continue this review of basic mathematics, a few definitions and identities pertaining to *vectors* and *vector calculus* will be presented. Consider vectors **A** and **B** to be expressed in terms of their Cartesian components as

$$\mathbf{A} = A_x\mathbf{i} + A_y\mathbf{j} + A_z\mathbf{k}$$

and

$$\mathbf{B} = B_x\mathbf{i} + B_y\mathbf{j} + B_z\mathbf{k},$$

where **i, j,** and **k** are the normal set of **unit vectors** that are parallel to the $X$, $Y$, and $Z$ axes, respectively. The **scalar product** (sometimes called the *dot*, or *inner* product) of **A** and **B** is defined by

Scalar Product

$$\mathbf{A} \cdot \mathbf{B} \equiv AB \cos \theta,$$

where $A$ and $B$ represent the absolute magnitude of vectors **A** and **B**, as given by

$$A \equiv (A_x^2 + A_y^2 + A_z^2)^{1/2}$$

and

$$B \equiv (B_x^2 + B_y^2 + B_z^2)^{1/2}.$$

Since the unit vectors have the properties

$$\mathbf{i} \cdot \mathbf{i} = \mathbf{j} \cdot \mathbf{j} = \mathbf{k} \cdot \mathbf{k} = 1$$

and

$$\mathbf{i} \cdot \mathbf{j} = \mathbf{j} \cdot \mathbf{k} = \mathbf{k} \cdot \mathbf{i} = 0,$$

the the *scalar product* can also be expressed by

Scalar Product

$$\mathbf{A} \cdot \mathbf{B} = A_xB_x + A_yB_y + A_zB_z$$

From these definitions it should be obvious that the **commutative law,**

Commutative Law

$$\mathbf{A} \cdot \mathbf{B} = \mathbf{B} \cdot \mathbf{A},$$

and **distributive law,**

Distributive Law

$$\mathbf{A} \cdot (\mathbf{B} + \mathbf{C}) = \mathbf{A} \cdot \mathbf{B} + \mathbf{A} \cdot \mathbf{C}$$

are valid.

The **vector product** (sometimes called the *cross* or *outer* product) of vectors **A** and **B** is defined by

Vector Product

$$\mathbf{A} \times \mathbf{B} \equiv AB \sin \theta \, \mathbf{n},$$

where **n** is a unit vector perpendicular to the plane of vectors **A** and **B.** With the unit vectors having the properties

$$\mathbf{i} \times \mathbf{i} = \mathbf{j} \times \mathbf{j} = \mathbf{k} \times \mathbf{k} = 0,$$
$$\mathbf{i} \times \mathbf{j} = \mathbf{k}, \quad \mathbf{j} \times \mathbf{k} = \mathbf{i}, \quad \mathbf{k} \times \mathbf{i} = \mathbf{j},$$
$$\mathbf{k} \times \mathbf{j} = -\mathbf{i}, \quad \mathbf{j} \times \mathbf{i} = -\mathbf{k}, \quad \mathbf{i} \times \mathbf{k} = -\mathbf{j},$$

the vector product of **A** and **B** becomes

$$\mathbf{A} \times \mathbf{B} = (A_yB_z - A_zB_y)\,\mathbf{i} + (A_zB_x - A_xB_z)\,\mathbf{j} + (A_xB_y - A_yB_x)\,\mathbf{k}.$$

Further, the properties

$$\mathbf{A} \times \mathbf{B} = -\mathbf{B} \times \mathbf{A}$$

and

$$\mathbf{A} \times (\mathbf{A} \times \mathbf{B}) = \mathbf{B} \times (\mathbf{A} \times \mathbf{B}) = 0$$

are immediate consequences of the definition.

Up to this point, the calculus review is consistent with an introductory treatment; however, an interesting and slightly more advanced observation is now possible with regards to the equation representing the *total derivative*. That equation can now be represented as a *scalar product*, as given by

$$\frac{dw}{dt} = \left(\frac{\partial w}{\partial x}\mathbf{i} + \frac{\partial w}{\partial y}\mathbf{j} + \frac{\partial w}{\partial z}\mathbf{k}\right) \cdot \left(\frac{dx}{dt}\mathbf{i} + \frac{dy}{dt}\mathbf{j} + \frac{dz}{dt}\mathbf{k}\right).$$

The first vector on the right-hand side is called the **gradient** of $w$ and normally denoted as

Gradient

$$\nabla w = \frac{\partial w}{\partial x}\mathbf{i} + \frac{\partial w}{\partial y}\mathbf{j} + \frac{\partial w}{\partial z}\mathbf{k}.$$

The *inverted delta* symbol $\nabla$ is normally considered to be an operator, as defined by

Del Operator

$$\nabla \equiv \frac{\partial}{\partial x}\mathbf{i} + \frac{\partial}{\partial y}\mathbf{j} + \frac{\partial}{\partial z}\mathbf{k},$$

which can be applied to any scalar function to produce a *gradient vector*. The scalar and vector product of the *del operator* with a vector **A** defines the **divergence**

Divergence

$$\nabla \cdot \mathbf{A} = \frac{\partial A_x}{\partial x} + \frac{\partial A_y}{\partial y} + \frac{\partial A_z}{\partial z}$$

and the **curl**

Curl

$$\nabla \times \mathbf{A} = \left(\frac{\partial A_z}{\partial y} - \frac{\partial A_y}{\partial z}\right)\mathbf{i} + \left(\frac{\partial A_x}{\partial z} - \frac{\partial A_z}{\partial x}\right)\mathbf{j} + \left(\frac{\partial A_y}{\partial x} - \frac{\partial A_x}{\partial y}\right)\mathbf{k},$$

respectively. The scalar product of the del operator with itself is just

Laplacian Operator

$$\nabla^2 = \frac{\partial^2}{\partial x^2} + \frac{\partial^2}{\partial y^2} + \frac{\partial^2}{\partial z^2},$$

which is called the **Laplacian operator.** Although the vector operations given above are used only occasionally in this textbook, the *del* and *Laplacian* operators will be most useful in condensing expressions in quantum mechanics.

## A.10 Definite Integrals

A few rather commonly used *definite integrals* are listed below for easy reference. These integrals will be most useful in the application of quantum and statistical mechanics.

$$\int_{-\infty}^{+\infty} e^{-au}\,du = \frac{2}{a}$$

$$\int_{-\infty}^{+\infty} e^{-au^2}\,du = \sqrt{\frac{\pi}{a}}$$

$$\int_{-\infty}^{+\infty} u^{2n}\,e^{-au^2}\,du = \sqrt{\frac{\pi}{a}}\left(\frac{1}{2}\cdot\frac{3}{2}\cdot\frac{5}{2}\cdots\frac{2n-1}{2}\right)a^{-n}$$

$$\int_{-\infty}^{+\infty} e^{-au^2}\,e^{cu}\,du = \sqrt{\frac{\pi}{a}}\,e^{c^2/4a}$$

$$I_n(a) \equiv \int_0^{\infty} u^n e^{-au^2}\,du$$

| n | $I_n(a)$ |
|---|---|
| 0 | $\frac{1}{2}\sqrt{\frac{\pi}{a}}$ |
| 1 | $\frac{1}{2a}$ |
| 2 | $\frac{1}{4}\sqrt{\frac{\pi}{a^3}}$ |
| 3 | $\frac{1}{2a^2}$ |
| 4 | $\frac{3}{8}\sqrt{\frac{\pi}{a^5}}$ |
| 5 | $\frac{1}{a^3}$ |

## Problems

**A.1** Derive the identity

$$\tan(\alpha \pm \beta) = \frac{\tan\alpha \pm \tan\beta}{1 \mp \tan\alpha \tan\beta},$$

by using the *addition formulas* for the sine and cosine functions given in section A.6.

*Solution:*

$$\begin{aligned}\tan(\alpha \pm \beta) &= \frac{\sin(\alpha \pm \beta)}{\cos(\alpha \pm \beta)} \\ &= \frac{\sin\alpha\cos\beta \pm \sin\beta\cos\alpha}{\cos\alpha\cos\beta \mp \sin\alpha\sin\beta} \\ &= \frac{\tan\alpha\cos\beta \pm \sin\beta}{\cos\beta \mp \tan\alpha\sin\beta} \\ &= \frac{\tan\alpha \pm \tan\beta}{1 \mp \tan\alpha\tan\beta},\end{aligned}$$

where we have multiplied numerator and denominator of the second equality by $1/\cos\alpha$ and of the third equality by $1/\cos\beta$.

**A.2** Derive the *addition formula* for the $\cot(\alpha \pm \beta)$, by using the addition formulas for the sine and cosine functions.

*Answer:* $\cot(\alpha \pm \beta) = \dfrac{\cot\alpha\cot\beta \mp 1}{\cot\beta \pm \cot\alpha}$

**A.3** Show that (a) $\sin(-\alpha) = -\sin\alpha$, (b) $\cos(-\alpha) = \cos\alpha$, (c) $\sin(\alpha \pm \pi/2) = \pm\cos\alpha$, and (d) $\cos(\alpha \pm \pi/2) = \mp\sin\alpha$, by using the appropriate sine and cosine addition formulas.

*Solution:*

(a) $\sin(-\alpha) = \sin(0° - \alpha) = \sin 0° \cos\alpha - \sin\alpha\cos 0° = -\sin\alpha$

(b) $\cos(-\alpha) = \cos(0° - \alpha) = \cos 0° \cos\alpha + \sin 0° \sin\alpha = \cos\alpha$

(c) $\sin\left(\alpha \pm \dfrac{\pi}{2}\right) = \sin\alpha\cos\dfrac{\pi}{2} \pm \sin\dfrac{\pi}{2}\cos\alpha = \pm\cos\alpha$

(d) $\cos\left(\alpha \pm \dfrac{\pi}{2}\right) = \cos\alpha\cos\dfrac{\pi}{2} \mp \sin\alpha\sin\dfrac{\pi}{2} = \mp\sin\alpha$

**A.4** By using the addition formulas for the sine, cosine, and tangent functions, show that (a) $\sin 2\alpha = 2 \sin \alpha \cos \alpha$, (b) $\cos 2\alpha = 1 - 2 \sin^2 \alpha = 2 \cos^2 \alpha - 1$, and (c) $\tan 2\alpha = 2 \tan \alpha/(1 - \tan^2 \alpha)$.

**A.5** Consider a triangle of sides $a$, $b$, and $c$ and interior angles $\alpha$, $\beta$, and $\gamma$, where $\gamma > 90°$. Show how the *law of cosines* is expressed in terms of the exterior angle $\theta$, where $\theta + \gamma = \pi$.

*Solution:*
In this case the $\cos \gamma$ can be expressed as

$$\begin{aligned}\cos \gamma &= \cos(\pi - \theta) = \cos \pi \cos \theta + \sin \pi \sin \theta \\ &= (-1) \cos \theta + 0 \cdot \sin \theta \\ &= -\cos \theta.\end{aligned}$$

Consequently, the law of cosines becomes

$$c^2 = a^2 + b^2 + 2ab \cos \theta.$$

**A.6** Express the quantity $(1 - z^2)^{-1/2}$ as a series using the Binomial expansion.

*Answer:* $(1 - z^2)^{-1/2} = 1 + \dfrac{z^2}{2} + \dfrac{3}{8} z^4 + \cdots$

**A.7** Using the Binomial expansion, find the first few terms in the series of $(1 - z)^{-1}$.

*Solution:*
By analogy with the Binomial expansion,

$$(x + y)^n = x^n + nx^{n-1}y + \frac{n(n-1)}{2!} x^{n-2}y^2 + \cdots,$$

with $x = 1$, $y = -z$, and $n = -1$, we have

$$\begin{aligned}(1 - z)^{-1} &= 1^{-1} + (-1)\, 1^{-2} (-z) + \frac{-1(-2)}{2} (1)^{-3} (-z)^2 + \cdots. \\ &= 1 + z + z^2 + z^3 + \cdots.\end{aligned}$$

**A.8** Find the series expansion for $(1 - z)^{-2}$, using the Binomial expansion.

*Answer:* $(1 - z)^{-2} = 1 + 2z + 3z^2 + 4z^3 + \cdots$

**A.9** Expand $e^x$ using the general form of the MacLaurin series.

*Solution:*
With the MacLaurin series,

$$f(x) = f(0) + xf'(0) + \frac{x^2}{2!} f''(0) + \cdots,$$

we immediately obtain

$$
\begin{aligned}
e^x &= e^0 + xe^0 + \frac{x^2}{2}e^0 + \frac{x^3}{6}e^0 + \cdots \\
&= 1 + x + \frac{x^2}{2} + \frac{x^3}{6} + \cdots .
\end{aligned}
$$

**A.10** Derive the complex exponential representation of (a) sin α and (b) cos α, by using Euler's relations for any angle α.

*Answer:* (a) $\sin \alpha = -\frac{1}{2}j(e^{j\alpha} - e^{-j\alpha})$, (b) $\cos \alpha = \frac{1}{2}(e^{j\alpha} + e^{-j\alpha})$

**A.11** Using the results of Problem A.10, verify the identity $\sin^2 x + \cos^2 x = 1$.

*Solution:*
Direct substitution into the identity yields

$$
\begin{aligned}
\sin^2 x + \cos^2 x &= \left[\frac{1}{2j}(e^{jx} - e^{-jx})\right]^2 + \frac{1}{4}(e^{jx} + e^{-jx})^2 \\
&= \left(-\frac{1}{4}\right)(e^{2jx} - 2e^{jx}e^{-jx} + e^{-2jx}) \\
&\quad + \frac{1}{4}(e^{2jx} + 2e^{jx}e^{-jx} + e^{-2jx}) \\
&= \frac{1}{4}(4e^{jx}e^{-jx}) = e^0 = 1.
\end{aligned}
$$

**A.12** Show that $2 \sin \alpha \cos \alpha = \sin 2\alpha$, by direct substitution of the results from Problem A.10.

**A.13** What generalization results from the successive application of the *differential operator d/dx* to the complex exponential $e^{jx}$?

*Solution:*
Taking successive derivatives of $e^{jx}$ yields

$$
\begin{aligned}
\frac{d}{dx}e^{jx} &= je^{jx}, \\
\frac{d^2}{dx^2}e^{jx} &= \frac{d}{dx}je^{jx} = j^2e^{jx}, \\
\frac{d^3}{dx^3}e^{jx} &= \frac{d}{dx}j^2e^{jx} = j^3e^{jx},
\end{aligned}
$$

which is easily generalized to the $n$th derivative as

$$\frac{d^n}{dx^n} e^{jx} = j^n e^{jx}.$$

**A.14** Derive the identity for the first order derivative of sec $u$, where $u = u(x)$.

*Answer:* $\frac{d}{dx} \sec u = \sec u \tan u \frac{du}{dx}$

**A.15** Derive the identity for $d \csc u/dx$, where $u = u(x)$.

*Solution:*
Using the notation $D \equiv d/dx$, we have

$$\begin{aligned} D \csc u &= D \sin^{-1} u \\ &= (-1) \sin^{-2} u D \sin u \\ &= -\sin^{-2} u \cos u \, Du \\ &= \frac{1}{\sin u} \frac{\cos u}{\sin u} Du \\ &= -\csc u \cot u \frac{du}{dx} \end{aligned}$$

**A.16** Derive the identity for $d \tan \alpha/d\alpha$.

*Answer:* $\frac{d}{d\alpha} \tan \alpha = \sec^2 \alpha$

**A.17** Derive the identity $\int \cos x \, dx = \sin x$ by differentiating $\sin x$.

*Solution:*
The first order derivative of $\sin x$ is

$$\frac{d}{dx} \sin x = \cos x.$$

Multiplying both sides of this equation by $dx$ and integrating yields

$$\int d(\sin x) = \int \cos x \, dx,$$

which reduces to

$$\sin x = \int \cos x \, dx.$$

**A.18** Derive the identity for $\int v du$ by differentiating $uv$ with respect to $x$.

*Answer:* $\int v \, du = uv - \int u dv$

# APPENDIX B

# Properties of Atoms in Bulk

| Element | Symbol | At. No. Z | Electron Configuration | Chemical Atomic Weight | Mass Density $10^3$ kg/m$^3$ at 20° C | Melting Point (° C) | Boiling Point (° C) |
|---|---|---|---|---|---|---|---|
| Actinium | Ac | 89 | [Rn]$6d7s^2$ | 227.0278 | (10.1) | 1050 | 3200 |
| Aluminum | Al | 13 | [Ne]$3s^2 3p$ | 26.98154 | 2.699 | 660 | 2467 |
| Americium | **Am** | 95 | [Rn]$5f^7 7s^2$ | (243) | 13.7 | 994 | 2607 |
| Antimony | Sb | 51 | [Pd]$5s^2 5p^3$ | 121.75 | 6.69 | 631 | 1950 |
| Argon | *Ar* | 18 | [Ne]$3s^2 3p^6$ | 39.948 | $1.784 \times 10^{-3}$ | −189.2 | −185.7 |
| Arsenic | As | 33 | [Ar]$3d^{10} 4s^2 4p^3$ | 74.9216 | 5.73 | 817 | 617 |
| Astatine | At | 85 | [Xe]$4f^{14} 5d^{10} 6s^2 6p^5$ | (210) | ... | (302) | 337 |
| Barium | Ba | 56 | [Xe]$6s^2$ | 137.33 | 3.5 | 725 | 1640 |
| Berkelium | **Bk** | 97 | [Rn]$5f^9{*}7s^2$ | (247) | (14) | 986 | ... |
| Beryllium | Be | 4 | [He]$2s^2$ | 9.01218 | 1.85 | 127 | 2970 |
| Bismuth | Bi | 83 | [Xe]$4f^{14} 5d^{10} 6s^2 6p^3$ | 208.9804 | 9.75 | 271 | 1560 |
| Boron | B | 5 | [He]$2s^1 2p$ | 10.81 | 2.34 | 2079 | 2550 |
| Bromine | **Br** | 35 | [Ar]$3d^{10} 4s^2 4p^5$ | 79.904 | 3.12 | −7.2 | 58.8 |
| Cadmium | Cd | 48 | [Pd]$5s^2$ | 112.41 | 8.65 | 320.9 | 765 |
| Calcium | Ca | 20 | [Ar]$4s^2$ | 40.08 | 1.55 | 839 | 1484 |
| Californium | **Cf** | 98 | [Rn]$5f^{10} 7s^2$ | (251) | ... | ... | ... |
| Carbon | C | 6 | [He]$2s^2 2p^2$ | 12.01 | 2.25 | 3367 | 4827 |
| Cerium | Ce | 58 | [Xe]$4f^2{*}6s^2$ | 140.12 | 6.66 | 798 | 3257 |
| Cesium | **Cs** | 55 | [Xe]$6s$ | 132.9054 | 1.873 | 28.4 | 669 |
| Chlorine | *Cl* | 17 | [Ne]$3s^2 3p^5$ | 35.453 | $3.214 \times 10^{-3}$ | 101.0 | −34.6 |
| Chromium | Cr | 24 | [Ar]$3d^5{*}4s$ | 51.996 | 7.19 | 1857 | 2672 |
| Cobalt | Co | 27 | [Ar]$3d^7 4s^2$ | 58.9332 | 8.9 | 1495 | 2870 |
| Copper | Cu | 29 | [Ar]$3d^{10}{*}4s$ | 63.546 | 8.96 | 1083 | 2567 |
| Curium | **Cm** | 96 | [Rn]$5f^7 6d7s^2$ | (247) | (13.5) | 1340 | ... |
| Dysprosium | Dy | 66 | [Xe]$4f^{10} 6s^2$ | 162.50 | 8.55 | 1412 | 2567 |
| Einsteinium | **Es** | 99 | [Rn]$5f^{11} 7s^2$ | (252) | ... | ... | ... |
| Erbium | Er | 68 | [Xe]$4f^{12} 6s^2$ | 167.26 | 9.07 | 1529 | 2868 |
| Europium | Eu | 63 | [Xe]$4f^7 6s^2$ | 151.96 | 5.24 | 822 | 1529 |
| Fermium | **Fm** | 100 | [Rn]$5f^{12} 7s^2$ | (257) | ... | ... | ... |
| Florine | *F* | 9 | [He]$2s^2 2p^5$ | 18.998403 | $1.69 \times 10^{-3}$ | −219.6 | −188.1 |
| Francium | **Fr** | 87 | [Rn]$7s$ | (223) | ... | 27 | 677 |
| Gadolinium | Gd | 64 | [Xe]$4f^7 5d6s^2$ | 157.25 | 7.90 | 1313 | 3273 |
| Gallium | **Ga** | 31 | [Ar]$3d^{10} 4s^2 4p$ | 69.72 | 5.904 | 29.8 | 2403 |
| Germanium | Ge | 32 | [Ar]$3d^{10} 4s^2 4p^2$ | 72.59 | 5.32 | 947.4 | 2830 |
| Gold | Au | 79 | [Xe]$4f^{14} 5d^{10} 6s$ | 196.9665 | 18.9 | 1064 | 3080 |

| Element | Symbol | At. No. Z | Electron Configuration | Chemical Atomic Weight | Mass Density $10^3$ kg/m$^3$ at 20° C | Melting Point (° C) | Boiling Point (° C) |
|---|---|---|---|---|---|---|---|
| Hafnium | Hf | 72 | $[Xe]4f^{14}5d^2 6s^2$ | 178.49 | 13.3 | 2227 | 4600 |
| Helium | *He* | 2 | $1s^2$ | 4.00260 | $0.1785 \times 10^{-3}$ | −272.2 | −268.9 |
| Holmium | Ho | 67 | $[Xe]4f^{11}6s^2$ | 164.9304 | 8.80 | 1474 | 2700 |
| Hydrogen | *H* | 1 | $1s$ | 1.00794 | $0.0899 \times 10^{-3}$ | −259.1 | −252.9 |
| Indium | In | 49 | $[Pd]5s^2 5p$ | 114.82 | 7.31 | 156.6 | 2080 |
| Iodine | I | 53 | $[Pd]5s^2 5p^5$ | 126.9045 | 4.93 | 113.5 | 184 |
| Iridium | Ir | 77 | $[Xe]4f^{14}5d^7 6s^2$ | 192.22 | 22.4 | 2410 | 4130 |
| Iron | Fe | 26 | $[Ar]3d^6 4s^2$ | 55.847 | 7.86 | 1535 | 2750 |
| Krypton | *Kr* | 36 | $[Ar]3d^{10}4s^2 4p^6$ | 83.8 | $3.743 \times 10^{-3}$ | −157 | −152 |
| Lanthanum | La | 57 | $[Xe]5d6s^2$ | 138.9055 | 6.145 | 920 | 3454 |
| Lawrencium | **Lw** | 103 | $[Rn]5f^{14}6d7s^2$ | (260) | ... | ... | ... |
| Lead | Pb | 82 | $[Xe]4f^{14}5d^{10}6s^2 6p^2$ | (207.2) | 11.4 | 327.5 | 1740 |
| Lithium | Li | 3 | $[He]2s$ | 6.941 | 0.534 | 180.5 | 1342 |
| Lutetium | Lu | 71 | $[Xe]4f^{14}5d6s^2$ | 174.967 | 9.84 | 1663 | 3402 |
| Magnesium | Mg | 12 | $[Ne]3s^2$ | 24.305 | 1.738 | 649 | 1090 |
| Manganese | Mn | 25 | $[Ar]3d^5 4s^2$ | 54.9380 | 7.43 | 1244 | 1962 |
| Mendelevium | **Md** | 101 | $[Rn]5f^{13}7s^2$ | (258) | ... | ... | ... |
| Mercury | **Hg** | 80 | $[Xe]4f^{14}5d^{10}6s^2$ | (200.59) | 13.5 | −38.9 | 357 |
| Molybdenum | Mo | 42 | $[Kr]4d^5 5s$ | 95.94 | 10.2 | 2617 | 4612 |
| Neodymium | Nd | 60 | $[Xe]4f^4 6s^2$ | 144.24 | 7.00 | 1016 | 3127 |
| Neon | *Ne* | 10 | $[He]2s^2 2p^6$ | 20.179 | $0.8999 \times 10^{-3}$ | −248.7 | −246.0 |
| Neptunium | **Np** | 93 | $[Rn]5f^4 6d7s^2$ | 237.0482 | 20.2 | 640 | (3900) |
| Nickel | Ni | 28 | $[Ar]3d^8 4s^2$ | 58.69 | 8.90 | 1453 | 2730 |
| Niobium | Nb | 41 | $[Kr]4d^4{*}5s$ | 92.9064 | 8.57 | 2468 | 4742 |
| Nitrogen | N | 7 | $[He]2s^2 2p^3$ | 14.0067 | $1.25 \times 10^{-3}$ | −209.9 | −195.8 |
| Nobelium | **No** | 102 | $[Rn]5f^{14}7s^2$ | (259) | ... | ... | ... |
| Osmium | Os | 76 | $[Xe]4f^{14}5d^6 6s^2$ | 190.2 | 22.6 | 3045 | 5030 |
| Oxygen | *O* | 8 | $[He]2s^2 2p^4$ | 15.9994 | $1.429 \times 10^{-3}$ | −218.4 | −183.0 |
| Palladium | Pd | 46 | $[Kr]4d^{10}{*}$ | 106.42 | 12.0 | 1554 | 3140 |
| Phosphorus | P | 15 | $[Ne]3s^2 3p^3$ | 30.97376 | 1.82 | 44.1 | 280 |
| Platinum | Pt | 78 | $[Xe]4f^{14}5d^9 6s$ | 195.08 | 21.4 | 1772 | 3827 |
| Plutonium | **Pu** | 94 | $[Rn]5f^6 7s^2$ | (244) | 19.8 | 641 | 3232 |
| Polonium | Po | 84 | $[Xe]4f^{14}5d^{10}6s^2 6p^4$ | (209) | (9.32) | 254 | 962 |
| Potassium | K | 19 | $[Ar]4s$ | 39.0983 | 0.86 | 63.25 | 760 |
| Praseodymium | Pr | 59 | $[Xe]4f^3 6s^2$ | 140.9077 | 6.77 | 931 | 3017 |
| Promethium | **Pm** | 61 | $[Xe]4f^5 6s^2$ | (145) | 7.26 | 1042 | 3000 |
| Protactinium | Pa | 91 | $[Rn]5f^2{*}6d7s^2$ | 231.0359 | (15.4) | 1570 | 4000 |
| Radium | Ra | 88 | $[Rn]7s^2$ | 226.0254 | 5.0 | 700 | 1140 |
| Radon | *Rn* | 86 | $[Xe]4f^{14}5d^{10}6s^2 6p^6$ | (222) | $9.73 \times 10^{-3}$ | (−71) | (−61.8) |
| Rhenium | Re | 75 | $[Xe]4f^{14}5d^5 6s^2$ | 186.207 | 21.0 | 3180 | (5600) |
| Rhodium | Rh | 45 | $[Kr]4d^8 5s$ | 102.9055 | 12.4 | 1966 | 3727 |
| Rubidium | Rb | 37 | $[Kr]5s$ | 85.4678 | 1.532 | 38.9 | 686 |
| Ruthenium | Ru | 44 | $[Kr]4d^7 5s$ | 101.07 | 12.4 | 2310 | 3900 |
| Samarium | Sm | 62 | $[Xe]4f^6 6s^2$ | 150.36 | 7.52 | 1074 | 1794 |

| Element | Symbol | *At. No.* Z | *Electron Configuration* | *Chemical Atomic Weight* | *Mass Density* $10^3$ kg/m$^3$ *at 20° C* | *Melting Point* (° C) | *Boiling Point* (° C) |
|---|---|---|---|---|---|---|---|
| Scandium | Sc | 21 | $[Ar]3d4s^2$ | 44.9559 | 2.99 | 1541 | 2832 |
| Selenium | Se | 34 | $[Ar]3d^{10}4s^2 4p^4$ | 78.96 | 4.79 | 217 | 685 |
| Silicon | Si | 14 | $[Ne]3s^2 3p^2$ | 28.0855 | 2.33 | 1410 | 2355 |
| Silver | Ag | 47 | $[Pd]5s$ | 107.8682 | 10.5 | 962 | 2212 |
| Sodium | Na | 11 | $[Ne]3s$ | 22.98977 | 0.971 | 97.8 | 883 |
| Strontium | Sr | 38 | $[Kr]5s^2$ | 87.62 | 2.54 | 769 | 1384 |
| Sulfur | S | 16 | $[Ne]3s^2 3p^4$ | 32.06 | 2.07 | 112.8 | 444.7 |
| Tantalum | Ta | 73 | $[Xe]4f^{14}5d^3 6s^2$ | 180.9479 | 16.6 | 2996 | 5425 |
| Technetium | **Tc** | 43 | $[Kr]4d^6 5s$ | (98) | (11.5) | (2172) | 4877 |
| Tellurium | Te | 52 | $[Pd]5s^2 5p^4$ | 127.60 | 6.24 | 449.5 | 989.8 |
| Terbium | Tb | 65 | $[Xe]4f^9{*}6s^2$ | 158.9254 | 8.23 | 1365 | 3230 |
| Thallium | Tl | 81 | $[Xe]4f^{14}5d^{10}6s^2 6p$ | 204.383 | 11.85 | 303 | 1457 |
| Thorium | Th | 90 | $[Rn]6d^2 7s^2$ | 232.0381 | 11.7 | 1750 | (4790) |
| Thulium | Tm | 69 | $[Xe]4f^{13}6s^2$ | 168.9342 | 9.32 | 1545 | 1950 |
| Tin | Sn | 50 | $[Pd]5s^2 5p^2$ | 118.69 | 7.31 | 232.0 | 2270 |
| Titanium | Ti | 22 | $[Ar]3d^2 4s^2$ | 47.86 | 4.54 | 1660 | 3287 |
| Tungsten | W | 74 | $[Xe]4f^{14}5d^4 6s^2$ | 183.85 | 19.3 | 3410 | 5660 |
| Unnilhexium | **Unh** | 106 | $[Rn]5f^{14}6d^4 7s^2$ | 263 | ... | ... | ... |
| Unniplentium | **Unp** | 105 | $[Rn]5f^{14}6d^3 7s^2$ | 262 | ... | ... | ... |
| Unnilquadium | **Unq** | 104 | $[Rn]5f^{14}6d^2 7s^2$ | 261 | ... | ... | ... |
| Uranium | U | 92 | $[Rn]5f^3 6d7s^2$ | 238.0289 | 19.0 | 1132 | 3818 |
| Vanadium | V | 23 | $[Ar]3d^3 4s^2$ | 50.9415 | 6.11 | 1890 | 3380 |
| Xenon | *Xe* | 54 | $[Pd]5s^2 5p^6$ | 131.29 | $5.887 \times 10^{-3}$ | −111.8 | −107.1 |
| Ytterbium | Yb | 70 | $[Xe]4f^{14}6s^2$ | 173.04 | 6.97 | 819 | 1196 |
| Yttrium | Y | 39 | $[Kr]4d5s^2$ | 88.9059 | 4.47 | 1523 | 3337 |
| Zinc | Zn | 30 | $[Ar]3d^{10}4s^2$ | 65.38 | 7.13 | 419.6 | 906 |
| Zirconium | Zr | 40 | $[Kr]4d^2 5s^2$ | 91.22 | 6.51 | 1852 | 4377 |

*An asterisk in the electron configuration denotes an irregularity.

*Source:* According to the Commission on Atomic Weights and Isotopic Abundances, International Union of Pure and Applied Chemistry

*Notes:* Atomic symbols in normal type (e.g., Ac, Al, etc.) are solid elements.

Atomic symbols in bold type (e.g., **Br, Cs,** etc.) are liquids at 20° C.

Atomic symbols in italics (e.g., *Ar, Cl,* etc.) are gases.

Atomic symbols in double wide type (e.g., **Am, Bk,** etc.) are synthetic elements.

Chemical atomic weight values in parentheses are for the most stable radioactive isotopes.

Melting points, boiling points, and mass density in parentheses are uncertain values.

APPENDIX C

# Partial List of Nuclear Masses

| Element | Symbol | Z | A | Relative Atomic Mass (u) | Relative Abundance (%) | No. of Isotopes Stable | Unstable |
|---|---|---|---|---|---|---|---|
| Neutron | n | 0 | 1(R) | 1.008665 | | 0 | 1 |
| Hydrogen | H | 1 | 1 | 1.007825 | 99.985 | 2 | 1 |
| | D | 1 | 2 | 2.014102 | 0.015 | | |
| | T | 1 | 3(NR) | 3.016052 | | | |
| Helium | He | 2 | 3 | 3.016029 | 0.0001 | 2 | 3 |
| | | | 4 | 4.002603 | 99.9999 | | |
| Lithium | Li | 3 | 6 | 6.015123 | 7.42 | 2 | 3 |
| | | | 7 | 7.016004 | 92.58 | | |
| Beryllium | Be | 4 | 9 | 9.012182 | 100 | 1 | 5 |
| Boron | B | 5 | 10 | 10.012938 | 19.78 | 2 | 4 |
| | | | 11 | 11.009305 | 80.22 | | |
| Carbon | C | 6 | 12 | 12.000000 | 98.89 | 2 | 5 |
| | | | 13 | 13.003355 | 1.11 | | |
| Nitrogen | N | 7 | 14 | 14.003074 | 99.63 | 2 | 6 |
| | | | 15 | 15.000109 | 0.37 | | |
| Oxygen | O | 8 | 16 | 15.994915 | 99.759 | 3 | 5 |
| | | | 17 | 16.999131 | 0.037 | | |
| | | | 18 | 17.999159 | 0.204 | | |
| Fluorine | F | 9 | 19 | 18.998403 | 100 | 1 | 5 |
| Neon | Ne | 10 | 20 | 19.992439 | 90.51 | 3 | 5 |
| | | | 21 | 20.993845 | 0.27 | | |
| | | | 22 | 21.991384 | 9.22 | | |
| Sodium | Na | 11 | 23 | 22.989770 | 100 | 1 | 6 |
| Magnesium | Mg | 12 | 24 | 23.985045 | 78.99 | 3 | 5 |
| | | | 25 | 24.985839 | 10.00 | | |

| Element | Symbol | Z | A | Relative Atomic Mass (u) | Relative Abundance (%) | No. of Isotopes Stable | No. of Isotopes Unstable |
|---|---|---|---|---|---|---|---|
| | | | 26 | 25.982595 | 11.01 | | |
| Aluminum | Al | 13 | 27 | 26.981541 | 100 | 1 | 7 |
| Silicon | Si | 14 | 28 | 27.976928 | 92.23 | 3 | 5 |
| | | | 29 | 28.976496 | 4.67 | | |
| | | | 30 | 29.973772 | 3.10 | | |
| Phosphorus | P | 15 | 31 | 30.973763 | 100 | 1 | 6 |
| Sulfur | S | 16 | 32 | 31.972072 | 95.02 | 4 | 6 |
| | | | 33 | 32.971459 | 0.75 | | |
| | | | 34 | 33.967868 | 4.21 | | |
| | | | 36 | 35.967079 | 0.017 | | |
| Chlorine | Cl | 17 | 35 | 34.968853 | 75.77 | 2 | 7 |
| | | | 37 | 36.965903 | 24.23 | | |
| Argon | Ar | 18 | 36 | 35.967546 | 0.337 | 3 | 5 |
| | | | 38 | 37.962732 | 0.063 | | |
| | | | 40 | 39.962383 | 99.60 | | |
| Potassium | K | 19 | 39 | 38.963708 | 93.26 | 2 | 7 |
| | | | 40(NR) | 38.963999 | 0.01 | | |
| | | | 41 | 40.961825 | 6.73 | | |
| Calcium | Ca | 20 | 40 | 39.962591 | 96.94 | 6 | 8 |
| | | | 42 | 41.958622 | 0.647 | | |
| | | | 43 | 42.958770 | 0.135 | | |
| | | | 44 | 43.955485 | 2.09 | | |
| | | | 46 | 45.953689 | 0.0035 | | |
| | | | 48 | 47.952532 | 0.187 | | |
| Scandium | Sc | 21 | 45 | 44.955914 | 100 | 1 | 10 |
| Titanium | Ti | 22 | 46 | 45.952633 | 8.25 | 5 | 4 |
| | | | 47 | 46.951765 | 7.45 | | |
| | | | 48 | 47.947947 | 73.7 | | |
| | | | 49 | 48.947871 | 5.4 | | |
| | | | 50 | 49.944786 | 5.2 | | |
| Vanadium | V | 23 | 50(NR) | 49.947161 | 0.25 | 1 | 8 |
| | | | 51 | 50.943962 | 99.75 | | |
| Chromium | Cr | 24 | 50 | 49.946046 | 4.35 | 4 | 5 |
| | | | 52 | 51.940510 | 83.79 | | |
| | | | 53 | 52.940651 | 9.50 | | |

| Element | Symbol | Z | A | Relative Atomic Mass (u) | Relative Abundance (%) | No. of Isotopes | |
|---|---|---|---|---|---|---|---|
| | | | | | | *Stable* | *Unstable* |
| | | | 54 | 53.938882 | 2.36 | | |
| Manganese | Mn | 25 | 55 | 54.938046 | 100 | 1 | 8 |
| Iron | Fe | 26 | 54 | 53.939612 | 5.8 | 4 | 6 |
| | | | 56 | 55.934939 | 91.8 | | |
| | | | 57 | 56.935396 | 2.1 | | |
| | | | 58 | 57.933278 | 0.3 | | |
| Cobalt | Co | 27 | 59 | 58.933198 | 100 | 1 | 9 |
| Nickel | Ni | 28 | 58 | 57.935347 | 68.3 | 5 | 7 |
| | | | 60 | 59.930789 | 26.1 | | |
| | | | 61 | 60.931059 | 1.1 | | |
| | | | 62 | 61.928346 | 3.6 | | |
| | | | 64 | 63.927968 | 0.9 | | |
| Copper | Cu | 29 | 63 | 62.929599 | 69.2 | 2 | 9 |
| | | | 65 | 64.927792 | 30.8 | | |
| Zinc | Zn | 30 | 64 | 63.929145 | 48.6 | 5 | 8 |
| | | | 66 | 65.926035 | 27.9 | | |
| | | | 67 | 66.927129 | 4.1 | | |
| | | | 68 | 67.924846 | 18.8 | | |
| | | | 70 | 69.925325 | 0.6 | | |
| Gallium | Ga | 31 | 69 | 68.925581 | 60.1 | 2 | 12 |
| | | | 71 | 70.924701 | 39.9 | | |
| Germanium | Ge | 32 | 70 | 69.924250 | 20.5 | 5 | 9 |
| | | | 72 | 71.922080 | 27.4 | | |
| | | | 73 | 72.923464 | 7.8 | | |
| | | | 74 | 73.921179 | 36.5 | | |
| | | | 76 | 75.921403 | 7.8 | | |
| Arsenic | As | 33 | 75 | 74.921596 | 100 | 1 | 12 |
| Selenium | Se | 34 | 74 | 73.922477 | 0.9 | 6 | 10 |
| | | | 76 | 75.919207 | 9.0 | | |
| | | | 77 | 76.919908 | 7.6 | | |
| | | | 78 | 77.917304 | 23.5 | | |
| | | | 80 | 79.916520 | 49.8 | | |
| | | | 82 | 81.916709 | 9.2 | | |
| Bromine | Br | 35 | 79 | 78.918336 | 50.7 | 2 | 15 |
| | | | 81 | 80.916290 | 49.3 | | |

| Element | Symbol | Z | A | Relative Atomic Mass (u) | Relative Abundance (%) | No. of Isotopes Stable | No. of Isotopes Unstable |
|---|---|---|---|---|---|---|---|
| Krypton | Kr | 36 | 78 | 77.920397 | 0.35 | 6 | 15 |
| | | | 80 | 79.916375 | 2.25 | | |
| | | | 82 | 81.913483 | 11.6 | | |
| | | | 83 | 82.914134 | 11.5 | | |
| | | | 84 | 83.911506 | 57.0 | | |
| | | | 86 | 85.910614 | 17.3 | | |
| Rubidium | Rb | 37 | 85 | 84.911800 | 72.2 | 1 | 16 |
| | | | 87(NR) | 86.909184 | 27.8 | | |
| Strontium | Sr | 38 | 84 | 83.913428 | 0.6 | 4 | 12 |
| | | | 86 | 85.909273 | 9.8 | | |
| | | | 87 | 86.908890 | 7.0 | | |
| | | | 88 | 87.905625 | 82.6 | | |
| Yttrium | Y | 39 | 89 | 88.905856 | 100 | 1 | 14 |
| Zirconium | Zr | 40 | 90 | 89.904708 | 51.5 | 5 | 13 |
| | | | 91 | 90.905644 | 11.2 | | |
| | | | 92 | 91.905039 | 17.1 | | |
| | | | 94 | 93.906319 | 17.4 | | |
| | | | 96 | 95.908272 | 2.8 | | |
| Niobium | Nb | 41 | 93 | 92.906378 | 100 | 1 | 13 |
| Molybdenum | Mo | 42 | 92 | 91.906809 | 14.8 | 7 | 11 |
| | | | 94 | 93.905086 | 9.3 | | |
| | | | 95 | 94.905838 | 15.9 | | |
| | | | 96 | 95.904675 | 16.7 | | |
| | | | 97 | 96.906018 | 9.6 | | |
| | | | 98 | 97.905405 | 24.1 | | |
| | | | 100 | 99.907473 | 9.6 | | |
| Technetium | Tc | 43 | 99(R) | 98.906252 | | 0 | 16 |
| Ruthenium | Ru | 44 | 96 | 95.907596 | 5.5 | 7 | 9 |
| | | | 98 | 97.905287 | 1.9 | | |
| | | | 99 | 98.905937 | 12.7 | | |
| | | | 100 | 99.904217 | 12.6 | | |
| | | | 101 | 100.905581 | 17.0 | | |
| | | | 102 | 101.904347 | 31.6 | | |
| | | | 104 | 103.905422 | 18.7 | | |
| Rhodium | Rh | 45 | 103 | 102.905503 | 100 | 1 | 14 |
| Palladium | Pd | 46 | 102 | 101.905609 | 1.0 | 6 | 12 |

| Element | Symbol | Z | A | Relative Atomic Mass (u) | Relative Abundance (%) | No. of Isotopes Stable | No. of Isotopes Unstable |
|---|---|---|---|---|---|---|---|
| | | | 104 | 103.904026 | 11.0 | | |
| | | | 105 | 104.905075 | 22.2 | | |
| | | | 106 | 105.903475 | 27.3 | | |
| | | | 108 | 107.903894 | 26.7 | | |
| | | | 110 | 109.905169 | 11.8 | | |
| Silver | Ag | 47 | 107 | 106.905095 | 51.8 | 2 | 14 |
| | | | 109 | 108.904754 | 48.2 | | |
| Cadmium | Cd | 48 | 106 | 105.906461 | 1.3 | 8 | 9 |
| | | | 108 | 107.904186 | 0.9 | | |
| | | | 110 | 109.903007 | 12.5 | | |
| | | | 111 | 110.904182 | 12.8 | | |
| | | | 112 | 111.902761 | 24.1 | | |
| | | | 113 | 112.904401 | 12.2 | | |
| | | | 114 | 113.903361 | 28.7 | | |
| | | | 116 | 115.904758 | 7.5 | | |
| Indium | In | 49 | 113 | 112.904056 | 4.3 | 1 | 18 |
| | | | 115(NR) | 114.903875 | 95.7 | | |
| Tin | Sn | 50 | 112 | 111.904823 | 1.0 | 10 | 11 |
| | | | 114 | 113.902781 | 0.7 | | |
| | | | 115 | 114.903344 | 0.4 | | |
| | | | 116 | 115.901743 | 14.7 | | |
| | | | 117 | 116.902954 | 7.7 | | |
| | | | 118 | 117.901607 | 24.3 | | |
| | | | 119 | 118.903310 | 8.6 | | |
| | | | 120 | 119.902199 | 32.4 | | |
| | | | 122 | 121.903440 | 4.6 | | |
| | | | 124 | 123.905271 | 5.6 | | |
| Antimony | Sb | 51 | 121 | 120.903824 | 57.3 | 2 | 20 |
| | | | 123 | 122.904222 | 42.7 | | |
| Tellurium | Te | 52 | 120 | 119.904021 | 0.1 | 7 | 14 |
| | | | 122 | 121.903055 | 2.5 | | |
| | | | 123(NR) | 122.904278 | 0.9 | | |
| | | | 124 | 123.902825 | 4.6 | | |
| | | | 125 | 124.904435 | 7.0 | | |
| | | | 126 | 125.903310 | 18.7 | | |
| | | | 128 | 127.904464 | 31.7 | | |
| | | | 130 | 129.906229 | 34.5 | | |
| Iodine | I | 53 | 127 | 126.904477 | 100 | 1 | 22 |

| Element | Symbol | Z | A | Relative Atomic Mass (u) | Relative Abundance (%) | No. of Isotopes | |
|---|---|---|---|---|---|---|---|
| | | | | | | Stable | Unstable |
| Xenon | Xe | 54 | 124 | 123.906120 | 0.1 | 9 | 16 |
| | | | 126 | 125.904281 | 0.1 | | |
| | | | 128 | 127.903531 | 1.9 | | |
| | | | 129 | 128.904780 | 26.4 | | |
| | | | 130 | 129.903509 | 4.1 | | |
| | | | 131 | 130.905076 | 21.2 | | |
| | | | 132 | 131.904148 | 26.9 | | |
| | | | 134 | 133.905395 | 10.4 | | |
| | | | 136 | 135.907219 | 8.9 | | |
| Cesium | Cs | 55 | 133 | 132.905433 | 100 | 1 | 20 |
| Barium | Ba | 56 | 130 | 129.906277 | 0.1 | 7 | 13 |
| | | | 132 | 131.905042 | 0.1 | | |
| | | | 134 | 133.904490 | 2.4 | | |
| | | | 135 | 134.905668 | 6.6 | | |
| | | | 136 | 135.904556 | 7.9 | | |
| | | | 137 | 136.905816 | 11.2 | | |
| | | | 138 | 137.905236 | 71.7 | | |
| Lanthanum | La | 57 | 138 | 137.907114 | 0.1 | 2 | 17 |
| | | | 139 | 138.906355 | 99.9 | | |
| Cerium | Ce | 58 | 136 | 135.907140 | 0.2 | 4 | 13 |
| | | | 138 | 137.905996 | 0.2 | | |
| | | | 140 | 139.905442 | 88.5 | | |
| | | | 142 | 141.909249 | 11.1 | | |
| Praseodymium | Pr | 59 | 141 | 140.907657 | 100 | 1 | 13 |
| Neodymium | Nd | 60 | 142 | 141.907731 | 27.2 | 6 | 8 |
| | | | 143 | 142.909823 | 12.2 | | |
| | | | 144(NR) | 143.910096 | 23.8 | | |
| | | | 145 | 144.912582 | 8.3 | | |
| | | | 146 | 145.913126 | 17.2 | | |
| | | | 148 | 147.916901 | 5.7 | | |
| | | | 150 | 149.920900 | 5.6 | | |
| Promethium | Pm | 61 | 147(R) | 146.915148 | | 0 | 13 |
| Samarium | Sm | 62 | 144 | 143.912009 | 3.1 | 4 | 12 |
| | | | 147(NR) | 146.914907 | 15.1 | | |
| | | | 148(NR) | 147.914832 | 11.3 | | |
| | | | 149(NR) | 148.917193 | 13.9 | | |
| | | | 150 | 149.917285 | 7.4 | | |
| | | | 152 | 151.919741 | 26.7 | | |

| Element | Symbol | Z | A | Relative Atomic Mass (u) | Relative Abundance (%) | No. of Isotopes Stable | No. of Isotopes Unstable |
|---|---|---|---|---|---|---|---|
| | | | 154 | 153.922218 | 22.6 | | |
| Europium | Eu | 63 | 151 | 150.919860 | 47.9 | 2 | 15 |
| | | | 153 | 152.921243 | 52.1 | | |
| Gadolinium | Gd | 64 | 152(NR) | 151.919803 | 0.2 | 6 | 11 |
| | | | 154 | 153.920876 | 2.1 | | |
| | | | 155 | 154.922629 | 14.8 | | |
| | | | 156 | 155.922130 | 20.6 | | |
| | | | 157 | 156.923967 | 15.7 | | |
| | | | 158 | 157.924111 | 24.8 | | |
| | | | 160 | 159.927061 | 21.8 | | |
| Terbium | Tb | 65 | 159 | 158.925350 | 100 | 1 | 17 |
| Dysprosium | Dy | 66 | 156 | 155.924287 | 0.1 | 7 | 12 |
| | | | 158 | 157.924412 | 0.1 | | |
| | | | 160 | 159.925203 | 2.3 | | |
| | | | 161 | 160.926939 | 19.0 | | |
| | | | 162 | 161.926805 | 25.5 | | |
| | | | 163 | 162.928737 | 24.9 | | |
| | | | 164 | 163.929183 | 28.1 | | |
| Holmium | Ho | 67 | 165 | 164.930332 | 100 | 1 | 19 |
| Erbium | Er | 68 | 162 | 161.928787 | 0.1 | 6 | 9 |
| | | | 164 | 163.929211 | 1.6 | | |
| | | | 166 | 165.930305 | 33.4 | | |
| | | | 167 | 166.932061 | 22.9 | | |
| | | | 168 | 167.932383 | 27.1 | | |
| | | | 170 | 169.935476 | 14.9 | | |
| Thulium | Tm | 69 | 169 | 168.934225 | 100 | 1 | 15 |
| Ytterbium | Yb | 70 | 168 | 167.933908 | 0.1 | 7 | 7 |
| | | | 170 | 169.934774 | 3.2 | | |
| | | | 171 | 170.936338 | 14.4 | | |
| | | | 172 | 171.936393 | 21.9 | | |
| | | | 173 | 172.938222 | 16.2 | | |
| | | | 174 | 173.938873 | 31.6 | | |
| | | | 176 | 175.942576 | 12.6 | | |
| Lutetium | Lu | 71 | 175 | 174.940785 | 97.4 | 1 | 13 |
| | | | 176(NR) | 175.942694 | 2.6 | | |
| Hafnium | Hf | 72 | 174(NR) | 173.940065 | 0.2 | 5 | 11 |

| Element | Symbol | Z | A | Relative Atomic Mass (u) | Relative Abundance (%) | No. of Isotopes Stable | No. of Isotopes Unstable |
|---|---|---|---|---|---|---|---|
| | | | 176 | 175.941420 | 5.2 | | |
| | | | 177 | 176.943233 | 18.6 | | |
| | | | 178 | 177.943710 | 27.1 | | |
| | | | 179 | 178.945827 | 13.7 | | |
| | | | 180 | 179.946561 | 35.2 | | |
| Tantalum | Ta | 73 | 180(NR) | 179.947489 | 0.01 | 1 | 14 |
| | | | 181 | 180.948014 | 99.99 | | |
| Tungsten | W | 74 | 180 | 179.946727 | 0.1 | 5 | 12 |
| | | | 182 | 181.948225 | 26.3 | | |
| | | | 183 | 182.950245 | 14.3 | | |
| | | | 184 | 183.950953 | 30.7 | | |
| | | | 186 | 185.954377 | 28.6 | | |
| Rhenium | Re | 75 | 185 | 184.952977 | 37.4 | 1 | 15 |
| | | | 187(NR) | 186.955765 | 62.6 | | |
| Osmium | Os | 76 | 184 | 183.952514 | 0.02 | 7 | 8 |
| | | | 186 | 185.953852 | 1.6 | | |
| | | | 187 | 186.955762 | 1.6 | | |
| | | | 188 | 187.955850 | 13.3 | | |
| | | | 189 | 188.958156 | 16.1 | | |
| | | | 190 | 189.958455 | 26.4 | | |
| | | | 192 | 191.961487 | 41.0 | | |
| Iridium | Ir | 77 | 191 | 190.960603 | 37.3 | 2 | 15 |
| | | | 193 | 192.962942 | 62.7 | | |
| Platinum | Pt | 78 | 190(NR) | 189.959937 | 0.1 | 4 | 24 |
| | | | 192(NR) | 191.961049 | 0.79 | | |
| | | | 194 | 193.962679 | 32.9 | | |
| | | | 195 | 194.964785 | 33.8 | | |
| | | | 196 | 195.964947 | 25.3 | | |
| | | | 198 | 197.967879 | 7.2 | | |
| Gold | Au | 79 | 197 | 196.966560 | 100 | 1 | 17 |
| Mercury | Hg | 80 | 196 | 195.965812 | 0.2 | 7 | 15 |
| | | | 198 | 197.966760 | 10.0 | | |
| | | | 199 | 198.968269 | 16.8 | | |
| | | | 200 | 199.968316 | 23.1 | | |
| | | | 201 | 200.970293 | 13.2 | | |
| | | | 202 | 201.970632 | 29.8 | | |
| | | | 204 | 203.973481 | 6.9 | | |

| Element | Symbol | Z | A | Relative Atomic Mass (u) | Relative Abundance (%) | No. of Isotopes | |
|---|---|---|---|---|---|---|---|
| | | | | | | Stable | Unstable |
| Thallium | Ti | 81 | 203 | 202.972336 | 29.5 | 2 | 18 |
| | | | 205 | 204.974410 | 70.5 | | |
| Lead | Pb | 82 | 204 | 203.973037 | 1.4 | 4 | 17 |
| | | | 206 | 205.974455 | 24.1 | | |
| | | | 207 | 206.975885 | 22.1 | | |
| | | | 208 | 207.976641 | 52.4 | | |
| Bismuth | Bi | 83 | 209 | 208.980388 | 100 | 1 | 16 |
| Uranium | U | 92 | 234(NR) | 234.040947 | 0.005 | 0 | 14 |
| | | | 235(NR) | 235.043925 | 0.720 | | |
| | | | 238(NR) | 238.050786 | 99.275 | | |

*Source:* Relative atomic masses are taken from J. H. E. Mattauch, W. Thiele, and A. H. Wapstra, *Nuclear Physics 67*, no. 1 (1965). Relative abundances are from the *Chart of the Nuclides*, 12th ed., General Electric Co, 1977. Number of ground state isotopes are from the *Handbook of Chemistry and Physics*, 64th ed., CRC Press, 1984.

*Note:* The symbol (NR) beside a mass number stands for Naturally Radioactive, and the symbol (R) stands for Radioactive (not found in nature).

# Index

Note: numbers in **bold** type indicate the location for the definition of an item. Because many fundamental quanities (acceleration, momentum, Newton's second law, kinetic energy, etc.) are used so frequently, they are referenced by only the page number where they are originally introduced and defined.

A 8
B 9
C 0
D 1
E 2
F 3
G 4
H 5
I 6
J 7

## COMMON DERIVATIVES

$$\frac{dx}{dx} = 1$$

$$\frac{d}{dx} nu = n \frac{du}{dx}$$

$$\frac{d}{dx} (u \pm v) = \frac{du}{dx} \pm \frac{dv}{dx}$$

$$\frac{d}{dx} uv = \frac{du}{dx} v + u \frac{dv}{dx}$$

$$\frac{df}{dx} = \frac{df}{du} \frac{du}{dx}$$

$$\frac{d}{dx} u^n = nu^{n-1} \frac{du}{dx}$$

$$\frac{d}{dx} e^u = e^u \frac{du}{dx}$$

$$\frac{d}{dx} \ln|u| = \frac{1}{u} \frac{du}{dx}$$

$$\frac{d}{dx} \sin u = \cos u \frac{du}{dx}$$

$$\frac{d}{dx} \cos u = -\sin u \frac{du}{dx}$$

$$\frac{d}{dx} \tan u = \sec^2 u \frac{du}{dx}$$

$$\frac{d}{dx} \cot u = -\csc^2 u \frac{du}{dx}$$

## COMMON INDEFINITE INTEGRALS

$$\int du = u$$

$$\int nu \, dx = n \int u \, dx$$

$$\int (u + v) \, dx = \int u \, dx + \int v \, dx$$

$$\int u \, dv = uv - \int v \, du$$

$$\int x^n \, dx = \frac{x^{n+1}}{(n + 1)}, \quad (n \neq -1)$$

$$\int \frac{1}{x} \, dx = \ln|x|$$

$$\int e^x \, dx = e^x$$

$$\int \ln x \, dx = x \ln x - x$$

$$\int \sin x \, dx = -\cos x$$

$$\int \cos x \, dx = \sin x$$

$$\int \tan x \, dx = \ln|\sec x|$$

$$\int \cot x \, dx = \ln|\sin x|$$